Alcoholic beverages

Related titles:

Consumer-driven innovation in food and personal care products
(ISBN 978-1-84569-567-5)
Sensory and consumer scientists require a range of knowledge and skills to contribute successfully to the innovation process. With contributions from experts from a variety of backgrounds, this book reviews significant advances in consumer and sensory research and developments in industry practice. The first section focuses on changes in the marketplace impacting on product innovation. Parts II and III cover emerging research methodologies for innovation and advances in data analysis. Chapters in the following section discuss topics associated with consumer behaviour. The final section reports the personal perspectives of respected experts on innovation and the use of research in innovation.

Sensory analysis for food and beverage quality control
(ISBN 978-1-84569-476-0)
Producing products of reliable quality is vitally important to the food and beverage industries. In particular, companies often fail to ensure that the sensory quality of their products remains consistent, leading to the manufacture of goods which fail to meet the desired specifications or are rejected by the consumer. This book is a practical guide for all those tasked with sensory quality control in the food and beverage industries. Chapters in Parts I and II review the design, implementation and running of an effective sensory quality control program. Contributions in the last part of the book illustrate the application and results of such a program in different industry sectors.

Flavour in food
(ISBN 978-1-85573-960-4)
The flavour of a food is one of its most important qualities. Edited by two leading authorities in the field, and with a distinguished international team of contributors, this important collection summarises the wealth of recent research on how flavour develops in food and is then perceived by the consumer. The first part of the book reviews ways of measuring flavour. Part II looks at the ways flavour is retained and released in food. It considers the way flavour is retained in particular food matrices, how flavour is released during the process of eating, and the range of influences governing how flavour is perceived by the consumer. *Flavour in food* guides the reader through a complex subject and provides the essential foundation in both understanding and controlling food flavour.

Details of these books and a complete list of Woodhead's titles can be obtained by:

- visiting our web site at www.woodheadpublishing.com
- contacting Customer Services (e-mail: sales@woodheadpublishing.com; fax: +44 (0) 1223 832819; tel.: +44 (0) 1223 499140 ext. 130; address: Woodhead Publishing Limited, 80, High Street, Sawston, Cambridge CB22 3HJ, UK)
- contacting our US office (e-mail: usmarketing@woodheadpublishing.com; tel.: (215) 928 9112; address: Woodhead Publishing, 1518 Walnut Street, Suite 1100, Philadelphia, PA 19102-3406, USA).

If you would like to receive information on forthcoming titles, please send your address details to: Francis Dodds (address, tel. and fax as above; e-mail: francis.dodds@woodheadpublishing.com). Please confirm which subject areas you are interested in.

Woodhead Publishing Series in Food Science, Technology and Nutrition:
Number 225

Alcoholic beverages

Sensory evaluation and consumer research

Edited by
John Piggott

WP
WOODHEAD PUBLISHING

Oxford Cambridge Philadelphia New Delhi

Published by Woodhead Publishing Limited,
80 High Street, Sawston, Cambridge CB22 3HJ, UK
www.woodheadpublishing.com

Woodhead Publishing, 1518 Walnut Street, Suite 1100, Philadelphia,
PA 19102-3406, USA

Woodhead Publishing India Private Limited, G-2, Vardaan House,
7/28 Ansari Road, Daryaganj, New Delhi – 110002, India
www.woodheadpublishingindia.com

British Library Cataloguing in Publication Data
A catalogue record for this book is available from the British Library.

Library of Congress Control Number: 2011938829

ISBN 978-0-08-101652-7 (print)
ISBN 978-0-85709-517-6 (online)
ISSN 2042-8049 Woodhead Publishing Series in Food Science, Technology and Nutrition (print)
ISSN 2042-8057 Woodhead Publishing Series in Food Science, Technology and Nutrition (online)

The publisher's policy is to use permanent paper from mills that operate a sustainable forestry policy, and which has been manufactured from pulp which is processed using acid-free and elemental chlorine-free practices. Furthermore, the publisher ensures that the text paper and cover board used have met acceptable environmental accreditation standards.

Typeset by RefineCatch Limited, Bungay, Suffolk

Contents

Contributor contact details

(* = main contact)

Editor

J. R. Piggott
2 Snowdon Place
Stirling
FK8 2NH
UK

Email: m.j.piggott@talk21.com

Chapter 1

W. L. P. Bredie* and P. Møller
Sensory Science Section
Department of Food Science
University of Copenhagen
Rolighedsvej 30
1958 Frederiksberg C
Denmark

Email: wb@life.ku.dk

Chapter 2

E. Chambers IV*
Sensory Analysis Center
Kansas State University
Dept HN, Justin Hall
Manhattan, KS 66506-1407
USA

Email: eciv@ksu.edu

D. McGrew
Product Development
Beam Global Spirits and Wine
300 High Rise Drive
Louisville, KY 40213
USA

Email: don.mcgrew@beamglobal.com

Chapter 3

M. G. O'Sullivan
School of Food & Nutritional Sciences
University College Cork
Cork
Ireland

Email: maurice.osullivan@ucc.ie

Chapter 4

E. Monteleone
Dipartimento di Biotecnologie Agrarie
Università degli Studi di Firenze
Via Donizetti 6
Firenze 50144
Italy

Email: erminio.monteleone@unifi.it

Chapter 5

B. Plutowska* and W. Wardencki
Department of Analytical Chemistry
Chemical Faculty
Gdańsk University of Technology
ul. Narutowicza 11/12
80-233 Gdańsk
Poland

Email: bplutowska@gmail.com

Chapter 6

D. K. Parker
Campden BRI
Coopers Hill Road
Nutfield
Surrey
RH1 4HY
UK

Email: debbie.parker@bri-advantage.com

Chapter 7

A. C. Silva Ferreira*
CBQF / Escola Superior de Biotecnologia
Universidade Católica Portuguesa
Rua Dr. António Bernardino de Almeida
4200-072 Porto
Portugal

Email: acferreira@esb.ucp.pt

A. G. J. Tredoux
Institute for Wine Biotechnology
University of Stellenbosch
Private Bag X1
Matieland 7602
Stellenbosch
South Africa

Email: atredoux@sun.ac.za

Chapter 8

S. Furukawa
Kizakura Co., Ltd
223 Shioyamachi
Fushimi-ku
Kyoto 612-8046
Japan

Email: furukawa@kzkappa.co.jp

Chapter 9

R. S. Jackson
Cool Climate Oenology and Viticulture Institute
Brock University
500 Glenridge Ave
St. Catharines
Ontario L2S 3A1
Canada

Email: r.s.jackson@globetrotter.net

Chapter 10

I. Zabetakis
Laboratory of Food Chemistry
Department of Chemistry
University of Athens
Panepistimioupolis
15771 Athens
Greece

Email: izabet@chem.uoa.gr

Chapter 11

L. Lurton*, G. Ferrari and G. Snakkers
Station Viticole du BNIC
69 Rue de Bellefonds
16100 Cognac
France

Email: station@cognac.fr

Chapter 12

M. Riu Aumatell
Departament de Nutrició i Bromatologia
Facultat de Farmàcia
Universitat de Barcelona
Av Joan XXIII s/n
08028 Barcelona
Spain

Email: montseriu@ub.edu

Chapter 13

L. Louw and M. G. Lambrechts*
Distell Ltd
PO Box 184
Stellenbosch 7599
South Africa

Email: LELouw@distell.co.za; MLambrechts@distell.co.za

Chapter 14

C. Da Porto
Department of Food Science
University of Udine
Via Sondrio 2/A
33100 Udine
Italy

Email: carla.daporto@uniud.it

Chapter 15

Y. Xu*
Centre for Brewing Science and Enzyme Technology
Jiangnan University
1800 Lihu Ave
Wuxi City
Jiangsu 214122
China

Email: yxu@jiangnan.edu.cn

K. Ji
Moutai Distillery Co., Ltd
2 Guojiu Ave
Maotai Town
Renhuai City
Guizhou 564500
China

Chapter 16

E. Bordeu*
Departmento de Fruticultura y Enología
Facultad de Agronomía e Ingeniería Forestal
Pontificia Universidad Católica de Chile
4860 Vicuña Mackenna
Santiago 7820436
Región Metropolitana
Chile

Email: ebordeu@uc.cl

E. Agosín
Departmento de Ingeniería Química y Bioprocesos
Facultad de Ingeniería
Pontificia Universidad Católica de Chile
4860 Vicuña Mackenna
Santiago 7820436
Región Metropolitana
Chile

Email: agosin@ing.puc.cl

G. Casaubon
Centro de Aromas y Sabores
DICTUC S. A.
Pontificia Universidad Católica de Chile
4860 Vicuña Mackenna
Santiago 7820436
Región Metropolitana
Chile

Email: gcasaubo@puc.cl

Chapter 17

J. B. Faria
Departamento de Alimentos e Nutrição
Faculdade de Ciências Farmacêuticas – UNESP
Rod. Araraquara-Jaú, km 01 – Campus Universitário
14.801-902 Araraquara-SP
Brazil

Email: fariajb@fcfar.unesp.br

Chapter 18

S. Villanueva-Rodriguez
Centro de Investigación y Asistencia en Tecnología y Diseño del Estado de Jalisco A. C.
Av. Normalistas 800
Colinas de la Normal
44270 Guadalajara
Jalisco
Mexico

Email: svillanueva@ciatej.net.mx

H. Escalona-Buendia*
Departamento de Biotecnologia
Universidad Autonoma Metropolitana – Iztapalapa
Av. San Rafael Atlixco 186
Vicentina
09340 Iztapalapa D. F.
Mexico

Chapter 19

F. R. Jack
The Scotch Whisky Research Institute
The Robertson Trust Building
Research Avenue North
Riccarton
Edinburgh
EH14 4AP
UK

Email: frances.jack@swri.co.uk

Chapter 20

L. Ettinger Lieberman,
D. I. Moskowitz and H. R.
Moskowitz*
Moskowitz Jacobs Inc.
1025 Westchester Ave
White Plains
NY 10604
USA

Email: llieberman@mji-designlab.com; david.moskowitz@mji-designlab.com; mjihrm@sprynet.com

Chapter 21

H. J. H. MacFie
43 Manor Road
Keynsham
Bristol
BS31 1RB
UK

Email: hal@halmacfie.com

J. R. Piggott*
2 Snowdon Place
Stirling
FK8 2NH
UK

Email: m.j.piggott@talk21.com

Woodhead Publishing Series in Food Science, Technology and Nutrition

1 **Chilled foods: a comprehensive guide** *Edited by C. Dennis and M. Stringer*
2 **Yoghurt: science and technology** *A. Y. Tamime and R. K. Robinson*
3 **Food processing technology: principles and practice** *P. J. Fellows*
4 **Bender's dictionary of nutrition and food technology Sixth edition** *D. A. Bender*
5 **Determination of veterinary residues in food** *Edited by N. T. Crosby*
6 **Food contaminants: sources and surveillance** *Edited by C. Creaser and R. Purchase*
7 **Nitrates and nitrites in food and water** *Edited by M. J. Hill*
8 **Pesticide chemistry and bioscience: the food-environment challenge** *Edited by G. T. Brooks and T. Roberts*
9 **Pesticides: developments, impacts and controls** *Edited by G. A. Best and A. D. Ruthven*
10 **Dietary fibre: chemical and biological aspects** *Edited by D. A. T. Southgate, K. W. Waldron, I. T. Johnson and G. R. Fenwick*
11 **Vitamins and minerals in health and nutrition** *M. Tolonen*
12 **Technology of biscuits, crackers and cookies Second edition** *D. Manley*
13 **Instrumentation and sensors for the food industry** *Edited by E. Kress-Rogers*
14 **Food and cancer prevention: chemical and biological aspects** *Edited by K. W. Waldron, I. T. Johnson and G. R. Fenwick*
15 **Food colloids: proteins, lipids and polysaccharides** *Edited by E. Dickinson and B. Bergenstahl*
16 **Food emulsions and foams** *Edited by E. Dickinson*
17 **Maillard reactions in chemistry, food and health** *Edited by T. P. Labuza, V. Monnier, J. Baynes and J. O'Brien*
18 **The Maillard reaction in foods and medicine** *Edited by J. O'Brien, H. E. Nursten, M. J. Crabbe and J. M. Ames*
19 **Encapsulation and controlled release** *Edited by D. R. Karsa and R. A. Stephenson*
20 **Flavours and fragrances** *Edited by A. D. Swift*
21 **Feta and related cheeses** *Edited by A. Y. Tamime and R. K. Robinson*

22 **Biochemistry of milk products** *Edited by A. T. Andrews and J. R. Varley*
23 **Physical properties of foods and food processing systems** *M. J. Lewis*
24 **Food irradiation: a reference guide** *V. M. Wilkinson and G. Gould*
25 **Kent's technology of cereals: an introduction for students of food science and agriculture Fourth edition** *N. L. Kent and A. D. Evers*
26 **Biosensors for food analysis** *Edited by A. O. Scott*
27 **Separation processes in the food and biotechnology industries: principles and applications** *Edited by A.S. Grandison and M. J. Lewis*
28 **Handbook of indices of food quality and authenticity** *R.S. Singhal, P. K. Kulkarni and D. V. Rege*
29 **Principles and practices for the safe processing of foods** *D. A. Shapton and N. F. Shapton*
30 **Biscuit, cookie and cracker manufacturing manuals Volume 1: ingredients** *D. Manley*
31 **Biscuit, cookie and cracker manufacturing manuals Volume 2: biscuit doughs** *D. Manley*
32 **Biscuit, cookie and cracker manufacturing manuals Volume 3: biscuit dough piece forming** *D. Manley*
33 **Biscuit, cookie and cracker manufacturing manuals Volume 4: baking and cooling of biscuits** *D. Manley*
34 **Biscuit, cookie and cracker manufacturing manuals Volume 5: secondary processing in biscuit manufacturing** *D. Manley*
35 **Biscuit, cookie and cracker manufacturing manuals Volume 6: biscuit packaging and storage** *D. Manley*
36 **Practical dehydration Second edition** *M. Greensmith*
37 **Lawrie's meat science Sixth edition** *R. A. Lawrie*
38 **Yoghurt: science and technology Second edition** *A. Y Tamime and R. K. Robinson*
39 **New ingredients in food processing: biochemistry and agriculture** *G. Linden and D. Lorient*
40 **Benders' dictionary of nutrition and food technology Seventh edition** *D. A. Bender and A. E. Bender*
41 **Technology of biscuits, crackers and cookies Third edition** *D. Manley*
42 **Food processing technology: principles and practice Second edition** *P. J. Fellows*
43 **Managing frozen foods** *Edited by C. J. Kennedy*
44 **Handbook of hydrocolloids** *Edited by G. O. Phillips and P. A. Williams*
45 **Food labelling** *Edited by J. R. Blanchfield*
46 **Cereal biotechnology** *Edited by P. C. Morris and J. H. Bryce*
47 **Food intolerance and the food industry** *Edited by T. Dean*
48 **The stability and shelf-life of food** *Edited by D. Kilcast and P. Subramaniam*
49 **Functional foods: concept to product** *Edited by G. R. Gibson and C. M. Williams*
50 **Chilled foods: a comprehensive guide Second edition** *Edited by M. Stringer and C. Dennis*
51 **HACCP in the meat industry** *Edited by M. Brown*
52 **Biscuit, cracker and cookie recipes for the food industry** *D. Manley*
53 **Cereals processing technology** *Edited by G. Owens*
54 **Baking problems solved** *S. P. Cauvain and L. S. Young*
55 **Thermal technologies in food processing** *Edited by P. Richardson*
56 **Frying: improving quality** *Edited by J. B. Rossell*
57 **Food chemical safety Volume 1: contaminants** *Edited by D. Watson*

58 **Making the most of HACCP: learning from others' experience** *Edited by T. Mayes and S. Mortimore*
59 **Food process modelling** *Edited by L. M. M. Tijskens, M. L. A. T. M. Hertog and B. M. Nicolaï*
60 **EU food law: a practical guide** *Edited by K. Goodburn*
61 **Extrusion cooking: technologies and applications** *Edited by R. Guy*
62 **Auditing in the food industry: from safety and quality to environmental and other audits** *Edited by M. Dillon and C. Griffith*
63 **Handbook of herbs and spices Volume 1** *Edited by K. V. Peter*
64 **Food product development: maximising success** *M. Earle, R. Earle and A. Anderson*
65 **Instrumentation and sensors for the food industry Second edition** *Edited by E. Kress-Rogers and C. J. B. Brimelow*
66 **Food chemical safety Volume 2: additives** *Edited by D. Watson*
67 **Fruit and vegetable biotechnology** *Edited by V. Valpuesta*
68 **Foodborne pathogens: hazards, risk analysis and control** *Edited by C. de W. Blackburn and P. J. McClure*
69 **Meat refrigeration** *S. J. James and C. James*
70 **Lockhart and Wiseman's crop husbandry Eighth edition** *H. J. S. Finch, A. M. Samuel and G. P. F. Lane*
71 **Safety and quality issues in fish processing** *Edited by H. A. Bremner*
72 **Minimal processing technologies in the food industries** *Edited by T. Ohlsson and N. Bengtsson*
73 **Fruit and vegetable processing: improving quality** *Edited by W. Jongen*
74 **The nutrition handbook for food processors** *Edited by C. J. K. Henry and C. Chapman*
75 **Colour in food: improving quality** *Edited by D MacDougall*
76 **Meat processing: improving quality** *Edited by J. P. Kerry, J. F. Kerry and D. A. Ledward*
77 **Microbiological risk assessment in food processing** *Edited by M. Brown and M. Stringer*
78 **Performance functional foods** *Edited by D. Watson*
79 **Functional dairy products Volume 1** *Edited by T. Mattila-Sandholm and M. Saarela*
80 **Taints and off-flavours in foods** *Edited by B. Baigrie*
81 **Yeasts in food** *Edited by T. Boekhout and V. Robert*
82 **Phytochemical functional foods** *Edited by I. T. Johnson and G. Williamson*
83 **Novel food packaging techniques** *Edited by R. Ahvenainen*
84 **Detecting pathogens in food** *Edited by T. A. McMeekin*
85 **Natural antimicrobials for the minimal processing of foods** *Edited by S. Roller*
86 **Texture in food Volume 1: semi-solid foods** *Edited by B. M. McKenna*
87 **Dairy processing: improving quality** *Edited by G Smit*
88 **Hygiene in food processing: principles and practice** *Edited by H. L. M. Lelieveld, M. A. Mostert, B. White and J. Holah*
89 **Rapid and on-line instrumentation for food quality assurance** *Edited by I. Tothill*
90 **Sausage manufacture: principles and practice** *E. Essien*
91 **Environmentally-friendly food processing** *Edited by B. Mattsson and U. Sonesson*
92 **Bread making: improving quality** *Edited by S. P. Cauvain*
93 **Food preservation techniques** *Edited by P. Zeuthen and L. Bøgh-Sørensen*
94 **Food authenticity and traceability** *Edited by M. Lees*

95 **Analytical methods for food additives** *R. Wood, L. Foster, A. Damant and P. Key*
96 **Handbook of herbs and spices Volume 2** *Edited by K. V. Peter*
97 **Texture in food Volume 2: solid foods** *Edited by D. Kilcast*
98 **Proteins in food processing** *Edited by R. Yada*
99 **Detecting foreign bodies in food** *Edited by M. Edwards*
100 **Understanding and measuring the shelf-life of food** *Edited by R. Steele*
101 **Poultry meat processing and quality** *Edited by G. Mead*
102 **Functional foods, ageing and degenerative disease** *Edited by C. Remacle and B. Reusens*
103 **Mycotoxins in food: detection and control** *Edited by N. Magan and M. Olsen*
104 **Improving the thermal processing of foods** *Edited by P. Richardson*
105 **Pesticide, veterinary and other residues in food** *Edited by D. Watson*
106 **Starch in food: structure, functions and applications** *Edited by A-C Eliasson*
107 **Functional foods, cardiovascular disease and diabetes** *Edited by A. Arnoldi*
108 **Brewing: science and practice** *D. E. Briggs, P. A. Brookes, R. Stevens and C. A. Boulton*
109 **Using cereal science and technology for the benefit of consumers: proceedings of the 12th International ICC Cereal and Bread Congress, 24-26th May, 2004, Harrogate, UK** *Edited by S. P. Cauvain, L. S. Young and S. Salmon*
110 **Improving the safety of fresh meat** *Edited by J. Sofos*
111 **Understanding pathogen behaviour in food: virulence, stress response and resistance** *Edited by M. Griffiths*
112 **The microwave processing of foods** *Edited by H. Schubert and M. Regier*
113 **Food safety control in the poultry industry** *Edited by G. Mead*
114 **Improving the safety of fresh fruit and vegetables** *Edited by W. Jongen*
115 **Food, diet and obesity** *Edited by D. Mela*
116 **Handbook of hygiene control in the food industry** *Edited by H. L. M. Lelieveld, M. A. Mostert and J. Holah*
117 **Detecting allergens in food** *Edited by S. Koppelman and S. Hefle*
118 **Improving the fat content of foods** *Edited by C. Williams and J. Buttriss*
119 **Improving traceability in food processing and distribution** *Edited by I. Smith and A. Furness*
120 **Flavour in food** *Edited by A. Voilley and P. Etievant*
121 **The Chorleywood bread process** *S. P. Cauvain and L. S. Young*
122 **Food spoilage microorganisms** *Edited by C. de W. Blackburn*
123 **Emerging foodborne pathogens** *Edited by Y. Motarjemi and M. Adams*
124 **Bender's dictionary of nutrition and food technology Eighth edition** *D. A. Bender*
125 **Optimising sweet taste in foods** *Edited by W. J. Spillane*
126 **Brewing: new technologies** *Edited by C. Bamforth*
127 **Handbook of herbs and spices Volume 3** *Edited by K. V. Peter*
128 **Lawrie's meat science Seventh edition** *R. A. Lawrie in collaboration with D. A. Ledward*
129 **Modifying lipids for use in food** *Edited by F. Gunstone*
130 **Meat products handbook: practical science and technology** *G. Feiner*
131 **Food consumption and disease risk: consumer-pathogen interactions** *Edited by M. Potter*
132 **Acrylamide and other hazardous compounds in heat-treated foods** *Edited by K. Skog and J. Alexander*

133 **Managing allergens in food** *Edited by C. Mills, H. Wichers and K. Hoffman-Sommergruber*
134 **Microbiological analysis of red meat, poultry and eggs** *Edited by G. Mead*
135 **Maximising the value of marine by-products** *Edited by F. Shahidi*
136 **Chemical migration and food contact materials** *Edited by K. Barnes, R. Sinclair and D. Watson*
137 **Understanding consumers of food products** *Edited by L. Frewer and H. van Trijp*
138 **Reducing salt in foods: practical strategies** *Edited by D. Kilcast and F. Angus*
139 **Modelling microorganisms in food** *Edited by S. Brul, S. Van Gerwen and M. Zwietering*
140 **Tamime and Robinson's Yoghurt: science and technology Third edition** *A. Y. Tamime and R. K. Robinson*
141 **Handbook of waste management and co-product recovery in food processing Volume 1** *Edited by K. W. Waldron*
142 **Improving the flavour of cheese** *Edited by B. Weimer*
143 **Novel food ingredients for weight control** *Edited by C. J. K. Henry*
144 **Consumer-led food product development** *Edited by H. MacFie*
145 **Functional dairy products Volume 2** *Edited by M. Saarela*
146 **Modifying flavour in food** *Edited by A. J. Taylor and J. Hort*
147 **Cheese problems solved** *Edited by P. L. H. McSweeney*
148 **Handbook of organic food safety and quality** *Edited by J. Cooper, C. Leifert and U. Niggli*
149 **Understanding and controlling the microstructure of complex foods** *Edited by D. J. McClements*
150 **Novel enzyme technology for food applications** *Edited by R. Rastall*
151 **Food preservation by pulsed electric fields: from research to application** *Edited by H. L. M. Lelieveld and S. W. H. de Haan*
152 **Technology of functional cereal products** *Edited by B. R. Hamaker*
153 **Case studies in food product development** *Edited by M. Earle and R. Earle*
154 **Delivery and controlled release of bioactives in foods and nutraceuticals** *Edited by N. Garti*
155 **Fruit and vegetable flavour: recent advances and future prospects** *Edited by B. Brückner and S. G. Wyllie*
156 **Food fortification and supplementation: technological, safety and regulatory aspects** *Edited by P. Berry Ottaway*
157 **Improving the health-promoting properties of fruit and vegetable products** *Edited by F. A. Tomás-Barberán and M. I. Gil*
158 **Improving seafood products for the consumer** *Edited by T. Børresen*
159 **In-pack processed foods: improving quality** *Edited by P. Richardson*
160 **Handbook of water and energy management in food processing** *Edited by J. Klemeš, R. Smith and J-K Kim*
161 **Environmentally compatible food packaging** *Edited by E. Chiellini*
162 **Improving farmed fish quality and safety** *Edited by Ø. Lie*
163 **Carbohydrate-active enzymes** *Edited by K-H Park*
164 **Chilled foods: a comprehensive guide Third edition** *Edited by M. Brown*
165 **Food for the ageing population** *Edited by M. M. Raats, C. P. G. M. de Groot and W. A Van Staveren*
166 **Improving the sensory and nutritional quality of fresh meat** *Edited by J. P. Kerry and D. A. Ledward*

167 **Shellfish safety and quality** *Edited by S. E. Shumway and G. E. Rodrick*
168 **Functional and speciality beverage technology** *Edited by P. Paquin*
169 **Functional foods: principles and technology** *M. Guo*
170 **Endocrine-disrupting chemicals in food** *Edited by I. Shaw*
171 **Meals in science and practice: interdisciplinary research and business applications** *Edited by H. L. Meiselman*
172 **Food constituents and oral health: current status and future prospects** *Edited by M. Wilson*
173 **Handbook of hydrocolloids Second edition** *Edited by G. O. Phillips and P. A. Williams*
174 **Food processing technology: principles and practice Third edition** *P. J. Fellows*
175 **Science and technology of enrobed and filled chocolate, confectionery and bakery products** *Edited by G. Talbot*
176 **Foodborne pathogens: hazards, risk analysis and control Second edition** *Edited by C. de W. Blackburn and P. J. McClure*
177 **Designing functional foods: measuring and controlling food structure breakdown and absorption** *Edited by D. J. McClements and E. A. Decker*
178 **New technologies in aquaculture: improving production efficiency, quality and environmental management** *Edited by G. Burnell and G. Allan*
179 **More baking problems solved** *S. P. Cauvain and L. S. Young*
180 **Soft drink and fruit juice problems solved** *P. Ashurst and R. Hargitt*
181 **Biofilms in the food and beverage industries** *Edited by P. M. Fratamico, B. A. Annous and N. W. Gunther*
182 **Dairy-derived ingredients: food and neutraceutical uses** *Edited by M. Corredig*
183 **Handbook of waste management and co-product recovery in food processing Volume 2** *Edited by K. W. Waldron*
184 **Innovations in food labelling** *Edited by J. Albert*
185 **Delivering performance in food supply chains** *Edited by C. Mena and G. Stevens*
186 **Chemical deterioration and physical instability of food and beverages** *Edited by L. H. Skibsted, J. Risbo and M. L. Andersen*
187 **Managing wine quality Volume 1: viticulture and wine quality** *Edited by A.G. Reynolds*
188 **Improving the safety and quality of milk Volume 1: milk production and processing** *Edited by M. Griffiths*
189 **Improving the safety and quality of milk Volume 2: improving quality in milk products** *Edited by M. Griffiths*
190 **Cereal grains: assessing and managing quality** *Edited by C. Wrigley and I. Batey*
191 **Sensory analysis for food and beverage quality control: a practical guide** *Edited by D. Kilcast*
192 **Managing wine quality Volume 2: oenology and wine quality** *Edited by A. G. Reynolds*
193 **Winemaking problems solved** *Edited by C. E. Butzke*
194 **Environmental assessment and management in the food industry** *Edited by U. Sonesson, J. Berlin and F. Ziegler*
195 **Consumer-driven innovation in food and personal care products** *Edited by S.R. Jaeger and H. MacFie*
196 **Tracing pathogens in the food chain** *Edited by S. Brul, P.M. Fratamico and T.A. McMeekin*

197 **Case studies in novel food processing technologies: innovations in processing, packaging, and predictive modelling** *Edited by C. J. Doona, K Kustin and F. E. Feeherry*
198 **Freeze-drying of pharmaceutical and food products** *T-C Hua, B-L Liu and H Zhang*
199 **Oxidation in foods and beverages and antioxidant applications Volume 1: understanding mechanisms of oxidation and antioxidant activity** *Edited by E. A. Decker, R. J. Elias and D. J. McClements*
200 **Oxidation in foods and beverages and antioxidant applications Volume 2: management in different industry sectors** *Edited by E. A. Decker, R. J. Elias and D. J. McClements*
201 **Protective cultures, antimicrobial metabolites and bacteriophages for food and beverage biopreservation** *Edited by C. Lacroix*
202 **Separation, extraction and concentration processes in the food, beverage and nutraceutical industries** *Edited by S. S. H. Rizvi*
203 **Determining mycotoxins and mycotoxigenic fungi in food and feed** *Edited by S. De Saeger*
204 **Developing children's food products** *Edited by D. Kilcast and F. Angus*
205 **Functional foods: concept to product Second edition** *Edited by M. Saarela*
206 **Postharvest biology and technology of tropical and subtropical fruits Volume 1:** Fundamental issues *Edited by E. M. Yahia*
207 **Postharvest biology and technology of tropical and subtropical fruits Volume 2:** Açai to citrus *Edited by E. M. Yahia*
208 **Postharvest biology and technology of tropical and subtropical fruits Volume 3:** Cocona to mango *Edited by E. M. Yahia*
209 **Postharvest biology and technology of tropical and subtropical fruits Volume 4:** Mangosteen to white sapote *Edited by E. M. Yahia*
210 **Food and beverage stability and shelf life** *Edited by D. Kilcast and P. Subramaniam*
211 **Processed Meats: improving safety, nutrition and quality** *Edited by J. P. Kerry and J. F. Kerry*
212 **Food chain integrity: a holistic approach to food traceability, safety, quality and authenticity** *Edited by J. Hoorfar, K. Jordan, F. Butler and R. Prugger*
213 **Improving the safety and quality of eggs and egg products Volume 1** *Edited by Y. Nys, M. Bain and F. Van Immerseel*
214 **Improving the safety and quality of eggs and egg products Volume 2** *Edited by F. Van Immerseel, Y. Nys and M. Bain*
215 **Animal feed contamination: effects on livestock and food safety** *Edited by J. Fink-Gremmels*
216 **Hygienic design of food factories** *Edited by J. Holah and H. L. M. Lelieveld*
217 **Manley's technology of biscuits, crackers and cookies Fourth edition** *Edited by D. Manley*
218 **Nanotechnology in the food, beverage and nutraceutical industries** *Edited by Q. Huang*
219 **Rice quality: A guide to rice properties and analysis** *K. R. Bhattacharya*
220 **Advances in meat, poultry and seafood packaging** *Edited by J. P. Kerry*
221 **Reducing saturated fats in foods** *Edited by G. Talbot*
222 **Handbook of food proteins** *Edited by G. O. Phillips and P. A. Williams*
223 **Lifetime nutritional influences on cognition, behaviour and psychiatric illness** *Edited by D. Benton*

224 **Food machinery for the production of cereal foods, snack foods and confectionery** *L.-M. Cheng*
225 **Alcoholic beverages: sensory evaluation and consumer research** *Edited by J. Piggott*
226 **Extrusion problems solved** *M. N. Riaz and G. J. Rokey*
227 **Handbook of herbs and spices Volume 1 Second edition** *Edited by K. V. Peter*
228 **Handbook of herbs and spices Volume 2 Second edition** *Edited by K. V. Peter*
229 **Bread making: improving quality Second edition** *Edited by S. P. Cauvain*
230 **Emerging food packaging technologies: principles and practice** *Edited by K. L. Yam and D. S. Lee*
231 **Infectious disease in aquaculture: prevention and control** *Edited by B. Austin*
232 **Diet, immunity and inflammation** *Edited by P. C. Calder and P. Yaqoob*
233 **Natural food additives, ingredients and flavourings** *Edited by D. Baines and R. Seal*
234 **Food decontamination: novel methods and applications** *Edited by A. Demirci and M. Ngadi*
235 **Chemical contaminants and residues in foods** *Edited by D. Schrenk*
236 **Robotics and automation in the food industry: current and future technologies** *Edited by D. G. Caldwell*
237 **Fibre-rich and wholegrain foods: improving quality** *Edited by J. A. Delcour and K. Poutanen*
238 **Computer vision technology in the food and beverage industries** *Edited by D.-W. Sun*
239 **Encapsulation technologies and delivery systems for food ingredients and nutraceuticals** *Edited by N. Garti and D. J. McClements*

Preface

Our aim in compiling this book was to describe the basic principles of sensory analysis and consumer research methods, and to show how they can be applied to alcoholic beverages. The production of fermented beverages is an ancient industry, but it was probably not until the twelfth century (in Europe at least) that distillation was applied to the production of potable spirits. It was not until the nineteenth century that the next major scientific advances were made, in the development of the Coffey still, and the beginnings of modern microbiology. Towards the end of the nineteenth century, scientific methods began to be applied more systematically to the study and improvement of alcoholic beverages, with the establishment, for example, of the research station in Cognac, and of the Institute of Brewing in the UK. In parallel with these developments ran an increasing interest in the quality of the industry's products, in terms of purity, safety, and sensory quality. Alcoholic beverages present unique challenges for sensory analysts, and a major part of the book describes how various sectors of the industry developed methods to overcome these challenges. However, these methods were largely developed in isolation, and were heavily dependent on the expertise of experienced individuals. Such individuals still make an invaluable contribution to the success of the industry, but since the middle of the twentieth century the contribution which formal sensory analysis can make has been increasingly recognised. However, the more elaborate methods of laboratory sensory analysis are not always appropriate. It is not practicable for the smallest distilleries, wine makers or microbreweries to sustain large panels of tasters. They must still rely on the skill of the distiller, oenologist or brewer to maintain product quality, but even so an understanding of the scientific basis of sensory analysis can help to ensure that their judgement is securely based.

To achieve this aim, the chapters of the book have been grouped into sections. In the first part the chapters describe the principles of sensory evaluation and their application to alcoholic beverages in general terms. While all the methods here will not be applicable in all circumstances, these chapters will provide information and ideas for all sectors of the industry, and producers large and small. In the second part, industry practices for fermented products are described, and in the third part for distilled products. These chapters necessarily vary in format, depending on the degree of understanding of the products, and the extent of research into their composition and properties. In some cases the relationships between chemistry and flavour are well understood, but in others less so. It would not have been possible to include all alcoholic beverages, but major products in terms of domestic production and volume of trade are included. There are others which we would like to have included, but lack of space, and the difficulty of finding willing authors with suitable expertise, precluded these. The final part describes the principles of some consumer research methods and their application to alcoholic beverages. In a book such as this it cannot be a complete description of consumer research; rather we have illustrated some methods with wide potential application which can help provide an understanding of consumer choice.

John Piggott

Part I

Sensory evaluation: principles and application to alcoholic beverages

1

Overview of sensory perception

W. L. P. Bredie and P. Møller, The University of Copenhagen, Denmark

Abstract: Perception of foods and beverages is mediated by the human senses, of which the five common senses play an important role. Sensory function at physiological and genetic levels is now understood in detail. This chapter provides details about the chemical senses, important for the perception of flavour. Chemical receptors for taste, smell and pain are widespread in the human body and are believed to have other functions, as in homeostasis and food intake regulation. Food perception is intertwined with the way humans select foods, food intake and learning of new food preferences. Sensory physiology undergoes changes and food preferences vary through life; perception of the sensory quality of foods is stable from early adolescence to old age.

Key words: food perception, common senses, sensory physiology, food interoception, preferences.

1.1 Introduction

The appreciation of foods and beverages is strongly influenced by the sensory properties of the appearance, smell, flavour and texture. The sensory properties of foods are mediated through all our senses including the common senses of sight, touch, hearing, taste, and smell, as well as pain and temperature. Many of the perceptions humans have with foods become more familiar and liked upon repeated exposure and are thus learned throughout their lifespan. The sound of opening a can of beer will create expectations regarding the freshness of the product, and even such a perception that is less central to our awareness is not trivial for the total sensory quality experienced with the product. Although the common senses are quite central to the understanding of the perceived sensory quality, the sensations and feelings related to the desires to eat and satisfaction with the consumption of food should not be neglected. These feelings occur in the alimentary tract and are important drivers for food intake and termination of eating and drinking. As with perception of foods, which deals with the interaction of the senses with the environment, feelings in the body can be referred to as interoception.

Our senses of sight and touch can set up expectations of the overall flavour of food which can be very hard to ignore. For example, the presentation of a prestigious cognac in a cheap polystyrene container would ruin not only the expected quality, but also the perceived sensory quality. The interplay of the presentation, expectation and situation in which the food is made available plays an important role in the experienced sensory quality. Equally, the colour of food can affect our perception of the flavour both by enhancing the magnitude of perception, for example the fruity and strawberry flavours in red drinks (e.g. Small and Prescott, 2005).

Between all the senses, the most significant for our appreciation of foods and drinks remain the chemical senses, which encompass taste, smell and chemesthesis (pain, touch and temperature). Taste or gustation detects compounds dissolved in liquids using delicately organized chemosensors, mostly in the mouth. Smell or olfaction detects air borne compounds, both from the ambient air, but also from those emitted from the food in the oral cavity and throat to the nose. Chemesthesis mediates information about irritants through nerve endings in the skin as well as other borders between us and the environment, including the epithelia in the nose, mouth, eyes and in the alimentary tract.

In this chapter we will give a closer look at how our senses interact with foods and beverages. Special attention will be given to the role and function of the senses, especially the chemical senses, which are central in the perception of flavour. Furthermore, the chapter describes what happens in the mouth when drinking a beverage and how the sensory signals that arise are transduced and integrated in the human brain. Several mechanisms of human perception will be explained and described with the aim to give a better understanding about why sensory properties and preferences are so different for different foods and beverages.

1.2 The common senses

Humans are endowed with many senses which make it possible to obtain information about 'what' is 'where' in the environment. A fair number of these different senses, or sensory modalities, are of crucial importance for perception and affective evaluation of foods and drinks, as well as for choice behaviour, which have been demonstrated to rely far more on the sensory properties than upon any other parameters.

In English, and many other languages, the overall sensory impression of food and drink is usually described by the word 'taste'. When we use the word in this sense it will be written in quotation marks, 'taste'; without quotes we refer to the separate sense of taste. Besides helping us to choose and create expectations, visual perception also contributes to preparing the gastrointestinal system for digestion via a number of so-called cephalic reflexes, such as increasing the flow of saliva in the mouth. This aids the chewing process and facilitates transport of tastants to taste cells, increasing release of insulin in the pancreas and increasing release of acid in the stomach (Giduck *et al.*, 1987; Mattes, 1997). Perception of food smells has also been demonstrated to have cephalic consequences. These effects take place via so-called orthonasal olfaction, or olfaction obtained by sniffing behaviour. This is different from the way the sense of smell contributes to the 'taste' of food and drink, which is referred to as retronasal olfaction. Recent work has suggested that retronasal and orthonasal olfaction might be distinct in many ways (Small *et al.*, 2005), possibly reflecting the different functional roles they play in food and other behaviours.

Finally, interoception, the sensation of the internal physiological state of the body, plays a large role in food behaviour, probably in particular with respect to how preferences and habits change, seemingly outside our conscious control (Craig, 2002; Capaldi, 2001).

1.2.1 General definitions of taste and flavour

It is customary to talk of five *basic* tastes: sweet, salty, sour, bitter and umami. The last one, umami, is less well known than the other four. It is often described as the taste of bouillon or glutamate. Umami was discovered by Japanese scientists and the name is meant to indicate 'pleasurable taste', and is associated with seaweed, fermented soy and fish products. The concept of basic tastes comprises the idea that sweet, salty, sour, bitter and umami tastes are basic, in the sense that no one of them can be obtained by any combination of the other four, and that any taste sensation possible can be created by an appropriate mixture of the five basic tastes. The basic tastes are usually defined in terms of sucrose (sweet), quinine (bitter), sodium chloride (salty), citric acid (sour) and monosodium glutamate (umami).

As already mentioned, the five basic tastes are not sufficient to create all of the thousands of different 'tastes' available to us from different foods. Smelling a food or drink gives an impression of its so-called aroma. The sense of smell would therefore seem to be an important sense for the perception of foods and drinks. Think about how food tastes when you have a common cold; or how wine tastes if you block your nose when drinking it. If smell was not important for the 'taste' of food and drinks it should be possible to 'create' the 'taste' of, e.g., an orange, by a particular mixture of the basic tastes. The sensation of smell produces an almost infinite number of possible 'tastes' when combined with the other senses important for food and drink perception. The dimensionality of the space describing the different possible tastes increases enormously from the five dimensions that the sense of taste provides on its own. In the scientific literature the central (i.e. in the brain) integration of taste, smell and trigeminality is often referred to as flavour. For the total experience of a food or a drink, it would make sense to also include tactile sensation (touch, mouth-feel) and hearing among the senses that contribute to the flavour of a food or drink. In common language, taste is often confused with flavour, as the tastes that people describe are merely derived from olfactory inputs or other sensations, such as touch and temperature.

Despite the fact that we only rarely doubt whether we have added too much salt to a dish or too much tonic in our gin and tonic, or if our coffee is too bitter, the underlying neurophysiological processes that immediately allow us to evaluate these questions are very complicated, as will be elaborated upon later in this chapter. Suffice it to mention here that the sensation of a taste starts with stimulation of taste cells in the mouth. Different neurophysiological mechanisms have been identified to code the different basic tastes. Smells are coded by receptor cells in the olfactory epithelium, which is located in the upper part of both nostrils. The epithelium occupies 4–5 cm^2 in each nostril and consists of about 350 different types of receptor cells. Most odorants activate many of these 350 different cells, but different smells activate different subsets to different degrees in patterns of activation to be disentangled by the brain.

Trigeminal stimuli are also occasionally referred to as 'irritants', since the sensations they give rise to can be unpleasant or even painful. Besides allowing us to perceive hot spices (chilli, garlic and ginger, etc.) and CO_2 in fizzy drinks, most chemical substances at sufficiently high concentration will also activate the trigeminal system. The receptors of the trigeminal system consist of so-called free nerve endings, and many of the functional properties of this system are very different from properties of olfaction (smell) and gustation (taste), as we will return to later in the chapter.

1.3 Oro-sensory systems

The oro-sensory area can be viewed as the total group of sensory receptors present in our lips, tongue, palate, soft palate and areas in the upper throat (pharynx and laryngopharynx). Although sensations of temperature, touch and pain can be sensed throughout the oro-sensory area, the sense of taste appears to be comparatively better organised. Even at the lips, temperature and pain are clearly sensed and temperatures at 32 °C are sensed as cool, whereas temperatures above 34 °C are sensed as warm and above 43 °C as painful (Heft and Robinson, 2010). Thus the drink or food touching the lips is already primed by a signal of suitability for consumption.

1.3.1 Sense of taste

In its scientific meaning, taste is mainly a function of the taste buds in the mouth, most of which are organised in structurally-defined units also called taste papillae. The receptor cells for taste are chemoreceptors as they have receptor proteins that respond to compounds put into the mouth (Fig. 1.1). The chemoreceptors in the oral cavity detect a range of volatile and non-volatile compounds leading to basic taste sensations. The perception of basic tastes has probably evolved to provide information about foods in accord with desires, and often in accord with metabolic needs for specific nutritive substances. Another function is to alert us to toxins or intake of a potential toxic dose of a food. It is therefore not surprising that very bitter

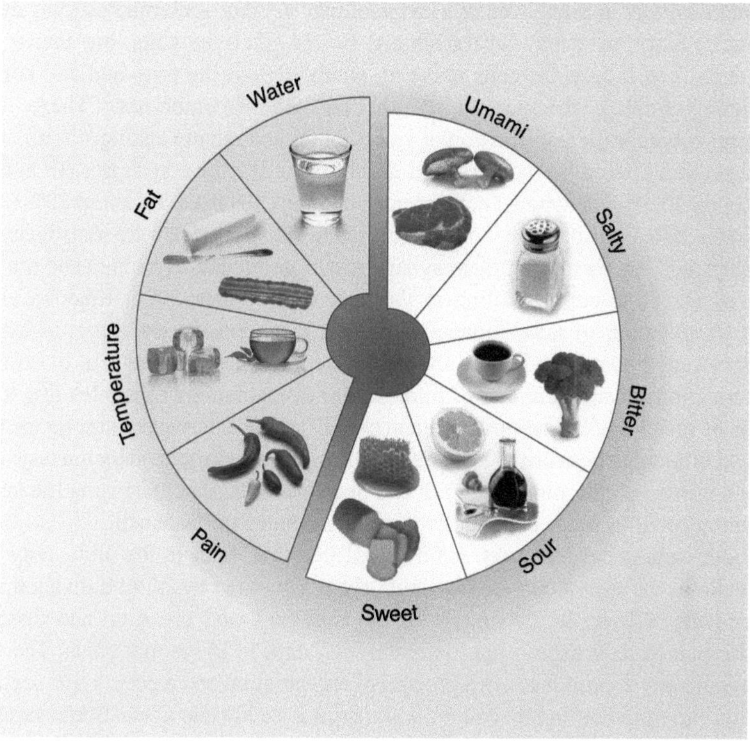

Fig. 1.1 Chemo- and thermoreceptors in the mouth extending the common view of the five basic taste qualities (sweet, sour, bitter, salty and umami) to other receptors that are also believed to be important for the overall taste quality of foods and beverages.

substances or high concentrations of sugar or alcohol are difficult to ingest in large portions. Beside the basic tastes, there is some evidence for another taste quality, namely fat. At present the architecture and sensory function of fat receptors in the mouth is a topic of scientific debate. The chemoreceptors for taste reside in the taste buds, which are mostly located in fungiform, foliate and circumvallate papillae on the tongue. About two-third of the taste buds are located in the taste papillae, whereas the remainder are situated on the palate and on the epiglottis (Miller, 1995). The filiform papillae in the middle part of the tongue have mainly trigeminal functions, like the burning sensations of CO_2 and capsaicin.

The taste bud is composed of up to 100 epithelial cells, some of which are supporting cells, called sustentacular cells, and others function as taste cells. The outer tips (apical) of the taste cells are organised around a minute taste pore. From the tip of the taste cells several taste hairs or microvilli protrude into the taste pore towards the cavity of the mouth. These microvilli contain the receptor proteins that interact with tastants. At present there are three major classes of taste cells. Type I cells (glial-like) have a support function for clearance of the taste bud for neurotransmitters and may remove extracellular K^+ that accumulates after action potentials. Salty taste may be transduced by the glial-like cells, but this is still uncertain. Type II receptor cells make up about 30% of the taste bud and contain G-protein coupled receptors (GPCRs) in their membranes. These cells are responsible for the transduction of sweet, bitter and umami tasting stimuli. They associate with afferent nerve fibres but do not form definitive synapses (Chaudhari and Roper, 2010). The type III cells or intermediate cells make up about 20% of the cells in the taste bud and have well-differentiated synapses. They are transducers for sour stimuli. They form prominent synapses with afferent nerve fibres and are also involved in cell-to-cell signalling in the taste bud (Roper, 2006; Kataoka *et al.*, 2008). Application of taste molecules on the taste cell causes a partial loss of the membrane potential, which is also called depolarization of the cell. In turn, the depolarization causes a number of intracellular events among others leading to the release of vesicles with neurotransmitter (type III cells). Interwoven among the taste cells is a branching network of taste nerve fibres that are stimulated by the taste cells through release of neurotransmitter. It has been observed that increasing the tastant concentration often leads to a greater depolarization of the taste cell.

Healthy adults have between 3000 and 10000 taste buds in the oral cavity. The taste cells in the taste buds are continuously regenerated by mitotic division. The mature taste cells lie in the middle of the taste bud, and break up and dissolve. The lifespan of taste cells is approximately ten days in lower mammals. The taste organ is thereby continuously in a process of regeneration and recovers from cellular damage, e.g. from burning the tongue when drinking a hot cup of tea. Some variation in the number of taste buds has been correlated to sensitivity to 6-propylthiouracil (PROP) and phenylthiocarbamide (PTC). Studies have shown that people with a high sensitivity to these substances have a greater number of fungiform papillae on the anterior part of the tongue (Miller and Reedy, 1990; Bartoshuk *et al.*, 1994).

Taste receptors and transduction

Taste components that cannot penetrate the taste cell membrane interact with taste receptor proteins located in the cell membrane. These proteins are a subclass of the super family of G-protein coupled receptors (GPCRs). The human taste receptors have been classified as T1R1, T1R2, T1R3 and T2Rs, and interact with bitter, sweet and umami substances. Other taste stimuli penetrate cell membranes including sodium, protons, undissociated acids and some bitter and sweet compounds. These compounds may interact with intracellular targets to activate the taste receptor cell. The definition of what would be a taste receptor for such ligands is therefore less clear (Bachmanov and Beauchamp, 2007). The activation of GPCRs by an external stimulus is the starting point of a succession of interactions between multiple proteins in the cell, leading to the release of chemical substances in the cell, also called second messengers. Although the cellular signal cascade is a general pattern of GPCRs, the large variety of each protein involved renders these mechanisms very complex and they are not completely understood. A schematic representation of the transduction cascades in taste cells is shown in Fig. 1.2.

(a)

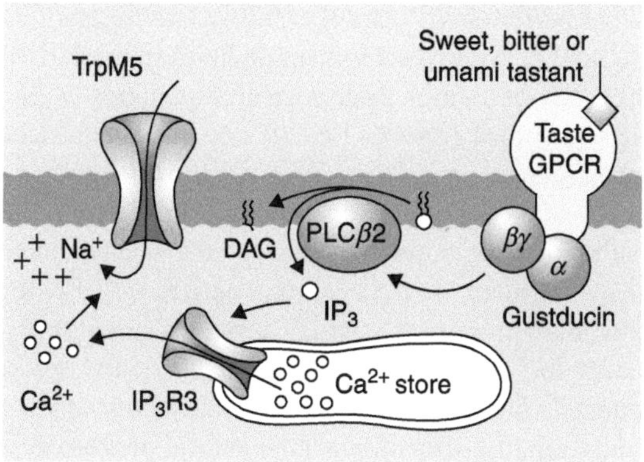

(b)

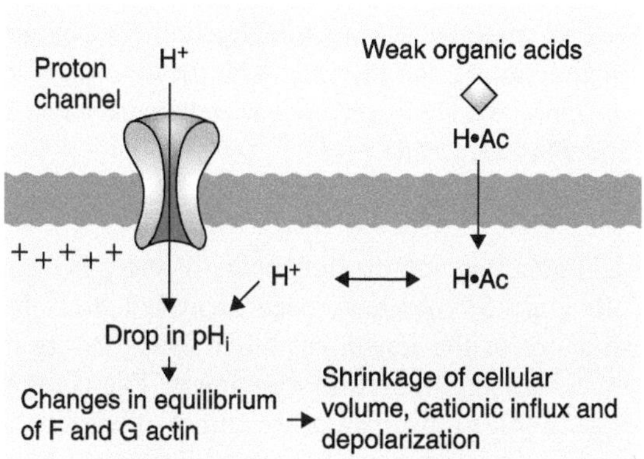

(c)

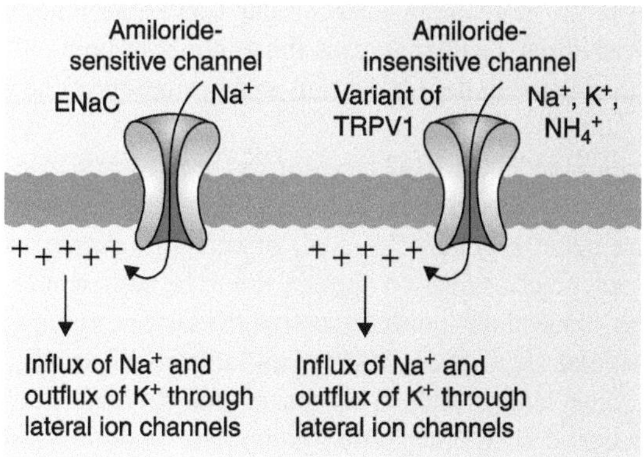

Fig. 1.2 Schematic overview of the apical transduction mechanisms for the five basic taste qualities. (a) The sweet, bitter and umami tastants are mediated by interactions with the taste receptors (GPCR) of the type II taste cells. The sour and salty tastants are mediated by ion channels causing different intracellular events. (b) The proton channels are presumably present in type III cells. (c) Type II or type III cells carry the channels for salty tastants. (Source: Chaudhari and Roper, 2010; Firestein and Beauchamp, 2008).

Sweet

The human receptor for sweet taste is a single heterodimeric T1R2/T1R3 receptor (Li *et al.*, 2002). Even though this receptor responds well to known sweet compounds, there may also be some low affinity of sweet tastants to its monomer or homodimers (Zhao *et al.*, 2003). Transduction of sweet taste is, however, generally accepted as proceeding via activation of the heterodimeric T1R2/T1R3 receptor with subsequent activation of the G protein gustducin and/or transducin and possibly $G_{i(2)}$. The gustucin $G_\beta G_\gamma$ subunit activates $PLC_{\beta 2}$ to act on membrane phosphatidylinositol to produce IP_3, which acts at its receptor IP_3R on intracellular Ca^{2+} storage sites to release Ca^{2+}. Finally, Ca^{2+} gates the transient receptor potential ion channel, TRPM5, enabling the inward flow of Na^+, depolarizing the taste bud cell and initiating signalling to the central nervous system (CNS). Even though only one receptor for sweetness exists, a plurality of binding sites is accessible as well as multiple agonist-binding loci (Dubois *et al.*, 2008). This means that the interaction of the receptor with the sweetener or combination of sweeteners can be complex. Hydrophobic sweeteners may also enter the taste cell to initiate the intracellular cascade.

Bitter

The receptors for bitter taste are also members of the GPCR family. They have been named T2Rs and 25 types have been identified from the human genome analysis. The structure of the functional bitter receptors are not known, but it seems unlikely that they are limited to monomers. The combinations of homo- and heterodimers easily lead to a manifold number of bitter receptors in humans. Many bitter tastants bind directly to GCPRs that activate PLC and the production of IP_3, leading to an increase in intracellular Ca^{2+}. The binding of intracellular Ca^{2+} to the ion channel TRPM5 causes the channel to open, allowing cations to enter the cell, thus causing depolarization and excitation of the taste receptor cell (Dubois *et al.*, 2008).

The genotype of the PTC/PROP receptor has been identified as T2R38 with PAV and AVI as halotypes. Accordingly, nontasters carry the recessive alleles AVI/AVI, while tasters are heterozygous (PAV/AVI) or carry the dominant AVI/AVI alleles. However, this genetic classification appears not to be sufficient to account for PTC/PROP sensitivity alone since it only expresses the taste receptor and not the number of fungiform papillae (Kim *et al.*, 2003). Furthermore, bitter receptors appear to be less specific to bitter compounds. This means that several T2Rs can interact with the same bitter ligand. The bitter taste system has, therefore, some analogy to the olfactory system, which is also broadly tuned with respect to odorants.

Umami

The receptor for umami taste is T1R1/T1R3. Both T1R1 and T1R3 are members of class C GPCRs, and T1R3 is the same 7-TMD protein present in the sweetener receptor T1R2/T1R3. In humans the umami receptor is specific for glutamate and aspartate, and exhibits strong enhancements upon addition of nucleotides such as inosine monophosphate. Beside the heteromeric receptor, several intracellular effectors mediate the umami taste. Umami taste is coded at the cellular level, just like sweet and bitter taste. Thus, within the taste bud some taste bud cells are specific sensors for umami taste stimuli (Dubois *et al.*, 2008).

Salt

Sodium salts are detected by the taste buds by permeating Na^+ through apical ion channels and thereby depolarizing the taste cells in the taste bud. The amiloride-sensitive epithelial Na channel (ENaC) has been associated with salt taste detection and is found in type I cells. In sensory studies scaling the intensity of salty taste in drinking water showed a good positive correlation with the concentration of Na^+, but Cl^- seemed also to play a role for salty taste perception in drinking water (Marcussen *et al.*, 2011, unpublished results).

Sour

The decrease of oral pH as a result of intake of an acid would suggest that acids leading to the same pH_{oral} have the same sourness; however, this is not the case. At a fixed pH, acetic acid is always more potent sour than an equivalent strong acid, e.g. HCl. This indicates that a decrease in pH_{oral} is not the signal for sour taste transduction. Changes in intracellular pH (pH_i) of taste bud cells correlate, however, very well with neural responses. This indicates that the pH_i is the proximal signal for sour taste transduction (Dubois *et al.*, 2008). Sour taste transduction is mediated by multiple pathways, including transport of H^+ through proton sensitive channels and cross membrane diffusion of undissociated acids.

Fat

There is growing evidence in support of taste receptors for dietary fat in humans. Although the sensory function to detect fat as a taste has been controversial, sham feeding experiments show that oral fat alone stimulates digestive responses in the gastro-intestinal tract (Crystal and Teff, 2006). Animal studies indicate that the lipid-binding protein CD36 is expressed in the apical regions, especially the circumvallate papillae. CD36 can transport long-chain fatty acids across the cellular membrane, and in mice it has been linked to preference of fatty acids and the cephalic phase response (Laugerette *et al.*, 2005). The mRNA of GPR 120, a fatty acid receptor expressed in the endocrine cells of the colon, has also recently been found in the circumvallate taste buds (Matsumara *et al.*, 2007). The G-protein coupled receptors GPR40 (Ffar1) and GPR120 are activated by medium- and long-chain fatty acids, and are expressed in mainly type I and type II cells of taste buds. Compared with wild-type mice, GPR120 knock-out and GPR40 knock-out mice show a diminished preference for linoleic acid and oleic acid, indicating a gustatory function of these oral fat receptors (Cartoni *et al.*, 2010).

Water

Although water is not normally considered to have a taste, it is well established that gustatory nerve fibres respond to water applied to the tongue. Hypo-osmotic solutions lead to an increase in cell capacitance accompanied by activation of a Cl^- conductance. Such a current is likely to play a role in the regulation of taste cell volume in response to varying osmotic taste stimuli (Gilbertson, 2002). Recently, the water-permeable ion channels involved in the hypo-osmotic swelling were identified (Watson *et al.*, 2007). Several aquaporins are expressed in taste cells, including AQP1 and AQP2, which label the basolateral membrane; and AQP5, which labels both the apical and basolateral membranes. AQP5 is the most likely candidate for water taste transduction in humans (Gilbertson *et al.*, 2006). Taste perception in humans largely depends on the state of adaptation in the oral cavity and thus affects the way water is perceived. Water elicits a strong sweet aftertaste when the oral cavity is exposed to sweetness blockers, which can be explained by interactions at the sweet taste receptor level (Galindo-Cuspinera *et al.*, 2006). The perceived freshness of pure, unfiltered drinking water from 20 different reservoirs in Denmark was negatively correlated with the total mineral content. The higher the total mineral level, the lower the perceived freshness was observed (Marcussen *et al.*, 2011, unpublished results). Future studies are likely to elucidate more clearly the physiology of water taste, and find explanations for the perceptions related to thirst.

The overall picture which emerges from these admittedly complicated details is that taste cells have a variety of transduction mechanisms. In general, individual taste cells respond to several types of chemosensory stimuli. Nevertheless, taste cells also exhibit gustatory selectivity. The lower the threshold concentration for detecting a single tastant, the greater the selectivity of the relevant taste cell (Bachmanov and Beauchamp, 2007). Taste receptor mechanisms also adapt to the ongoing presence of a stimulus. If a tastant is left in the mouth for a sufficient amount of time the taste will vanish, especially when the stimulus intensity is low. Thus, to obtain the full taste of foods, one must either frequently change the types of foods placed in the mouth or wait for a while between servings.

1.3.2 Sensing of pain and temperature

Temperature and pain sensations are very important to how people drink and eat their food. They set the range of physical temperatures and chemical hotness/coolness of foods and beverages that are acceptable for human consumption. There are three different types of stimuli that can excite pain receptors, namely mechanical, thermal and chemical. In all cases the receptors for pain sensations on the lips and in the mouth are free nerve endings of the so-called nociceptors. One major family of sensory transducers in nociceptors belong to the transient receptor potential (TRP) family of cation channels (Clapham *et al.*, 2005; Montell, 2003). Their unique response to temperature has given them the name thermoTRPs. These include members of the subfamily vanilloid TRPV (TRPV1, 2, 3, and 4), melastatin TRPM (TRPM8), and ankyrin transmembrane proteins TRPA (TRPA1). Between them, response to noxious heat is mediated by TRPV1 and TRPV2, innocuous warm temperature by TRPV3 and TRPV4, innocuous cool temperature by TRPM8, and noxious cold by TRPA1 (Dhaka *et al.*, 2006). These nociceptors are found in the trigeminal ganglia, and TRPM8 is also found in taste papillae. Several flavour compounds and other compounds found in foods have been shown to be specific targets and antagonists for these nociceptors (Mandadi and Roufogalis, 2008) as shown in Table 1.1.

Table 1.1 Thermoreceptors (TRPs) with some known food components as ligands and antagonists

TRP channel	Temperature (°C)	Activators (ligands)	Antagonists	Tissue/cell expression
TRPV1	>41	Vanilloids (capsaicin, piperine, eugenol, gingerols, capsiate); non-vanilloids (ginsenosides, unsaturated 1,4-dialdehyde, terpenes, camphor)	E.g. oleanolic acid	Sensory neurons and fibres, urinary bladder epithelium
TRPA1	<10	Isothiocyanates, cinnamaldehyde, acrolein, eugenol, methylsalicylate, gingerol, allicin, acrolein	E.g. camphor, menthol	Sensory neurons, inner ear hair cells
TRPM8	18–24	Terpenes (menthol, menthone, geraniol, linalool, cis/trans-p-menthane-3,8-diol); non-terpenes (WS3, cooling agent 10)	E.g. cinnamaldehyde	Sensory neurons
TRPV3	>33	Carvacrol, thymol, eugenol, dihydrocarveol, carveol, camphor, menthol, cinnamaldehyde, unsaturated fatty acids (sensitizer)	GRC15133 (structure not public)	Sensory neurons (low level)

Source: Cortright and Szallasi, 2009; Mandadi and Roufogalis, 2008

The chemosensory substances are especially important in stimulating the slow, persistent type of pain as is observed for capsaicin. In contrast to most other sensory receptors of the body, the pain receptors adapt very little and sometimes not at all. The pain sensation is linked to cellular changes in tissues and the release of the peptide bradykinin (Guyton and Hall, 1997).

Physical temperature of a drink is often perceived painful at temperatures above 45 °C (Hardy, 1956). This is also the temperature at which cellular tissues begin to be damaged by heat. It is therefore not surprising that the firing rate of the thermoreceptors peaks at approximate 45 °C in the transition from perceiving warm to hot. The thermoreceptors for the cold sensation have a much broader temperature range of activation and sensations of cold to freezing cold take place at temperatures below 10 °C (Guyton and Hall, 1997). The physical temperature of a drink also influences taste perception. Several studies have shown that basic tastes like sucrose, NaCl and quinine are more easily detected between 25–35 °C. A similar effect has also been observed for monosodium glutamate (MSG) (Jensen and Nielsen, 2011, unpublished data). The reason for this effect is not clearly understood and may be due to physiological and perceptual interaction effects. At high intensity of a basic taste, the effect of temperature on taste perception becomes negligible. Still, it seems not trivial at which temperature a drink is served, and to utilise specific components to, for instance, compensate for a higher physical serving temperature.

1.4 Sense of smell

The receptors for the sense of smell are located in the nose and occupy 4–5 cm^2 in each nostril. It is estimated that humans have about 350 different types of receptor cells. It is noteworthy that this number of different cells is dramatically different from the number of different receptor cells we have in the eye (three types (cones) for day vision and one type (rod) for night vision/vision under low ambient luminance conditions). This difference strongly suggests that coding and processing of visual and olfactory information use different strategies. Most odorants activate many of these 350 different cells, but different smells activate different subsets to different degrees. From the receptor cells in the nose signals are transmitted to neurons in the olfactory bulbs, bulb-formed neural structures located behind the eyes. After signal transformations in the bulbs, olfactory information is relayed via the olfactory tract to the piriform cortex, the so-called primary olfactory cortex and amygdaloid areas. The amygdalae (one in each side of the brain) are part of the limbic system, which is strongly implicated in emotional processing. This means that olfactory information reaches emotional brain areas after only two relays. Signals in other senses need to pass more relay stations before reaching the amygdalae and other emotional brain areas. This strongly suggests that the sense of smell has particular salience as a conveyer of emotions. This is a suggestion supported by a number of investigations (Herz *et al.*, 1995; Herz *et al.*, 2004). From the piriform cortex and amygdaloid areas, signals are relayed to other areas in the brain, both in the midbrain and in the frontal areas in the brain. Various frontal areas have been implicated in olfactory processing, most notably the orbitofrontal cortex.

The olfactory brain is ipsilaterally organized, as opposed to the other senses, which are contralaterally organized. Ipsilateral means that signals from the right nostril are transmitted to the same (ipsi) side, i.e. the right side; and signals from the left nostril are transmitted to the left side of the olfactory brain. In a contralateral sense (contra meaning opposite) such as vision, signals from the left visual field are transmitted to the right part of the visual brain and vice versa for signals from the right visual field. After having been processed in the right half of the brain, signals from the right nostril can eventually influence activity in the left half of the brain by signals transmitted via the corpus callosum, which is a bundle of nerve fibres connecting the two halves of the brain.

Furthermore, olfactory signals do not have to pass through a midbrain relay in the thalamus, as signals in the other senses do, before they are transmitted to limbic areas and to areas in the orbitofrontal cortex. From a structural anatomical point of view, olfaction is, thus, very differently wired compared with the other senses. This, together with the dramatic differences in number of different types of receptor cells (in e.g. vision and olfaction) might suggest that olfaction could also have very different properties from, e.g. vision.

1.4.1 Common function of the sense of smell

Smell serves us in two ways when we deal with foods and drinks. Before we ingest a food or a drink, we often smell it, by sniffing at it. This sniffing behaviour allows us to judge if the food or drink is safe to ingest, and we form anticipations of the quality and taste of it. When a smell is estimated or appreciated by sniffing, odorants enter the nose through the nostrils. This type of olfactory perception is orthonasal perception. Once we have decided that the food or drink we estimate by smelling is safe and (probably) well-tasting, we ingest it and chew it (if food) followed by swallowing. During chewing and swallowing, aroma molecules are released from the food or drink and these reach the olfactory epithelium via the connection of the mouth and nose. Olfactory perception via this route is called retronasal olfactory perception. These different functions of olfactory perception seem to be reflected in the neural underpinnings of anticipatory and consummatory (actually eating or drinking) food chemosensation. Small and colleagues have found separable and overlapping representations of anticipatory and consummatory chemosensation (Small *et al.*, 2008). It has also been demonstrated that the human brain responds differentially to odorants presented orthonasally and retronasally (Small *et al.*, 2005).

There is no strict inter-subjectivity in olfaction. Olfactory experiments often require at least 20 subjects, whereas it is perfectly acceptable to only perform vision measurements on 3–5 subjects. Most olfactory judgements are learned, i.e.

we are not born with certain preferences. This is particularly important for the perception and evaluation of foods and drinks, and explains why very different culinary traditions have developed in different parts of the world, based upon what nature has to offer at particular places. We come to like what we have or to which we have access.

Also of importance to food and drinking behaviour is the fact that olfactory adaptation is very strong and sometimes complete. Adaptation means that the sensitivity to a stimulus is reduced when we are exposed to it. These effects can be quite quick, rendering an odorant to become unnoticeable within a few seconds. For other odorants, adaptation might take minutes. There are both peripheral (in the nose) adaptation mechanisms and central (in the brain) mechanisms. Furthermore, an odorant does not only adapt the olfactory system's ability to detect the presence of itself (auto-adaptation) but may also affect the system's perception of other odorants (cross-adaptation). These effects strongly affect what we perceive when we engage in continual eating and drinking. If a substance with an unwanted flavour is present in a drink, for example, but is not perceived because it is masked by another flavour component, then adaptation to the masker could lead to an off-flavour perception caused by the substance with the unwanted flavour. This might be why the second beer often tastes different from the first that you drink.

Smells are seldom in our focus of attention, but this does not mean that they do not serve to guide our behaviours. A number of recent experiments have revealed that odours and flavours that are not attended to at all are nevertheless encoded and remembered (Møller *et al.*, 2004; Møller *et al.*, 2007; Sulmont-Rossé *et al.*, 2008). This type of learning is referred to as 'incidental' learning, as opposed to intentional learning where people learn a piece of information intentionally. Intentional learning is characteristic of what we do when we attend school. The memories we build are of an explicit nature and what we learn can be declared (declarative, explicit memories). This is quite different from ordinary life and in our dealings with food and drink, where we seldom have any intention of encoding what we encounter. The memories we form of such events are of a much more implicit nature and learning is incidental. We nevertheless do remember events and objects even without any intention to do so. The memories we build under such circumstances seem to have very different properties from more explicit types of memories, often involving semantic, visual and verbal material. It has been found that incidentally-learned information is not based on actual recognition of a certain target, but rather is based on detection of novelty. That is, when prompted, we detect that a certain stimulus is not identical to the one we encountered previously. When presented with the same stimulus as previously (the target) we are often at chance level at detecting that it is the same stimulus. These effects have also been found in other lower senses than smell (Mojet *et al.*, 2005; Köster *et al.*, 2004; Møller *et al.*, 2007; Laureati *et al.*, 2008). Furthermore, opposed to what is usually found in memory experiments, that ability to remember decays with age, in the incidental learning experiments on the lower senses referred to above, elderly people remember stimuli as well as young people. These results are highly relevant for the understanding of eating behaviour as well as choice (of foods and drinks) and the changes observed in these choices. They might also help explain the relative constancy of food and drink preferences over life despite rather dramatic changes in the sensory systems that underlie food perception.

Relying on memory systems with these properties in conjunction with senses that are highly sensitive to expectation and anticipation effects, might be sufficient to explain the relative constancy of food preferences as well as why a change in a food or drink occasionally turns out to be disastrous. Since olfactory stimuli are rather meagre in information content compared to, e.g. visual stimuli, one might expect that olfaction could be more prone to influences from top-down effects. Such effects have indeed been found in olfaction. It has been demonstrated how different expectations to an odour dramatically changes the activities in the olfactory system (de Araujo *et al.*, 2005; Grabenhorst *et al.*, 2008; Grabenhorst and Rolls, 2008). These effects might go some way in explaining why advertising actually might work under appropriate circumstances.

1.5 Integration of sensory information

As explained above, many senses contribute to the taste of foods and drinks. Even though a number of separate senses contribute to each food sensation, we do not perceive foods as a number of individual sensations, but rather as a coherent (integrated) whole. Integration of different sensory modalities is commonplace in the sensory system. When you watch a film in a movie theatre you perceive the sounds and speech as coming from the actors, mouths, even though the sound is produced by loudspeakers located elsewhere in the movie theatre. Information from your visual and auditory systems is integrated centrally in the brain to bind the different types of sensory information into coherent meaningful wholes.

Much the same happens when we perceive one of the many tastes (flavours) available to us. The binding together of different sensations into coherent wholes take place both within a particular sense as well as between different senses. Most food aromas (smells) are mixtures of hundreds of different types of molecules with their own smells. These individual smells are in general completely disguised by the integration into an overall aroma (smell) of the food. It has been demonstrated that humans are not capable of identifying more than two to three components correctly in mixtures of smells (Jinks *et al.*, 1999). That is, if a mixture consists of more than three components, we cannot reliably report what these components are without making a mistake. The mixture, of course, still has a characteristic smell, but it is a synthetic perception, which is the result of olfactory integration processes which abolish perception of the smell of the individual components. These integration processes are of crucial importance for the perception of foods and drinks as well as fragrances. A smell which is unpleasant on its own, can, nevertheless, when added to a mixture of other odorants, turn this into a more pleasantly-smelling mixture. An odorant present in a mixture in such a small concentration that it cannot be detected if it was presented alone can change and improve the overall pleasantness of the mixture. At present we do not have a good understanding of these highly non-linear phenomena. A number of models have, however, been developed, but these are linear and can only capture very few of the effects that are important for mixtures in the real world (foods and drinks) (Lapid *et al.*, 2008).

Integration of the different senses into an overall flavour percept has been studied both by psychophysical and neuroimaging means. Verhagen and Engelen have collected and described many important effects in multimodal food perception in a recent review (Verhagen *et al.*, 2006). Neuroimaging studies conducted on flavour perception have revealed a number of brain areas involved in the integration of signals from the different food-related senses. Most notable among these are the so-called orbito-frontal cortices, located in the front of the brain over the orbits of the eyes. These areas receive input from smell, taste, touch and vision, and are part of a putative flavour circuit in the human brain (Rolls, 1999). The orbito-frontal cortex has also been implicated in human brain imaging as the area that computes and represents sensory specific satiety, the well-known phenomenon that liking of a particular food or drink decreases with the amount eaten or drunk, without affecting appreciably the liking of other foods or drinks.

1.6 Mechanisms of perception, cognition and emotion

Each of the senses has its own neural system in the brain, and only later along the chain of processing is there massive integration of signals from the different sensory systems. The perception of flavour constitutes a prime example of integration of information from different senses (Kandel *et al.*, 1985; Smock, 1999). From a functional point of view the brain's tasks are often subdivided into categories of perception, cognition, emotion and action. Each of the senses performs tasks from each category. Perception allows the perceiver to obtain knowledge about what is where and when. Typical perceptual tasks are detecting the presence of a certain stimulus (a smell, a taste, etc.) and discrimination between different stimuli. Determining the perceived intensity of a stimulus is also a perceptual task. The 'higher senses' (vision and audition) also provide information about the layout of the environment as well as the movement of objects. The higher senses are sometimes referred to as 'far senses', because they provide information about the environment that is 'far away' and they do not rely on direct interactions with the physical materials in the environment. Such types of information are not provided by the 'lower senses' (smell, taste, trigeminality and touch) which 'only' provide information about the 'quality' and quantity and rely on direct interactions between physical materials and receptor cells. These senses are, therefore, also referred to as 'near senses'.

Cognition is responsible for higher mental states (e.g. rational thinking, planning), which are often of a kind that allow for introspection and consciousness. Even though some cognitive processes are embedded in each of the senses, they are much more central to the so-called higher senses, vision and audition, but, as we discussed previously, some types of memories seem to be very important for food and drink behaviour.

Emotional processes, on the other hand, seem to be more central to the lower senses of taste, smell, touch and trigeminality. Any judgement of the pleasantness or hedonic quality of a stimulus is of an emotional nature, and such processes are central to the senses that are strongly involved in the perception and evaluation of foods and drinks.

1.7 Sensory perception, satiation and food reward

When we eat and drink we become satiated, and intake normally ends. Much of satiety is controlled by a set of so-called homeostatic processes, which are negative feedback loop processes that help keep appropriate balances of nutrients in the body. Besides the homeostatic satiety processes, humans also possess so-called sensory-specific satiety mechanisms. Sensory-specific satiety denotes the decline in liking of a food eaten to satiety without significant effects on the appreciation on other foods. An animal endowed with such mechanisms will tend to eat a varied diet, which in turn will counteract the risk of malnutrition. These mechanisms obviously have much influence in guiding an eater/drinker in their choices during a meal. These mechanisms are not innate, as demonstrated recently (Olsen *et al.*, 2011). It was found that sensory-specific satiety in children is bound strongly to the whole product eaten, whereas clear transfer effects were found for adults. That is, for adults, eating or drinking a product with a certain sensory profile will lead to some transfer of drop of liking to other products that share one or more of the same characteristic sensory attributes present in the product eaten to satiation.

These mechanisms also highlight the importance of reward for food intake. Reward is a term that covers a set of different psychological and neurobiological processes. Reward is often divided into three categories: liking, wanting and learning (Berridge, 1996; Berridge *et al.*, 2003). All three types of processes come in two different types: some of which we have awareness of and some, the core processes, of which we have no conscious awareness. Liking describes the immediate appreciation of a stimulus (e.g. food or drink) whereas wanting covers a set of motivational processes which capture how much we will work to obtain a certain stimulus. Liking and wanting, of course, often co-vary, but there are numerous examples of cases where a certain food or drink is liked but not wanted and therefore not chosen. It is reasonable to include learning processes in the set of processes we need to consider to understand reward. This is particularly clear with respect to foods and drinks where almost all rewards are learned via a number of mechanisms.

1.8 Sensory perception and learning of preferences

Some foods and drinks are liked more than others, and thus we prefer one food or drink rather than another. Preferences vary with culinary culture, whereas the perceptual system does to a much lesser degree. Preferences are considered as learned behaviours and it has been demonstrated that learning takes place in the foetal state (Schaal *et al.*, 2000) and first months of life (Hausner *et al.*, 2010). This is believed to take place via transfer of flavour substances eaten by the mother to the amniotic fluid and from there to the developing sense organs of the foetus. After birth, and probably all through life, a number of so-called conditional learning mechanisms act to change our preferences. Conditional learning means that we learn, or change, on the condition that another unconditional stimulus is present together with the stimulus we learn about, the conditional stimulus. In the case of preference learning an already well-liked stimulus plays the role of the unconditional stimulus, which is a stimulus to which we have an unconditional positive response. The response to the conditional stimulus presented alone is called the conditional response. For a new food or drink not resembling anything ever encountered previously the conditional response will most often be a rejection.

If, however, the new food is presented together with an already liked stimulus, it turns out that after a sufficient number of presentations of the conditional and unconditional stimuli together we will have a changed response to the conditional stimulus presented alone. What initially was rejected and not liked will now be accepted and liked. It is as if the properties of the unconditional stimulus have been taken over by the conditional stimulus. If a known liked flavour is paired with a novel flavour the conditional learning is called flavour–flavour learning. Another type of conditional learning important for changes of preferences of foods and drinks is the so-called flavour–nutrient learning effect. In this type of learning it is the nutrient value/energy in an unconditional stimulus which, when paired with the novel flavour of a conditional stimulus, will increase the liking of the novel flavour. Finally, the so-called ‘mere exposure effect’, where it is a sufficient number of exposures to a new flavour which changes our appreciation of it, is an important mechanism for preference development (Capaldi, 2001; Hausner *et al.*, 2011). Mere exposure might be a result of conditional learning, but in this interpretation, it is the absence of adverse effects, after being exposed to a novel flavour, which plays the role of the unconditional stimulus (Zajonc, 2001).

These forms of preference formation mechanisms are believed to be important in forming children’s food and drink preferences. They play undoubtedly also a role in the changes of preferences experienced by adults, but for this population it is not as well-defined to talk of novel flavours, since such a flavour might have certain resemblances to other flavours that have been perceived previously.

1.9 Conclusion and future trends

In this chapter we have given a comprehensive overview and update on interdisciplinary aspects of sensory perception providing the reader with a background on the topic. Sensory perception and appreciation of foods and beverages is of great importance for the enjoyment of our life and the human well-being in social contexts. The underlying scientific mechanisms within molecular biology, physiology, psychology, and food science and nutrition have only been discovered over the last decades. This has given many insights and inspiration for product developers and other professionals to make foods that give us rewarding experiences at a reasonable cost. It is hoped that the advancement and further integration of the diverse disciplines dealing with the acceptability of foods by the consumer will keep on stimulating the advancements for the needs of foods in the future.

1.10 Sources of further information

DOTY, R.L. (ed.) (2003), *Handbook of Olfaction and Gustation*, Marcel Dekker, New York, 2nd edition.

FIRESTEIN, S. and BEAUCHAMP, G.K. (eds) (2008), *The Senses: a Comprehensive Reference, Vol. 4, Olfaction and Taste*, Academic Press, Elsevier, Oxford.

SMOCK, T.K. (1999), *Physiological Psychology. A Neuroscience Approach*, Prentice Hall, New York.

1.11 References

DE ARAUJO, I.E.T., ROLLS, E.T., VELASCO, M.I., MARGOT, C. and CAYEUX, I. (2005), 'Cognitive modulation of olfactory processing', *Neuron*, **46**, 671–679.

BACHMANOV, A.A. and BEAUCHAMP, G.K. (2007), 'Taste and receptor genes', *Annual Review of Nutrition*, **27**, 389–414.

BARTOSHUK, L.M., DUFFY, V.B. and MILLER, I.J. (1994), 'PTC/PROP tasting: Anatomy, psychophysics, and sex effects', *Physiology and Behavior*, **56**, 1165–1171.

BERRIDGE, K.C. (1996), 'Food reward', *Neuroscience and Biobehavioral Reviews*, **20**, 1–25.

BERRIDGE, K.C. and ROBINSON, T.E. (2003), 'Parsing reward', *Trends in Neurosciences*, **26**, 507–513.

CARTONI, C., YASUMATSU, K., OHKURI, T., SHIGEMURA, N., YOSHIDA, R. *et al.* (2010), 'Taste preference for fatty acids is mediated by GPR40 and GPR120', *The Journal of Neuroscience*, **30**, 8376–8382.

CAPALDI, E.D. (2001), 'Conditioned food preferences'. In: *Why We Eat What We Eat – The Psychology of Eating*, Capaldi, E.D. (ed.), Washington DC: American Psychological Association, 53–80.

CHAUDHARI, N. and ROPER, S.D. (2010), 'The cell biology of taste', *Journal of Cell Biology*, **190**, 285–296.

CLAPHAM, D.E., JULIUS, D., MONTELL, C. and SCHULTZ, G. (2005), 'International Union of Pharmacology. XLIX. Nomenclature and Structure–function relationships of transient receptor potential channels', *Pharmacological Reviews*, **57**, 427–50.

CORTRIGHT, D.N. and SZALLASI, A. (2009), 'TRP channels and pain', *Current Pharmaceutical Design*, **15**, 1736–1749.

CRAIG, A.D. (2002), 'How do you feel? Interoception: the sense of the physiological condition of the body', *Nature Reviews Neuroscience*, **3**, 655–66.

CRYSTAL, S.R. and TEFF, K.L. (2006), 'Tasting fat: cephalic phase hormonal responses and food intake in restrained and unrestrained eaters', *Physiology and Behavior*, **89**, 213–220.

DHAKA, A., VISWANATH, V. and PATAPOUTIAN, A. (2006), 'Trp ion channels and temperature sensation', *Annual Review Neuroscience*, **29**, 135–61.

DUBOIS, G.E., DESIMONE, J. and LYALL, V. (2008), 'Chemistry of gustatory stimuli'. In: *The Senses: A Comprehensive Reference, Vol. 4, Olfaction and Taste*, Firestein, S. and Beauchamp, G.K. (eds), Academic Press, Elsevier, Oxford.

GALINDO-CUSPINERA, V., WINNING, M., BUFE, B., MEYERHOF, W. and BRESLIN, P.A. (2006), 'A TAS1R receptor-based explanation of sweet "water-taste" ', *Nature*, **441**, 354–357.

GIDUCK, S.A., THREATTE, R.M. and KARE, M.R. (1987), 'Cephalic reflexes: their role in digestion and possible roles in absorption and metabolism', *Journal of Nutrition*, **117**, 1191–1196.

GILBERTSON, T.A. (2002), 'Hypoosmotic stimuli activate a chloride conductance in rat cells', *Chemical Senses*, **27**, 383–394.

GILBERTSON, T.A., BAQUERO, A.F. and SPRAY-WATSON, K.J. (2006), 'Water taste: the importance of osmotic sensing in the oral cavity', *Journal of Water and Health*, **4**, 35–40.

GRABENHORST, F., ROLLS, E.T. and BILDERBECK, A. (2008a), 'How cognition modulates affective responses to taste and flavour: Top-down influences on the orbitofrontal and pregenual cingulated cortices', *Cerebral Cortex*, **18**, 1549–1559.

GRABENHORST, F. and ROLLS, E.T. (2008b), 'Selective attention to affective value alters how the brain processes taste stimuli', *European Journal of Neuroscience*, **27**, 723–729.

GUYTON, A.C. and HALL, J.E. (1997), *Human Physiology and Mechanisms of Disease*, 6th edition, W.B. Saunders Company, Philadelphia.

HARDY, J.D. (1956), 'The nature of pain', *Journal of Chronic Diseases*, **4**, 22–51.

HAUSNER, H., NICKLAUS, S., ISSANCHOU, S., MØLGAARD, C. and MØLLER, P. (2010), 'Breastfeeding facilitates acceptance of a novel dietary flavour compound', *Clinical Nutrition*, **29**, 141–148.

HAUSNER, H., HARTVIG, D., REINBACH, H., WENDIN, K. and BREDIE, W.L.P. (2011), 'Effects of repeated exposure on acceptance of initially disliked and liked Nordic snack bars in 9 to 11 year-old school children', *Clinical Nutrition* (in press).

HEFT, M.W. and ROBINSON, M.E. (2010), 'Age differences in orofacial sensory thresholds', *Journal of Dental Research*, **89**, 1102–1105.

HERZ, R.C. and CUPCHIK, G.C. (1995), 'The emotional distinctiveness of odor-evoked memories', *Chemical Senses*, **20**, 517–528.

HERZ, R.C., ELIASSEN, J., BELAND, S. and SOUZA, T. (2004), 'Neuroimaging evidence for the emotional potency of odor-evoked memory', *Neuropsychologia*, **42**, 371–378.

JENSEN, S. and NIELSEN, B. (2011), 'The influence of temperature on the threshold of umami taste'. MSc thesis (in Danish), University of Copenhagen.

JINKS, A. and LAING, D.G. (1999), 'A limit in the processing of components in odour mixtures', *Perception*, **28**, 39–404.

KANDEL, E.R. and SCHWARTZ, J.H. (1985), *Principles of Neural Science*, Elsevier, London, New York.

KATAOKA, S., YANG, R., ISHIMARU, Y., MATSUNAMI, H., SEVIGNY, J. *et al.* (2008), 'The candidate sour taste receptor, PKD2L1, is expressed by type III taste cells in the mouse', *Chemical Senses*, **33**, 243–254.

KIM, U.K., JORGENSON, E., COON, H., LEPPERT, M., RISCH, N. *et al.* (2003), 'Positional cloning of the human quantitative trait locus underlying taste sensitivity to phenylthiocarbamide', *Science*, **299**, 1221–1225.

KÖSTER, M.A., PRESCOTT, J. and KÖSTER, E.P. (2004), 'Incidental learning and memory for three basic tastes in food', *Chemical Senses*, **29**, 441–453.

LAPID, H., HAREL, D. and SOBEL, N. (2008), 'Prediction models for the pleasantness of binary mixtures in olfaction', *Chemical Senses*, **33**, 599–609.

LAUGERETTE, F., PASSILLY-DEGRACE, P., PATRIS, B., NIOT, I., FEBBRAIO, M. *et al.* (2005), 'CD36 involvement in orosensory detection of dietary lipids, spontaneous fat preference, and digestive secretions', *Journal of Clinical Investigation*, **115**, 3177–3184.

LAUREATI, M., MORIN-AUDEBRAND, L., PAGLIARINI, E., SULMONT-ROSSÉ, C., KÖSTER, E.P. *et al.* (2008), 'Food memory and its relation with age and liking: An incidental learning experiment with children, young and elderly people', *Appetite*, **51**, 273–282.

LI, X., STASZEWSKI, L., XU, H., DURICK, K., ZOLLER, M. *et al.* (2002), 'Human receptors for sweet and umami taste', *Proceedings of the National Academy of Sciences USA*, **99**, 4692–4696.

MANDADI, S. and ROUFOGALIS, B.D. (2008), 'ThermoTRP channels in nociceptors: Taking a lead from capsaicin receptor TRPV1', *Current Neuropharmacology*, **6**, 21–38.

MATSUMURA, S. MIZUSHIGE, T., YONEDA, T., IWANAGA, T., TSUZUKI, S., INOUE, K. and EUSHICKI, T. (2007), 'GPR expression in the rat taste bud relating to fatty acid sensing', *Biomedical Research*, **28**, 49–55.

MARCUSSEN, H., BREDIE, W.L.P., NIELSEN, S.S., BRÜSCH, W., HOLM, P.E. *et al.* (2011), 'Taste of drinking water as a function of inorganic chemical content'. Internal report, University of Copenhagen, Denmark.

MATTES, R.D. (1997), 'Physiologic responses to sensory stimulation by food: nutritional implications', *Journal of the American Dietetic Association*, **97**, 406–413.

MILLER, I.J. and REEDY, F.E. (1990), 'Variations in human taste bud density and taste intensity perception', *Physiology and Behavior*, **47**, 1213–1219.

MILLER, I.J. (1995), 'Anatomy of the peripheral taste system'. In: Handbook of Olfaction and gustation, R.L. Doty (ed.). Mareel Dekker. New York. 1st edition, pp. 521–547.

MOJET, J. and KÖSTER, E.P. (2005), 'Sensory memory and food texture', *Food Quality and Preference*, **16**, 251–266.

MONTELL, C. (2003), 'Thermosensation: hot findings make TRPNs very cool', *Current Biology*, **13**, R476–8.

MØLLER, P., MOJET, J. and KÖSTER, E.P. (2007), 'Incidental and intentional flavor memory in young and older subjects', *Chemical Senses*, **32**, 557–567.

MØLLER, P., WULFF, C. and KÖSTER, E.P. (2004), 'Do age differences in odor memory depend on differences in verbal memory?', *Neuroreport*, **15**, 915–917.

OLSEN, A., RITZ, C., HARTVIG, D.L. and MØLLER, P. (2011), 'Comparison of sensory specific satiety and sensory specific desires to eat in children and adults', *Appetite* **57**, 6–31.

ROLLS, E.T. (1999), *The Brain and Emotion*, Oxford University Press, Oxford.

ROPER, S.D. (2006), 'Cell communication in taste buds', *Cellular and Molecular Life Sciences*, **63**, 1494–1500.

SCHAAL, B., MARLIER, L. and SOUSSIGNAN, R. (2000), 'Human foetuses learn odours from their pregnant mother's diet', *Chemical Senses*, **25**, 729–737.

SMALL, D.M., GERBER, J.C., MAK, Y.E. and HUMMEL, T. (2005), 'Differential neural responses evoked by orthonasal versus retronasal odorant perception in humans', *Neuron*, **47**(4), 593–605.

SMALL, D.M. and PRESCOTT, J. (2005), 'Odor/taste integration and the perception of flavour', *Experimental Brain Research*, **166**, 345–357.

SMALL, D.M., VELDHUIZEN, M.G., FELSTED, J., MAK, Y.E. and MCGLONE, F. (2008), 'Separable substrates for anticipatory and consummatory food chemosensation', *Neuron*, **57**, 786–797.

SULMONT-ROSSÉ, C., MØLLER, P., ISSANCHOU, S. and KÖSTER, E.P. (2008), 'Effect of age and food novelty on food memory', *Chemosensory Perception*, **1**, 199–209.

VERHAGEN, J.V. and ENGELEN, L. (2006), 'The neurocognitive bases of human multimodal food perception: Sensory integration', *Neuroscience and Biobehavioral Reviews*, **30**, 613–650.

WATSON, K-J., KIM, I., BAQUERO, A.F., BURKS, C.A., LUI, L. *et al.* (2007), 'Expression of aquaporin water channels in rat taste buds', *Chemical Senses*, **32**, 411–421.

ZAJONC, R.B. (2001), 'Mere exposure: A gateway to the subliminal', *Current Directions in Psychological Science*, **10**, 224–228.

ZHAO, G.Q., ZHANG, Y., HOON, M.A., CHANDRASHEKAR, J., ERLENBACH, I. *et al.* (2003), 'The receptors for mammalian sweet and umami taste', *Cell*, **115**, 255–266.

2

Sensory quality control and assurance of alcoholic beverages through sensory evaluation

D. McGrew, Beam Global Spirits & Wine, USA and E. Chambers IV, Kansas State University, USA

Abstract: Tools are available, through various reference books, to develop a purposeful sensory quality program. Some companies already have a strong sensory program in place; others may require a cultural change to facilitate the implementation. This chapter indicates some of the challenges to be overcome, covers some current quality control (QC) sensory practices, and addresses advantages and disadvantages of expert tasters. Some specific issues regarding sensory evaluations of alcohol beverages are discussed, and critical factors in production are reviewed with discussion on the potential for off-taint development.

Key words: alcoholic beverages, sensory quality, expert tasters, off odors, quality control.

2.1 Introduction

There are numerous reference materials on sensory quality and its place within the quality control (QC)/quality assurance (QA) function. These reference materials give explicit detail on how to plan, set up, and implement a sensory program for QC. Most current sensory analysis books devote at least a chapter to the subject. The American Society for Testing and Materials (ASTM), for example, published a 52-page manual detailing the intricacies of implementing a sensory program in a manufacturing plant's QC environment. 'to bring sensory testing for quality control . . . out of the "touchy/feely" and value judgment realm into the world of quality control data reporting' (Yantis, 1992).

Despite the availability of such reference materials, the integration of sensory evaluation into the areas of QC/QA has achieved only moderate success throughout the food and beverage industries. As Stone and Sidel (1993) mention, 'most QC sensory programs have not been as successful as anticipated'. Even after nearly two decades that statement holds true; and the success of these programs in the alcohol beverage industry is no exception.

Stone and Sidel (1993) speculated that the lack of success in the implementation of sensory programs is likely to be a result of the organizational environment. This is reasonable because many manufacturing plants operate under the old axiom, 'if it's not broke, don't fix it'. The challenge is for QC sensory professionals who understand the benefits of a well-run sensory program to continue to enlighten industry management by demonstrating the efficiency and practicality of a sensory quality model. The difficulty of this challenge is in describing the benefits of a sensory program and how it can save time and money, while concurrently increasing value for consumers. Phrased in this manner the business case in support of a QC sensory program is more meaningful to management than scientific jargon about statistical power, repeatability or reproducibility.

2.2 Sensory quality concerns and issues in the alcohol beverage industry

Most alcohol beverage manufacturers perform some form of flavor evaluation on their products. Some companies, typically the larger ones, have established sensory groups/departments responsible for the overall sensory programs used in the production facilities' QC departments. In these instances, the sensory group will communicate, as needed, standard operating procedures, conduct and monitor panel training and interpret results for corrective actions. Other companies perform sensory testing on an *ad hoc* basis where the evaluations are conducted to meet the needs of a specific situation. On this point, it should be understood that many recognizable brands in the alcohol beverage industry have been producing products much longer than current organized sensory practices have been in use. The implication is not that sensory quality is an unimportant aspect in these products' production. In fact, the United States Bottled-in-Bond Act of 1897 was enacted to protect whiskey consumers from illicit and often lethal productions of blended whiskey. The Bottled-in-Bond Act is considered a precursor to the US Pure Food and Drug Act of 1906. Despite providing no direct guidance for sensory quality, whiskey labeled as Bottled-in-Bond came to be known as 'the good stuff' because the Act did provide assurance to the consumer (Veach, undated).

Meiselman and Schutz (2003) note that the old Seagram's™ company was one the first companies, in any industry, to embrace sensory techniques established in the 1940s by the US Army's Quartermaster Corps. Throughout the 1900s, Seagram's™ was one of, if not the, largest distilled spirits companies in the world. The Seagram's™ brand still exists, but is currently owned by Diageo™. Meiselman and Schutz (2003) describe the hiring of David Peryam and Norman Giradot, two of Seagram's™ quality control scientists, by the US Army's laboratory in 1949 as a critical step that led to the development of many sensory analysis techniques still in use today. It is apparent that the alcohol beverage industry has made an important impact in the development of sensory evaluation methods and the field of sensory analysis as a whole. Despite this link in the development of sensory evaluation methods to the alcohol beverage industry, many companies still lack proper support for, and use of, sensory quality practices. Munoz (2002) stated, 'Current QC/sensory programs in industry cover the whole gamut of "quality", from "expert" measures similar to those used a century ago to state-of-the-art programs'.

At some point in the history of QC evaluations of alcohol beverages the phrase 'organoleptic testing' crept into the industry's vocabulary. Though organoleptic testing does refer to the use of the senses for product evaluations, that is where the comparison to sensory evaluations stops. Sensory evaluation involves greater attention to details such as methodological assessor training, sample preparation and recording and reporting steps, that are useful to the scientific evaluation of products. Therefore, organoleptic tests are a more qualitative, subjective method; whereas sensory evaluations typically have a more scientific basis. In this context it is not surprising that most sensory texts avoid discussing the use of organoleptic testing. Since at least the early 1980s, sensory professionals have been trying to eliminate the use of the word organoleptic from the practice of product evaluations (Jellinek, 1985; McDaniel, 1985). The problem is that many people continue to use the words organoleptic and sensory in the same context. The misuse of these words is especially prevalent in the alcohol beverage industry. It is important for the laboratory analyst, sensory scientist, or whoever is charged with the function of product quality assessment, to fully understand these differences, and the deficiencies related to organoleptic testing. It is also important to understand that the terms organoleptic testing and sensory evaluation are not synonymous, and should never be used interchangeably.

A similar but perhaps more damaging effect to the integrity of sensory practices has been misuse of methods. It is not at all uncommon to find QC staff combining difference tests, e.g. combining triangle tests or paired comparisons with a preference test. Many researchers have commented on the misapplication or combining of test methods and the resulting problems from this practice (Stone and Sidel, 1993; O'Mahony, 1986). Stone and Sidel explained the inherent problems (1993) as follows:

- Odd sample bias: panelists tend to prefer samples that are not perceived as different (or odd in the case of a triangle test) from a reference. Since the initial test is to determine sample difference, panelists are now predisposed to a preference bias.
- Skewed results: panelists for a discrimination test are recruited based on their acuity and usually specific experience in looking for differences in a product.
- Small sample sizes: discrimination tests typically use 20–30 panelists while preference/acceptance tests require much larger panel sizes.
- Transition of tasks: it is more difficult than it might appear to mentally 'flip the switch' from trained discriminator to consumer, which potentially leads to odd sample bias.

- Making sense of the results: do you only include those preferences from subjects that were correct on the difference test? Do you include the preferences from all subjects?

Additionally, the use of preference tests as part of plant sensory quality does not make practical sense, because a primary role of QC is to make a product consistently, irrespective of personal likes or dislikes. Any manufacturer would be horrified to know that based on preference or acceptance tests alone, they could be bottling another manufacturer's beverage just because it tasted good – even though it did not match the flavor of their own product. Other common misuses of sensory practices include the use of discrimination test methods (i.e. triangle or duo-trio tests) with too few respondents for statistical relevance, repeated test panels to achieve a desired result, and having panelists evaluate more than one set of samples on the same test to increase the number of responses.

2.2.1 Review of current sensory practices in the alcohol beverage industry

Because the goal of QC is to maintain a consistent and standard quality product from one production to another, the use of difference tests (including triangle, duo-trio, and paired comparisons) are often used. These types of tests are more likely to be conducted by a QA or corporate sensory group using their own set of trained panelists than by a plant's QC group. These tests typically require a larger number of panelists and more time than a plant QC lab has available. Difference tests are useful to evaluate the acuity (the ability to discriminate) of panelists on a particular product or attribute (or sometimes a defect) of a product and can be used appropriately as a final check when there is a question about whether a product matches a control or previous production.

It is more common to find simple 'go/no go' comparison tests, using smaller panel sizes, often 3–6 panelists, for quality control evaluations. In these tests a sample is compared to a reference standard of the same product; the panel will then determine whether or not the sample is comparable to the reference sample. If it is determined to be comparable by the panel, the product is released for use. If it is not approved, then the decision as to what corrective action is required must be made. A key aspect of this type of determination is what is comparable. Usually, 'comparable' means that there is an expectation that consumers would identify or accept the product in the package as typical or representative of the expected product. This type of testing requires that the panelists have some training and background in typical or representative products and then use good judgment to determine whether finding small, sometimes inherent, differences result in the product being comparable.

When the product is not found to be in compliance with standards, the corrective action may involve blending of the unapproved product in small amounts with approved batches and retesting to ensure that the new blend is not compromised in quality. At other times the decision may be to re-work the product by going through additional processing, i.e. additional filtration.

Occasionally, when a panel cannot reach a consensus to accept or reject, the questionable product will go to a larger difference panel, i.e. triangle, duo-trio or paired comparison. This panel will determine whether there is a significant difference between two samples – the test sample and a control batch selected as typical or representative. It is critical in these cases to have large enough panels so that decisions of no-difference can correctly be interpreted.

It is also common to find multiple sample comparison tests in use by QC areas. These types of tests are useful when multiple production samples of the same product are available; usually 3–5 products compared to a reference. This test type can be conducted as either a scaled test whose results are analyzed using analysis of variance (ANOVA), or a ranking of samples for difference by Friedman analysis (Meilgaard *et al.*, 2007). These types of tests often require 8–12 panelists, so their use may be restricted in plant QC situations if not enough panelists are trained and available.

Descriptive analysis methods for QA/QC are less common throughout the alcohol beverage industry. Some producers, most notably the larger beer and wine companies, use descriptive analysis methods to define product flavor profiles. The information gathered from descriptive testing on a product's flavor profile can be used to train panelists in the identification and intensity of critical product characteristics. Such information also may help the sensory staff understand sensitivities that particular individuals on QA/QC panels have to specific flavor compounds that may influence their decisions. Research also suggests that trained sensory panel results may be useful as predictive indicators of consumer acceptance (Rousseau, 2010; Lattey *et al.*, 2010; Drake and Civille, 2003) although such predictions would need to be determined on a product-by-product basis. Again, it is important to emphasize that a QC/QA test typically should not be designed simply to test 'overall acceptability' or 'liking', but rather to ensure that products meet agreed-upon standards of typicality and representativeness.

2.2.2 Advantages and disadvantages of expert tasters

Master brewer, cellar master, enologist, winemaker, master distiller and master blender are some of the titles used in the alcohol beverage industry that indicate experts in their field. The expertise of these individuals is not only about their acuity as part of a difference panel or their descriptive abilities, but also refers to their knowledge of the production process and the products (Hughson and Boakes, 2002; Lawless, 1985).

Use of the term 'expert' can be misleading, in regards to who is responsible for the sensory quality of a product. Is it the individual or the panel that has the responsibility for sensory quality? It is true, at least in the not too distant past, that experts would be the final decision makers regarding a product's quality. This way of working has changed in recent years in many companies. Although there are individuals who have in-depth knowledge of and experience with specific products, these individuals rarely make sensory quality decisions exclusive of others' input. In many cases where experts make a final decision, they also participate on panels and use the panel's results to help form their decisions of a particular product's quality. These panels usually evaluate products on a go/no go basis. Training of new panelists in this setting often takes place in what can be described as an apprenticeship with the expert and the panel. There obviously are problems with this training method, i.e. lack of standardized training protocols, understanding individual's sensitivities and anosmias, and the effect of panel influences among others.

Training and maintenance protocols for panelists are what set sophisticated sensory programs apart from other programs. All companies want to produce the most consistent quality products possible. However, companies that have sensory procedures in place without a sensory program strategy and strong management support typically have fewer resources (i.e. time, money and people), made available to them than their counterparts in other companies. Granitto *et al.* (2008) stated, 'the most difficult and time consuming stage of sensory profiling is probably the selection and training of a panel, in the attempt to reach agreement in the use of scales and in the meaning of each attribute'.

Thorough training and maintenance protocols for the development and assessment of panelists have been implemented by some companies, but there are many who still rely on expert tasters to evaluate their products. There have been many published reports on improved reliability of trained sensory panels over the expert tasters (Guinard *et al.*, 1999; Latreille *et al.*, 2006; Gawel and Godden, 2008). There is no argument that training of panelists improves reliability of results. The research by Guinard *et al.* (1999, p 59), for example, suggests that expert tasters' judgments differ significantly from person to person when asked to evaluate the quality of a variety of beers.

2.2.3 Basic issues in sensory evaluation of alcoholic beverages

All food and beverage categories have their own inherent sensory quality issues related to storage, serving, handling and evaluation. One of the key aspects of importance with alcoholic beverages generally is the alcohol content; most of the world uses the terminology 'percent alcohol' or 'percent alcohol by volume' (% abv) to express alcohol content. In the US, and with primary regard to distilled spirits, the term "proof" is used to express alcohol content. US proof units are double the percent alcohol of a given product. For clarity and consistency in this chapter's discussion, only the term percent alcohol will be used.

Reducing alcohol content for sensory evaluations

Alcohol can desensitize the palate and nose of the assessor if evaluations are conducted on products with high percent alcohol. Even when assessors expectorate, they may still become adapted and desensitized by evaluating samples with high alcohol content. This is caused, in part, by the ease with which alcohol can be absorbed through the soft tissue of the mouth unless proper caution and common sense is employed.

Another reason for reducing alcohol content by dilution is that it can be much easier to expose off-taint characteristics that otherwise may have gone undetected.

For example, geosmin, an aromatic compound caused by blue-green algae growth, exhibits an earthy, musty character. The sources of geosmin contamination typically include the production water and/or the grain, fruit, and other raw material substrates from which the alcohol is initially derived. It is much easier to detect geosmin when alcohol content is diluted.

One problem with dilution is that delicate flavor notes, particularly in some highly aged products, can be diluted to the point that they are lost or overwhelmed by other characteristics. This presents a problem that must be addressed on a case-by-case basis by the company in determining whether dilution is necessary and appropriate.

Most wines and beers are of low enough alcohol content that alcohol reduction is not required before testing. However, for those distilled spirit products with greater than 35% alcohol content, it is common practice to dilute the alcohol content using de-ionized or distilled water. Depending on the producer of the product, the range for evaluation can be from 15–25% alcohol content. It is important that a single diluted alcohol content is adhered to for quality assessments of a product. Varying the alcohol content from one panel to the next can cause confusion and erroneous results. There is a simple formula to use when diluting alcohol content of a product (Eq. 2.1):

$$\frac{\textit{desired product volume} \times \textit{desired alcohol content}}{\textit{current alcohol content}}$$

$$= \text{required product volume at current alcohol content} \qquad [2.1]$$

The remaining desired volume would be filled with de-ionized water to bring the product to the desired alcohol content.

Selection of serving ware for sensory evaluations

In conventional practice most QC labs use tulip-shaped glasses, often a short stemmed glass with a bulbous body (narrower at the opening than at the base), for their evaluations. However, as Piggott and Macleod (2010) note there is no scientific research to demonstrate improved flavor detection using this type of glassware. Some companies have opted to move away from glass altogether for their product evaluations, citing several factors for this change:

- Time and energy/water savings from washing,
- Avoidance of sample-to-sample contamination, because even after washing, some products leave residual flavors behind (e.g. compounded gin with very flavorful essential oils is difficult to neutralize),
- Avoidance of residuals from detergents as well as the rinse water; residual calcium from rinse water can inhibit the detection of geosmin (Schrader and Blevins, 2001),
- Replacement, both from a cost perspective but also in sourcing glass from the same glass manufacturer and glass mold.

The requirement remains that whatever servingware is selected it must be vetted for product compatibility, meaning that the vessel will remain inert and will not impact the product evaluation. An outline of how to screen raw materials; including transfer and containment vessels is presented in Section 2.4, pertaining to processing and product transfer.

Other special concerns related to the sensory evaluations of alcoholic beverages
Along with the previously discussed issue regarding alcohol content, Piggott and Macleod (2010) also provide insight into other areas of special concern when evaluating alcoholic beverage. These areas of concern include:

- Time from sample preparation to serving and evaluation,
- Sample serving temperature,
- The topic of nosing as compared to tasting.

The first of these, time from preparation to serving, suggests serving as close to a fresh preparation as possible. Nonetheless those authors recognize that this is not always possible and conclude that most sample preparations (e.g. that are not temperature-sensitive or time-sensitive in the case of carbonated beverages) may be pre-poured and held for evaluation up to eight hours without noticeable differences in results. Next is sample serving temperature, with the basic concern being that, although products generally may be served at a wide range of temperatures, product comparisons of a given product should always be served at the same temperature. The last topic, nosing, is a common practice given the number of samples that a QC lab may evaluate daily. In many labs the appearance and smell of the product are determined during nosing and used to determine whether the product will pass QC sensory tests. In those cases, the product may not actually be tasted. This presents a conundrum that must be understood no matter what testing decision finally is made. It may be important to control the amount of beverage that is tasted, because the alcohol content can result both in adaptation and desensitization and can produce physiological effects if over-consumed. Even if assessors do not swallow the samples being evaluated, the alcohol can still be absorbed through the mouth. This information suggests that nosing, without actual tasting, is preferred. However, tasting the sample can release flavors or heighten certain flavors that would not be detected in the same way if the sample is only smelled. Again, companies need to make decisions based on informed discussion before deciding on a specific set of protocols. Regardless, if the alcoholic beverage samples are being tasted, expectorating almost always is encouraged for QA/QC purposes when numerous samples (more than three or four depending on alcohol content) are to be evaluated in a single session.

Two other areas of concern for quality evaluations to be considered include sample order and the sample number and serving size. It is not uncommon for the quality assessor to be involved in multiple evaluations of a variety of products. If this is the case, samples often should be organized in an order of least flavorful, i.e. most neutral, such as vodka, to more heavily flavored products such as gin or a full-bodied red wine. Flavor impact is not the only concern because mouth-feel and texture also come into play when considering sample order. More viscous types of products may leave a mouth coating that can influence subsequent evaluations.

The number of samples and serving size is critical to sensory quality assessment and even more so if consumers are involved in the QA process, as they might be when standards are first established or are checked to determine if modifications are necessary. The Distilled Spirits Council (DISCUS) of the US lists one of its main goals as the promotion of responsible drinking. They have published guidelines comparing serving sizes of distilled spirits, wine and beer. In a press release, DISCUS proposed that 'serving sizes are 1.5 fluid ounces for 80-proof distilled spirits, 12 fluid ounces for regular beer, 5 fluid ounces of wine' (DISCUS, 2010). These guidelines are important for all researchers and participants involved with alcohol beverage evaluations to understand. Likewise, when conducting consumer research it is the responsibility of the manufacturer to understand the laws regarding total allowable serving volume, which affect the number of samples that may be presented as well as the serving size that each respondent can receive. Tables 2.1 and 2.2 show equivalent number of servings based on serving size and alcohol content. It should be noted these are not recommended serving sizes for individual samples for testing purposes, but rather are included only to

Table 2.1 Serving sizes (in ozs) for consumer tests

	Number of samples presented					
Proof	1	2	3	4	5	6
5	24.00	12.00	8.00	6.00	4.80	4.00
10	12.00	6.00	4.00	3.00	2.40	2.00
15	8.00	4.00	2.67	2.00	1.60	1.33
20	6.00	3.00	2.00	1.50	1.20	1.00
25	4.80	2.40	1.60	1.20	0.96	0.80
30	4.00	2.00	1.33	1.00	0.80	0.67
35	3.43	1.71	1.14	0.86	0.69	0.57
40	3.00	1.50	1.00	0.75	0.60	0.50
45	2.67	1.33	0.89	0.67	0.53	0.44
50	2.40	1.20	0.80	0.60	0.48	0.40
55	2.18	1.09	0.73	0.55	0.44	0.36
60	2.00	1.00	0.67	0.50	0.40	0.33
65	1.85	0.92	0.62	0.46	0.37	0.31
70	1.71	0.86	0.57	0.43	0.34	0.29
75	1.60	0.80	0.53	0.40	0.32	0.27
80	1.50	0.75	0.50	0.38	0.30	0.25
85	1.41	0.71	0.47	0.35	0.28	0.24
90	1.33	0.67	0.44	0.33	0.27	0.22
95	1.26	0.63	0.42	0.32	0.25	0.21
100	1.20	0.60	0.40	0.30	0.24	0.20

Note: 1 fluid ounce = 29.57 ml

Table 2.2 Serving sizes (in ml) for consumer tests

	Number of samples presented					
% ABV	1	2	3	4	5	6
2.5	709.68	354.84	236.56	177.42	141.94	118.28
5.0	354.84	177.42	118.28	88.71	70.97	59.14
7.5	236.56	118.28	78.85	59.14	47.31	39.43
10.0	177.42	88.71	59.14	44.36	35.48	29.57
12.5	141.94	70.97	47.31	35.48	28.39	23.66
15.0	118.28	59.14	39.43	29.57	23.66	19.71
17.5	101.38	50.69	33.79	25.35	20.28	16.90
20.0	88.71	44.36	29.57	22.18	17.74	14.79
22.5	78.85	39.43	26.28	19.71	15.77	13.14
25.0	70.97	35.48	23.66	17.74	14.19	11.83
27.5	64.52	32.26	21.51	16.13	12.90	10.75
30.0	59.14	29.57	19.71	14.79	11.83	9.86
32.5	54.59	27.30	18.20	13.65	10.92	9.10
35.0	50.69	25.35	16.90	12.67	10.14	8.45
37.5	47.31	23.66	15.77	11.83	9.46	7.89
40.0	44.36	22.18	14.79	11.09	8.87	7.39
42.5	41.75	20.87	13.92	10.44	8.35	6.96
45.0	39.43	19.71	13.14	9.86	7.89	6.57
47.5	37.35	18.68	12.45	9.34	7.47	6.23
50.0	35.48	17.74	11.83	8.87	7.10	5.91

Note: 29.57 ml = 1 fluid ounce

help make comparisons for total alcohol consumption during testing based on total amount consumed during a test.

2.3 Similarities and differences in alcoholic beverages

The most basic similarity of beer, wine and distilled spirit products is quite obvious: they all contain alcohol in the form of ethanol. The alcohol content of these products arguably affects their sensory quality more than any other single factor. For example, the fermentation process that produces ethanol also contributes to the formation of several volatile aromatic compounds including acetaldehyde, diacetyl, acetic acid and numerous other compounds. The equilibrium of ethanol and other compounds produce further aromatic compounds; an example would be the formation of ethyl acetate through the esterification of acetic acid in the presence of ethanol. Additionally, alcohol influences the sensory quality of these products by the fact that it is an active solvent. This ability to serve as a solvent for other flavor fractions has an impact on the sensory testing of the products. As products become more complex and more flavorful, QC testing becomes more difficult both from the standpoint of the typicality of the product and the timing and serving protocols that must be considered.

The alcohol content of a product also impacts the maturation process of products that are aged in wood, the order of addition and solubility of ingredients in compounded products (e.g. cordials, blended whiskies), and shelf stability.

The differences that affect the sensory quality for beer, wine and distilled spirits are numerous. The first difference is the fermentable substrate(s) used in the production of each type of product. Next, the production processes for wine, beer and distilled spirits are quite different. Some products are carbonated, some are aged in wood and others require filtration. The operational controls that must be included for sensory testing of the various products are product-specific and must be considered based on the total product and its intended use.

2.4 Factors influencing the development of off-flavors and taints

2.4.1 Raw materials

Water is a critical component in the production of all alcoholic beverages. The quality of water used in production often goes unnoticed when there are no problems, but when quality issues arise in production, water is one of the primary suspects. The use of water is ubiquitous throughout production. Whether use of water is limited to rinsing and cleaning production vessels, or it is added as an ingredient in the production process (e.g. beer and distilled spirit productions), the quality of water is essential to the quality of the finished product. Routine evaluations to monitor water quality should be included as part of a sensory quality program, because conventional water treatment procedures are not always adequate to remove off-odors (Srinivasan and Sorial, 2011). Geosmin and 2-methyl isoborneol (2-MIB), which have been described as having earthy and musty notes respectively, are two of the most problematic taints found in water.

All alcohol starts from some sort of fermentable substrate (e.g. grains, fruits or tubers). Any material that contains simple sugars, or complex carbohydrates that can be converted into simple sugars has the potential to produce alcohol through fermentation. Establishing sensory procedures to evaluate these materials as they come into the production facility adds assurance that new shipments meet predetermined quality criteria for flavor and consistency. A common method used for grain evaluations involves making a 'tea' where hot water is poured into a glass containing the grain. This sample is compared side-by-side with a reference sample of the specific grain, prepared in the same way. Often it is the grain operators/receivers who are charged with this responsibility, which works only if they are given procedural support and training to conduct these evaluations.

Common off-notes that can be detected by this method include geosmin, 2-MIB, and diesel fuel (used commonly in machinery to dry grains).

The yeast variety used in fermentation also impacts the flavor of the finished product (Lilly *et al.*, 2000; Lee *et al.*, 1995). Many companies choose to use dry yeast, because it allows for more consistent yeast application and is less costly to maintain. Though use of dry yeast does not prevent cross-contamination from wild strains, there are fewer opportunities for this interaction. Companies that choose to propagate their own proprietary strain of yeast must do so with care. Wild yeasts are everywhere and can mutate a particular strain of yeast if proper control conditions are not in place. Proper sensory testing to ensure that no extraneous flavors are formed by unexpected yeast contamination is essential.

2.4.2 Operational controls

Time, temperature and pH are critical operational control points in the production of alcoholic beverages. Beer and many spirits products begin with a mashing process, where grains are cooked with water (i.e. the mash) to gelatinize the starch content of the grains. To cook the mash, the temperature is elevated. However, prior to the addition of enzymes, or malt in the case of straight whiskies, used to convert the starches to fermentable sugar, the temperature must be reduced so as not to inhibit enzymatic action. As the mixture is transferred into a fermenter it must continue to cool before the yeast is added. In some products an additional cooling step takes place prior to the enzyme addition. For bourbon production, for example, milled rye is added to mash. If the rye is added at too high a temperature it will form 'rye balls' (i.e. the rye meal will clump together) and the interior portion of the rye ball does not cook. This uncooked rye meal can act as a medium for bacterial growth, which can contaminate the fermenter. Increased levels of phenolic compounds (e.g. ethyl phenol and ethyl guaiacol) are associated with contaminated fermentations due to the presence of rye balls. In addition to proper temperature control, the yeast activity in the fermenter is influenced by the pH of the mash. Each yeast strain operates in its own optimal pH range. Control of pH in the fermenter will impact the yield and quality of the alcohol produced (Bra'nyik *et al.*, 2004; Davis *et al.*, 1985). Sensory testing during the production process, often by nosing intermediate products of production, can help to determine if problems are occurring before additional processing steps compound the problems.

There are other operational controls to minimize the risk of product contamination. Proper sanitation and cleaning of all equipment is critical. Cleaning minimizes bacteria and mold growth potential as well as reducing the chance of cross-contamination from shared equipment. In plants with stills, keeping copper portions of the still clean reduces the occurrence of sulfur compounds. Clean, active copper sites will bind sulfurous compounds to minimize their migration through the still. Many companies have adopted closed-topped stainless steel fermenting tanks. In the past, fermenters often were open-topped wooden containers. Stainless steel is easier to clean and sanitize than wood. The closed system also reduces contamination from air borne wild yeast and bacteria (Bra'nyik *et al.*, 2005).

2.4.3 Maturation/storage

Many wines and spirits are aged, often in oak barrels. This process is known as maturation. The sensory impact from maturation can be huge depending on the length of time the product ages, the type and size of the barrels, alcohol content in the barrel, and whether the barrel is a new charred or reused barrel. It is possible to underage and overage products. Not enough time and the product may appear immature. Too much time the product may be considered over-aged. Product in smaller barrels will mature (i.e. gain more wood character) more quickly than product aged in larger barrels, because the ratio of surface area to volume is greater in the small barrels. French oak provides a different flavor (drier, more tannic) profile than American white oak (more vanilla notes). Product with higher alcohol content will extract more tannins than a lower alcohol content product. More delicate distilled spirits (e.g. rum, Canadian whiskey and malted whiskies) are aged in reused barrels. Straight whiskies (e.g. bourbon and rye whiskies) are aged in new charred oak barrels (this is required by US law). Research also suggests that barrel ageing can contribute to increased levels of phenolic compounds in wine (Rayne and Eggers, 2007). The difficulty in sensory testing of these products is that the product changes considerably during the ageing process and the initial product going into the barrel is quite different from the end product. However, it is clear that poor quality, tainted product entering the ageing process will not improve with age.

2.4.4 Processing and product transfer

All materials that come into contact with product must be tested to ensure that the material will not alter the flavor of the product. These materials include, but are not limited to, filter pads, filter media, hoses and gaskets. A 72-hour test should be sufficient to evaluate the integrity of these materials, since they are not intended to be in constant contact with product for an extended period of time. A common method for doing this is to use grain neutral spirits, at 95% abv, and place a sample, at least 2.5 cm^2 in size, of the material into a beaker. Grain spirits (150 ml) is added and the beaker covered. When evaluating hose materials it is important to receive a long enough length of hose that it can be bent into a u-shape. The grain spirits are then placed inside the hose. Be sure to note the start time of the test and retain enough of the original spirits to use as a reference for evaluation at 2, 6, 24 and 72-hour intervals. At each interval dilute samples of the reference and test material to 20% abv for evaluation.

2.4.5 Packaging materials

Finished product containers (e.g. cans, glass and PET) and closures also need to be evaluated for product compatibility. The difference here is that these materials will be in contact for a much longer time period. Packaging materials can initially be reviewed using the 72-hour test as described above, but for approval these materials must be placed into long-term storage testing containing product similar to that which it is intended to hold. Test increments can vary; but intervals of 1, 3, 6, 9 . . . 24 months often are recommended.

If closures are being tested it is suggested that the container be laid on its side to maximize product contact. Historically, natural cork has been used as a closure with many alcohol beverage products. Over the past 20 years, many companies have replaced natural corks with synthetic cork, plastics or metallic closures. Natural corks have the disadvantage of occasionally imparting an off-odor, 2,4,6-trichloroanisole (TCA), with odor characteristics similar to geosmin or 2-MIB.

2.5 The use of instrumental methods to aid sensory quality evaluations

Instrumentation, namely gas chromatography-flame ionization detector (GC-FID) and high-performance liquid chromatography (HPLC) has been useful in the quantification of volatile compounds and barrel extractives, respectively. Until recently, instrumentation has been limited by its detection capabilities. Many taint compounds, such as geosmin, 2-MIB, TCA and sulfurous compounds, have best been detected by the human nose, but have low detection limits, in the parts per billion and parts per trillion range. Advancements in instrumental detection due to advancements such as stir bar sorptive extraction (SBSE) have made it possible to quantify compounds at these very low levels (Nakamura *et al.*, 2001). SBSE technology requires the use of gas chromatography mass spectroscopy (GC-MS). Most QC labs do not contain these instruments because the cost of the equipment and the associated maintenance and staff to run the equipment is high. However, if such equipment is available in a research and development laboratory it can be used to identify and quantify potential training standards for a sensory quality panel, in addition to the other advantages these equipment have as research tools.

2.6 Future trends in quality evaluations

The food and beverage industry continues to lead the way in the advancement of biomimetic technology. The goal of biomimetics is to mimic biological systems. In this instance, the biological systems being mimicked are specifically the biochemical response mechanisms for odor and taste. Electronic noses (e-nose) have been in development since the early 1980s. Since that time e-tongue devices have also emerged, along with combined e-nose and e-tongue devices, to provide a more robust instrumental comparison to the human experience. As related to application in quality control, the benefits of such instruments include the speed and consistent accuracy of off-flavor detection. These devices make on-line inspections for quality and safety possible (Ghasemi-Varnamkhasti *et al.*, 2010).

These systems also have an advantage over traditional analytical instruments, like gas chromatography (GC), in their ease of operator use and portability. The manufacturers of zNose™, an alternative e-nose utilizing acoustic sensors, offer both a portable bench top model as well as a hand-held device. The manufacturers, Electronic Sensor Technology, contend that the zNose™ can detect 2,4,6-trichloroanisole (TCA) in the part-per-trillion (ppt) range even in complex products, such as wine (Staples, 2000). Electronic nose systems have been shown to be able to differentiate a control beer from its "tainted" beer counterpart within the same brand (Ragazzo-Sanchez *et al.*, 2009).

It is essential to remember that electronic detection systems are only as good as the "training" they are given. If the system is trained or set to detect ten different compounds that result in off-flavors, but then is given a sample with a different off-flavor compound, the system will pass or "OK" the product, even though it would clearly fail on a human panel.

Research comparing beer analysis to that of red wine, using e-nose technology, concludes that the complexity of the product being analyzed does impact the ability of the device to recognize the defect. This effect is compounded particularly where there are variables unaccounted for, such as origin of the wine. While e-noses are more precise in the recognition of off-flavor compounds, they are not as sensitive as humans in the detection of overall flavor differences (Arroyo *et al.*, 2009). The implication is that as biomimetic technology continues to improve, this means of analysis is not a replacement for sensory evaluation, but rather a compliment to it.

2.7 Additional sources of information

A quick search for "sensory analysis" or "sensory evaluation" on the ASTM International website (http://www.astm.org/) will lead to a number of international sensory standards, books and other publications. To gain both a historical and technical perspective on a wide range of products, the *Handbook of Alcoholic Beverages* (Buglass, 2011) is a helpful resource.

The trade association links compiled by Piggott and Macleod (2010) provide information related to various spirits and include: the Scotch Whisky Association (http://www.scotch-whisky.org.uk), the Bureau National Inter-professional du Cognac (http://www.cognac.fr/), the Gin and Vodka Association (http://www.ginvodka.org/), and the West Indies Rum and Spirits Producers Association (http://www.truerum.com/) among others. In addition to these links there are similar links for wine, beer and spirits from the US. The Wine Institute is an advocacy group for California wines (http://www.wineinstitute.org/), the American Society of Brewing Chemists' site provides information about the science of beer making (http://www.asbcnet.org/), and the Kentucky Distillers Association's website has recently been renovated and is now dedicated to bourbon whiskey (http://www.kybourbon.com/).

In addition to books and articles specific to alcoholic beverages, current books on sensory methods and journals such as *Journal of Sensory Studies* or *Food Quality and Preference* and other food science journals often have articles on quality control and assurance that are adaptable for alcoholic beverage systems.

2.8 Conclusions

Great strides have been made in the alcohol beverage industry to improve sensory quality programs, but room for further improvement exists. Continued training is the key to the successful implementation of sensory programs as well as a means of retraining employees to an enhanced understanding of quality. Training should include discussions with management to establish quality goals, but should integrate other functional groups within the company, especially those groups involved in activities that can influence the development of off-odors and taints.

2.9 References

ARROYO T, LOZANO J, CABELLOS JM, GIL-DIAZ M, SANTOS JP *et al.* (2009), 'Evaluation of wine aromatic compounds by a sensory panel and an electronic nose', *Journal of Agricultural and Food Chemistry*, **57**, 11543–11549.

BRA'NYIK T, VICENTE AA, CRUZ JM and TEIXERIA JA (2004), 'Continuous primary fermentation of beer with yeast immobilized as spent grain: the effect of operational conditions', *Journal of the American Society of Brewing Chemist*, **62**(1), 29–34.

BRA'NYIK T, VICENTE AA, DOSTA'LEK P and TEIXEIRA JA (2005), 'Continuous beer fermentation using immobilized yeast cell bioreactor systems', *Biotechnology Progress*, **21**, 653–663.

BUGLASS AJ, ed (2011), *Handbook of Alcoholic Beverages: Technical, Analytical and Nutritional Aspects, Volume II*, Chichester, John Wiley and Sons Ltd.

DAVIS CR, WIBOWO D, ESCHENBRUCH R, LEE TH and FLEET GH (1985), 'Practical implications of malolactic fermentation: A review', *American Journal of Enology and Viticulture*, **36**(4), 290–301.

DISCUS (2010), Distilled Spirits Council urges government support for standard drink information on alcohol labels, 22 April 2010, Washington, DC, Distilled Spirits Industry Council. Available from: http://www.discus.org/media/press/article.asp?NEWS_ID=589 (accessed 19 November 2010).

DRAKE MA and CIVILLE GV (2003), 'Flavor lexicons', *Comprehensive Reviews in Food Science and Food Safety*, **2**(1), 33–40.

GAWEL R and GODDEN PW (2008), 'Evaluation of the consistency of wine quality assessments from expert wine tasters', *Australian Journal of Grape and Wine Research*, **14**, 1–8.

GHASEMI-VARNAMKHASTI M, MOHTASEBI SS and SIADAT M (2010), 'Biomimetic-based odor and taste sensing systems to food quality and safety characterization: An overview on basic principle and recent achievements', *Journal of Food Engineering*, **100**(3), 377–387.

GRANITTO PM, BIASIOLI F, ENDRIZZI I and GASPERI F (2008), 'Discriminant models based on sensory evaluations: Single assessors versus panel average', *Food Quality and Preference*, **19**, 589–595.

GUINARD J, YIP D, CUBERO E and MAZZUCCHELLI R (1999), 'Quality ratings by experts, and relation with descriptive analysis ratings: A case study with beer', *Food Quality and Preference*, **10**, 59–67.

HUGHSON A and BOAKES R (2002), 'The knowing nose: the role of knowledge in wine expertise', *Food Quality and Preference*, **13**, 463–472.

JELLINEK G (1985), *Sensory Evaluation of Food: Theory and Practice*, Chichester, Ellis Horwood Ltd.

LATREILLE J, MAUGER E, AMBROISINE L, TENENHAUS M, VINCENT M *et al.* (2006), 'Measurement of the reliability of sensory panel performances', *Food Quality and Preference*, **17**, 369–375.

LATTEY KA, BRAMLEY BR and FRANCIS IL (2010), 'Consumer acceptability, sensory properties and expert quality judgments of Australian Cabernet Sauvignon and Shiraz wines', *Australian Journal of Grape and Wine Research*, **16**, 189–202.

LAWLESS H (1985), 'Psychological perspectives on wine tasting and recognition of volatile flavours', in BIRCH G and LINDLEY M, *Alcoholic Beverages*, London, Elsevier Applied Science, 97–113.

LEE S, VILLA K and PATINO H (1995), 'Yeast strain development for enhanced production of desirable alcohols/esters in beer', *Journal of the American Society of Brewing Chemist*, **53**, 153–156.

LILLY M, LAMBRECHTS MG and PRETORIUS IS (2000), 'Effect of increased yeast alcohol acetyltransferase activity on flavor profiles of wine and distillates', *Applied and Environmental Microbiology*, **66**(2), 744–753.

MCDANIEL MR (1985), 'Sensory evaluation of food flavors', in BILLS D *et al.* (eds), *Characterization and Management of Flavor Compounds*, Washington DC, American Chemical Society, 1–9.

MEILGAARD M, CIVILLE G and CARR TB (2007), *Sensory Evaluation Techniques*, Boca Raton, Fl, CRC Press.

MEISELMAN H and SCHUTZ H (2003), 'History of food acceptance research in the US Army', *Appetite*, **40**, 199–216.

MUNOZ A (2002), 'Advances in sensory evaluation for quality control', *Food Quality and Preference*, **13**, 327–328.

NAKAMURA S, NAKAMURA N and ITO S (2001), 'Determination of 2-methylisoborneol and geosmin in water by gas chromatography-mass spectrometry using stir bar sorptive extraction', *Journal of Separation Science*, **24**(8), 674–677.

O'MAHONY M (1986), *Sensory Evaluation of Food*, New York, Marcel Dekker.

PIGGOTT JR and MACLEOD S (2010), 'Sensory quality control of distilled beverages', in KILCAST D (ed.), *Sensory Analysis for Food and Beverage Quality Control: A Practical Guide*, Cambridge, UK, Woodhead Publishing, 262–275.

RAGAZZO-SANCHEZ JA, CHALIER P, CHEVALIER-LUCIA D, CALDERON-SANTOYO M and GHOMMIDH C (2009), 'Off-flavors detection in alcoholic beverages by electronic nose coupled to GC', *Sensors and Actuators B*, **140**, 9–34.

RAYNE S and EGGERS NJ (2007), '4-ethylphenol and 4-ethylguaiacol in wines: Estimating non-microbial sourced contributions and toxicological considerations', *Journal of Environmental Science and Health Part B – Pesticides Food Contaminants and Agricultural Waste*', **42**(8), 887–897.

ROUSSEAU B (2010), 'Action standards in a successful sensory discrimination program', *IFPress*, **13**(4), 3–4.

SCHRADER K and BLEVINS W (2001), 'Effects of carbon source, phosphorous concentration, and several micronutrients on biomass and geosmin production by *Streptomyces halstedii*', *Journal of Industrial Microbiology & Biotechnology*, **26**, 241–247.

SRINIVASAN R and SORIAL G, (2011), 'Treatment of taste and odor causing compounds 2-methyl isoborneol and geosmin in drinking water: A critical review', *Journal of Environmental Sciences*, **23**(1), 1–13.

STAPLES EJ (2000), Acoustical Society of America: paper number 2aEA4, 'The zNose™, a new electronic nose using acoustic technology', 1–8, Available from: http://www.aromako.co.kr/aroma/tech/pdf/asa2000.pdf (accessed 7 March 2011).

STONE H and SIDEL JL (1993), *Sensory Evaluation Practices*, 2nd ed., San Diego, CA, Academic Press.

VEACH M (ed) (2008), The Filson Historical Society: The Filson Newsmagazine, **7**:4. Available from: http://www.filsonhistorical.org/archive/news_v7n4_distilling.html (accessed 19 November 2010).

YANTIS J (ed) (1992), *The Role of Sensory Analysis in Quality Control*, ASTM Manual Series: MNL 14, Baltimore, MD, ASTM International.

3

Principles of sensory shelf-life evaluation and its application to alcoholic beverages

M. G. O'Sullivan, University College Cork, Ireland

Abstract: As this book is principally concerned with sensory aspects of alcoholic beverages, the sensory and the instrumental methods which can be used as sensory measurement indices will be primarily presented in this chapter. In addition, accelerated shelf-life testing will be presented, including the discussion of limitations of these methods. Microbiological shelf life is discussed because of its importance for beer shelf-life evaluation. However, this is less important for other alcoholic beverages with higher alcohol levels due to the antimicrobial nature of ethanol and where sensory quality is the primary means of defining shelf life.

Key words: shelf life, sensory, alcoholic beverages, testing.

3.1 Introduction: principles of sensory evaluation in shelf-life testing

Most food and beverage products have a defined shelf life that is governed by firstly microbiological safety and then product sensory quality. At the end of shelf life undesirable characteristics develop which can be microbiological, chemical or physical. The Codex Alimentarius defines shelf life as the period during which a food product maintains its microbiological safety and suitability at a specified storage temperature and where appropriate, under specified storage and handling conditions. Also, shelf life is defined in European legislation as the 'date of minimum durability' (Food Safety Authority of Ireland, 2011). The UK Institute of Food Science and Technology (IFST) defines shelf life as 'the period of time during which the food product will remain safe, retain its desired sensory, chemical and microbiological characteristics, and comply with any label declaration of nutrition data' (IFST, 1993).

Microbiological safety is determined by manufacturers by testing periodically, over the shelf life, statistically sufficient quantities of the product that also reflect the batch-to-batch variation with respect to component ingredients, or perhaps variations in the manufacturing process and packaging. Inadequate shelf-life determination will lead to consumer dissatisfaction or complaints. At best, such dissatisfaction will eventually affect the acceptance and sales of brand name products and, at worst, it can lead to malnutrition or illness. For these reasons, food processors pay great attention to adequate storage stability or shelf life (Freitas and Costas, 2006).

Sensory analysis is generally performed in parallel with monitoring of specific sensory attributes that are used as indices of sensory quality. These attributes may include the loss of particular fresh sensory notes and the development of oxidised or stale sensory notes in the case of beers or the development of fruity notes in cream liqueurs. With new beverage products, the determination of product shelf life usually starts at the product development stage. Product development involves defining the new product concept, followed by pilot scale production. Microbial and sensory evaluations are included at this stage to estimate product shelf life. Consumer sensory evaluation or preference studies may also be undertaken to determine the commercial impact of the new product prior to the production of a finished product on a commercial scale for consumer purchase. However, irrespective of the stage a food product is at in its development, it is important to ensure that shelf life is considered at each stage and determined accurately using all available data (FSAI, 2010).

The procedure leading to the definition of the acceptance limit strictly depends on the nature of the deteriorative reaction most limiting food quality. For all microbiologically stable foods with no health-specific claims, sensory evaluation is the key factor for determining the acceptance limit, and hence shelf life (Manzocco and Lagazio, 2009). This is particularly the case with most alcoholic beverages due to the antimicrobial properties of ethanol where spoilage or pathogen growth is prevented. All food and beverage products will change with respect to flavour profile over the course of their respective shelf lives. Some products like vintage wine, due to changes in tannin and phenolic compounds or whisky, due to flavour concentration from evaporation while in the cask, can actually improve in flavour as they age. However, most alcoholic beverages have a finite shelf life, which is dictated by sensory changes during storage. Cream liqueurs stored at ambient temperatures can have shelf lives of anywhere between six months and two years depending on the quality of the raw ingredients and the homogenisation technology employed. These beverages tend also to become more 'fruity' with age due to the formation of ethyl esters from the reaction of the alcohol with the cream component of the beverage. This can be viewed as a negative sensory attribute in cream liqueurs. Another limitation to cream liqueur shelf life can be increases in viscosity, phase separation and plugging of the neck of product bottles.

Beer typically has a shelf life of between six months and one year depending on the type and whether it is canned, bottled or kegged. The limiting factor to beer shelf life is the development of stale or oxidised flavours such as E-2-nonenal with a parallel reduction in fresh fruity or hop notes. Oxidation can also occur in wine, limiting shelf life, but can be prevented with the addition of or naturally occurring levels of the anti-oxidant sulphur dioxide.

This chapter will discuss the sensory and instrumental methods employed in the shelf-life determination of alcoholic beverages as well as other useful tests such as accelerated storage analysis. The advantages and disadvantages of each will be discussed and specific examples of alcoholic beverages will be included where appropriate.

3.2 General principles of shelf-life estimation

According to Walker (1994), prior to determining the shelf life of a food, it is essential to determine which factor(s) limit the shelf life, which may cause physical, chemical or biological changes that result in a sensory change in the food. However, from a food safety perspective it is important to determine how long a product is safe to consume within the designed storage parameters. In order to achieve this the product is stored at the recommended storage temperature and conditions until the maximum legally allowable limit for numbers of defined micro-flora are reached and then manufacturers subtract a further period of time to allow an additional safety margin by which the product can be consumed. This margin of safety should be determined and applied by the manufacturer after examining all reasonably foreseeable conditions of processing and use (FSAI, 2010). With microbial growth it is often useful to consider safety and spoilage separately although the controlling factors for both may be identical (Walker, 1994).

The growth of spoilage or pathogenic microorganisms in a product will depend on a number of factors which include the prevailing hygienic conditions of the product during manufacture; the physiochemical composition of the product, such as moisture content, ethanol concentration and pH; the heat treatment used in manufacture such as pasteurisation; the packaging materials used; and finally storage temperatures. Storage conditions may include defining the temperature of storage such as refrigeration temperatures or perhaps defining an upper temperature limit or exposure to light or humidity.

Food spoilage organisms can be readily identifiable by a change in appearance, off-odour development or perhaps changes in texture. The principal danger of pathogenic species of micro-organism, such as *Salmonella* or *Listeria,* is the lack of sensory changes to the product after growth of these bacteria. Appearance, odour, flavour or texture sensory cues may be absent and thus do not indicate to the individual consuming the product that growth has occurred, presenting a health risk.

High osmotic potential conditions occur in cream liqueurs (Fig. 3.1), due to a high sucrose content (10–20%) which provides an additional hurdle effect, but the relatively high intrinsic ethanol content, 10–15% abv, prevents the growth of pathogenic organisms. The ethanol content will have an antimicrobial effect, which is enhanced with increases in concentration. Food-poisoning

Fig. 3.1 Baileys® Cream liqueur with a label configuration *circa* 2005.

microorganisms are of less concern to alcoholic beverages because of the antimicrobial effect of ethanol. Additionally low pH, as is the case with wine or beer, will contribute a hurdle effect along with the ethanol content and increase the antimicrobial properties combined with such components as organic acids in wine or hop bitters in beer.

Beer is a tough medium for bacterial growth, due to the presence of ethanol (0.5–10% w/w), hop bitter compounds (approximately 17–55 parts per million of isoacids), high CO_2 (approximately 0.5% w/w), low pH (3.8–4.7), extremely reduced content of oxygen (< 0.1 ppm) and the presence of only traces of the nutritive substances glucose, maltose and maltotriose (Sakamoto and Konings, 2003). Bacteria of the genera *Lactobacillus* and *Pediococcus* are a significant problem for the brewing industry, and account for the majority of spoilage incidents (Sakamoto and Konings, 2003; Yasui and Yoda, 1997; Simpson and Fernandez, 1994). These bacteria not only produce off-flavour compounds and increase acidity (which alters taste) but they also affect the clarity of the final beer (Fernandez and Simpson, 1995). Aerophilic lactic acid bacteria can grow in beer because of their relative indifference to free oxygen and acquisition of tolerances to hop antiseptics, ethanol up to about 6% and to low pH (Rainbow, 1981).

The strictly anaerobic gram-negative beer spoilage bacteria of the genera *Pectinatus* and *Megasphaera* have increased because of the improved technology in modern breweries which results in a significant reduction of O_2 content in the final products. Wild yeasts cause less serious spoilage problems than bacteria but are considered a serious nuisance to brewers because of the difficulty of discriminating them from brewing yeasts (Sakamoto and Konings, 2003).

3.3 Microbial analysis

In quality control of packaged beers, enrichment cultivation is the only practical method for the presence/absence testing of the strictly anaerobic beer spoilers due to their oxygen sensitivity. In this method, the development of turbidity in a bottled beer mixed with a concentrated growth medium is followed for 2–6 weeks (Haikara and Helander, 2002). Traditional incubation on culture media is used for the detection of beer spoilage bacteria. Bacteria will form visible colonies on plates or increase the turbidity in nutrient broths after a week or longer. This presents a problem for the brewery as the products are often already released for sale before the microbiological results become available (Sakamoto and Konings, 2003).

In the wine industry, spoilage yeasts are rarely sought during wine fermentation, but during storage or ageing and during the bottling process (Loureiro and Malfeito-Ferreira, 2003). The surface of the grapes contains a large variety of moulds, bacteria and yeasts. In particular, a great variety of yeasts are present on grapes, but only a minor portion can participate in alcoholic fermentation (Romano *et al.*, 2003). Spontaneous fermentation is still used in numerous wineries, but in the last twenty years, yeast-manufacturing companies have developed selected starter cultures (Marullo *et al.*, 2004). The spread plate technique can be used to isolate and enumerate yeasts with the aim of preventing the growth of all innocent yeast and promote the growth of all spoilage yeasts (Loureiro and Malfeito-Ferreira, 2003).

3.4 Sensory discrimination tests

Discrimination tests are perceived as one of the easiest classes of sensory testing to apply in an industrial environment, and are consequently heavily used (Kilcast, 2000). Discrimination or difference tests include overall difference tests and attribute-specific directional difference tests. Difference tests provide a method to determine if a sensory difference is detectable between samples, whereas attribute difference tests determine if a specified attribute is perceived as being different between samples.

Commonly used difference tests are paired comparison, duo-trio test and the triangle test (Lawless and Heymann, 1998a; Piggott *et al.*, 1998). In the paired comparison test the assessors are presented with two coded samples and asked to assess a specific criterion. The panellists are usually instructed to make a choice (forced-choice procedure), even if they have to make a guess, or they may be allowed to record a 'no-difference' response (Kilcast, 2000). For the duo-trio test, the assessors are presented with two coded samples and also a sample that is identified as a reference. One of the coded samples is the same as the reference and the other is the test sample. With triangle tests, three coded samples are presented to the assessors: two are the same, the third is different. Assessors are instructed to pick the odd one out or score the sample as not being different.

3.5 Quantitative descriptive tests

Descriptive analysis uses sensory panellists who are trained to quantify specific sensory attributes for appearance, flavour, texture and aftertaste. The procedure involves the use of sensory terms that are produced in collaboration with the panellists and the panel leader, which is called the quantitative descriptive analysis (QDA) method, or the Spectrum method, which uses a strict technical sensory vocabulary using reference materials. The method has a number of advantages over difference testing in that it is quantitative and can be used to describe differences between products and the main sensory drivers, be they positive or negative, identified within products or especially when combined with objective consumer testing and objective multivariate data analysis (O'Sullivan *et al.*, 2011). However, the method can be expensive and time-consuming because of the necessity to train and profile individual panellists over extended periods of time; days or even weeks. It is also not a method that can be readily used for routine analysis (O'Sullivan *et al.*, 2011). Free choice profiling (FCP) can also be used and this involves panellists developing their own descriptive terms (Delahunty *et al.*, 1997). Assessors are allowed to develop their own individual vocabularies to describe sensory perceptions and to use these to score sets of samples. As a consequence of removing the need to agree vocabularies, free choice profiling requires little training – just instruction in the use of the chosen scale (Kilcast, 2000).

Descriptive terms can be determined from lexicons. The initial list of sensory terms can be compiled by experts with product knowledge (Johnson and Civille, 1986; Johnsen *et al.* 1988; Ordonez *et al.*, 1998), perhaps using lexicons as a starting point along with any additional terms that are deemed relevant or useful. Descriptors may also be incorporated from information from consumers (Sawyer *et al.*, 1988). Sensory lexicons have been developed for a number of alcoholic beverages. The beer industry has a standardised terminology wheel of mouth-feel and taste terms (Meilgaard *et al.*, 1979). Meilgaard *et al.* (1979) showed how the principal flavour notes for beer may be represented as spokes in a 'flavour wheel'. The taste of beer is described in terms of the four basic tastes, from acidic to bitter, plus certain other taste impressions that are grouped under mouth-feel and fullness. The system divides the more complex part of beer flavour, the odour (or aroma), into eight classes in an approximate order from pleasant to unpleasant (Meilgaard *et al.*, 1982). For whisky, Lee *et al.* (2000) reviewed the origins of the flavours in whiskies and developed a revised flavour wheel for Scotch whisky.

Noble *et al.* (1984) proposed a list of standardised wine aroma terminology. From an initial list composed of nearly all possible wine descriptors, terms were selected which were analytical and free of hedonic or value-judgement connotations.

After a list of appropriate sensory terms has been compiled the training of panellists may commence. Usually sensory descriptive analysis (QDA) is performed using between 8–12 judges (Machado *et al.*, 2010). Machado used 22 well-trained judges to evaluate 17 red wines using 24 attributes, in three repetitions. They found that clearly the standard 8–12 judges were sufficient, but it is also important to note, that with well-trained judges and with replication it may be possible to use as few as six judges. This information could be important for timing and cost constrains.

A subset of the full complement of samples to be profiled which reflects the sensory variation in these samples is generally used. These could include samples at the beginning or end of the shelf-life period along with intermediate samples, depending on the design. This meta list of terms can then be reduced by panel discussion and sensory term reduction protocols, as discussed below.

3.6 Sensory term reduction

There are certain rules or criteria by which a sensory term can be included in a sensory profile. The sensory terms chosen should fill the purpose of the analysis, correspond to the samples and have relevant definitions (Civille and Lawless, 1986; Civille, 1987; Piggott, 1991). Also the descriptive term must allow differentiation between sensations, identification of the object it describes, and recognition of the object by others seeing the term (Harper *et al.*, 1974). In summary, the sensory terms selected must:

- Be relevant to the samples,
- Discriminate between the samples,
- Have cognitive clarity,
- Be non-redundant.

(Byrne *et al.*, 1999a, 1999b, Byrne *et al.*, 2001; O'Sullivan *et al.*, 2002a, 2002b; O'Sullivan *et al.*, 2003).

Various means can be employed in this term reduction process and these have included principal component analysis (PCA) in conjunction with assessor suggestions (Byrne *et al.*, 1999a, 1999b, 2001; O'Sullivan *et al.*, 2003). Free choice profiling (FCP) can also be used and this involves panellists developing their own descriptive terms (Delahunty *et al.*, 1997). The problem with this method is the subjective correlation of terms derived by different assessors may not, in reality, be related.

Generalised Procrustes analysis (GPA) may also been used in term reduction (Byrne *et al.*, 2001) as well as descriptor leverage (O'Sullivan *et al.*, 2002a). These are similar methods as they involve a level and range correction for assessors, but the descriptor leverage method does not include the rotational aspect of GPA. The descriptor leverage method is rapid and unambiguous and was investigated by O'Sullivan *et al.* (2002a) to determine whether it provided any discriminative improvement in the term reduction process. These authors concluded that the use of descriptor leverage provided a greater amount of information regarding the elimination of sensory terms as opposed to PCA and assessor suggestions alone (O'Sullivan *et al.*, 2002a).

Detailed descriptions of sensory terminology and procedural guidelines for the identification and selection of descriptors for establishing a sensory profile by a multidimensional approach have been described in ISO (1992) and ISO (1994). Defined methodologies have been proposed for sensory descriptive analysis of food products using the Spectrum method and QDA (Byrne and Bredie, 2002).

The Spectrum method provides tools with which to design a descriptive procedure for a given product category. The panellist scores perceived intensities with reference to pre-learned 'absolute' intensity scales. The purpose is to make the resulting profiles universally understandable. The method provides 'lexicons' which are an array of standard attribute names, each with its set of standards that define a scale of intensity (Muñoz and Civille, 1992; Meilgaard *et al.*, 1999).

The QDA method first proposed by Stone *et al.* (1974) requires the use of product references to stimulate the generation of terminology during panel training. This method also uses statistical analysis to determine the appropriate terms, procedures and panellists to be used for the analysis of a specific product. The panel leader does not influence the group, but acts as a facilitator, where feedback is provided to panellists based on the statistical analysis of the taste session data and facilitator discretion (Lawless and Heymann, 1998b; Meilgaard *et al.*, 1999; Miller, 1994). Panellists are not allowed to discuss data, terminology or samples after each taste session.

3.7 Sensory profile changes in alcoholic beverages

Sensory profile changes in a selection of alcoholic beverages are discussed briefly below. Beers, cream liqueurs, wines and spirits will have very different sensory profiles and change very significantly from the time of initial production to the extreme end of the respective recommended shelf life.

3.7.1 Sensory profile changes in beer

The sensory profiles of beers are as diverse as the large variety and types of beers produced and consumed worldwide. Each beer, be it lager, stout, boch or wheat beer has its own unique characteristic sensory profiles and sensory attributes that define quality, freshness or the degree to which they have aged. This section will discuss the sensory profile changes that occur in beer in the broadest sense. However, generally speaking there are some common sensory attributes which are affected by ageing. In contrast to some wines, beer ageing is usually considered negative for flavour quality. The main focus of research on beer ageing has been the study of the cardboard-flavoured component (E)-2-nonenal and its formation by lipid oxidation with other stale compounds (Vanderhaegen *et al.*, 2006) or increases in the compound furfural with beer ageing. The limiting factor to beer shelf life is the development of these stale, oxidised flavours such as E-2-nonenal, with a parallel reduction in fresh fruity or hop notes. Positive flavour attributes of beer, such as fruity/estery and floral aroma tend to decrease in intensity during ageing. The most important flavour-active esters in beer are ethyl acetate (solvent-like aroma), isoamyl acetate (fruity), banana acetate (solvent-like aroma), isoamyl acetate (fruity, banana aroma), ethyl caproate and ethyl caprylate (sour apple), and phenyl ethyl acetate (flowery, roses, honey) (Verstrepen, 2003). For the overall impression, the decrease of positive flavours, such as esters, may be just as important as development of stale flavours (Bamforth, 1999; Whitear *et al.*, 1979; Vanderhaegen *et al.*, 2006).

A constant decrease in bitterness is also observed during beer ageing. This is partly due to sensory masking by an increasing sweet taste. Bitterness in beer arises when the α-acids in hops are isomerised during boiling with wort (Hough *et al.*, 1982).

In contrast to an initial acceleration of sweet aroma development, the formation of caramel, burnt sugar and toffee-like aromas (also called leathery) coincides with the sweet taste increase (Vanderhaegen *et al.*, 2006). Contact with oxygen causes a rapid deterioration of the flavour of beer and is dependent on the oxygen content of beer (Clapperton, 1976).

'Light struck skunk odour' can also develop which results when light causes riboflavin to react with and break down isohumulones, a bitter compound derived from hops which results in formation of the compound 3-methyl-2-butene-1-thiol. This principally occurs in beers packaged in clear or green bottles; brown bottles, however, offer a certain level of protection.

Many beer manufacturers will perform regular sensory profiling of their beer products over the course of real time shelf life. Panels are trained as described in the above section on descriptor profiling and sensory changes are tracked over time. Producers may also employ triangle tests to differentiate a product at a point in its shelf life from a fresh or gold standard product.

3.7.2 Sensory profile changes in cream liqueurs

Cream liqueurs are typically composed of cream, sodium caseinate, sugar, alcohol, flavours, colours and low-molecular-weight surfactants (Banks and Muir, 1988; Lynch and Mulvihill, 1997). Various food grade thickeners may be added to contribute to the mouth-feel of cream liqueurs (Banks *et al.*, 1981a). Corn syrups, molasses, maltose, ribose, galactose, honey, lactose, sucrose, dextrin, modified starch and glucose have been proposed for use as the carbohydrate source in cream/alcohol-containing beverages (Rule, 1983). Typical manufacture of a cream liqueur involves the preparation of a caseinate trisodium–citrate blend at 55–85 °C, followed by addition of cream and molten glycerol monostearate (GMS) with continuous high-speed mixing to give the cream base. An aqueous-ethanol-sucrose solution is then added to the cream base and mixed thoroughly. The blended pre-emulsion is normally homogenised twice at 45–55 °C and 20–30 MPa (two-stage) using a standard radial diffuser homogeniser, and cooled to 20 °C (Muir and Banks, 1985, 1986; Dickinson *et al.*, 1989; Lynch and Mulvihill, 1997).

The most important step in the manufacture of cream liqueurs is homogenisation, which is necessary to produce a finely dispersed emulsion and to ensure the long-term stability of cream liqueurs (Banks and Muir, 1988). The homogenisation technology employed has a profound effect on shelf life of the resulting cream liqueur. The more stable the fat globules in the aqueous phase of the emulsion, the less likely they are to coalesce and result in creaming or the unsightly neck plug that occurs in the bottle necks as seen in extremely aged cream liqueurs. During homogenisation, the size of the fat globules is decreased and sodium caseinate transfers from the serum phase to the newly-exposed fat globule surface to stabilise the product. Droplet size distribution is one of the most important characteristics of an emulsion, influencing stability (Heffernan *et al.*, 2009).

The viscosity of a fluid emulsion is an important characteristic since it influences such aspects as the rate of creaming, the physical shelf life of the product and the organoleptic properties of the product (McClements, 1999). Viscosity of cream liqueurs will increase over time and, as such, viscosity can be used as an index of shelf life. Eventually the viscosity will become a limiting factor to consumer acceptability. Viscosity can be quantified using a viscometer which will be discussed in more detail in the section covering instrumental analysis.

The flavour of cream liqueurs involves a complex interaction of volatile and non-volatile flavour compounds. These flavours originate from the liqueur components such as cream, cocoa, vanilla and whiskey. The polarity of these flavour compounds determines how they partition in the continuous phase of the cream liqueur emulsion, within the fat globule itself or across the globule aqueous phase interface. Fat has a significant effect on the partition of volatile compounds between the food and the air phases with lipophilic aroma compounds being the most affected (Bayarri *et al.*, 2006).

This flavour partitioning also affects flavour release and ultimately the flavour profile of the liqueur with more polar flavours perceived first and the more non-polar perceived last perhaps even as after-taste sensations. Ethyl esters also contribute to the overall flavour balance of cream liqueurs, but at high concentrations they can develop a very fruity flavour which can become the dominant flavour in the liqueur. In beers and wines, ethyl esters contribute to positive sensory attributes; for beer they denote freshness, and decrease during ageing. In contrast, ethyl esters increase in concentration as cream liqueurs age. Sensory descriptors for fruity flavours may include pear-like, banana-like, pineapple-like, apple-like, strawberry-like, ester-like, ethereal or just fruity. Ethyl esters form due to the reaction of the alcohol and fatty acids and can result in the formation of excessive levels of short-chain fatty acids such as ethyl acetate, ethyl butanoate and ethyl hexanoate, which are the principal causes of this fruity defect. The level of fruitiness may become a limiting factor to consumer acceptance in aged liqueurs. Typically, the levels of these ethyl esters can be monitored and quantified using gas chromatography (GC) or gas chromatography-mass spectrometry (GC-MS) and liquid extraction of methylated esters or using solid-phase micro-extraction (SPME). Effective flavour partition can be measured under static equilibrium headspace conditions but does not always relate well to the release profile that is observed *in vivo* during consumption of a product (Doyen *et al.*, 2001).

Cream liqueur manufacturers will also perform regular sensory profiling of their products over the course of their real-time shelf life. As described above, panels are trained and sensory changes are tracked over time. Also, producers may employ triangle tests to differentiate a product at a point in its shelf life from a fresh or gold standard product.

3.7.3 Sensory profile changes in wines

All wines do not have the same ability to age well throughout long cellaring. Some wines maintain their integrity or even improve with cellaring, whilst other wines are elaborated for consumption when they are young (e.g. Beaujolais nouveau) (Jaffré *et al.*, 2009). Conversion of grape sugars to alcohol and other end-products by specific yeast populations may yield wines with distinct organoleptic quality (Romano, 2003).

Phenolic compounds constitute one of the most important quality parameters of wines since they contribute to their organoleptic characteristics, particularly colour, astringency and bitterness (Monagas *et al.*, 2006). During wine ageing these phenolic compounds undergo transformations, such as oxidation, condensation and polymerisation, and the modification of these phenolic compounds would lead to a decrease in astringency (Mateus *et al.*, 2004). Anthocyanins are progressively transformed into more stable oligomeric and polymeric pigments which give rise to important changes in the colour (from bright red to brick-red hues) and in the astringency of wines (Monagas *et al.*, 2006). Volatile esters are only trace compounds in wines, however they are extremely important for the flavour profile of these drinks. Essentially we know very little about the assessment of the ageing potential of a wine (Dubourdieu, 2005; Jaffré *et al.*, 2009). According to Jaffré *et al.* (2009) consumers should rely on professional experts' advice to identify which wines can or should be consumed immediately and which wines can be kept. Professional experts base their judgements regarding potential for ageing, the so-called 'Vin de garde', on tasting, but there is no explicit rule to determine whether a wine has an ageing potential or not.

3.7.4 Sensory profile changes in spirits

Extensive changes take place in the spirits during ageing, and the final product has a very different flavour and aroma from the newly distilled spirit as is the case with whisky (Nishimura and Matsuyama, 1989; Piggott *et al.*, 2000). In whisky, spicy, smooth, vanilla, woody and sweet aroma notes increase with time (Piggott *et al.*, 1993), which are caused primarily by materials such as vanillin, aromatic aldehydes and other materials extracted from the wood (Conner *et al.*, 1999). The taste and odour of freshly distilled spirits, particularly whisky, is rather raw and unpleasant; desired flavour components develop during years of ageing in wood (Freitas and Costa, 2006) (Fig. 3.2).

Rum is a cane spirit obtained by distillation of sugar cane molasses, after fermentation with yeast, and subsequent ageing in oak barrels, where the spirit acquires its special characteristics of flavour and aroma during the time it is in contact with the wood (Pino, 2007). Dark rum is aged in oak barrels, whereas white rum is aged in stainless steel tanks (Fig. 3.3). Some materials reduce in concentration in the spirit, through losses by evaporation through the cask wood; there is a general increase in concentration as water and ethanol are lost by evaporation (Piggott *et al.*, 2000).

Fatty acid ethyl esters are the main components of rum aroma and play an important role in the sensory quality of these distilled alcoholic beverages (Pino

Fig. 3.2 Two rare bottles of Baileys® Irish whisky, which had a limited production and were sold for a period of time in Dublin, Ireland around 2002.

Fig. 3.3 Caribbean rums from the island of St. Lucia: Chairman's Reserve Rum, considered one of the finest on the island; and Kweyol Spiced Rum infused with the bark of the Bois Bande tree, a West Indian aphrodisiac.

et al., 2002). Fusel alcohols, acetic acid and ethyl acetate are present in these beverages at relatively large amounts. Blending may be employed in the production process and caramel may be added to standardise the colour of dark rums.

3.8 Consumer acceptability testing

In order for any given food product to be commercially successful, consumer desires and demands must be addressed and met with respect to the sensory properties of such products, before other quality dimensions become relevant (Chambers and Bowers, 1993). Therefore, the consumer acceptability of food and beverage products must be taken into account not only in understanding product quality, but also in shelf-life analysis as virtually all products will change with respect to their sensory profile over the course of their respective shelf lives.

The choice of scientists and companies for food acceptance limits is still often food-driven, since it occurs independently of the actual attitude of the consumers to eat it or not. In fact, it is evident that, at the end of shelf life, the product is certainly still acceptable by some consumers, despite being rejected by others (Manzocco and Lagazio, 2009). Shelf life is the result of the interaction of foods with the consumers, since foods have no sensory shelf lives of their own (Giménez *et al.*, 2007; Hough *et al.*, 2003). Survival analysis of food acceptability has been applied to estimate shelf life of different foods and is focused on monitoring the risk of consumer's rejection of the product during storage. According to industrial policy, the food company can thus choose to be exposed to more or less risk by selecting, as acceptance limit, the proper percentage of consumers rejecting the product (Manzocco and Lagazio, 2009).

The most common procedure is to ask consumers representative of the target population to scale acceptability on a 9-point category scale, anchored from 'like extremely' to 'dislike extremely' (Kilcast, 2000). One of the most important parameters in consumer testing is determining the appropriate number of consumers to use. The greater the number of assessors the more expensive and time-consuming the test, but also the more accurate the data collected. Peryam (1964) states that a minimum of 50 consumers should be used, and preferably 100, although lower numbers (32–40) have been reported. On the other hand O'Sullivan *et al.* (2011) describe how 25 naïve assessors (untrained assessors) were used to evaluate ready meals by indicating their score on a continuous 10 cm line scale ranging from 0 (none) on the left to 10 (extreme) on the right for each sensory or hedonic descriptor. The descriptors included those for appearance, flavour, texture, off-flavour as well as product overall acceptability and packaging quality and the presentation order for all samples was randomised to prevent first-order and carry-over effects (MacFie *et al.*, 1989). O'Sullivan *et al.* (2011) showed how a consumer-driven optimisation strategy, using 25 naïve assessors, who also fitted the consumer demographic, could be used with success for product development and optimisations for products such as meat-based ready-to-eat meals, soups, savoury pies, convenience-type seafood products and alcoholic beverages. They argue that one may question using such a small segment in making such important commercial decisions, but that the reality is that small food businesses cannot afford large-scale consumer evaluations using perhaps 200, 300 or even 1000 consumers (O'Sullivan *et al.*, 2011). Larger-scale consumer evaluations take a considerably longer time to organise, conduct and data mine, and months can elapse before completion in which time the consumer market place or demographic can change, which questions the accuracy of the information accumulated. O'Sullivan *et al.* (2011) propose that effective, robust consumer-driven innovation strategies can assist companies in getting their products into the marketplace in a speedy fashion and also ensure the optimised products are not obsolete from a consumer-relevance perspective, but are specifically designed to meet consumer requirements. One way of validating small-scale consumer studies is to replicate them by repeating with a different consumer panel, but the same demographic segment.

Also, from a shelf-life evaluation perspective, the descriptors used to evaluate the product must capture the most important sensory quality changes over the course of shelf-life testing. Information on individual modalities (appearance, odour, flavour and texture) can also be obtained, together with attribute intensity information, but it is preferable to keep such tests simple and to focus on overall acceptability (Kilcast, 2000). For the evaluation of cream liqueurs, the reaction of fatty acids and alcohol over time results in the formation of ethyl esters which have fruity notes. These fruity notes must be included as a sensory descriptor for consumer evaluation. Similarly, a reduction of fresh notes and an increase in stale or oxidised notes can be used as indices to track the change in the consumer flavour profile of beer.

3.9 Instrumental analysis

Instrumental analysis is often employed by the food and beverage industry to measure sensory changes that occur in products over the course of shelf life. These methods may be quicker than sensory analysis, which traditionally can be quite expensive. Descriptive profiling can be particularly expensive and time-consuming because of the necessity to maintain a testing laboratory or facility, train and profile individual panellists over extended periods of time, days or even weeks. It is also not a method that can be readily used for routine analysis. However, instrumental methods are usually quick and should be directly correlated to a sensory or primary shelf-life criterion. Instrumental and sensory limits can be assessed using survival analysis results to determine what instrumental limits correspond to the appropriate sensory limit of acceptability. An example of this is the rheological measurement of texture or viscosity in the case of some alcoholic beverages. Emulsion-based alcoholic beverages such as cream liqueurs become more viscous over time and this can be a limiting factor with respect to shelf life. Some of the measurements that can be used to assess cream liqueur shelf life include droplet size distributions and viscosity. Droplet size distributions can be determined by laser diffraction using a Malvern Mastersizer S (Malvern Instruments, Malvern, Worcestershire, UK), which facilitates the measurement of particles in the size range of 0.050–880 μm (Heffernan *et al.*, 2009). A viscometer is an instrument used to measure the viscosity of a fluid. Typically rotational viscometers are used to quantify the viscosity of cream liqueurs, such as the 'cup and bob' type, which measure the torque required to rotate the bob in a fluid at a known speed, where this measured torque is a function of the viscosity of that fluid. Also, viscosity of cream liqueurs can be determined using a viscometer such as Carrimed CSL 100 rheometer (Carri-Med Ltd, Dorking, Surrey, UK). This is generally performed at a constant shear rate at a specific temperature of 20 °C (Power, 1996; Lynch and Mulvihill, 1997).

For standard beer analysis, European Brewing Convention (EBC) colour and bitterness units can be assessed by photometry. Organic acids are determined using enzymatic analyses or liquid chromatography. For a standard spirit drink analysis, higher alcohols and other volatile compounds are determined using gas chromatography. Therefore, by means of these traditional techniques, beverage analysis is time-consuming and expensive (Lachenmeier, 2007). Lachenmeier (2007) investigated Fourier Transform Infrared spectroscopy in combination with partial least squares (PLS) regression as a complete multi-component screening method for spirit drinks and beer in the context of official food control and found that the majority of samples were classified as being in conformance with legal and quality requirements and that FTIR spectroscopy is faster and only requires a simple sample preparation compared to conventional methods.

3.10 Accelerated storage tests

Real-time shelf-life studies have the potential of taking months, if not years, to complete because of the inherent durability of alcoholic beverages. Some of the premium brands of cream liqueur can have shelf lives of two years when stored in ambient conditions. Beers can have shelf lives of between six months and two years at ambient temperatures, but this is obviously increased when stored under refrigerated conditions. Also, some wines and spirits can have indefinite shelf lives and the importance of accelerated shelf-life testing may not be relevant.

Accelerated storage tests are designed to reduce shelf-life testing time by speeding up the deteriorative changes that occur in the product by exposure to factors such as high temperatures or humidity. Figure 3.4 shows a Binder (BINDER Gmbh, Tuttlingen, Germany) climatic cabinet (KBF 720) which is used for accelerated storage testing. Temperature, humidity and light parameters can be controlled with ranges from −10–100 °C without humidity and lighting, 10–99 °C with humidity and 20–90 °C with humidity and lighting. Photo stability testing can be undertaken with the doors incorporating Lux 4000 lighting. In the system shown in Fig. 3.4 a remote data recording system has been incorporated. These wireless recorders communicate by radio frequency with the server PC at defined regular intervals and record many different parameters such as temperature and humidity in this case, but can also record additional measurements, depending on the application such as hygrometry measurements or pressure if required.

Accelerated ageing traditionally involves use of temperature increases to speed reactions. The process of estimating ambient stability involves estimating the reaction rate at different temperatures, and then extrapolating to the desired temperature (Waterman and Adami, 2005). Accelerated tests involve a single test condition where it is assumed, for example, that if a product lasts one month at 40 °C, it will last four months at ambient temperature, 20 °C. Thus, it is assumed that the accelerating factor from 20–40 °C is equal to 4. Often this value is a company tradition with unknown origins, other times it is estimated from data on a previous similar product (Nelson, 1990).

Mathematical modelling can be used to take into account the differences, linear or non-linear between real-time shelf life and accelerated tests. Mathematical modelling can be applied to any deterioration process that has a valid kinetic

Fig. 3.4 Binder (BINDER Gmbh, Tuttlingen, Germany) KBF 720 climatic cabinet, used for accelerated storage testing. Included in this configuration is a SPYRF® (JRI, Argenteuil Cedex France) remote data recording system.

model, which could be chemical, physical, biochemical or microbial. Temperature is the most frequently used acceleration factor in the Arrhenius equation. The key parameter of this equation to estimate shelf life at different temperatures is the activation energy. Usually, the reaction rate is determined at three or more temperatures and the logarithm of these values are regressed versus 1/temperature to obtain the activation energy from the resulting slope. When applying regression techniques, statistical analysis is used to determine the 95% confidence limits of the activation energy. If only three reaction rates at their corresponding temperatures are available, the confidence range is usually wide (Taoukis *et al.*, 1997), therefore more temperature measurements may be required to obtain data that can be more accurately extrapolated, but more time effort and resources are required to achieve this. Mizrahi (2000) describes the 'initial rate approach' which may be applicable to cases where the deterioration process can be monitored by an extremely accurate and sensitive analytical method. This method should be capable of measuring minute changes in the extent of deterioration after a relatively short storage time at actual conditions.

Taoukis *et al.* (1997) also describe the one-step method, where the reaction rate equation is combined with the Arrhenius equation. Experimental data of physicochemical or sensory measures versus time for all tested temperatures are used, substantially increasing the degrees of freedom and hence giving much narrower confidence intervals for the estimated parameters. Activation energy is estimated by nonlinear regression.

The kinetic model approach is the most common method for accelerated shelf-life testing, which has been extensively described by Mizrahi (2000). The kinetic model approach involves the selection of the desired kinetically active factors for acceleration of the deterioration process to be measured. A kinetic study of the deterioration process is then run at such levels of the accelerating factors that the rate of deterioration is fast enough to be a useful indicator. The parameters of the kinetic model are evaluated and the data extrapolated to normal storage conditions. The extrapolated data or the kinetic model are then used to predict shelf life at actual storage conditions (Mizrahi, 2000).

Some indices which could be used as accelerated shelf-life indices in alcoholic beverages are discussed below. Heffernan *et al.* (2009) measured the stability of model cream liqueurs by evaluating changes in volume frequency distribution of fat droplet diameter and the viscosity of the cream liqueurs during a 28-day accelerated shelf-life storage period at 45 °C (Banks *et al.*, 1981b; Muir, 1987; Muir *et al.*, 1991; Power, 1996; Lynch and Mulvihill, 1997). In order to assess the stability of model cream liqueurs, accelerated shelf-life tests were carried out by incubating the liqueurs for 28 days at 45 °C and monitoring the changes in volume frequency distribution of fat droplets and liqueur viscosity (Heffernan *et al.*, 2009). Alternatively, the concentration of certain ethyl esters, perhaps short chain, could be measured, using GC or GCMS in an accelerated shelf-life storage period, also at 45 °C. These are directly correlated to the development of fruity notes which could be deleterious sensory quality attributes benchmarked against consumer sensory data. A factor may then be calculated to correlate accelerated storage to real-time storage at ambient temperatures. Similarly for beer, accelerated shelf-life storage temperature of 45 °C could also be used and, again, GC or GCMS used to quantify E-2-nonenal as an index of staling, perhaps also correlated to consumer sensory data.

There are certain problems that occur with accelerated shelf-life testing. The most difficult problem is to find a simple mathematical relation able to describe the evolution of the chosen attribute as a function of storage time (Guerra *et al.*, 2008). It can be difficult to predict shelf life from accelerated storage testing as the relationship to real-time shelf-life analysis is not always linear. The main problem is how to devise a reliable method, or methods, that will allow the extrapolation of the accelerated storage data in a manner that will guarantee a correct prediction of what will happen under normal storage and handling conditions (Corradini and

Peleg, 2007). The reason for this is that deteriorative mechanisms can prevail in the accelerative test that would not normally occur in the real-time situation under normal storage conditions. Accelerated shelf-life testing may be difficult to implement when no valid kinetic model is believed to exist for any accelerating kinetic factor, thus there is no test procedure available to make measurements. Also, testing may not be practical in cases where too many parameters need to be measured and thus the analysis becomes overly complicated (Mizrahi, 2000).

Certain processes follow models that have two or more temperature-dependent parameters, in which case the rate cannot be expressed by a single constant, as the Arrhenius equation demands. There are also situations where the deterioration kinetics are not known and there is insufficient evidence to conclude that it follows any of the traditional models (Corradini and Peleg, 2007). There is simply neither evidence nor theoretical reason for the effects of different parameters temperature, pH, aW, or any other factor for that matter, to be additive or multiplicative in accelerated shelf-life models (Peleg, 2006).

It is essential that the proper statistical methods be used in accelerated shelf-life testing. These methods test the sensitivity of the model by a cross-validation method by using part of the data to verify the validity of the model. This requires a wider range of accelerated storage conditions. The closer they are to the actual storage conditions the better. Such an approach costs more both in time and in experimental efforts (Mizrahi, 2000). Guerra *et al.* (2008) state that it is important to pay attention to the planning of test sessions, recruitment and training of the panel and, the phase of data analysis and the choice of processing methods, in order not to run into incorrect predictions. Given the number of arbitrary choices, they concluded that the shelf-life concept for microbiologically stable food products is more company or researcher driven than product or consumer dependent (Guerra *et al.*, 2008). There is also good reason to believe that many, or perhaps even most, of the deterioration processes in foods might not follow any of the standard kinetic models. The assumption that the process or reaction must follow a particular kinetic is therefore unnecessary (Corradini and Peleg, 2007).

3.11 Future trends

The principal criteria by which beverage product quality is judged are by sensory attributes. As such, sensory methods will continue to be the most important methods of assessing product quality over the course of alcoholic beverage shelf life. Many sensory methods are available as have been discussed above, but the industry will strive towards developing more reliable, cheaper and quicker methods so that the route to market is shortened with an effective evaluation of the sensory profile of a product over the course of shelf life. Kilcast (2000) proposes a potential adaptation of sensory methods, driven by the retail sector in the UK. This involves the development of sensory specifications for foods, and incorporates a simple assessment of product quality against specification. Although relatively crude, such systems offer the opportunity for low-cost sensory appraisal of shelf life on a qualitative or semi-quantitative basis. O'Sullivan *et al.* (2011), as described above, suggest that validated small-scale consumer studies may assist small food businesses who cannot afford large-scale consumer evaluations in assessing the consumer perception of their products.

Rapid instrumental analysis that effectively monitors sensory changes in alcoholic beverages may present a future opportunity for alcoholic beverage analysis. FTIR appears so far as the most advantageous screening method for spirits and beers because of the lower investment and operational costs, after the establishment of the FTIR calibrations (Lachenmeier, 2007). FTIR/PLS offers considerable advantages when measured against conventional methods of analysis and will acquire increasing importance as an efficient high-throughput tool for screening alcoholic beverages (30 samples/hour) (Lachenmeier, 2007).

3.12 Sources of further information and advice

Sensory Evaluation of Food, Principles and Practices (1998), H.T. Lawless and H. Heymann (eds). Chapman and Hall, New York, USA.

Smart Packaging Technologies, (2007), J.P. Kerry and P. Butler (eds). John Wiley & Sons, Chichester, West Sussex, UK, 211–230.

3.13 References

BAMFORTH, C.W. (1999), 'The science and understanding of the flavour stability of beer: a critical assessment', *Brauwelt International*, **17**(2), 98–110.

BANKS, W. and MUIR, D.D. (1988), 'Stability of alcohol-containing emulsions'. In Dickinson, E., Stainby, G. (eds), *Advances in Food Emulsions and Foams*, Elsevier Science Publishing Co., New York, pp. 257–283.

BANKS, W., MUIR, D.D. and WILSON, A.G. (1981b), 'Extension of the shelf-life of cream-based liqueurs at high ambient temperatures', *Journal of Food Technology*, **16**, 587–595.

BANKS, W., MUIR, D.D. and WILSON, A.G. (1981a), 'The formation of cream based liqueurs', *The Milk Industry*, **83**, 16–18.

BAYARRI, S., TAYLOR, A.J. and HORT, J. (2006), 'The role of fat in flavor perception: effect of partition and viscosity in model emulsions', *Journal of Agriculture and Food Chemistry*, **54**, 8862–8868.

BYRNE, D.V. and BREDIE, W.L.P. (2002), 'Sensory meat quality and warmed-over flavour: A Review.' In F. Toldrá, *Research Advances in the Quality of Meat and Meat Products. Volume Within Agriculture & Food Chemistry, Research Signpost*, 95–212.

BYRNE, D.V., BAK, L.S., BREDIE, W.L.P., BERTELSEN, G. and MARTENS, M. (1999a), 'Development of a sensory vocabulary for warmed-over flavour 1: in porcine meat', *Journal of Sensory Studies*, **14**, 47–65.

BYRNE, D.V., BREDIE, W.L.P. and MARTENS, M. (1999b), 'Development of a sensory vocabulary for warmed-over flavour: Part II. in chicken meat', *Journal of Sensory Studies*, **14**, 67–78.

BYRNE, D.V., O'SULLIVAN, M.G., DIJKSTERHUIS, G.B., BREDIE, W.L.P. and MARTENS, M. (2001), 'Sensory panel consistency during development of a vocabulary for warmed-over flavour', *Food Quality and Preference*, **12**, 171–187.

CASAMAYOR, P. (2001), *Les livrets du vin. La dégustation*. Hachette: Paris.

CHAMBERS, E. and BOWERS, J. (1993), 'Consumer perception of sensory quality in muscle foods: Sensory characteristics of meat influence consumer decisions', *Food Technology*, **47**, 116–120.

CIVILLE, G.V. (1987), 'Development of vocabulary for flavour descriptive analysis'. In *Flavor Science and Technology*, M. Martens, G.A. Dalen and H. Russwurm Jr., (eds.), John Wiley & Sons Ltd, Chichester, 357–368.

CIVILLE, G.V. and LAWLESS, H.T. (1986), 'The importance of language in describing perceptions', *Journal of Sensory Studies*, **1**, 203–215.

CLAPPERTON, J.F. (1976), 'Ribes flavour in beer', *Journal of the Institute of Brewing*, 82, 175–176.

CONNER, J.M., PATERSON, A., BIRKMYRE, L. and PIGGOTT, J.R. (1999), 'Role of organic acids in maturation of distilled spirits in oak casks', *Journal of the Institute of Brewing*, **105**, 287–291.

CORRADINI, M.G. and PELEG, M. (2007), 'Shelf-life estimation from accelerated storage data', *Trends in Food Science and Technology*, **18**, 37–47.

DELAHUNTY, C.M., MCCORD, A., O'NEILL, E.E. and MORRISSEY, P.A. (1997), 'Sensory characterisation of cooked hams by untrained consumers using free-choice profiling', *Food Quality and Preference*, **8**, 381–388.

DICKINSON, E., NARHAN, S.K. and STAINSBY, G. (1989), 'Stability of cream liqueurs containing low-molecular-weight surfactants', *Journal of Food Science*, **54**, 77–81.

DOYEN, K., CAREY, M., LINFORTH, R.S.T., MARIN, M. and TAYLOR, A.J. (2001), 'Volatile release from an emulsion: headspace and in-mouth studies', *Journal of Agriculture and Food Chemistry*, **49**, 804–810.

DUBOURDIEU, D. (2005), 'Comment la garde vient aux vins. Le point', *Special Vins*, **1721**, 168–170.

FERNANDEZ, J. and SIMPSON, W.J. (1995), 'Measurement and prediction of the susceptibility of lager beer to spoilage by lactic acid bacteria', *Journal of Applied Microbiology*, **78**, 419–425.

FREITAS, M.A. and COSTA, J.C. (2006), 'Shelf life determination using sensory evaluation scores: A general Weibull modeling approach', *Computers and Industrial Engineering*, **51**, 652–670.

FOOD SAFETY AUTHORITY OF IRELAND (2011), available at http://www.fsai.ie/faqs/best_before_and_use_by_dates.html#date of minimum

GIMÉNEZ, A., VARELA, P., SALVADOR, A., ARES, G., FISZMAN, S. *et al.* (2007), 'Shelf life estimation of brown pan bread: A consumer approach', *Food Quality and Preference*, **18**, 196–204.

GUERRA, S., LAGAZIO, C., MANZOCCO, L., BARNABÀ, M. and CAPPUCCIO, R. (2008), 'Risks and pitfalls of sensory data analysis for shelf life prediction: Data simulation applied to the case of coffee', *LWT – Food Science and Technology*, **41**, 2070–2078.

HAIKARA, A. and HELANDER, I. (2002), '*Pectinatus, Megasphaera* and *Zymophilus*'. In Dworkin, M. (ed.), *The Prokaryotes, An Evolving Electronic Database for the Microbiological Community*, 3rd ed. Springer Verlag, New York.

HARPER, R., BATE SMITH, E.C. and LAND, D.G. (1974), *Odour Description and Odour Classification: A Multidisciplinary Examination*, American Elsevier Publishing Co., New York.

HEFFERNAN, S.P, KELLY, A.L. and MULVIHILL, D.M. (2009), 'High-pressure-homogenised cream liqueurs: Emulsification and stabilization efficiency', *Journal of Food Engineering*, **95**, 525–531.

HOUGH, J.S., BRIGGS, D.E., STEVENS, R., and YOUNG, T.W. (1982), *Malting and Brewing Science* (Vol. 2, 2nd ed). Chapman and Hall: London, p. 389.

HOUGH, G., LANGOHR, K., GOMEZ, G., and CURIA, A. (2003), 'Survival analysis applied to sensory shelf life of foods', *Journal of Food Science*, **68**, 359–362.

IFST (1993), *Shelf Life of Foods – Guidelines for its Determination and Prediction.* Institute of Food Science and Technology: London.

ISO (1992), 'International Standard 5492. Sensory analysis-vocabulary', Ref. no. ISO 5492:1992 (E/F), International Organization for Standardization, Genève.

ISO (1994), 'International Standard 11035. Sensory analysis-identification and selection of descriptors establishing a sensory profile by a multidimensional approach', Ref. no. ISO 11035:1994 (E), International Organization for Standardization, Genève.

JAFFRÉ, J., VALENTIN, D., DACREMONT, C. and PEYRON, D. (2009), 'Burgundy red wines: representation of potential for aging', *Food Quality and Preference*, **20**, 505–513.

JOHNSON, P.B. and CIVILLE, G.V. (1986), 'A standardized lexicon of meat WOF descriptors', *Journal of Sensory Studies*, **1**, 99–104.

JOHNSEN, P.B., CIVILLE, G.V., VERCELLOTTI, J.R., SANDERS, T.H. and DUS, C.A. (1988), 'Development of a lexicon for the description of peanut flavor', *Journal of Sensory Studies*, **3**, 9–17.

KILCAST, D. (2000), 'CH4. Sensory evaluation methods for shelf-life assessment'. In David Kilcast and Persis Subramaniam (eds). *The Stability and Shelf-life of Food*, Woodhead Publishing Limited, Cambridge, 79–103.

LACHENMEIER, D.W. (2007), 'Rapid quality control of spirit drinks and beer using multivariate data analysis of Fourier transform infrared spectra', *Food Chemistry*, **101**, 825–832.

LAWLESS, H.T. and HEYMANN, H. (1998a), 'Sensory evaluation in quality control'. In H.T. Lawless and H. Heymann. *Sensory Evaluation of Food, Principles and Practices*. New York: Chapman and Hall, 548–584.

LAWLESS, H.T. and HEYMANN, H. (1998b), 'Descriptive analysis'. In H.T. Lawless and H. Heymann (eds), *Sensory Evaluation of Food, Principles and Practices*. New York: Chapman and Hall, 117–138, 341–378.

LEE, M., PATERSON, A., PIGGOTT, J.R. and RICHARDSON, G.D. (2000), 'Origins of flavour in whiskies and a revised flavour wheel: a review', *Journal of the Institute of Brewing*, **107**, 287–313.

LOUREIRO, V. and MALFEITO-FERREIRA, M. (2003), 'Spoilage yeasts in the wine industry', *International Journal of Food Microbiology*, **86**, 23–50.

LYNCH, A.G. and MULVIHILL, D.M. (1997), 'Effect of sodium caseinate on the stability of cream liqueurs', *International Journal of Dairy Technology*, **50**, 1–7.

MACFIE, H.J., BRATCHELL, N., GREENHOFF, K. and VALLIS, L.V. (1989), 'Designs to balance the effect of order of presentation and first-order carry-over effects in hall tests', *Journal of Sensory Studies*, **4**, 129–48.

MACHADO, B., HEYMANN, H., ROBINSON, A.L. and TORRI, L. (2010), 'How many judges are required for sensory descriptive analysis'. *Fourth European Conference on Sensory and Consumer Research: a Sense of Quality*, 5–8 Sept, Palacio Europa, Vitoria-Gasteiz, Spain.

MANZOCCO, L. and LAGAZIO, C. (2009), 'Coffee brew shelf life modelling by integration of acceptability and quality data', *Food Quality and Preference*, **20**, 24–29.

MARULLO, P., BELY, M., MASNEUF-POMAREDE, I., AIGLE, M. and DUBOURDIEU, D. (2004), 'Inheritable nature of enological quantitative traits is demonstrated by meiotic segregation of industrial wine yeast strains', *FEMS Yeast Research*, **4**, 711–719.

MCCLEMENTS, D.J. (1999), *Food Emulsions: Principles, Practices and Techniques*. CRC Press, Boca Raton, FL.

MEILGAARD, M.C. (1982), 'Prediction of flavor differences between beers from their chemical composition', *Journal of Agriculture and Food Chemistry*, **30**, 1009–1017.

MEILGAARD, M.C., CIVILLE, G.V. and CARR, B.T. (1999), *Sensory Evaluation Techniques*, 3rd ed, pp 54–55. Florida: Academic Press.

MEILGAARD, M.C., DALGLEISH, C.E. and CLAPPERTON, J.R. (1979), 'Beer flavour terminology', *Journal of the Institute of Brewing*, **85**, 38.

MILLER, R. (1994), 'Sensory methods to evaluate muscle foods'. In D.M. Klinsman, A.W. Kotula, B.C. Breidenstein (eds), *Muscle Foods: Meat Poultry and Seafood Technology* (333–360). Chapman and Hall: New York.

MIZRAHI, S. (2000), 'CH5, Accelerated shelf-life tests'. In D. Kilcast and P. Subramaniam (eds), *The Stability and Shelf-life of Food.* Woodhead Publishing Limited, Cambridge.

MONAGAS, M., GOMEZ-CORDOVES, C. and BARTOLOMÉ, B. (2006), 'Evolution of the phenolic content of red wines from *Vitis vinifera L.* during ageing in bottle', *Food Chemistry*, **95**(3), 405–412.

MUIR, D.D. (1987), 'Cream liqueur manufacture: assessment of efficiency of methods using a viscometric technique', *Dairy Industries International*, **52**, 38–40.

MUIR, D.D. and BANKS, W. (1985), 'From Atholl Brose to cream liqueurs: development of alcoholic milk drinks stabilized with trisodium caseinate'. In Galesloot, T.E., Tinbergen, B.J. (eds), *Proceedings of the International Congress on Milk Proteins*, Luxemburg, May 7–11, 1984, 120–128.

MUIR, D.D. and BANKS, W. (1986), 'Technical note: multiple homogenisation of cream liqueurs', *Journal of Food Technology*, **21**, 229–232.

MUIR, D.D., MCCRAE, C.H. and SWEETSUR, A.W.M. (1991), 'Characterization of dairy emulsions by forward lobe laser light-scattering – application to cream liqueurs', *Milchwissenschaft*, **46**, 691–694.

MUÑOZ, A.M. and CIVILLE, G.V. (1992), 'The spectrum descriptive analysis method'. In R.C. Hootman (ed), *ASTM Manual on Descriptive Analysis*, Pennsylvania: American Society for Testing and Materials.

NELSON, W. (1990), *Accelerated testing. Statistical Models, Test Plans and Data Analyses*, John Wiley & Sons: New York.

NISHIMURA, K., and MATSUYAMA, R. (1989), 'Maturation and maturationchemistry'. In J. R. Piggott, R. Sharp, and R. E. B. Duncan (eds), *The Science and Technology of Whiskies*, 235–263, Longman, Harlow.

NOBLE, A.C., ARNOLD, R.A., MASUDA, B.M., PECORE, S.D., SCHMIDT, J.O. *et al.* (1984), 'Progress towards a standardized system of wine aroma terminology', *American Society for Enology and Viticulture*, **35**, 107–109.

O'SULLIVAN, M.G., BYRNE, D.V. and MARTENS, M. (2002a), 'Data analytical methodologies in the development of a vocabulary for evaluation of meat quality', *Journal of Sensory Studies*, **17**, 539–558.

O'SULLIVAN, M.G., MARTENS, H., BOBERG, S., KRISTENSEN, L. and MARTENS, M. (2002b), 'The reliability of naïve assessors in sensory evaluation visualised by pragmatical multivariate analysis', *Journal of Food Quality*, **25**, 395–408.

O'SULLIVAN, M.G., BYRNE, D.V., NIELSEN, J.H., ANDERSEN, H.J. and MARTENS, M. (2003), 'Sensory and chemical assessment of pork supplemented with iron and vitamin E', *Meat Science*, **64**, 175–189.

O'SULLIVAN, M.G., KERRY, J.P. and BYRNE, D.V. (2011), 'Use of sensory science as a practical commercial tool in the development of consumer-led processed meat products'. In J. P. Kerry and J.F. Kerry *Processed Meats Improving Safety and Nutritional Quality.*. Woodhead Publishing Ltd, Cambridge.

ORDONEZ, A.I., IBANEZ, F.C., TORRE, P. and BARCINA, Y. (1998), 'Application of multivariate analysis to sensory characterisation of ewes' milk cheese', *Journal of Sensory Studies*, **13**, 45–55.

PELEG, M. (2006), *Advanced quantitative microbiology for food and biosystems: Models for predicting growth and inactivation.* CRC Press: Boca Raton, FL.

PERYAM, D.R. (1964), 'Consumer preference evaluation of the storage stability of foods', *Food Technology*, September, 214–17.

PIGGOTT, J.R. (1991), 'Selection of terms for descriptive analysis'. In (H.T. Lawless and B.P. Klein, eds) *Sensory Science Theory and Applications in Foods*, 339–351, Marcel Dekker, New York.

PIGGOTT, J.R., CONNER, J.M., PATERSON, A. and CLYNE, J. (1993), 'Effects on Scotch whisky composition and flavour of maturation in oak casks with varying histories', *International Journal of Food Science and Technology*, **28**, 303–318.

PIGGOTT, J.R., HUNTER, E.A. and MARGOMENOU, L. (2000), 'Comparison of methods of analysis of time-intensity data: application to Scotch malt whisky', *Food Chemistry*, **71**, 319–326.

PIGGOTT, J.R., SIMPSON, S.J. and WILLIAMS, S.A.R. (1998), 'Sensory analysis', *International Journal of Food Science and Technology*, **33**(1), 7–12.

PINO, J.A. (2007), 'Characterization of rum using solid-phase microextraction with gas chromatography–mass spectrometry', *Food Chemistry*, **104**, 421–428.

PINO, J., MARTÍ, M.P., MESTRES, M., PÉREZ, J., BUSTO, O. *et al.* (2002), 'Headspace solid-phase microextraction of higher fatty acid ethyl esters in white rum aroma', *Journal of Chromatography A*, **954**, 51–57.

POWER, P.C. (1996), 'The formulation, testing and stability of 16% fat cream liqueurs', Ph.D. Thesis, National University of Ireland, Cork.

RAINBOW, C. (1981), 'Beer spoilage microorganisms'. In (ed) *Brewing Science* Vol.2. J.R.A Pollock. Academic Press, 491–550.

ROMANO, P., FIORE, C., PARAGGIO, M., CARUSO, M. and CAPECE, A. (2003), 'Function of yeast species and strains in wine flavour', *International Journal of Food Microbiology*, **86**, 169–180.

RULE, C.E. 1983. Cream/alcohol containing beverages. US Patent No. 4,419,378.

SAKAMOTO, K. and KONINGS, W.N. (2003), 'Beer spoilage bacteria and hop resistance', *International Journal of Food Microbiology*, **89**, 105– 124.

SAWYER, F.M., CARDELLO, A.V. and PRELL, P.A. (1988), 'Consumer evaluation of the sensory properties of fish', *Journal of Food Science*, **53**, 12–18, 24.

SIMPSON, W.J. and FERNANDEZ, J.L. (1994), 'Mechanism of resistance of lactic acid bacteria to *trans*-isohumulone', *Journal of the American Society of Brewing Chemists*, **52**(1), 9–11.

STONE, H., SIDEL, J., OLIVER, S., WOOLSEY, A. and SINGLETON, R.C. (1974), 'Sensory evaluation by quantitative descriptive analysis', *Food Technology*, **28**, 24–34.

TAOUKIS, P.S., LABUZA, T.P. and SAGUY, I.S. (1997), 'Kinetics of food deterioration and shelf-life prediction'. In K. J. Valentas, E. Rotstein, and R. P. Singh (eds), *Handbook of Food Engineering Practice.* CRC Press: Boca Raton, Florida.

VANDERHAEGEN, B., NEVEN, H., VERACHTERT, H. and DERDELINCKX, G. (2006), 'The chemistry of beer aging – a critical review', *Food Chemistry*, **95**, 357–381.

VERSTREPEN, K.J., DERDELINCKX, G., DUFOUR, J.P., WINDERICKX, J., THEVELEIN, J.M. *et al.*, (2003), 'Flavor-active esters: adding fruitiness to beer', *Journal of Bioscience and Bioengineering*, **96**(2), 110–118.

WALKER, S.J. (1994), 'The principles and practice of shelf life prediction of microorganisms'. In: C.M.D. Man and A.A. Jones. *Shelf Life Evaluation of Foods*. (eds) Chapman and Hall, London, 40–51.

WATERMAN, K.C. and ADAMI, R.C. (2005), 'Accelerated aging: prediction of chemical stability of pharmaceuticals', *International Journal of Pharmaceutics*, **293**, 101–125.

WHITEAR, A.L., CARR, B.L., CRABB, D., and JACQUES, D. (1979), 'The challenge of flavour stability'. In: *Proceedings of the European Brewery Convention Congress*, 13–25.

YASUI, T. and YODA, K. (1997), 'Imaging of *Lactobacillus brevis* single cells and microcolonies without a microscope by an ultrasensitive chemiluminescent enzyme immunoassay with a photoncounting television camera', *Applied Environmental Microbiology*, **63**, 4528–4533.

4

Sensory methods for product development and their application in the alcoholic beverage industry

E. Monteleone, University of Florence, Italy

Abstract: This chapter deals with the role of sensory evaluation in product development. The aim is to present sensory methods to describe sensory similarities and differences among products suitable for alcoholic beverages. Particular relevance is given to methods appropriate to obtain so-called perceptual maps; besides the conventional descriptive analysis and the free choice profile, recent advances in sensory research on wine are discussed, from flash profile to sorting and napping. Temporal dominance of sensation, a new method to analyse temporal aspects of perception, is discussed because of its potential relevance in product development and optimization of alcoholic beverages. Finally, sensory procedures appropriate for setting up a predictive model of astringency induced by phenolic compounds in alcoholic beverages are reported.

Key words: perceptual maps, descriptive analysis, projective mapping, temporal dominance of sensations, astringency.

4.1 Introduction

No food or beverage is worth producing, distributing or marketing without at least an approximate idea that consumers accept its sensory quality. Modern product development and hard competition within the food industry require clear understanding of sensory aspects of foods, and adequate sensory techniques (Tuorila and Monteleone, 2009).

As described by Stone and Sidel (2007), there are both strategic and practical reasons to include sensory evaluation in food development activities. Usually, marketers do not know, on a sensory basis, the key dimensions which generate liking, or fit to specific end uses (Moscowitz, 1999). The lack of information about the relationship between sensory characteristics and liking is not just important at a technical level, but it is important because marketers need to define the most relevant sensory properties that their product possesses: given a marketing plan, sensory information contributes to understanding how consumers will perceive a new product. Moreover, from an operative point of view, sensory evaluation is essential to understand the marketplace since it provides information on how to manipulate the sensory properties of products in relation to consumers' hedonic responses. Analytical sensory evaluation serves to identify perceived product attributes and differences. In this capacity, its function is similar to chemical, physical and microbiological characterization of products. However, being closer to consumer perceptions than physicochemical and biological assays, sensory analysis helps to interpret consequences of process modifications, and the like. Also, using methodologies applicable to consumers, sensory research can directly acquire information about consumer response to products.

The need to combine the information on product characteristics and consumer behaviour in product development has led sensory scientists to adopt methods from other scientific disciplines. With wine as an example, Table 4.1 shows different lines of recent sensory research. Sensory analysis is widely applied in wine research to describe the effect of factors such as grape variety or processing on properties of wine, and to study the relationship between chemical and sensory characteristics, but there are many further approaches and targets of research.

Table 4.1 Recent advances in sensory research on wine

Research area	Topic	Reference
New method and tools in sensory description of wine	Vocabulary of terms for sensory evaluation of dessert port wine Descriptive wheel of mouth-feel characteristics of red wine Technique for descriptive profile validation Procedures for reducing carry-over effects in evaluation of red wines Projective mapping for describing differences among wines Temporal dominance of sensations (TDS) for illustrating wine perception pattern over time	Cristovam *et al.*, 2000 Gawel *et al.*, 2001 O'Neill *et al.*, 2003 Colonna *et al.*, 2004 Pagès, 2005 Meillon *et al.*, 2010
Sensory evaluation of wine by experts	Estimation of the consistency of experts in describing sensory properties of wine Study of differences between experts and consumers in discriminating wines	Sauvageot *et al.*, 2006 Hughson and Boakes, 2002 Parr *et al.*, 2004 Ballester *et al.*, 2005
Relationship between chemical and sensory characteristics	Role of polyphenols in determining bitterness and astringency of red wine Evaluation of key wine components on mouth-feel perception Multivariate approach for defining wine aroma model	Brossaud *et al.*, 2001 Vidal *et al.*, 2004 Lee and Noble, 2003

(*Continued overleaf*)

Table 4.1 Continued

Research area	Topic	Reference
Prediction of astringency induced by grape and wine phenols	*In vitro* assay for astringency prediction from grape and wine phenol extract	Monteleone *et al.*, 2004
Flavour perception in wine	Wine discrimination with orthonasal and retronasal profiles Detection thresholds for 2-isopropyl-3-methoxypyrazine in wine Consumer rejection threshold for cork taint in white wine	Aubry *et al.*, 1999 Pickering *et al.*, 2007 Prescott *et al.*, 2005
Consumer preference for wine	Effect of contextual factors on liking for wine Consumer reactions to different attributes of red and white wine Perspectives for consumer-driven winemaking and wine business strategies Effect of partial alcohol reduction in wine on consumer expectations and overall liking in real-life settings	Hersleth *et al.*, 2003 Hughson *et al.*, 2004 Lesschaeve, 2007 Meillon *et al.*, 2010a
Assessment of sensory quality of wine	Hedonic scaling method to wine quality assessment Estimation of the consistency of wine quality assessment	Lawless *et al.*, 1997 Parr *et al.*, 2006 Gawel and Godden, 2008
Wine and food combination	Effect of cheese on sensory description of wine	Nygren *et al.*, 2002 Madrigal-Galan and Heymann, 2006 Bastian *et al.*, 2010

Sensory work requires many skills, but this is not always appreciated. The importance of the field is so obvious that many parties, whether trained in sensory food science or not, are seeking its support for their decision-making. Sensory data are easy to collect, because human subjects respond to questions and perform tasks, even when a task is obscure or inappropriate. This creates an insidious potential for collecting data of poor validity.

It is generally assumed that there are two types of methods appropriate for product development and optimization: descriptive sensory analysis or so-called sensory profiling, and affective methods. The reason is quite obvious if we consider that in product development situations one of the main interests is to detect and understand differences between products and how these relate to consumer liking. The aim of this chapter is to present sensory methods to describe sensory similarities and differences among products suitable for alcoholic beverages. Consumer data are not considered. Particular relevance is given to methods appropriate to obtain the so-called perceptual maps, including alternatives to conventional descriptive analysis.

Temporal dominance of sensation, a new method to analyse temporal aspects of perception, is discussed because of its potential relevance in describing the effect of factory conditions on the sensory properties of alcoholic beverages.

Finally, sensory procedures appropriate for setting up a predictive model of astringency induced by phenolic compounds in alcoholic beverages are reported.

4.2 Perceptual maps

The use of perceptual maps is common in food and beverage development and optimization. A perceptual map is defined as a 'pictorial representation that capture[s] the relationships among a set of products' (Lawless and Heymann, 1998) and it is obtained by submitting sensory data to a variety of multivariate statistical techniques. These techniques can extract from complex data (a matrix with many products and variables) the most important information (how different are the products from each other and how much each variable contributes to the difference among samples) and present the results in a simplified picture or map in two or three dimensions that can be easily understood. Because of its characteristics, perceptual mapping is very attractive but also effective. In Fig. 4.1 an example of a perceptual map is reported. It shows the differences and similarities between a set of wines: products that are similar to one another are positioned close to one another. Furthermore, vectors corresponding to important attributes are projected to interpret directions through the space.

It is easy to understand that perceptual maps can be useful for many purposes in product development. They can be used in an early phase of an investigation as an explorative tool in order to provide an overview, but they can also be used to generate hypotheses and ideas for further experimentations (Naes *et al.*, 2010a) as well as to confirm previous hypotheses. Perceptual mapping is a necessary step to effectively relate sensory properties to both consumer hedonic responses by means of preference mapping and to chemo-physical characteristics of samples in multi-product studies.

In particular, by interpreting perceptual maps it is possible:

- To compare a product with its competitors,
- To show the effect of modifications in product composition,
- To define desired changes of the sensory properties of a product,
- To select products to be tested by consumers.

Perceptual maps can be obtained using different sensory methods and applying a variety of statistical analysis, as seen in Table 4.2. The table reports the methods that can be applied in laboratory small-panel studies to obtain a map. Descriptive analysis is probably the most important tool in sensory analysis for describing products and differences between products; however, alternative ways of capturing sensory differences between products, as free choice profile, flash analysis, sorting and napping are available for specific aims and needs. In this chapter these methods are illustrated considering their use in product development of alcoholic beverages, with particular reference to wine.

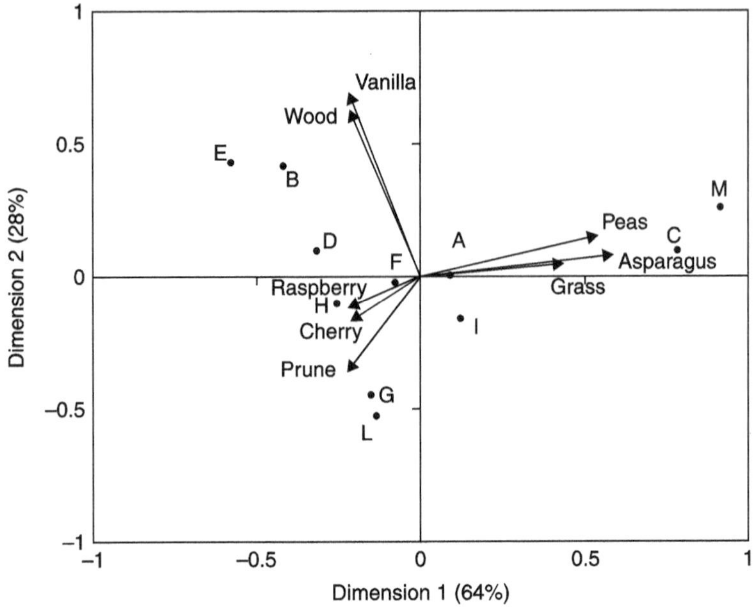

Fig. 4.1 Example of a perceptual map. The graph shows the bi-plot (scores and loadings) from principal component analysis (PCA) of sensory descriptive data (eight aroma attributes) from 11 wines (A-M) (Picchi *et al.*, 2010). The first two dimensions account for approximately 92% of the variation (64% and 28%). The first dimension, from left to right of the map, primarily discriminates sample M and C from the rest of the wines by contrasting 'herbaceous' or 'vegetative' attributes (asparagus, pea and grass) with 'fruity' (raspberry, cherry, dried prune) and wood (vanilla and wood) attributes. Visual inspection of the second dimension, from the bottom to the top of the map, indicates that wines E and B are separated from the rest by contrasting the wood notes with dried prune.

Table 4.2 Sensory methods applied in laboratory small-panel studies to obtain a perceptual map

Sensory response	Sensory method	Statistical method
Descriptive data	Descriptive analysis (conventional profile)	Principal Component Analysis (PCA)
	Free choice profile	Generalized Procrustes Analysis (GPA)
	Flash analysis	
Similarity data	Sorting	Multidimentional Scaling (MDS)
	Projective map or nappe method	MDS, GPA, PCA, Multifactorial Analysis (MFA)

4.2.1 Conventional descriptive analysis

Descriptive analysis is a term generally used to describe a sensory method by which identification, quantification and description of sensory attributes (the so-called sensory profile) of food by human subjects are obtained (Piggott *et al.*,

1998). Several descriptive methods are considered: the flavor profile (Cairncross and Sjostrom, 1950), the texture profile (Szczesniak, 1963), quantitative descriptive analysis (Stone *et al.*, 1974), and the sensory spectrum (Meilgaard *et al.*, 1991). However, the most widely used profile technique combines different aspects from the above mentioned methods and is generally named 'conventional descriptive analysis' (see ISO 11035, 1994; Lawless and Heymann, 1998). Free choice profiling (Williams and Langron, 1984), and flash profiling (Sieffermann, 2000) are further descriptive techniques that are presented below.

Conventional descriptive analysis has been widely used to provide qualitative as well as quantitative measures of food and beverages properties, including wine (see Carlucci and Monteleone, 2008) and other alcoholic beverages. It is based on independent judgments of panelists (trained subjects) and statistical testing rather than group discussion and consensus procedures. Thus it is capable of providing a picture of how products differ among themselves, implying a comparison among products. Several products are assessed together, and the descriptive profile of a single product is both placed in and compared with the context of other products (Murray *et al.*, 2001). Descriptive analysis can indicate exactly how, in the sensory dimension, the competitor's product is different from yours, and it is frequently used in product development to measure how close a new introduction is to the target or to assess suitability of prototype products (Lawless and Heymann, 1998). It is a multi-product test that uses a limited number of subjects. It requires a language development process, and subjects provide an intensity rating for each descriptive term. The data are averaged across the panel and it implies statistical evaluation of results.

Subjects

Usually, a descriptive panel involves between 10–15 trained assessors recruited according to their ability to detect differences in important product attributes. Subjects must be qualified prior to their participation. They must be users of the product class to be evaluated. It is also important that individuals participate in a series of difference test organized to represent a range of difficulty and to include relevant modalities (vision, taste, olfaction, etc.). In wine sensory research it is frequently the case (Campo *et al.*, 2010; Biasoto *et al.*, 2010) that subjects are qualified in relation to:

- Their sensibility to recognize the basic tastes (sweetness, bitter, saltiness, sourness) and to perceive astringency,
- Their ability to determine differences in the flavour of relevant wine samples in discrimination tests,
- Their ability to identify wine aromas present in the Wine Aroma Wheel® (Noble *et al.*, 1987), among others.

Language development, subject training and subjects reproducibility

Before assessment of products, assessors participate in a series of language sessions managed by a panel leader. Generally, a new panel develops the sensory language itself. This is a consensus-building process aimed at defining the attributes the panel utilizes to represent their perceptions. Subjects familiarize themselves with the product space and generate attributes that describe the differences among products. The way of presenting products to the subjects may vary. For example, subjects can be exposed to the entire range of products and asked to write down the descriptors that describe perceived differences. The number of training sessions is dependent on the number of the products but, in general, 6–10 one-hour sessions training are needed. The initial list of attributes is normally reduced to achieve a list that comprehensively and accurately describes the product space: redundant and/or less-cited terms are grouped on a semantic basis and/or eliminated according to the subjects' consensual decisions. To facilitate the consensus and to calibrate the subjects, reference standards are presented to the panel, discussed and modified if needed. Refining descriptive terms, reference standards and definitions continues until the panelists reach a consensus.

There are a number of scientific papers reporting attributes generated by descriptive panels to describe the sensory properties of wine. Tables 4.3 and 4.4 report the attributes (and related standards) generated by descriptive panels working

Table 4.3 Attributes and aroma reference standards employed in conventional descriptive analysis training. Quantities specified are those added to 60 ml of neutral red wine

Term	Definition	Composition
Red berries	Odor associated with strawberry and raspberry	2 ml each of strawberry and raspberry (Teisseire)
Black berries	Odor associated with blackcurrant and blackberry	2 ml each of blackcurrant and blackberry syrup (Teisseire)
Jam/cooked berries	Odor associated with red fruit jam	1 ml each raspberry, strawberry and blackberry marmalade (Bon Maman)
Kirsch	Odor associated to cherry in alcohol	5 ml cherry syrup (Védrene)
Prune	Odor associated with prune	One crushed pitted prune (Maître Prunille)
Vegetal	Odor associated to grass	6 chopped brines of fresh and dried grass
Bell pepper	Odor associated to bell pepper	5 g chopped green bell pepper
Earthy/mushroom	Odor associated to mushrooms	2 ml water from canned mushrooms
Wood/toasted	Odor associated to oak wood wine barrels	4 1 cm^2 pieces of oak wood (HM Arobois)
Coffee	Odor associated to coffee	1 ml liquid coffee aroma (Vahiné)
Vanilla	Odor associated to vanilla	1 ml liquid vanilla aroma (Vahiné)
Pepper	Odor associated to black pepper	2 small round grains (Amora)
Licorice	Odor associated to licorice candy	One chopped licorice candy (Haribo)
Butter	Odor associated to butter	1 g butter (President)
Alcohol	Odor associated to pure alcohol	3 ml of 95% ethanol

Source: Campo *et al.*, 2010

Table 4.4 Formulation of sensory reference standards used to anchor attributes. All aroma standards made up in 40 ml base wine

Term*	Composition
Berry (a,f)	2 slightly mashed frozen cherries + 2 slightly mashed frozen blackberries
Oak (a,f)	5 g oak chips
Mushroom (a)	½ fresh white mushroom, sliced
Mint (a)	1 fresh chopped mint leaf
Dried fruit (a,f)	1 prune + 5 raisins
Spice (a)	1 tsp pumpkin spice (McCormick)
Vegetal (a)	2 tbsp canned mixed vegetables (Veg-All)
Bell pepper (a)	1 tbsp chopped green bell pepper
Vanilla (a)	6 drops vanilla extract (McCormick)
Butter (a)	3 drops butter flavor popcorn oil (Orville Redenbacher's)
Chocolate (a,f)	1 tsp cocoa (Scharffenberger)
Leather (a)	2 cm leather shoelace
Ethanol (f)	5 ml rum (Island Pride Rum)
Bitter (t)	1 g caffeine/l, caffeine USP/FCC, anhydrous (Fisher Chemical)
Astringent (tx)	1.5 g tannic acid/l, tannic acid (Mallinckrodt)
Sour (t)	1.5 g citric acid/l, citric acid, monohydrate (J.T. Baker)

* Aroma (a), flavor by mouth (f), texture (tx), taste (t)

Source: Madrigal-Galan and Heymann, 2006

on red wine (Campo *et al.*, 2010; Madrigal-Galan and Heymann, 2006). Table 4.5 reports attributes and standards relative to white wine. A new panel may also adopt an already existing terminology. In this case, if the descriptive vocabulary was developed by another laboratory or in a different country or region, difficulties in understanding and interpreting the terms could arise. This latter problem could be bypassed providing full definitions and standards to panelists (Hunter and McEwan, 1998). At the end of the 1980s, Noble and co-workers (1987) developed a standardized system of aroma terminology that is still a milestone in the descriptive analysis of wine. The aroma wheel described above gives reference standards in order to clarify each descriptive term. Similar contributions are represented by the descriptive wheel of mouth-feel characteristics of red wine (Gawel *et al.*, 2001) and the vocabulary of terms for sensory evaluation of dessert port wine (Cristovam *et al.*, 2000).

During language development, subjects practice scoring products in order to familiarize themselves with them and the scale rating system. Different scaling methods are used by different researchers, and their efficiency seems to be similar. The unstructured, 10 cm linear scale anchored 'not perceived' at the left end and 'very intense' (other anchors are frequently used) at the right end is very common in wine research, but 9-point category and 15-point unlabeled box scales are used too.

In conventional descriptive analysis it is important to determine assessor reproducibility. Normally, panel performance is evaluated at the end of the training period by having the panel evaluate, in triplicate, a subset of samples to be used for the real study. Data are frequently analysed for each attribute by means of a

Table 4.5 Standard samples used for descriptors' identification and learning

Aromatic terms			Composition of standard†
First tier term	Second tier term	Third tier term	
Fruity	Citrus	Orange	2.10 ml orange extract
		Grapefruit	2.10 ml grapefruit extract
		Lemon	2.10 ml lemon extract
	Tree fruit	Apricot	50 ml apricot nectar
		Peach	5 g canned peach pieces + 65.5 ml syrup
		Apple	9 g apple peel + 10.5 pieces of apple pulp
	Tropical fruit	Pineapple	87 ml pineapple nectar
		Melon	20 g melon peel + 45 g pieces of melon pulp
		Banana	1.30 ml isoamyl acetate
		Coconut	4.5 g coconut flour
	Dried fruit	Fig	18 g fig pieces
Herbaceous or vegetative	Fresh	Cut grass	16 ml cis-2-exenol
		Eucalyptus	30 g crushed eucalyptus leaf
		Mint	2.5 ml mint extract
	Canned	Green beans	14 g canned green beans pieces
		Asparagus	13 g canned green asparagus pieces
		Green olive	14 g canned green olive pieces
		Black olive	10 g canned black olive pieces
		Artichoke	15 g canned artichoke pieces
	Dried	Hay/straw	15 g hay/straw
		Tea	3.5 g flakes of black tea
		Tobacco	3.4 g flakes of tobacco
Nutty	Nutty	Walnut	30 g walnuts
		Hazelnut	16 g hazelnuts
		Almond	120 ml almond extract
Caramelized	Caramel	Honey	12.5 g honey
		Butter	15 g butter
Wood	Phenolic	Vanilla	1.5 g vanillin
	Resinous	Cedar	100 g cedar pieces
		Oak	100 g oak pieces
Floral	Floral	Rose	200 ml 2-phenylethanol
		Orange blossom	210 ml orange blossom extract
Spicy	Spicy	Cloves	1.8 ml eugenol
		Licorice, anise	2.5 g anethole
		Fennel	1.5 g fennel seed

† Standards were prepared in 150 ml of white wine deodorized with activated carbon.

Source: Carlucci and Monteleone, 2008.

two (sample and assessor) or three-way analysis of variance (samples, assessors, replications) to determine whether there are significant panelists by sample interactions. The significant result of this interaction implies that the panel leader will determine which assessor should be further trained in the use of particular attributes. The need for solid and validated information from sensory data in decision-making processes is extremely clear to sensory scientists. It is not a case that user-friendly statistical tools for validating panel performance are now available for free. Panel check software developed at Nofima in Norway (www.nofima.no/en) allows the panel leader to control the quality of sensory profile data using both univariate and multivariate approaches (see Naes *et al.*, 2010a). The software provides a series of graphs for quick visual inspection of data:

- Mean and standard deviation and box plot for all assessors for each attribute,
- Line plot showing the panel average,
- Individual scores for all assessors for each product.

F-values of either two or three-way analysis of variance associated with sample, assessor, and their two-way interaction effects are plotted in different colours depending on the significance level, allowing a quick detection of significant effects. Furthermore, the comparison of assessors' ability to detect differences among products for each attribute can be checked using the p-MSE plot. The plot allows the analyst to easily detect subjects able to discriminate between samples and, at the same time, able to reproduce their own scores reliably. Finally, the software permits an overall assessment of assessor differences using all variables simultaneously by means of a methodology referred to as Tucker1. This is a multivariate method (consensus principal component analysis) able to detect assessors who differ from the rest, and attributes that are affected by poorly performing assessors.

Experimental design

The sample evaluation in descriptive analysis is run after defining an appropriate experimental design. The term indicates a series of experimental procedures that have been developed for providing as much information as possible in the most efficient way. A design can be defined only once the aim and the unique properties of the experiment are stated. In descriptive analysis the objective is to collect a sensory profile of a heterogeneous group of treatments (products) using a defined number of assessors and replicates. In general, the design should take into account assessor variation, presentation order effect, first order and carry-over effects, and any specific limitation associated with samples and assessors (Piggott *et al.*, 1998).

It is possible to say that when planning a profile study there are two important aspects to consider. The first is the initial choice of products (Naes *et al.*, 2010a). It is important to select product carefully. When the interest is to compare a product to similar products in the market, products that span the entire variation in the study must be included. Some sensory method applied to obtain perceptual maps (e.g. free choice profile, flash profile and projective map) can be very helpful for the purpose. Secondly, a further important aspect of an experimental design is how to present the samples to panelists. In sensory wine research it is common to use complete designs with replications involving three factors: sample, replicate and assessors. Randomization is a key principle in experimental design. A proper randomization ensures that the effects of extraneous factors are averaged out in the long run. The presentation order represents a source of variation of sensory data in itself, and a balanced design for first-order and carry-over effects is needed. This can be obtained by adopting modified Latin-square designs as reported by MacFie *et al.* (1989). Software for collecting sensory data and running tests normally allows the experimenter to easily design the presentation order in respect of this important requirement.

In descriptive analysis each assessor is asked to replicate the evaluation of samples. The number of replicates depends on the size of the differences that the experimenter is required to detect. Small expected differences require a higher number of replicates. However, the number of replicates in wine sensory studies frequently ranges between 2–4, with 10–12 panelists involved in the test.

Sensory procedure

In sensory evaluation of wine, assessors are frequently presented up to 12 samples in each session. Generally each sample (25–30 ml wine) is presented in a coded tulip wineglass covered with a plastic Petri dish. Samples are served monadically and panelists are required to expectorate the wines. A 5–10 minute break is taken after the evaluation of four samples. Drinking water and unsalted crackers are provided to clean the mouth between each sample. Specific rinsing procedures are often defined to limit astringency carry-over effects (see Section 4.4). Tests are conducted in isolated booths, under red lighting. Scores are frequently recorded directly on a computer system using dedicated software.

Data analysis

Analysis of variance (ANOVA) is the most used statistical method for testing the significance of effects included in descriptive analysis experimental designs. As reported above, three-way ANOVA involving sample, assessor and repetition as fixed factors and all first-order interactions are frequently computed for each attribute in order to evaluate panel performance. When replicate and interaction effects are not significant for the tested attributes, the panel is assumed to perform well, and data are analysed in order to study similarities and differences between products. Spider-plots are used to represent the sensory profile of samples.

In product development studies, mixed ANOVA models are often computed before submitting sensory data to a multivariate analysis in order to obtain a perceptual map. Two-way mixed models are used to test the importance of sensory attributes (sample effect) considering assessors at random. In these cases, Naes *et al.* (2010a) suggest to check for interaction effects in order to get a complete overview of differences between products. Non-significant sample effects identify unimportant attributes in discriminating between products. These attributes are not used in perceptual mapping.

Principal component analysis (PCA) is a very well-known multivariate statistical method and probably the most applied approach to obtain perceptual maps from descriptive data (see Martens and Martens, 2001). For the purpose, PCA models are computed on the product-by-attribute matrix, after having averaged out both replicates and assessors. The method is based on the computation of the most interesting directions of variability, called principal components. The main results are graphically presented in the score plot, which describes the relations between the products, and the loading plot, which describes the relations between the sensory attributes and the principal components. Martens and Martens (2001) proposed a modification of the loading plot that can facilitate the interpretation of the importance of variables in discriminating between products, particularly when the PCA model is computed on non-standardized sensory data: the correlation loading plot. This is a two-dimensional scatter plot of correlations between the principal components and the variables themselves. The advantage of this plot is that the researcher can directly obtain information about how much each variable is explained by each component. Moreover, the correlation plot gives the possibility of drawing circles in the plot corresponding to various degrees of explained variances. Typically, a circle for 100% explained variance and for 50% explained variance for the two components is drawn.

The graphical interpretation of the plots follows simple rules (Naes *et al.*, 2010a):

- Products which are close to each other have similar overall properties and samples which are far apart are very different,
- Attributes which are close are strongly and positively correlated, while those on opposite sides of the origin have negative correlation,
- Products to the left of the score plot are characterized by attributes to the left in the loading plot; products to the top of the score plot are characterized by attributes to the top in the loading plot, etc.
- The higher the explained variance, the more valid is the information obtained from the perceptual map. In PCA models each component describes as much variance as possible and each new component is orthogonal to the previous. The explained variance is largest for the first component, next largest for the second, etc. The problem is how many components can be looked at safely. Often, two or three components are computed in order to explain a substantial amount of variance in the data, but even with a small number of components there is an interest in evaluating how reliable the components are. Cross-validation is frequently used to test the goodness of PCA models. The procedure computes an explained 'validation' variance for each component. The explained validation variance can be plotted as function of the number of components. The point where the curve flattens out and becomes stable is where one should stop interpreting components. Figure 4.2 gives examples of all the plots mentioned above.

Use of perceptual maps from conventional descriptive analysis data in studying the combined effect of sensory properties and information on affective consumer responses to food and beverages

The study of the combined effect of sensory properties and information is an important task in developing or optimizing food product finalized to respond to

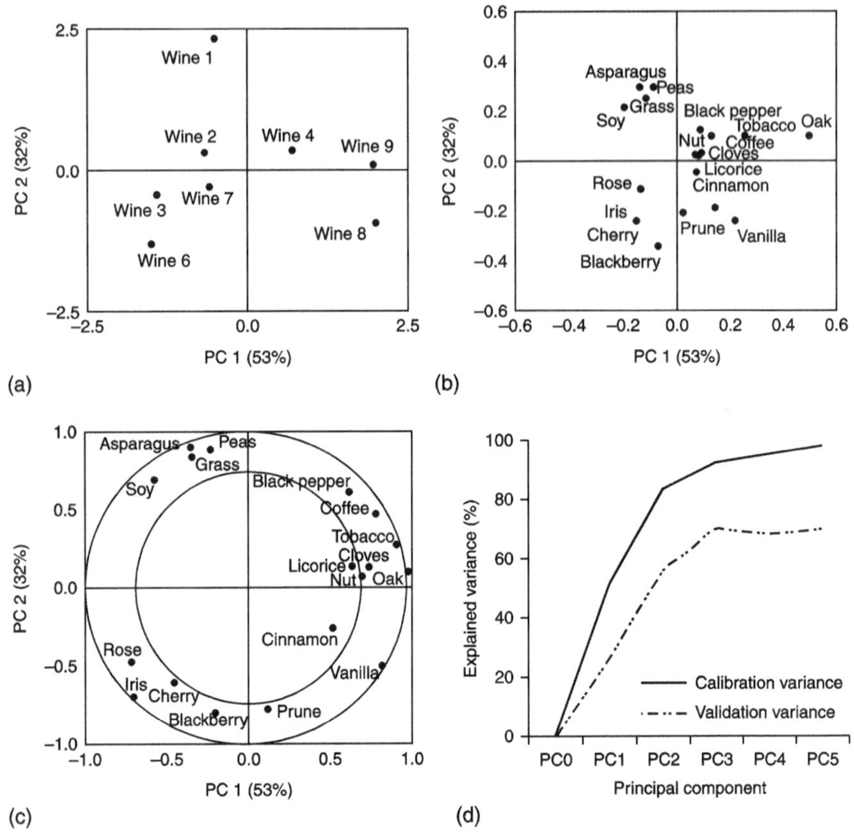

Fig. 4.2 PCA on descriptive sensory data: 8 red wines and 18 attributes. (a) Score plot: the plot gives information about patterns in the samples. (b) Loading plot: the loading plot shows which variables are important and which variables correlate. (c) Correlation loading plot: the importance of individual variables is visualized more clearly in the correlation loading plot compared to the standard loading plot. (d) Explained variance plot: this plot gives an indication of how much of the variation in the data is described by the different components.

consumer needs. In the last 15 years the common approach in these sort of studies has been characterized by the collection of three data sets from consumers (blind, expected and actual liking) and by a data analysis based on score subtractions to determine disconfirmation (positive or negative), assimilation, or contrast effect. Main limitations of this approach are:

- The collection of three sets of data requires time and money,
- Test conditions are very far away from real-life situations,
- Subtracting blind scores from expected and informed conditions might not be correct,
- It is difficult to estimate the combined effect of specific sensory properties and information.

Two recent papers have introduced a new way of approaching the study of the effect of sensory and information cues on consumer hedonic responses using a conjoint design (Naes *et al.*, 2010b; Bølling *et al.*, 2009). The authors pointed out that there are, of course, examples in the literature where product tasting was incorporated in a conjoint framework, but sensory properties were not specifically highlighted. Therefore, it is very difficult to identify what are the actual sensory drivers of liking, which is often of major concern in product development. Sensory analysis is needed both for ensuring that samples span the relevant sensory space and as part of the data analysis itself.

When combining intrinsic (specific sensory properties) and extrinsic attributes (information) in conjoint analysis, the question is how the samples should be selected and combined with the extrinsic attributes. The most important aspect is that the samples should span the relevant variability of the sensory attributes within realistic limits. Thus the first step of the proposed approach is the descriptive analysis of an appropriate set of products. The sensory data are subjected to PCA. The design samples are selected from the PCA score plot according to a geometric structure, corresponding to the two most important sensory dimensions; they can be thought of as new latent orthogonal variables, called 'meta-attributes' (Fig. 4.3).

The two meta-attributes are then combined with information variables (Fig. 4.4). This is done using a factorial design strategy. Subjects' test samples are

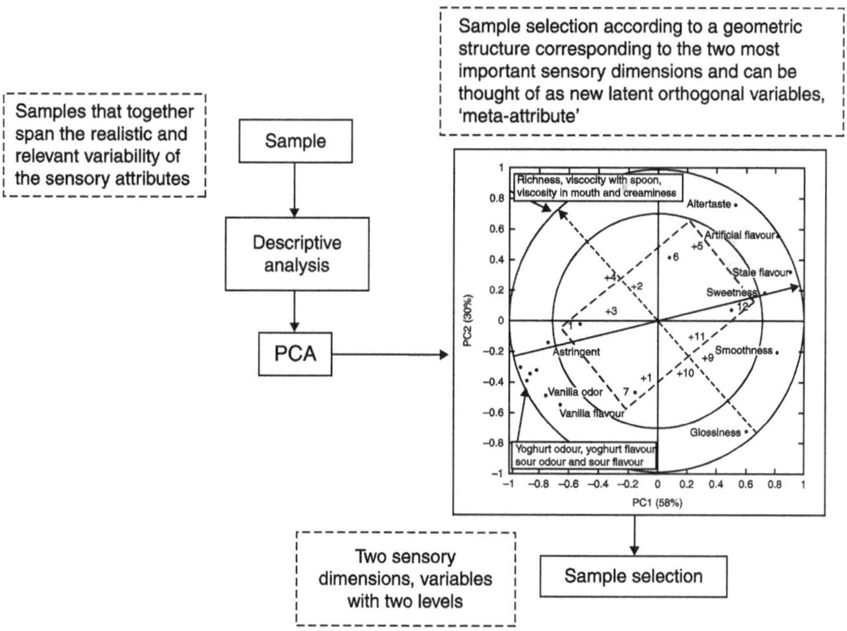

Fig. 4.3 Conjoint design to approach the study of the effect of sensory and information cues on consumer hedonic responses: design samples selection from PCA score plot (meta-attribute).

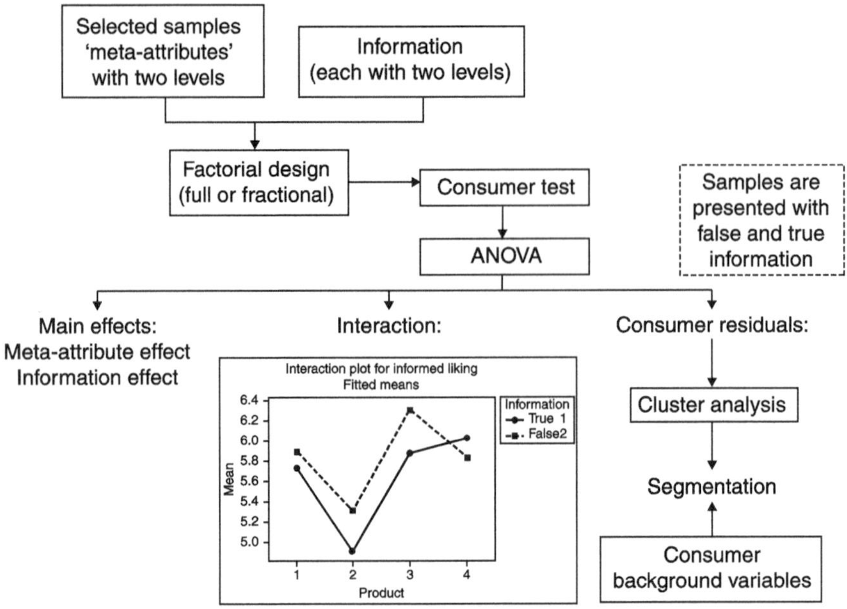

Fig. 4.4 Conjoint design to approach the study of the effect of sensory and information cues on consumer hedonic responses: factorial design strategy and consumer segmentation.

presented with true and false information. The ANOVA model applied to analyze the data allows one to estimate the effect of both meta-attribute and information as well as the combined effect of sensory and information (interaction effect). Furthermore, residuals can be used to segment consumers.

4.2.2 Alternative descriptive methods

Alternative and complementary methods to conventional descriptive analysis are often used to evaluate a set of products providing a quick access on a relative sensory positioning of a set of products. Thus the best use of these tools is for rapid market analyzes. Free choice profile (FCP) and flash analysis (FA) are the two methods reported here. These methods differ from conventional descriptive analysis for the following characteristics:

- Each panelist creates his/her own list of descriptive terms: panelists are not extensively trained to create a consensus vocabulary for the product,
- The data set obtained from these methods requires a specific statistical procedure to be analyzed in order to obtain a perceptual map: the Procrustes analysis.

In product development these methods can be seen as convenient sensory mapping tools for conducting preliminary phases of sensory studies. In fact they may provide a general overview of the sensory differences between products allowing the researcher to select samples that span the entire sensory space. Selected samples can be further studied by means of conventional descriptive analysis and in consumer studies. FCP and FA are also particularly useful when only a one-shot evaluation is possible. They can also be used for conducting preliminary phases of language development step in conventional descriptive analysis.

Free choice profile (FCP)

Free choice profile (FCP) (Williams and Langron, 1984) is a descriptive technique in which each assessor produces individual profiles of the products, using his or her own terms for describing products. The spatial configurations derived from individual profiles are analyzed by Generalized Procrustes analysis (GPA) and result in a consensus configuration (a perceptual map) revealing the interrelationships between the samples for the panel as a whole. The method has been largely used with product experts, sensory evaluation experts who have previously participated in several descriptive evaluation, and with consumers (see Jack and Piggott, 1991–1992). Here we do not consider consumer applications.

In FCP, panelists are allowed to term the sensations they perceive and rate them on a scale. Some panelists may use very few descriptors and others may use many. This technique can be applied for many purposes of perceptual mapping, particularly for explorative purposes when a representation of differences between products is quickly required. For instance, it can be used to check immediately how much and why a new product differs from competitors' or regular products. In this case, its application is suggested with panels composed either with product experts or subjects who have previously participated in several descriptive analyzes of the products. It should be recalled that FCP shows the main differences among samples and makes it possible to interpret them but does not indicate the stable differences among products that are often very important to product developers (Lawless and Heymann, 1998).

Data from FCP are treated by means of GPA (Gower, 1975). This procedure allows one to obtain a consensus space from individual spatial configurations (one for each panelist for each replication) by means of three steps: translation, scaling and rotation. The steps produce a perceptual space for each panelist which is matched as closely as possible with other panelists (Piggott and Watson, 1992). Then, the consensus map is calculated as average configuration for all panelists. This is usually simplified as a reduced dimensional plot by PCA. Thus, the consensus configuration is interpreted similarly to the PCA map. The residual errors, which means the distances between the panelists' individual configurations and the consensus, can be used to identify outliers or groups.

Figure 4.5 shows the consensus map describing differences among seven white wines. It can be noted that wines are represented by ellipses, which gives an empirical picture of the variation in the positions of the wines. If ellipses are superimposed, they are not significantly different. It is possible to plot the descriptors used by the panelists into the consensus space and interpret them in the same fashion as descriptors on a PCA plot (Fig. 4.6). The consensus configuration may be interpreted in terms of each individual panelist's vocabulary,

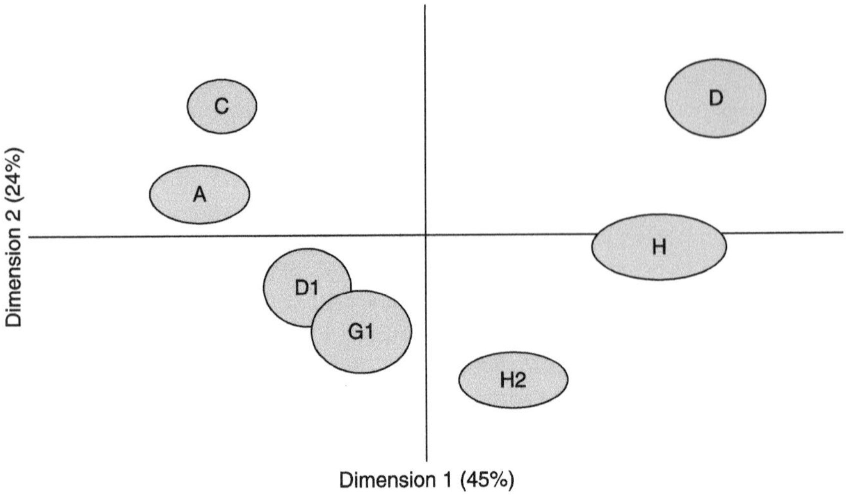

Fig. 4.5 Generalized Procrustes analysis: an example of a consensus plot.

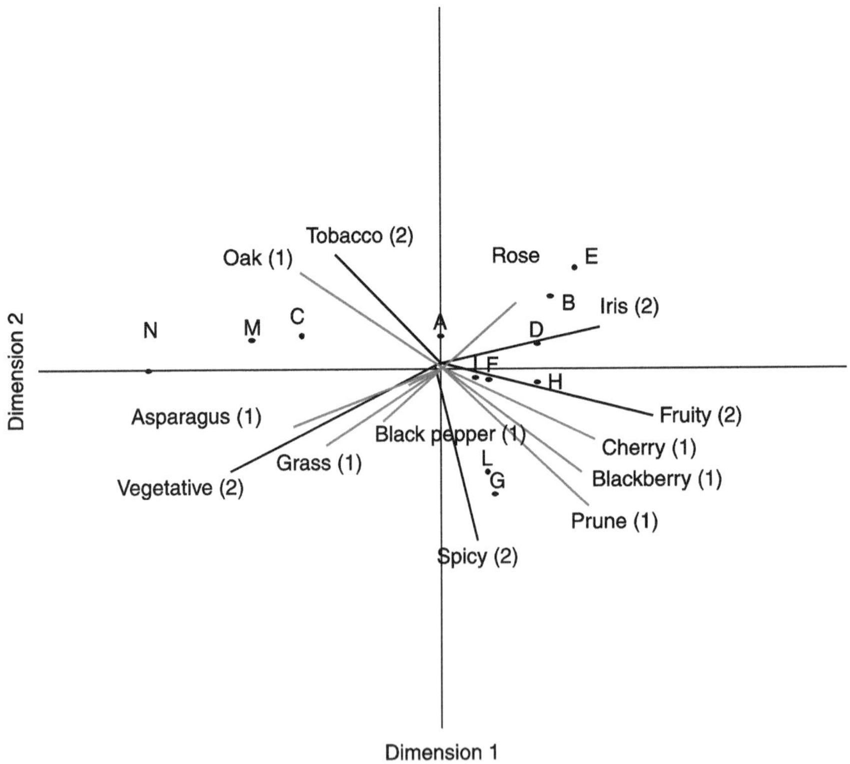

Fig. 4.6 Generalized Procrustes analysis: attribute vectors for subjects 1 and 2.

and the researcher can evaluate how different terms used by different subjects are related to each other.

Sensory procedures in FCP do not differ from those reported for the conventional profile. Two or three replicates should be run to verify both panelists and panel consistency. Replicate samples should be plotted together. When it is not possible to run replicates of the entire set of samples, duplicate samples should be included in the test and they should be plotted together on the consensus map.

In order to estimate the statistical validity of the dimensionality of the consensus model it is suggested that a permutation test is run on the data. This procedure rearranges at random each individual's rating scores and produces new permuted data sets in which the scores no longer correspond to the products to which they were originally given. The GPA is carried out on a number of permuted data sets (e.g. 50) and the percentage of variance accounted for is computed for each permuted data set. The distribution of these values reflects the probability that a variance accounted for occurs by chance alone. A variance accounted for in the real data higher than the 95th percentile of this distribution indicates a probability of less than 5% that the consensus generated has arisen by chance.

Flash analysis

Flash analysis (Sieffermann, 2000) has strong similarities with FCP. In fact, this method is based on the combination of free choice profiling and a comparative evaluation of the whole product set, which is presented simultaneously. It is applied with a small number of subjects that are allowed to use their own list of attributes to profile the presented products. For this reason, no panel training is needed. On the other hand, it is important to select sensory evaluation experts who have previously participated in several descriptive evaluations, even if they are not necessarily product experts. Furthermore, the method is specifically suitable for professionals such as oenologists (Delarue and Sieffermann, 2004).

Products (6–10) are presented anonymously. At the beginning of the session, subjects are provided with all the samples. They are asked to rank them (with ties allowed) along the sensory characteristics for which the samples differed. In fact, each expert produces his own attributes according to the major differences perceived among the products. Then, the panelist directly ranks products on the chosen descriptors, attribute per attribute. This forces them to focus on the perceived differences and to use discriminating attributes. The sessions are individual and could last from 40–75 minutes to about half a day. This is due to the fact that no limitations are given regarding the number of sensory characteristics they should evaluate, the number of re-tastings or the breaks needed. As a repeatability probe, a product can be replicated in the sample set; or the whole evaluation can be replicated. All sessions can be carried out individually, on separate days and separate locations, providing that all samples are available. GPA is frequently applied to the data from flash profiling to assess the consensus between assessors' sensory maps (Delarue and Sieffermann, 2004).

4.2.3 Perceptual maps from similarity data

Sorting and projective mapping (the nappe method) are methods frequently applied to obtain representations of the differences amongst products. Because of their immediateness and ease of application they are assumed to be useful tools in product development (Buck, 2007).

In the sorting method, subjects are asked to sort items into two or more groups: items that have something in common are placed in the same group, while items that differ from one another should be placed in different groups. In this method similarity is a group-derived estimate; in fact, similarity is inferred from the number of times two items are sorted into the same group across a panel of participants. Data from the sorting task for each sample are converted into a similarity matrix by summing over all participants the number of times each pair of wines is sorted into the same group. This matrix is analyzed with multidimensional scaling to obtain a map. After completing their sorting task, subjects can be asked to give a few words or descriptors that provided the criteria on which they had based their sorting. This information is used to interpret the differences among products.

It is reported that the sorting task is easy to perform with sets of 10–20 samples. However, in wine studies the sorting technique has been applied with a number of samples ranging from 6–12 (see Maitre *et al.*, 2010).

Sorting can be usefully applied to study differences amongst products and to verify the correspondence of some prototypes to specific profiles. This is the reason why in wine research the method has been used with product experts to study the perceived typicality of products. In the sorting task wine professionals use personal criteria based on their technical knowledge; if the dominant criterion seems to be the individuation of grape variety (Ballester *et al.*, 2008), inside the same grape variety the sorting dimensions can be linked to perceived typicality (Perrin *et al.* 2008) or correlated to sensory dimensions (Parr *et al.*, 2010). Results of the sorting task depend mainly on previous knowledge of participants, that is to say that the level of expertise influences the categorization. Concerning this, Maitre *et al.* (2010) wonder if we can actually consider the wine professionals as a homogeneous population; they suggest that it would be useful to characterize precisely the assessors' experience in order to classify them according to their actual knowledge of the wine type.

The sorting task is usually simple and easy to obtain data, but there are limitations. As observed by Nestrud and Lawless (2008), a limitation of sorting is that it provides only grouped or aggregate data on similarity, which means that it only puts similar products into nominal level categories. Furthermore, the stability of models needs to be checked. From a practical point of view the simplest way to check the reliability of results is to insert a blind duplicate of one of the products to see whether they are plotted together on the map.

Projective mapping may provide more graded information than sorting, because it is based on the individuation of similarities and differences using a graphic representation and not a nominal categorization (Pagès, 2005; Nestrud and Lawless, 2008). In fact, this is a simple user-friendly technique which allows subjects (naive consumers, trained subjects and experts) to express perceptual similarities and groupings among a set of products by placing them on a two-dimensional surface. Subjects are simply required to place products on the surface according to the similarities and differences they perceive, so that similar products appear near to one another and different products are farther apart. Placing items on a two-dimensional surface provides a potentially richer amount of information than the simple categorical information from sorting tasks: measuring the distance on the surface amongst products for each person, it is possible to calculate the greater and smaller similarities (and differences) amongst products.

This method was originally developed by Risvik and colleagues (Risvik *et al.*, 1994) as a possible alternative to profiling and (dis)similarity scaling methods to obtain perceptual maps. From its first application, this method has been viewed as a technique useful for the study of consumer perception. In a further work from Risvik and colleagues the mapping dimensions from naive consumers were compared to those from the profile data. Results showed that 'the best similarity was found when comparing the first dimension, thus suggesting good agreement on the obvious aspects of the product' (Risvik *et al.*, 1997).

In earlier studies more than one multidimensional method has been applied to analyze projective map data, such as multidimensional scaling (MDS), generalized Procrustes analysis (GPA), and principal component analysis (PCA). More recently, Pagès (2003, 2005) renewed the projective mapping technique, calling it the 'nappe' method (or napping). This method consists of collecting the perceived differences amongst products by positioning the products on a sheet of blank paper: in fact, in French language the word nappe means 'tablecloth'. Obtained data (the coordinates for each product) are treated by multiple factor analysis (MFA), which considers the individual subject's matrices more fully.

In a further work (Morand and Pagès, 2006), results from GPA and MFA, applied to the same nappe data set, are compared showing that the two approaches give very similar configurations. These findings were recently confirmed by Nestrud and Lawless (2008).

In wine research the nappe procedure was proposed as a method particularly relevant to wine professionals (see Maitre *et al.*, 2010). In the first napping application, Pagès (2005) examined wine professionals' perceptions of 10 white wines from the Loire Valley, showing the differences and similarities from one subject to another. With napping it is possible to get a representation of the products, which integrates the relative importance for the subjects of the characteristics of the products; however, this does not characterize the product itself. Thus, the nappe method is presented as a complementary method and not as an alternative to descriptive analysis. Perrin *et al.* (2008) compared the napping procedure to a conventional profile elaborated by a trained panel, taken as reference, and two simplified profiles provided by wine professionals (ultra-flash profile and free choice profile). While results provided by the three descriptive methods were similar (even if not identical, certainly), nappe method appeared to be less fine, but it revealed only the criteria important for wine discrimination according to wine professionals. It is also pointed out that combining nappe procedure with ultra-flash profiling it is possible to take advantage of the great knowledge of wine professionals.

As with any other method applied in food sensory science, nappe procedure shows some weaknesses, as reported in most of the papers cited above. One important issue is certainly the reliability of the results from this method: results are strongly dependent from the starting question that needs to be attentively and consciously determined. Another open issue is the real contribution of the second dimension of the maps in discriminating between samples; subjects seem to discriminate on the basis of only one dimension.

When applying projective mapping the following rules are suggested:

- Define the objective of the test and select an appropriate set of samples;
- Give instructions about principle and procedure of the method to the subjects, as disposed by Pagès (2005) and Nestrud and Lawless (2008):
  - Principle. You are asked to evaluate the similarities (or dissimilarities) between several wines. You have to do this according to your own criteria, those that are significant for you. You do not have to indicate your criteria. There is no good or bad answer.
  - Procedure. First, it is important to taste each sample and make any notes or comments about the sample according to your own criteria in the space provided below and on the back of the sheet. Be sure to save enough wine for three or four tastes. Cleanse your palate as needed with water. Then, after tasting all the samples, you have to position the wines on the paper in such a way that two wines are very near if they seem identical to you and that two wines are distant to one another if they seem different to you. This must be done according to your own criteria. Do not hesitate to express strongly the differences you perceive by using most of the sheet. When the operation is finished, write down on the sheet the number of the wine in the place it occupies.

    Notes or comments about the samples can be directly plotted on the consensus map by using MFA (Pagès, 2005). Alternatively they can be used to arrange a follow-up session to re-tasting the samples and rate the generated attributes (Nestrud and Lawless, 2008; Perrin *at al.*, 2008);
- Run the test with a number of subjects from 15–20;
- Run a pilot test to check the number of products to taste. According to Pagès (2005), in the case of wines this number cannot be higher than 12 in order to avoid the weariness of tasters. It is important to add duplicated samples to the group of examined products;
- Before submitting data to multidimensional methods (GPA or MFA) a visual inspection of data is strongly suggested in order to remove non-reliable subjects from the data set. In Fig. 4.7, 4.8 and 4.9 examples of individual maps are reported. Cases as those reported in Fig. 4.8 and 4.9 are not infrequent in projective map tests. Figure 4.9 is an example of a subject that should be

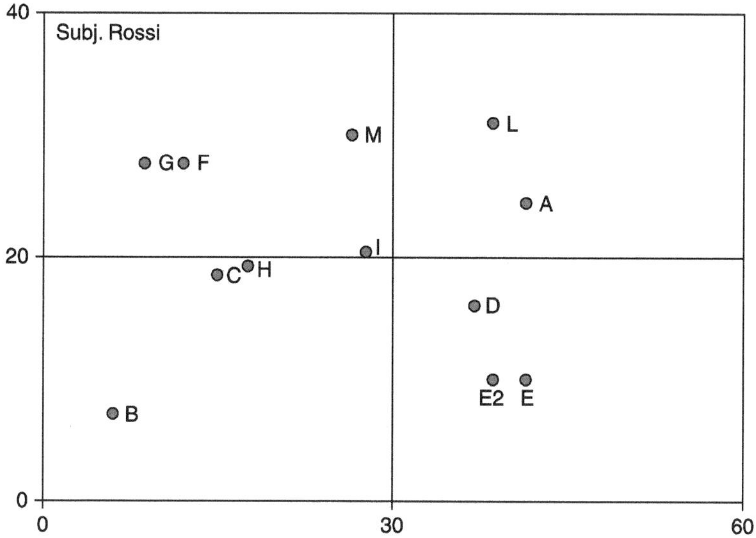

Fig. 4.7 Projecting mapping: raw data visual inspection. Samples are distributed in a two-dimensional space.

removed from the analysis. From Fig. 4.8 a more general question arises: do subjects really use two dimensions in segregating samples? This question is related to the evidence that often the contribution of the second dimension of the consensus map in discriminating between samples is poor.

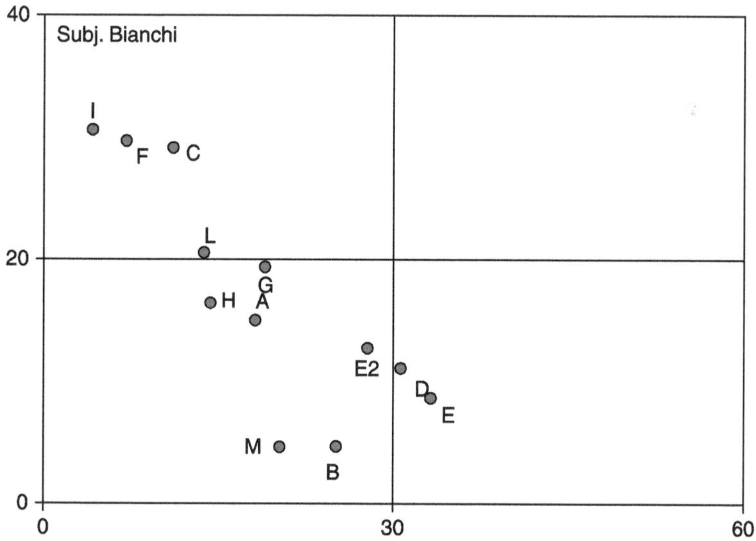

Fig. 4.8 Projecting mapping: raw data visual inspection. Samples are distributed along one dimension.

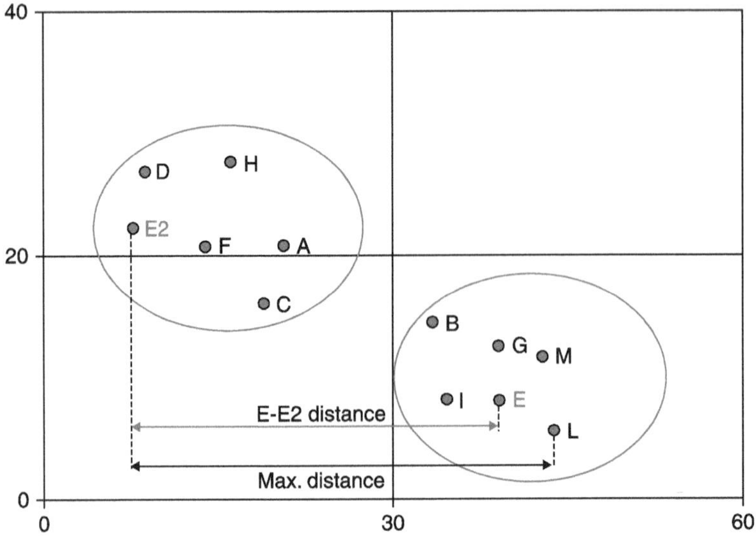

Fig. 4.9 Projecting mapping: raw data visual inspection. Samples are discriminated in two groups along one dimension; the maximum distance among products is very close to the one between the duplicated samples.

4.3 Temporal dominance of sensation (TDS)

Time intensity is a very well-known method of evaluating the temporality of the sensory perception in food products. It has been largely used in wine research as a means of studying the persistence of tastes such as bitterness and astringency (see Lesschaeve and Noble, 2005). Although the method has been extensively and successfully used to study the temporal aspects of the perception of a given sensory attribute in a product, when it is used for several attributes it quickly becomes very time-consuming because one run is needed per attribute. This limitation was clear to Rose Marie Pangborn (1989) who predicted increasing sensory research activities in analyzing temporal aspects of perception. Temporal dominance of sensation (TDS) (see Pineau *et al.*, 2009) is certainly the most innovative method alternative to time intensity. TDS was developed in 1999 at the Centre Européen des Sciences du Goût, France. It consists of presenting to the panelist the complete list of attributes on a computer screen. The panelist is then asked to assess which of the attributes is perceived as dominant (i.e. the most striking perception at a given time). During the testing of a product, the panelist is free to select an attribute several times. Conversely, another attribute may not be selected at all. In the course of the evaluation, when the panelist considers that the dominant attribute has changed, they select the new dominant attribute, and so on, until the perception ends. For each run, this method enables the collection of a sequence of sensory attributes at different times during the tasting.

The TDS method is applied with a trained panel which means that, as for descriptive panels, qualified subjects develop a vocabulary to describe the sensations they perceive, considering the temporal evolution of the sensations. They also establish a consensual definition of each attribute and are trained to use the method itself. For the training sessions, a dominant attribute is defined as the attribute associated with the sensation catching the attention at a given time. Thus the dominant attribute is not necessarily the one with the highest intensity.

The size of a TDS panel is similar to that of descriptive panels. Sample presentation follows the same rules of descriptive analysis. In order to avoid any order effect from the list of attributes, the order of descriptors should be balanced among subjects using Williams Latin squares. The application of the method requires dedicated software for data acquisition, which is now commercially available. This software also allows analysis of the data and generation of the TDS curves for each product. TDS curve computation is well explained in Pineau *et al.* (2009). The procedure considers each attribute separately. For each point of time, the proportion of runs (subject × replication) for which the given attribute is assessed as dominant is computed. These proportions, smoothed using appropriate procedures, are plotted against time and are called TDS curves (Fig. 4.10). On the plot it is possible to draw a line corresponding to the 'chance level', which is the dominance rate that an attribute can obtain by chance. Its value is equal to 1/p, p being the number of attributes. It is also possible to draw a line corresponding to the 95% significance level.

This can be obtained using a relatively standard application of the binomial distribution. The distribution is well characterized and one can, for any value of P0 (chance level), calculate the probability that the observed percentage is higher

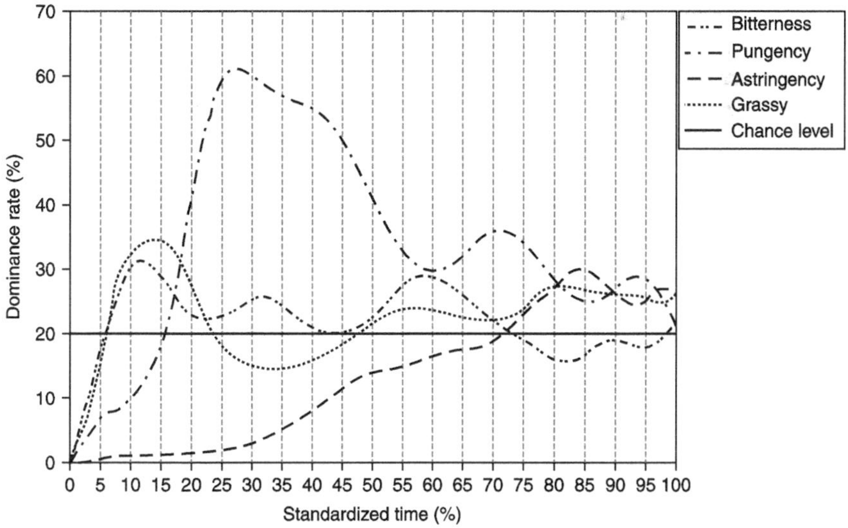

Fig. 4.10 TDS curves.

than a certain threshold. This means that the smallest value C for which P(DR ≥ C) < 0.05 (where DR is the observed dominance rate) can always be found (given the value of P0). This procedure is slightly more cumbersome than the procedure advocated by Pineau *et al.* (2009), but the advantage is that it can always be used.

TDS has recently been applied in wine studies to test the effect of dealcoholization on sensory properties of wine in relation to the perceived complexity (Meillon *et al.*, 2010b).

TDS seems to be more appropriate than descriptive analysis to study sensory interactions occurring when tasting food (such as suppression, release from suppression and enhancement).

For this reason TDS also seems to be a promising method to develop new approaches when studying wine–food pairing. However, further developments of TDS will require paying attention to panel training, attributes selection, intensity scores collection and data analysis and validation.

4.4 Prediction of perceived astringency induced by phenolic compounds in alcoholic beverages

Sensations perceived in the act of tasting are important determinants of consumer response to wine and alcoholic beverages. Astringency descriptors account for more than half of the total terms in the mouth-feel wheel for red wine (Gawel *et al.*, 2000). Studies addressing acceptance of astringency indicate that this is perceived as a negative attribute and is cited as a reason for food and wine rejection (Drenowski and Gomez-Carneros, 2000; Lesschaeve and Noble, 2005). On the other hand, this sensation enhances the complexity and palate-length of high-quality red wines.

Astringency is a tactile sensation perceived as a diffuse stimulus in the mouth, and is commonly described as a puckering, roughing and drying of the oral surface (Lee and Lawless, 1991).

Various substances such as multivalent metallic cations (particularly aluminium salts), dehydrating agents, mineral and organic acids and plant phenol compounds are able to induce this sensation. Dietary tannins form complexes with salivary proteins resulting in their aggregation and/or precipitation and the loss of their lubricating properties (Kallithraka *et al.*, 2001). A friction-based mechanism for astringency development, due to precipitation of salivary proteins, has been suggested (de Wijk and Prinz, 2006; Prinz and Lucas, 2000).

Polymeric flavan-3-ols, which occur either as galloylated species, conjugated with anthocyanins or in free form, are believed to be largely responsible for red wine astringency, to which they contribute in a different manner depending on their chemical structure and composition (Herderich and Smith, 2005; Kennedy *et al.*, 2006; Vidal *et al.*, 2004). Ethanol, acidity, pH and polysaccharides are also involved in the elicitation and perception of wine astringency.

The simplification of complex physiological phenomena in suitable *in vitro* assays has already proved to be a useful approach (Garrett *et al.*, 1999) whenever the main purpose of research is of practical importance and aimed at finding ways of optimizing food production or developing new products.

The possibility of the prediction of the strength of astringency induced by polyphenols could help processing conditions and facilitate the selection of prototypes in product development. The gelatin index is probably the most popular index amongst winemakers (Glories, 1978). The method provides an index of the potential reactivity of wine phenols with a model protein, but index values are not related to the perceived astringency in wine (Monteleone *et al.*, 2007). In several studies the reactivity of phenols with different proteins was considered as an index of phenolic compound astringency (Edelmann and Lendl, 2002; Hofmann *et al.*, 2006; Kennedy *et al.*, 2006; Poncet-Legrande *et al.*, 2007; Wang *et al.*, 2007). Among several techniques that have been used to investigate the interaction between dietary polyphenols and protein, nephelometry has proved to be very useful (Carvalho *et al.*, 2006; de Freitas *et al.*, 2003; de Freitas and Mateus, 2002). Main limitations of the proposed approaches and methods can be summarized in the following points:

- Phenol concentration ranges are chosen in order to maximize the analytical response and are not related to induced astringency at all. In other words, the relationship between analytical responses and perceived astringency is assumed but is not demonstrated;
- Most of the proposed methods ignore that sensory responses to astringent stimuli are described by curves in which the intensity steeply increases with the concentration of astringent stimuli in a range from 0.32–1.63 g/l (Monteleone *et al.*, 2004) while, at higher levels of concentration (which are very often relevant to red wine), the strength of the perceived sensation is less affected by increasing astringent levels and curves tend to plateau. Because of this, the proposed methods cannot have any practical application in product development and optimization.

However, nephelometry has results sensitive enough to determine the reactivity of protein with phenolic samples in a range of physiologically active phenol concentrations (Horne *et al.*, 2002). Based on this evidence, an *in vitro* assay called the astringency mucin index (AMI) was defined (Monteleone *et al.*, 2004; Condelli *et al.*, 2006; Fia *et al.*, 2009). The proposed assay tends to reproduce the polyphenol–protein reaction which takes place in the oral cavity. Phenolic solutions in ethanol are mixed with mucin solutions under appropriate conditions of pH and temperature.

The main characteristics of this predictive method can be summarized as the following:

- The use of mucin as a model protein; in fact, it represents the major glycoprotein synthesized and secreted by mucous cells and serves to lubricate the oral cavity and to protect it from the external environment; it has already been reported to form complexes with tannins;
- The selection of a protein concentration *in vitro* corresponding to the average saliva protein content in humans;

- The selection of phenol concentrations varying from 0.5–3 g/l within which the elicited astringency ranges from weak to strong.

This chapter does not aim to present in detail the *in vitro* assay and the reader can refer to the bibliography to find out more (see Fia *et al.*, 2009). The aim here is to report what are the most appropriate sensory procedures to adopt in order to build up a predictive model of astringency induced by phenolic compounds and to validate it.

4.4.1 Selection of subjects

Recent findings in understanding individual differences in astringency perception induced by phenolic compounds clearly indicate that high (HR) and low (LR) respondents to astringent stimuli exist in the population (Condelli *et al.*, 2006; Dinnella *et al.*, 2009, 2010). Based on this evidence, it is reasonable to select subjects on the basis of their saliva characteristics in order to reduce the effect of individual differences in panel performance when astringency is evaluated. Different procedures can be applied for the purpose; however, from a practical point of view, the simplest way is to measure, for each subject, two saliva parameters: flow rate and haze form capacity. Extreme subjects for both parameters should not be selected for panel composition, as suggested by Condelli *et al.* (2006).

4.4.2 Panel training

Phenolic compounds in wine and other alcoholic beverages can elicit more than one oral sensation. In fact, phenolic stimuli can be perceived as bitter, sour and astringent at the same time. For this reason, subjects should be trained to recognize and rate the perceived intensity of the three sensations using the following standards: citric acid (0.25, 0.38, 0.50 g/l) for sourness; quinine monohydrochloride dehydrate (0.025, 0.037, 0.050 g/l) for bitterness; aluminium potassium sulphate (0.3, 0.6, 0.9 g/l) for astringency (Monteleone *et al.*, 2004). For each standard the three concentrations tend to represent low, medium and strong intensity of the sensation, respectively. Experience in the sensory lab at Florence University over the last eight years indicate that two training sessions are normally sufficient to obtain 90% of the presented standards correctly recognized and scored.

4.4.3 Sample evaluation

The perceived intensity of an astringent stimulus increases with repeated ingestion (Lyman and Green, 1990). Because of this well known 'carry-over' effect, the evaluation of astringent products such as red wine cannot be made in typical side-by-side comparison (Lesschave and Noble, 2005). A complete control of product presentation is necessary. Sensory data necessary to calibrate a predictive model of astringency induced by phenolic compounds in wine are often acquired from water/ethanol solutions of phenolic extracts (from wine or grape), while validating the quality of the predictive model requires astringency ratings from wines.

In each session subjects can easily evaluate up to eight wine or phenolic solution samples, with a 30 minute interval between the evaluations of two subsets, each consisting of four samples. Within each session the presentation order of samples needs to be balanced for first-order and carry-over effects (MacFie *et al.*, 1989). The amount of sample to be tested should range from 10–15 ml. In order to evaluate the astringency of target stimuli, subjects should be asked to hold each sample in their mouth for 10 s, spit it out, wait for a further 20 s, and rate the intensity of astringency, bitterness and sourness on a scale (Dinnella *et al.*, 2010). It is recommended that subjects are asked to rate the intensity of the three sensations to avoid the risk of false enhancements in collecting astringency data. In fact, as reported by Lawless and Heymann (1998), 'respondents may choose to dump some of their impressions into the only allowable response if the attribute they perceive is otherwise unavailable on the ballot'. Along with product order, the attribute order should be balanced among subjects. Either 7- or 9-point category scale (from extremely weak to extremely strong) has been used to rate astringency intensities (Monteleone *et al.*, 2004; Condelli *et al.*, 2006); however when a panel is asked to evaluate a set of samples with an increasing amount of phenols it could be worth considering either the 15-point category scale (Horne *et al.*, 2002) or the labeled magnitude scale (Dinnella *et al.*, 2010) to better discriminate the intensities of samples at higher phenol concentration.

In order to control possible risk of carry-over effect, a break and a rinsing procedure should be planned between the evaluations of two samples. A break of 30 s reduces the carry-over effect, while a break of 60 s controls it (Lesschave and Noble, 2005). The effectiveness of several rinsing procedures in controlling the carry-over effect has been demonstrated. Colonna *et al.* (2004) reported that the combination water – cracker – water efficiently reduced the carry-over effect, an effect similar to a 5g/l pectin water solution. Based on this evidence, it is suggested that subjects are asked to rinse their mouths with distilled water for 45 s, to have some plain crackers for 30 s and finally to rinse their mouths with water for a further 45 s (Monteleone *et al.*, 2004). A minimum number of three replicates is recommended.

4.4.4 Data analysis

An overview of panel performance can be obtained by computing a two-way ANOVA and estimating sample, assessor and sample × assessor effect.

It has been recently shown (Dinnella *et al.*, 2011) that the pronounced variation in sensory perception, due to the physiologically different salivary responses, may lead to different hedonic responses during exposure to foods and beverages containing phenolic compounds.

4.4.5 Recent sensory research advances in understanding astringency induced by phenolic compounds

Phenolic compounds are found in a wide variety of fruits and vegetables. Their presence in human food is of nutritional interest since they may be responsible for beneficial as well as detrimental effects on health (Chobot et al., 2009). Phenolic compounds can be divided into 15 classes depending on their chemical structure. Flavonoids constitute the largest and most diverse family of phenolic compounds. Their strong radical scavenging activity probably accounts for their role in preventing diseases related to oxidative stress, such as coronary heart disease and various forms of cancer (Arts and Hollman, 2005). Tannins constitute a complex group of flavonoid-based polymers capable of binding and eventually precipitating proteins. The antioxidant activity has been demonstrated for tannins, but harmful effects have also been reported for these compounds, such as inhibition of digestive enzymes, lowering of dietary protein digestibility, suppressed growth in rats, altered food consumption and acute hepatotoxicity (Mueller-Harvey, 2006).

Sensory properties of flavonoids and tannins can be summarized by two main descriptors: bitter and astringent, both eliciting negative consumer reactions when perceived at high intensities (Lesschaeve and Noble, 2005). Astringency arises from the interaction of dietary tannins with lubricating salivary proteins (Nayak and Carpenter, 2008; Dinnella *et al.*, 2010). It has been proposed that the sensation of astringency represents a warning cue to discourage animals from ingesting foods too high in tannins and thus protecting them from their potential harmful anti-nutritional effects (Shimada, 2006).

However, the oral environment appears to be a complex and dynamic reaction medium in which the protein/tannin interactions take place. These cannot be considered a simple alert mechanism based on the perception of an increased astringency (friction), but instead may be considered to be a more sophisticated system aimed at defending the organism by inactivating these potentially dangerous compounds. Astringency arises from rupture of the lubricating saliva film that lines the oral cavity (de Wijk and Prinz, 2006). A two-step molecular model of salivary protein/dietary phenol interaction has been hypothesized for astringency elicitation (Nayak and Carpenter, 2008). The first step might involve a dynamic parotid protein film (prolin rich proteins, amylases, cystatins and hystatins) exerting a phenol-sequestering role, thus protecting against the negative effects of phenolic compounds on nutritional uptake. The second step might be based on phenol interactions with an adsorbed glycoprotein layer with the consequent oral cavity delubrication and astringency elicitation. Individual differences in perceived astringency was shown to be correlated to the extent of the change in salivary protein content (SPs D-value) and in haze development in saliva-tannic acid mixtures (HFC D-value) after repeated stimulation (Dinnella *et al.*, 2009). Based on subjects' ability to react to oral stimulation, groups with different responsiveness to astringent phenolic stimuli can be identified. The LR group shows low SPs D-values, indicating a high ability to constantly maintain salivary protein concentration. Negative D-values characterize the HR group, since a strong salivary glycoprotein depletion takes place in this subject group after repeated oral stimulation. A decrease of parotid protein secretion is involved in the depletion of whole saliva protein concentration induced by oral stimulation in the HR group, thus resulting in a temporary lowering of the overall saliva defence level against phenolic compounds. Thereby, an increased responsiveness to astringent stimuli provides a warning cue (Dinnella *et al.*, 2010).

The capacity of LR subjects to maintain constant salivary characteristics, in terms of both concentration and composition, may parallel their capacity to intercept and inactivate phenols; these subjects are then better protected and less sensitive to astringent stimuli. The higher sensitivity to astringent stimuli of HR subjects seems to be related to a decrease of saliva phenol-binding capacity after repeated stimulation conditions compared with basal conditions, and a modification of the prolin-rich proteins (PRP) profile seems to be involved in this. These findings can also be relevant for applied research studies conducted with the aim of setting up astringency predictive methods.

4.5 Future trends

In traditional sensory evaluation, wines are evaluated blind and all efforts are made to minimize the impact of external cues. This is assumed to yield unbiased information about perceived characteristics of the samples, without interference from the brand or labeling. Marketing research has been left with the task of finding out what contributions the brand, package or other information may have to the purchase or acceptance of a wine. Recently, consumer-oriented sensory research focuses on understanding wine preference and choice as a whole. Potential interactions of sensory properties and other factors affecting consumer responses have become relevant to wine research (see Tuorila and Monteleone, 2009). This interest is still growing, as proved by several papers published as a result of the Eighth Pangborn Sensory Science Symposium (Bertuccioli and Monteleone, 2010).

New approaches to assess the relative impact of sensory properties and information cues on overall liking have been recently proposed to investigate, in real-life settings, consumer expectation and overall liking for wine. In general it is possible to say that the exchange of expertise and methods between sensory scientists who work in wine evaluation and those who study human behavior will be incremented in the near future.

It is detrimental to food and wine industries that the only information available on wine–food pairing are provided by wine and food writers, sommeliers and culinary experts according to their personal sensitivity (Bastian *et al.*, 2010). This is a field of research that surely deserves attention from sensory scientists. In fact, while it seems clear that consumer behavior data indicate that wine consumption without the accompaniment of food is not highly practiced, a very limited number of scientific papers examine wine–food sensory interaction. It is possible that in the near future wine sensory studies will focus on these aspects optimizing the use of existing methods like descriptive analysis and TDS.

4.6 References

ARTS ICW and HOLLMAN PCH (2005), 'Polyphenols and disease risk in epidemiological studies', *American Journal of Clinical Nutrition*, **81**(1), 317–325.

AUBRY V, SCHLICH P, ISSANCHOU S and ETIÉVANT P (1999), 'Comparison of wine discrimination with orthonasal and retronasal profilings', *Food Quality and Preference*, **10**, 253–259.

BALLESTER J, DACREMONT C, LE FUR Y and ETIÉVANT P (2005), 'The role of olfaction in the elaboration and use of the Chardonnay wine concept', *Food Quality and Preference*, **16**, 351–359.

BALLESTER J, PATRIS B, SYMONEAUX R, and VALENTIN D (2008), 'Conceptual vs. perceptual wine spaces: Does expertise matter?', *Food Quality and Preference*, **19**(3), 267–276.

BASTIAN SEP, COLLINS C and JOHNSON TE (2010), 'Understanding consumer preferences for Shiraz wine and Cheddar cheese pairings', *Food Quality and Preference*, **21**(7), 668–678.

BERTUCCIOLI M and MONTELEONE E (eds) (2010), Eighth Pangborn Sensory Science Symposium – Wine Special Issue, *Food Quality and Preference* **21**(7) October 2010.

BIASOTO ACT, CATHARINO RR, SANVIDO GB, EBERLIN MN and DA SILVA MAAP (2010), 'Flavour characterization of red wines by descriptive analysis and ESI mass spectrometry', *Food Quality and Preference*, **21**(7), 755–762.

BØLLING JS, NAES T, ØYAAS J and HERSLETH M (2009), 'Acceptance of calorie-reduced yoghurt: Effect of sensory characteristics and product information', *Food Quality and Preference*, **21**(1), 13–21.

BROSSAUD F, CHEYNIER V and NOBLE AC (2001), 'Bitterness and astringency of grape and wine polyphenols', *Australian Journal of Grape and Wine Research*, **7**, 33–39.

BUCK D (2007), 'Methods to understand consumer attitudes and motivations in food product development', in MacFie H (ed.), *Consumer-led Food Product Development*, Woodhead Publishing, Cambridge, 141–157.

CAIRNCROSS SE and SJOSTROM LB (1950), 'Flavor profiles – A new approach to flavor problems', *Food Technology*, **4**(8), 308–311.

CAMPO E, BALLESTER J, LANGLOIS J, DACREMONT C and VALENTIN D (2010), 'Comparison of conventional descriptive analysis and a citation frequency-based descriptive method for odor profiling: an application to Burgundy Pinot Noir wines', *Food Quality and Preference*, **21**, 44–55.

CARLUCCI A and MONTELEONE E (2008), 'A procedure of sensory evaluation for describing the aroma profile of single grape variety wines', *Journal of Sensory Studies*, **23**, 817–834.

CARVALHO E, POVOAS MJ, MATEUS N, and DE FREITAS V (2006), 'Application of flow nephelometry to the analysis of the influence of carbohydrates on protein–tannin interactions', *Journal of the Science of Food and Agriculture*, **86**(6), 891–896.

CHOBOT V, HUBER C, TRETTENHAHN G and HADACEK F (2009), 'Catechin: chemical weapon, antioxidant, or stress regulator?', *Journal of Chemical Ecology*, **35**, 980–996.

COLONNA AE, ADAMS DO and NOBLE A (2004), 'Comparison of procedures for reducing astringency carry-over effects in evaluation of red wines', *Australian Journal of Grape and Wine Research*, **10**(1), 26–31.

CONDELLI N, DINNELLA C, CERONE A, MONTELEONE E and BERTUCCIOLI M (2006), 'Prediction of perceived astringency induced by phenolic compounds II: criteria for panel selection and preliminary application on wine sample', *Food Quality and Preference*, **17**, 96–107.

CRISTOVAM E, PATERSON A and PIGGOTT JR (2000), 'Development of a vocabulary of terms for sensory evaluation of dessert port wines', *Italian Journal of Food Science*, **12**, 129–142.

DE FREITAS V and MATEUS N. (2002), 'Nephelometric study of salivary protein-tannin aggregates', *Journal of the Science of Food and Agriculture*, **82**, 113–119.

DE FREITAS V, CARVALHO E and MATEUS N (2003), 'Study of carbohydrate influence on protein-tannin aggregation by nephelometry', *Food Chemistry*, **81**, 503–509.

DE WIJK RA and PRINZ JF (2006), 'Mechanisms underlying the role of friction in oral texture', *Journal of Texture Studies*, **37**, 413–427.

DELARUE J and SIEFFERMANN J-M (2004), 'Sensory mapping using flash profile comparison with a conventional descriptive method for the evaluation of the flavour of fruit dairy products', *Food Quality and Preference*, **15**(4), 383–392.

DINNELLA C, RECCHIA A, FIA G, BERTUCCIOLI M and MONTELEONE E (2009), 'Saliva characteristics and individual sensitivity to phenolic astringent stimuli', *Chemical Senses*, **34**, 295–304.

DINNELLA C, RECCHIA A, TUORILA H and MONTELEONE E (2011), 'Individual astringency responsiveness affects the acceptance of phenol-rich foods', *Appetite*, in press.

DINNELLA C, RECCHIA A, VINCENZI S, TUORILA H and MONTELEONE E (2010), 'Temporary modification of salivary protein profile and individual responses to repeated phenolic astringent stimuli', *Chemical Senses*, **35**, 75–85.

DREWNOWSKI A and GOMEZ-CARNEROS C. (2000), 'Bitter taste, phytonutrients, and the consumer: a review', *American Journal of Clinical Nutrition*, **72**, 1424–1435.

EDELMANN A and LENDL B (2002), 'Towards the optical tongue: flow-through sensing of tannin–protein interactions based on FTIR spectroscopy', *Journal of the American Chemical Society*, **124**, 14741–14747.

FIA G, DINNELLA C, BERTUCCIOLI M and MONTELEONE E (2009), 'Prediction of grape polyphenol astringency by means of a fluorimetric micro-plate assay', *Food Chemistry*, **113**, 325–330.

GARRETT DA, FAILLA ML and SARAMA RJ (1999), 'Development of an *in vitro* digestion method to assess carotenoid bioavailability from meals', *Journal of Agriculture and Food Chemistry*, **47**, 10, 4301–4309.

GAWEL R and GODDEN PW (2008), 'Evaluation of the consistency of wine quality assessments from expert wine tasters', *Australian Journal of Grape and Wine Research*, **14**, 1–8.

GAWEL R, ILAND PG and FRANCIS IL (2001), 'Characterizing the astringency of red wine: a case study', *Food Quality and Preference*, **12**, 83–94.

GAWEL R, OBERHOLSTER A and FRANCIS I (2000), 'A mouth-feel wheel: terminology for communicating the mouth-feel characteristics of red wine', *Australian Journal of Grape and Wine Research*, **6**, 203–207.

GLORIES Y (1978), *Recherches sur la matière colorant des vin rouges*, PhD Thesis, University of Bordeaux II.

GOWER J C (1975), 'Generalized Procustes analysis', *Psychometrika*, **20**, 33–51.

HERDERICH MJ and SMITH PA (2005), 'Analysis of grape and wine tannins: methods, application and challenges', *Australian Journal of Grape and Wine Research*, **11**, 205–214.

HERSLETH M, MEVIK BH, NAES T and GUINARD JX (2003), 'Effect of contextual factors on liking for wine – use of robust design methodology', *Food Quality and Preference*, **14**, 345–349.

HOFMANN T, GLABASNIA A, SCHWARZ B, WISMAN KN, GANGWER KA *et al.*, (2006), 'Protein binding and astringent taste of polymeric procyanidin, 1,2,3,4,6-Penta-O-galloil b-D-glucopyranose, castalagin and grandinin', *Journal of Agricultural and Food Chemistry*, **54**, 9503–9509.

HORNE J, HAYES J and LAWLESS HT (2002), 'Turbidity as a measure of salivary proteins reactions with astringent substances', *Chemical Senses*, **27**, 653–659.

HUGHSON AL, ASHMAN H, DE LA HUERGA V and MOSKOWITZ H (2004), 'Mind-sets of the wine consumer', *Journal of Sensory Studies*, **19**, 85–105.

HUGHSON AL and BOAKES RA (2002), 'The knowing nose: the role of knowledge in wine expertise', *Food Quality and Preference*, **13**, 463–472.

HUNTER EA and MCEWAN JA (1998), 'Evaluation of an international ring trial for sensory profiling of hard cheese', *Food Quality and Preference*, **9**(5), 343–354.

ISO 11035 (1994), 'Sensory analysis. Identification and selection of descriptors for establishing a sensory profile by a multidimensional approach'.

JACK FR and PIGGOTT J R (1991–1992), 'Free choice profiling in consumer research', *Food Quality and Preference*, **3**(3), 129–134.

KALLITHRAKA S, BAKKER J, CLIFFORD MN and VALLIS L (2001), 'Correlations between saliva protein composition and some T-I parameters of astringency', *Food Quality and Preference*, **12**, 145–152.

KENNEDY JA, FERRIER J, HARBERTSON JF and PEYROT DES GACHONS C (2006), 'Analysis of tannins in red wine using multiple methods: correlation with perceived astringency', *American Journal of Enology and Viticulture*, **57**(4), 481–485.

KENNEDY JA, SAUCIER C and GLORIES Y (2006), 'Grape and wine phenolics: history and perspective', *American Journal of Enology and Viticulture*, **57**(3), 239–248.

LAWLESS H, LIU YF and GOLDWYN C (1997), 'Evaluation of wine quality using a small-panel hedonic scaling method', *Journal of Sensory Studies*, **12**, 317–332.

LAWLESS HT and HEYMANN H (1998), *Sensory evaluation of foods: principles and practices*, Chapman & Hall, New York [2nd ed., 1999, Springer, New York].

LEE CB and LAWLESS HT (1991), 'Time-course of astringent materials', *Chemical Senses*, **16**, 225–238.

LEE SJ and NOBLE AC (2003), 'Characterization of odor-active compounds in Californian Chardonnay wines using GC-olfactometry and GC-mass spectrometry', *Journal of Agricultural and Food Chemistry*, **51**, 8036–8044.

LESSCHAEVE I (2007) 'Sensory evaluation of wine and commercial realities: review of current practices and perspectives', *American Journal of Enology and Viticulture*, **58**, 252–258.

LESSCHAEVE I and NOBLE AC (2005), 'Polyphenols: factors influencing their sensory properties and their effects on food and beverage preference', *American Journal of Clinical Nutrition*, **81**, 330–335.

LYMAN BJ and GREEN BG (1990), 'Oral astringency: effects of repeated exposure and interactions with sweeteners', *Chemical Senses*, **15**, 151–164.

MACFIE HJ, BRATCHELL N, GREENHOFF K and VALLIS LV (1989), 'Designs to balance the effect of order of presentation and first-order carry-over effects in hall tests', *Journal of Sensory Studies*, **4**(2), 129–148.

MADRIGAL-GALAN B and HEYMANN H (2006), 'Sensory effects of consuming cheese prior to evaluating red wine flavor', *American Journal of Enology and Viticulture*, **57**(1), 12–22.

MAITRE I, SYMONEAUX R, JOURJON F and MEHINAGIC E (2010), 'Sensory typicality of wines: how scientists have recently dealt with this subject', *Food Quality and Preference*, **21**, 726–731.

MARTENS H and MARTENS M (2001), *Multivariate Analysis of Quality: An Introduction*, John Wiley & Sons, Ltd, Chichester.

MEILGAARD M, CIVILLE GV and CARR BT (1991), *Sensory Evaluation Techniques*, CRC Press Boca Raton, [2nd ed., 2000].

MEILLON S, URBANO C, GUILLOT G and SCHLICH P (2010a), 'Acceptability of partially dealcoholized wines – Measuring the impact of sensory and information cues on overall liking in real-life settings', *Food Quality and Preference*, **21**, 763–773.

MEILLON S, VIALA D, MEDEL M, URBANO C, GUILLOT G *et al.* (2010b), 'Impact of partial alcohol reduction in Syrah wine on perceived complexity and temporality of sensations and link with preference', *Food Quality and Preference*, **21**, 732–740.

MONTELEONE E, CONDELLI N, DINNELLA C and BERTUCCIOLI M (2004), 'Prediction of perceived astringency induced by phenolic compounds', *Food Quality and Preference*, **15**, 761–770.

MONTELEONE E, DINNELLA C and CONDELLI N (2007), 'Perceived astringency in wine: A predictive model', *Acta Horticolturae* (ISHS), **754**, 523–532.

MORAND E and PAGÈS J (2006), 'Procrustes multiple factor analysis to analyse the overall perception of food products', *Food Quality and Preference*, **17**, 36–42.

MOSCOWITZ HR (1999), 'On the intersection of products and concepts: opportunities for sensory analysis to improve the commercial development process', *Food Quality and Preference*, **10**(4–5), 333–342.

MUELLER-HARVEY I (2006), 'Unravelling the conundrum of tannins in animal nutrition and health', *Journal of the Science of Food and Agriculture*, **86**, 2010–2037.

MURRAY JM, DELAHUNTY CM and BAXTER IA (2001), 'Descriptive sensory analysis: past, present and future', *Food Research International*, **34**, 461–471.

NAES T, BROCKHOFF PB and TOMIC O (2010a), *Statistics for Sensory and Consumer Science*, John Wiley & Sons Ltd, Chichester.

NAES T, LENGARD V, BØLLING JS and HERSLETH M (2010b), 'Alternative methods for combining design variables and consumer preference with information about attitudes and demographics in conjoint analysis', *Food Quality and Preference*, **21**, 368–378

NAYAK and CARPENTER GH (2008), 'A physiological model for tea induced astringency', *Physiological Behaviour*, **95**, 290–294.

NESTRUD MA and LAWLESS HT (2008), 'Perceptual mapping of citrus juices using projective mapping and profiling data from culinary professionals and consumers', *Food Quality and Preference*, **19**, 431–438.

NOBLE AC, ARNOLD RA, BUECHSENSTEIN J, LEACH EJ, SCHMIDT JO *et al.* (1987), 'Modification of a standardized system of wine aroma terminology', *American Journal of Enology and Viticulture*, **38**(2), 143–146.

NYGREN IT, GUSTAFSSON IB and JOHANSSON L (2002), 'Perceived flavour changes in white wine after tasting blue mould cheese', *Food Service Technology*, **2**, 163–171.

O'NEILL L, NICKLAUS S and SAUVAGEOT FO (2003), 'A matching task as a potential technique for descriptive profile validation', *Food Quality and Preference*, **14**, 539–547.

PAGÈS J (2003), 'Recueil direct de distances sensorielles: Application à l'évaluation de dix vins blancs de Val de Loire', *Sciences des Aliments*, **23**, 679–888.

PAGÈS J (2005), 'Collection and analysis of perceived product inter-distances using multiple factor analysis: Application to the study of 10 white wines from the Loire Valley', *Food Quality and Preference*, **16**, 642–649.

PANGBORN RM (1989), 'The evolution of sensory science and its interaction with IFT', *Food Technology*, **43**(9), 248–256, 307.

PARR WV, GREEN JA and WHITE KG (2006), 'Wine judging, context and New Zealand Sauvignon Blanc', *European Review of Applied Psychology*, **56**, 231–238.

PARR WV, WHITE KG and HEATHERBELL DA (2004), 'Exploring the nature of wine expertise: what underlies wine experts' olfactory recognition memory advantage?', *Food Quality and Preference*, **15**, 411–420.

PARR W V, VALENTIN D, GREEN J and DACREMONT C (2010), 'Evaluation of French and New Zealand Sauvignon wines by experienced French wine assessors', *Food Quality and Preference*, **21**, 56–64.

PERRIN L, SYMONEAUX R, MAITRE I, ASSELIN C, JOURJON F *et al.* (2008), 'Comparison of three sensory methods for use with the Napping™ procedure: Case of ten wines from Loire valley', *Food Quality and Preference*, **19**, 1–11.

PICCHI M, RECCHIA AM, BERTUCCIOLI M, TUORILA H, NAES T. and MONTELEONE E (2010), 'Experts and consumers evaluation of odour (dis)-similarities among wines: behind the projective map approach', *Food Quality and Preference*, **21**, 793.

PICKERING GJ, KARTHIK A, INGLIS D, SEARS M and KER K (2007), 'Determination of ortho- and retronasal detection thresholds for 2-isopropyl-3-methoxypyrazine in wine', *Journal of Food Science*, **72**, S468–S472.

PIGGOTT JR, SIMPSON SJ and WILLIAMS SAR (1998), 'Sensory analysis', *International Journal of Food Science and Technology*, **33**, 7–18.

PIGGOTT JR and WATSON MP (1992), 'A comparison of free choice profiling and the repertory grid method in the flavour profiling of cider', *Journal of Sensory Studies*, **7**, 133–145.

PINEAU N, SCHLICH P, CORDELLE S, MATHONNIÈRE C, ISSANCHOU S *et al.* (2009), 'Temporal Dominance of Sensations: Construction of the TDS curves and comparison with time–intensity', *Food Quality and Preference*, **20**(6), 450–455.

PONCET-LEGRAND C, EDELMANN A, PUTAUX JL, CARTALADE D, SARNI-MANCHADO P *et al.* (2006) 'Poly(L-proline) interactions with flavan-3-ols units: influence of the molecular structure and the polyphenol/protein ratio', *Food Hydrocolloids*, **20**, 687–697.

PRESCOTT J, NORRIS L, KUNST M and KIM S (2005), 'Estimating a "consumer rejection threshold" for cork taint in white wine', *Food Quality and Preference*, **16**, 345–349.

PRINZ JF and LUCAS PW (2000), 'Saliva tannin interaction', *Journal of Oral Rehabilitation*, **27**, 991–994.

RISVIK E, MCEWAN J A and RODBOTTEN M (1997), 'Evaluation of sensory profiling and projective mapping data', *Food Quality and Preference*, **8**(1), 63–71.

RISVIK E, MCEWAN J A, COLWILL J S, ROGERS R and LYON D H (1994), 'Projective mapping: a tool for sensory analysis and consumer research', *Food Quality and Preference*, **5**, 263–269.

SAUVAGEOT FO, URDAPILLETA I and PEYRON D (2006), 'Within and between variations of texts elicited from nine wine experts', *Food Quality and Preference*, **17**, 429–444.

SHIMADA T (2006), 'Salivary proteins as a defense against dietary tannins', *Journal of Chemical Ecology*, **32**, 1149–1163.

SIEFFERMANN JM (2000), 'Le profil flash—un outil rapide et innovant d'évaluation sensorielle descriptive', in *AGORAL 2000*, XIIèmes rencontres « L'innovation: de l'idée au succès », Montpellier, France, 335–340.

STONE H and SIDEL JL (2007), 'Sensory research and consumer-led food product development', in MacFie H (ed.), *Consumer-led food product development*, Woodhead Publishing, Cambridge, 307–320.

STONE H, SIDEL J, OLIVER S, WOOLSEY A and SINGLETO RC (1974), 'Sensory evaluation by quantitative descriptive analysis', *Food Technology*, **28**(11), 24–34.

SZCZESNIAK AS (1963), 'Objective measurements of food texture', *Journal of Food Science*, **28**, 410–420.

TUORILA H and MONTELEONE E (2009), 'Sensory food science in the changing society: Opportunities, needs and challenges', *Trends in Food Science and Technology*, **20**, 54–62.

VIDAL S, LEIGH F, NOBLE A, KWIATKOWSKI M, CHEYNIER V *et al.* (2004), 'Taste and mouth-feel properties of different types of tannin-like polyphenolic compounds and anthocyanins in wine', *Analytica Chimica Acta*, **513**, 57–65.

WANG X, HO C and HAUNG Q (2007), 'Investigation of adsorption behavior of (-)-epigallocatechin gallate on bovine serum albumin surface using quartz crystal microbalance with dissipation monitoring', *Journal of Agricultural and Food Chemistry*, **55**, 4987–4992.

WILLIAMS AA and LANGRON S P (1984), 'The use of free-choice profiling for the evaluation of commercial ports', *Journal of the Science of Food and Agriculture*, **35**, 558–568.

5

Gas chromatography-olfactometry of alcoholic beverages

B. Plutowska and W. Wardencki, Gdansk University of Technology, Poland

Abstract: In recent years extensive studies have been performed concerning the relationship between aroma and the chemical composition of the volatile fraction of alcoholic beverages, using gas chromatography-olfactometry (GC-O). The chapter first describes the principles of GC-O analysis comprising such issues as sample preparation procedures, methods of data collecting, GC instrumentation and other analytical conditions. The chapter then discusses application of the GC-O technique in analysis and quality assessment of various alcoholic beverages such as beer, wine and spirits.

Key words: gas chromatography-olfactometry, alcoholic beverages, odour, volatile compounds.

5.1 Introduction

The quality of alcoholic products depends strongly on their composition and the levels of odour compounds they contain. The beverage industry considers aroma to be of great importance as it contributes to product quality and hence to consumer acceptance. The smell of an alcoholic beverage is the effect of a large number of chemical compounds with different properties (such as volatility or polarity) occurring at different concentrations. The chemical composition of the odour it produces depends on the quality and type of the raw materials used, as well as the conditions of the fermentation and ageing processes.

Alcoholic fermentation conducted under industrial conditions leads to a series of byproducts in addition to ethanol. The composition and concentration levels of typical byproducts such as carbonyl compounds, alcohols, esters, acids and acetals can vary widely. The influence of individual compounds on the odour profile can be very different. Quite frequently, compounds appearing in trace quantities (even as law as mg/l) have a greater influence on the sensory properties of alcoholic products than those which appear in high concentrations (such as hundreds of mg/l).

Conventional techniques for assessment of flavour quality, such as sensory analysis are often time-consuming, labour-intensive and not always objective. A breakthrough in aroma research was the introduction of a technique combining olfactometry and gas chromatography, which associates the resolution power of capillary gas chromatography (GC) with the selectivity and sensitivity of the human nose. Gas chromatography-olfactometry (GC-O) studies on the odours of alcoholic beverages usually aim to determine the relationship between the composition and the content of volatile compounds and the organoleptic properties of different alcoholic beverages such as beer, wines, agriculture distillates and spirit beverages. Another aim is to identify and compare the compounds that make up the aroma of different alcoholic beverages. Another goal might be the determination of compounds responsible for undesired odours.

The GC-O results are often correlated with the results of conventional sensory evaluation conducted in parallel, usually using typical methods. Sensory analysis of the individual samples yields descriptors characterizing their smell. Chromatographic analysis with olfactometric detection makes it possible to determine which compounds are responsible for the individual descriptors.

5.2 Principles of gas chromatography-olfactometry (GC-O)

GC-O, proposed for the first time by Fuller *et al.* (1964) is nowadays a valuable technique for the determination of odour-active components in complex mixtures such as alcoholic beverages. Its general principle is illustrated schematically in Fig. 5.1. Each GC-O analysis consists of four stages:

1. Preparation of the sample,
2. Separation of the extracted mixture,
3. Detection of the odorants,
4. Interpretation of the acquired data.

Many factors determine the quality of the results obtained, such as the choice of the extraction procedure and the method of data collection, GC instrumentation and other analytical conditions. Sample handling influences the composition of the final extract and in consequence the quality and quantity of aroma compounds perceived by human assessors. The correct detection and assessment of the volatile compounds eluting from the GC column depend also on the separation capability of the GC system. The usefulness of the olfactometric data obtained can be increased by appropriate configuration of the instruments as well as the simultaneous use of additional detectors. Finally, since GC-O analysis involves the measurement and interpretation of the human olfactory impression, it is important to ensure the assessors perform optimally and to apply a fit-for-purpose method of detection and quantification of odour compounds.

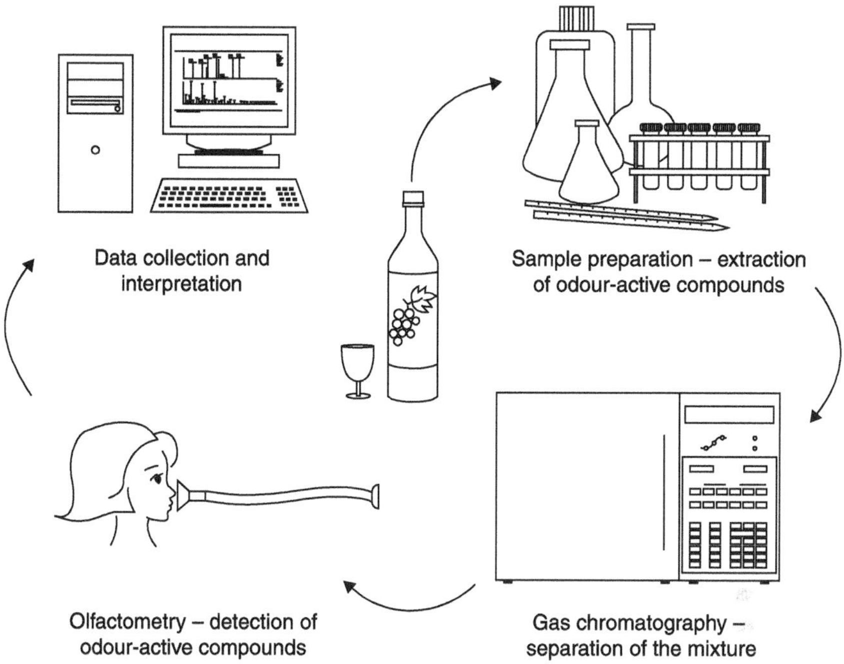

Fig. 5.1 Schematic of GC-O analysis.

While some of these factors are equally important for any other chromatographic technique, the replacement of the conventional detector with a human nose entails many additional aspects of analysis which must be considered in this specific context.

5.2.1 Sample preparation

The aroma composition of alcoholic beverages depends on the proportion of odorants in the gas phase available for perception. It must be remembered that only a portion of volatile components is released from a beverage matrix into the air, depending on their chemical properties and binding to nonvolatile components such as sugars or proteins. On the other hand, in a GC-O analysis the sample extract is completely vaporized and delivered to the assessor's nose in volatile form. Therefore, in order to relate chemical odour composition to the sensory properties of the sample, the isolate used for analysis should reflect the aroma of the beverage when drunk or smelled. The qualitative and quantitative representativeness of the sample extract depends on the sample preparation method. Numerous studies revealed that the use of different sample preparation techniques might affect the composition and contents of the isolated aroma profile (Bonino *et al.*, 2003; Falqué López and Fernández Gómez, 2000; Nonato *et al.*, 2001).

The choice of an appropriate extraction method is not easy. The volatile fraction of most alcoholic beverages is very complex (usually hundreds of compounds) and the individual components are found in a very wide range of concentrations (usually several orders of magnitude). An ideal method should be able to isolate the wide range of the odorous substances over the wide range of concentrations. In order to address the aforementioned issues, different analytical methods have been applied, including liquid–liquid extraction (LLE) (Gurbuz *et al.*, 2006; Guth, 1997a; Kishimoto *et al.*, 2006; Kotseridis and Baumes, 2000; Lee and Noble, 2003; Miranda-Lopez *et al.*, 1992; Sarrazin *et al.*, 2007a; Silva Ferreira *et al.*, 2003; Soares da Costa *et al.*, 2004; Torrens *et al.*, 2010), continuous liquid–liquid extraction (Escudero *et al.*, 2000a; Ferreira *et al.*, 1998; Komes *et al.*, 2006), demixtion and distillation (Bernet *et al.*, 2002; Escudero and Etievant, 1999; Priser *et al.*, 1997), steam distillation (SD) (Simpson *et al.*, 2004), solid phase extraction (SPE) (Aznar *et al.*, 2001; Cullere *et al.*, 2004; Escudero *et al.*, 2004; Ferreira *et al.*, 2001; Ferreira *et al.*, 2002a; Lermusieau *et al.*, 2001; Lopez *et al.*, 1999; Saenz-Navajas *et al.*, 2010), solvent-assisted flavour evaporation (SAFE) (Campo *et al.*, 2005; Fang and Qian, 2005; Fritsch and Schieberle, 2005; Garruti *et al.*, 2006; Kishimoto *et al.*, 2008; Xu *et al.*, 2007), static and dynamic headspace analysis (Campo *et al.*, 2005; Cullere *et al.*, 2008; Escudero *et al.*, 2007; Fan *et al.*, 2010; Ferreira *et al.*, 2009; Gomez-Miguez *et al.*, 2007; Poisson and Schieberle, 2008), headspace solid-phase micro-extraction (HS-SPME) (Ebeler *et al.*, 2000; Fan and Qian, 2005; Gurbuz *et al.*, 2006; Marti *et al.*, 2003; Plutowska and Wardencki, 2009), stir-bar sorptive extraction (SBSE) (Buettner, 2004; Buettner and Welle, 2004), etc. Their applications, advantages and disadvantages have been extensively reviewed (Plutowska and Wardencki, 2007, 2008).

All of these methods can be divided into two groups – solvent and solvent-free techniques. In principle, solventless techniques such as headspace analysis exemplify the liberation of volatile components from the matrix to the gaseous phase; therefore they can be easily correlated with the results of sensory analysis. On the other hand, solvent techniques produce liquid extracts which often can be directly checked for their representativeness using sensory analysis.

Solvent extraction methods yield a total extract of the sample which is not always close to the aroma profile. An analysis of alcoholic beverages requires the application of solvents with a low affinity to ethanol such as dichloromethane, hexane, pentane and diethyl ether or freon 11 (Pons *et al.*, 2008). Since the non-polar solvents may not extract polar compounds efficiently, a solid-phase extraction with Amberlite XAD resins is a common alternative to the liquid–liquid extraction. Continuous liquid–liquid extraction methods are considered to be the most sensitive; however, for many applications, a single extraction procedure is sufficient. In general, the extracts obtained by these techniques are very complex and the volatile compounds present at low concentrations, e.g. β-damascenone, may not be chromatographically resolved from those present at higher concentrations, e.g. the fusel alcohols (Delahunty *et al.*, 2006). Consequently, numerous co-elutions in GC-O occur and the subsequent preparative chromatography or other clean-up procedures of the crude isolate such as acidic/basic fractionation are needed (Benn and Peppard, 1996; Cullere *et al.*, 2003; Escudero *et al.*, 2000a; Fan and Qian, 2006a, 2006b; Fang and Qian, 2005; Ferreira *et al.*, 1999; Fritsch and Schieberle, 2005; Lee and Noble, 2003; Poisson and Schieberle, 2008; Rogerson *et al.*, 2001; Xu *et al.*, 2007). On the other hand, highly volatile compounds may be lost during solvent removal or they can co-elute with the solvent and so may not be detected (Fritsch and Schieberle, 2005). Another problem is the simultaneous extraction of non-volatile compounds. To prevent their presence in the extract, solvent-assisted flavour evaporation is being used. Further aspects that should be considered are the need for large sample amounts in order to attain a satisfactory concentration and the presence of impurities from the solvent in the extract.

A potential solution to the limitations of solvent techniques is to use headspace analysis. Both static and dynamic headspace methods can be used for this purpose; however, because of the possibility of analyte enrichment, dynamic methods are more often used (Pollien *et al.*, 1997). In these methods, the sample is purged with an inert gas (e.g. nitrogen or helium), which removes volatiles from the sample. The aroma compounds are then trapped from the gas stream. Most commonly adsorbent traps (e.g. Lichrolut EN, Porapak Q, Tenax or alternative polymer) and subsequent elution with solvents are applied (Ferreira *et al.*, 2009; Garruti *et al.*, 2006).

Another widely used solvent-free technique is SPME. This technique has already been linked with the detection frequency method (Ebeler *et al.*, 2000), the dilution to threshold method – aroma extract dilution analysis (AEDA) (Fan and Qian, 2005; Marti *et al.*, 2003) and time-intensity methods (Gurbuz *et al.*, 2006; Plutowska and Wardencki, 2009). The aroma profiles of the collected volatile compounds depend upon the type, thickness and length of the fibre, as well as the sampling conditions which are selected according to the type of matrix. Therefore, the utilization of this technique in GC-O analysis is limited due to its selectivity. One must be aware that not only SPME but all headspace analysis methods are selective to some extent and therefore may not disclose the total aroma composition of the sample. A good example is the study of Le Fur *et al.* (2003) showing that dynamic headspace sampling may emphasize the herbaceous character of wine while it does not reproduce its spicy attributes.

Although numerous aroma isolation methods are known, the most appropriate way to obtain an optimum aroma profile of an alcoholic beverage is to employ more than one extraction technique or to choose the technique fit-for-purpose. It is important to consider the matrix, analytes of interest, time consumption and objectives of the analysis in choosing analytical methodology (Reineccius, 2006). For example, Ebeler *et al.* (2000) reported that LLE with Freon 11 was more efficient for extracting the higher alcohols in brandy than SPME using polydimethylsiloxane (PDMS) fibre, while SPME was more selective for esters than LLE. In an analogous study on apple cider, Xu *et al.* (2007) showed that the SAFE technique revealed a higher recovery for acids and hydroxy-compounds, while the HS-SPME had a higher efficiency for esters and highly volatile compounds.

For identification of the key components of an aroma profile, a selective isolation method can be applied. Volatile thiols can be specifically extracted from wines by reversible combination of the thiols with sodium *p*-hydroxymercuribenzoate (*p*-HMB) (Bailly *et al.*, 2006; Sarrazin *et al.*, 2007b). If extraction of the total aroma profile is required, the similarity of extracts and corresponding samples should be evaluated (Barbe *et al.*, 2008). This is usually accomplished by sensory analysis using classical discriminative and matching tests such as duo-trio or triangle test (Aznar *et al.*, 2001; Escudero *et al.*, 2000b; Escudero *et al.*, 2004; Ferreira *et al.*, 1999, 2002a; Priser *et al.*, 1997).

5.2.2 Instrumentation

GC-O uses the human nose as a detection device, generally parallel to a standard chromatographic detector, i.e. flame ionization detector (FID) or more frequently a mass spectrometer (MS), and thus permits rapid identification of so-called odorant zones in a chromatogram. An analysis using the human sense of smell is carried out by an appropriately trained person or a group of evaluating persons who in the course of the test sniff the eluate from column and relate the aromatic impressions to the retention times. It should be mentioned that the human nose is able to detect and distinguish some volatiles compounds at the level of 10^{-17} g, while detectors commonly used in chromatography require, as a rule, at least 10^{-13} g to identify a compound (Acree and Barnard, 1984).

Many improvements have been applied to GC-O devices since the first description of a gas chromatograph modified for the sniffing of effluent to determine volatile odour activity was presented in 1964 (Fuller *et al.*, 1964). In most of the presently used systems, an organoleptic estimation of odour is facilitated with the use of a specially constructed attachment, the so-called olfactometric port, in most cases a PTFE or glass cone fitting the shape of the human nose, connected in parallel to conventional detectors (Fig. 5.2). Simultaneous detection is realized by dividing the eluent stream in an appropriate ratio so that it reaches both detectors. The most favourable is the simultaneous use of an olfactometric detector and mass spectrometer. Such an approach allows for both the description of odours and an identification of the compounds responsible for them, followed by a determination of which of them is characterized by the most intense aroma. The possibility of divergence between the retention times for analytes detected both by means of the mass spectrometer and the olfactometer (where the retention times are longer) can be overcome by installing a restrictor (thin capillary tube) and adjusting the flows of the carrier and auxiliary gas (Hochereau and Bruchet, 2004). When the tested extracts are sufficiently concentrated, they can be separated into a few streams which are simultaneously applied to separate olfactometric ports and analyzed by several evaluators operating in parallel. The results obtained in such situations are the most representative (Debonneville *et al.*, 2002). The eluate arriving at the olfactometric port may cause drying of the mucous membrane in the nose of the evaluating person, particularly during long-lasting analyses. To exclude this unfavourable

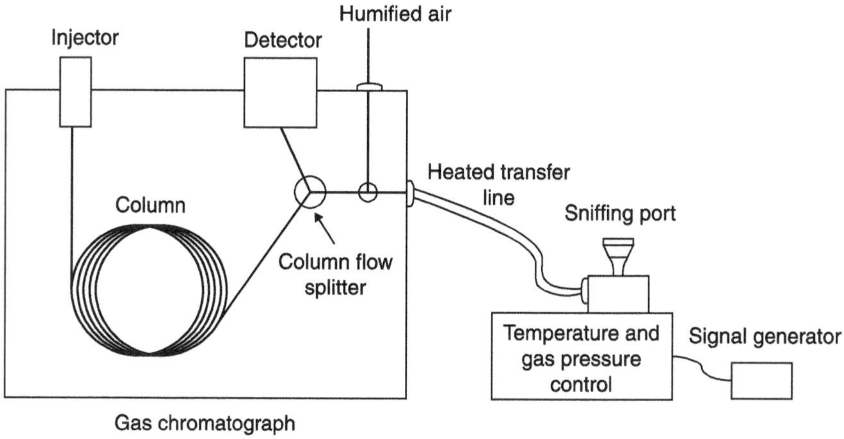

Fig. 5.2 Chromatograph equipped with olfactometric detector.

factor, affecting final results, an appropriate addition of an auxiliary gas, usually moist air, is used. The eluate from the column is fed into the olfactometric port through the transfer line. Its length must be adjusted so that the evaluating person assumes a comfortable position during the detection procedure and to avoid any discomfort resulting from the nearness of hot equipment. Moreover, the transfer line should be heated in the course of analysis, to prevent the condensation of medium or highly volatile analytes on capillary walls.

5.2.3 Methods of detection

Each odour-active compound that elutes from the GC column can be characterized on the basis of four properties: detection threshold (presence or absence of odour), quality and intensity of impression and the duration of perception. Various olfactometric (or 'sniffing') methods are available to collect and process these data. They can all be classified into three categories:

- Dilution to threshold methods: based on an estimation of the detection threshold of the odour using a series of extract dilutions e.g. Aroma Extract Dilution Analysis (AEDA) (Schieberle and Grosch, 1987; Ullrich and Grosch, 1987), Combined Hedonic Aroma Response Measurement (CHARM) (Acree and Barnard, 1984; Acree *et al.*, 1984),
- Detection frequency methods – based on the determination of the number of assessors over a group who detected a presence of the given odour (Linssen *et al.*, 1993; Pollien *et al.*, 1997; van Ruth *et al.*, 1996),
- Direct intensity methods:
    - Posterior intensity methods: based on the assessment of the perceived intensity of the odour recorded after elution from the column (Casimir and Whitfield, 1978),

- Time-intensity methods: based on an evaluation of the perceived intensity of the odour recorded simultaneously with the elution of the compound from the column e.g. Odour Specific Magnitude Estimation (OSME) (McDaniel *et al.*, 1990; Miranda-Lopez *et al.*, 1992; Sanchez *et al.*, 1992),
- Finger-Span Cross Matching (FSCM) (Etievant *et al.*, 1999; Guichard *et al.*, 1995).

The choice of an appropriate method is thus of great importance. The principles, quantitative objectives and the way to achieve these objectives differ among the different GC-O methodologies. While the first group of methods measures the potency of the aroma compound by stepwise dilution of the extract, the other two measure the relative importance or the intensities of the odour in a single extract. Therefore, application of different methods may lead to a different characterization of the same sample. Recent reviews report a detailed comparison of the commonly used approaches (d'Acampora Zellner *et al.*, 2008; Delahunty *et al.*, 2006; Plutowska and Wardencki, 2008; van Ruth, 2001).

In dilution analysis, an aroma extract is successively diluted and each dilution is sensory evaluated until no odour is perceived by the assessors. Up to now, the most frequently applied method in the analysis of alcoholic beverages is AEDA. In AEDA the relative impact of the odorant is expressed as its flavour dilution factor (FD) corresponding to the maximum dilution at which that compound can be detected by the judges. A hierarchy of odour-active compounds in the extract is presented simply by listing the FD factor values or in the form of an aromagram (Fig. 5.3). Sometimes FD factors are compared to those obtained for a well-known component of the matrix, usually present at similar concentrations – close to or above its odour threshold, e.g. 3-methyl-1-butyl acetate (isoamyl acetate) in the case of beers (Lermusieau *et al.*, 2001; Callemien *et al.*, 2006). FD factor values equal or higher than that of such a reference analyte indicate organoleptically active components in beer.

Dilution to threshold methods are considered time-consuming due to the numerous dilution steps (Delahunty *et al.*, 2006). The number of panelists is very limited (usually not more than three persons) which affects method sensitivity to differences between different samples as well as the precision of the method (Eyres *et al.*, 2007; Lopez *et al.*, 2003). They are also criticized for nonconformity with the psychophysical relationship between the concentration of the compound and the intensity of its odour (Delahunty *et al.*, 2006; van Ruth, 2001).

The detection frequency method in its pure form is very simple and does not require any training of the panelists since no intensity scale is used (Pollien *et al.*, 1997). However, the lack of scale is also recognized as the main drawback of this method (d'Acampora Zellner *et al.*, 2008; Delahunty *et al.*, 2006; Plutowska and Wardencki, 2008; van Ruth, 2001). Despite the intensity of the aroma increasing with the concentration, the detection frequency remains constant and equal to 100% above the compound detection threshold. So for two compounds, one odour may be just over the sensory threshold and the second may be very intense, but both of them will be viewed as being equal by this approach. Analogously, the

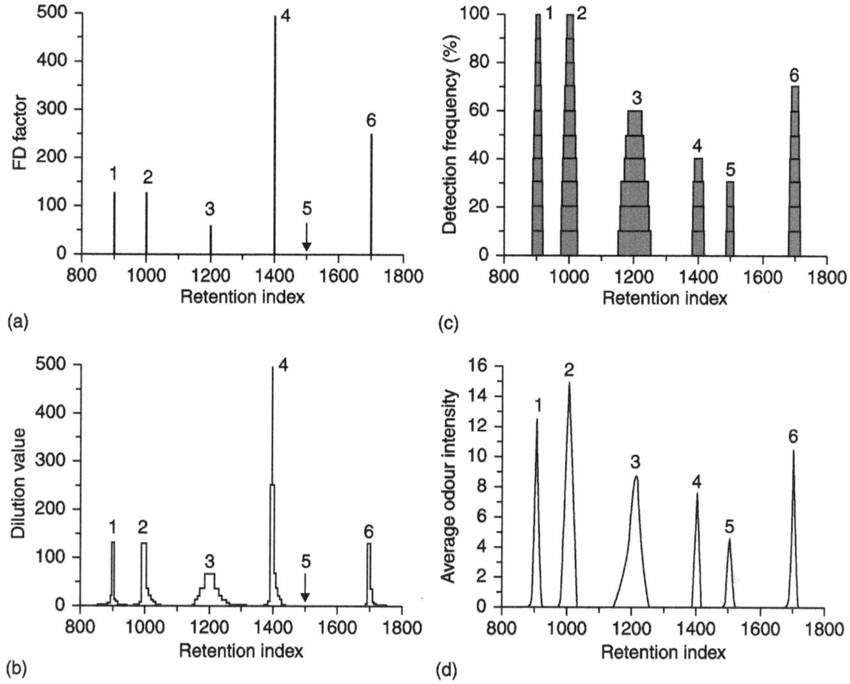

Fig. 5.3 Comparison of olfactograms simulated for six compounds using four different GC-O methods. (a) AEDA. (b) CHARM. (c) Detection frequency method. (d) OSME (adapted from Delahunty *et al.*, 2006).

detection frequency method will not discriminate between two concentrations of the same compound in two different samples if the odour is perceived by all assessors. To overcome the aforementioned disadvantage, several authors have modified the detection frequency approach, merging it with the posterior intensity measurement according to Dravnieks (Dravnieks, 1985) and used it in wine aroma studies (Escudero *et al.*, 2007; Ferreira *et al.*, 2009; Saenz-Navajas *et al.*, 2010; Torrens *et al.*, 2010).

All direct intensity methods are based on the psychophysical Steven's power law (Stevens, 1971) which explains the relationship between the perceived aroma intensity and the concentration of the compound. Among all direct intensity methods, posterior intensity procedures do not require any special equipment for data treatment. Up to now, this approach has been applied mainly in aroma and quality investigations of wine (Botelho *et al.*, 2007; Campo *et al.*, 2005; Cullere *et al.*, 2004; Gomez-Miguez *et al.*, 2007; Moyano *et al.*, 2010) and spirit liquors (Fan and Qian, 2006b).

Posterior methods assess only the maximum intensity of the aroma impression. However, an increase in the concentration in GC-O also increases the duration of the odour as the effect of the chromatographic separation process. For this reason, the most reliable results of intensity measurements can be achieved using methods based on the continuous recording of the intensity as a function of time. Data obtained in this way can be easily processed by any chromatography software and compared to a conventional detector signal. Although relatively new, time-intensity methods have become more and more popular in alcoholic beverage analysis for wines (Bernet *et al.*, 2002), beers (Callement *et al.*, 2001) or raw spirits (Plutowska and Wardencki, 2009) as well as in other aroma research areas; thus their advantages and drawbacks are well known. Unquestionably, the quality of the time-intensity methods is the quantity of information achieved.

5.2.4 Analytical conditions

Gas chromatography-olfactometry is, in fact, a combination of instrumental and sensory analysis techniques. Consequently, GC-O analytical conditions are analogous to the practices commonly used in sensory testing. For example, to minimize bias and distraction, the experiments should be conducted in a temperature- and pressure-controlled, quiet and odour-free environment. Similarly to classical sensory analysis, in GC-O tests, bias is related to the ability of a person to remember odour profiles and the ability to foresee or assume the composition and content of the odour compounds in later samples. The probability of making this type of error can be minimized by analysing the samples in random order and/or by randomly adding blank samples.

Unlike in conventional methods of sensory analysis, the person evaluating the sample has to deal with many different odours appearing for only a few seconds at irregular time intervals during a chromatographic analysis, which usually lasts tens of minutes at a time. Reduced detection sensitivity due to the tiredness of the assessors may be overcome by dividing the entire analysis into several parts, and changing the evaluator for each part (Bernet *et al.*, 2002). The variability of results within and between evaluators will be reduced if each measurement is replicated. All measurements should be performed by a group of assessors, carefully chosen and checked for olfactory abilities and specific anosmias (Berdague *et al.*, 2007; Pollien *et al.*, 1997). It is known that for some of the odour compounds commonly present in alcoholic beverages, like β-ionone, odour thresholds vary significantly within the population (Delahunty *et al.*, 2006).

Due to the specificity of GC-O where the stimuli last for only a few seconds, the detection results are strongly influenced by the assessor's respiratory cycle (Hanaoka *et al.*, 2001). Since the odours are perceived only when breathing in, some compounds may be missed during expiration, particularly when the odour is not intense and long-lasting, or when a subsequent odour appears within an insignificant time interval from the preceding one. If inspiration starts in the course of the appearance of an odour, only a part of the odour is registered and the real intensity of the odour is lowered; this can cause difficulties in the interpretation of results.

To achieve reproducible and mutually compatible results of analyses, it is indispensable that the evaluating team has some experience. Therefore, before commencing with testing real samples, appropriate training should be conducted

(Berdague *et al.*, 2007; Botelho *et al.*, 2007; Ferrari *et al.*, 2004; Ferreira *et al.*, 2001, 2002b). Training familiarizes assessors with the intensity scale, odour descriptors and the types of odours that may appear in the analyzed samples. Usually, mixtures of reference standard solutions are used for this purpose. Depending on the character of the investigation, different kinds of tests may be applied. In order to validate raw spirit differentiation by the fingerspan method, Plutowska and Wardencki (2009) evaluated panelists' reliability using two tests. During the repeatability test, standard samples of invariable composition were analyzed in replicates; thus both the areas and heights of peaks of individual aroma compounds should remain at a similar level. The second test checked the ability to discern between the intensity of aromas. This time, samples of the same composition but with varying concentrations of the tested aroma compounds were analyzed. Thus, with changing concentrations of the tested aroma compounds, the heights and surface areas of peaks ought to change. To eliminate errors that might result from changes in the sensory susceptibility of panel members in the course of investigations, a constant amount of an aroma compound hex-3-en-1-ol was always added, serving as an internal standard.

5.3 Applications of GC-O in the flavour analysis of alcoholic beverages

The volatile fraction of alcoholic beverages is a complex mixture of several hundreds of different compounds dissolved in an ethanol–water solution. It is widely known that not all of these volatiles contribute to the aroma or flavour perception. Some of them have a very high odour detection threshold and are sensorially inactive; others are odour-active but present at a concentration too low to be detected. Numerous studies show that usually only a few tens of volatile fraction components have a real impact on the bouquet of alcoholic beverages, and GC-O is an excellent screening tool to recognize them.

5.3.1 Identification of the impact aroma compounds in alcoholic beverages

The vast number of applications reported in the literature concern identification of the key aroma compounds responsible for the characteristic bouquet of various alcoholic beverages or characteristic hints and odour notes detectable by classical sensory analysis. Most studies have focused on the odour profiles of traditional alcoholic beverages typical of any given country or region, which are usually distinguished by individual aromas resulting from specific raw materials or unique technologies used during their production. A widely studied matrix is wine. Numerous examples regarding wines from different grape varieties such as Cabernet Sauvignon (Falcao *et al.*, 2008; Gurbuz *et al.*, 2006; Kotseridis and Baumes, 2000; Lopez *et al.*, 1999), Chardonnay (Lorrain *et al.*, 2006; Moio *et al.*, 1994), Devín (Petka *et al.*, 2006), Grenache (Ferreira *et al.*, 1998; Lopez *et al.*, 1999; Schneider *et al.*, 1998), Merlot (Gurbuz *et al.*, 2006; Kotseridis and Baumes,

2000; Lopez *et al.*, 1999), Pinot Noir (Fang and Qian, 2005; Miranda-Lopez *et al.*, 1992), Riesling (Chisholm *et al.*, 1994; Komes *et al.*, 2006), Scheurebe and Gewürztraminer (Guth, 1997b), Gual, Verdello, Marmajuelo, Listán and Malvasia (Lopez *et al.*, 2003), Kalecik Karasi (Selli *et al.*, 2004), Zalema (Gomez-Miguez *et al.*, 2007), Maccabeo (Escudero *et al.*, 2004), Petite Arvine (Fretz *et al.*, 2005) as well as wines produced in different regions of the world such as Bordeaux (Kotseridis and Baumes, 2000; Moio *et al.*, 1994), Sauternes (Bailly *et al.*, 2006) and Champagne (Escudero *et al.*, 2000b) from France, Port from Portugal (Silva Ferreira *et al.*, 2003), Rioja from Spain (Aznar *et al.*, 2001), Sauternes from France (Bailly *et al.*, 2006), Tannat from Uruguay (Boido *et al.*, 2003; Escudero *et al.*, 2007) are available. These studies often demonstrate that the differences in bouquet composition between wines are quantitative rather than qualitative, since regardless of the grape variety or the production conditions, the main and common part of the wine flavour is always derived from the mixture of the well-known grape-derived constituents and the fermentation by-products such as esters, acids, alcohols, lactones, volatile phenols, etc. However, in many wine samples, unusual aroma profiles can be explained by relatively few aroma compounds (Table 5.1).

Apart from wines, the application of GC–O to alcoholic beverages enabled the characterization of the impact flavour compounds of other matrices, including beers (Fritsch and Schieberle, 2005), spirit-based beverages such as: American Bourbon whisky (Poisson and Schieberle, 2008), Slovak brandies (Janacova *et al.*, 2008), Calvados (Ledauphin *et al.*, 2003), Cognac (Ferrari *et al.*, 2004; Lablanquie *et al.*, 2002), tequila (Benn and Peppard, 1996), Chinese Yanghe Daqu (Fan and Qian, 2005, 2006b) and Wuliangye and Jiannanchun (Fan and Qian, 2006a) as well as some unique liquors like palm wine (Lasekan *et al.*, 2007), orange wine (Selli *et al.*, 2008) and apple cider (Xu *et al.*, 2007).

Extensive research has been dedicated to the strong odour-active components responsible for the typical nuances experienced during classical sensory evaluation (Fan and Qian, 2006a). Wanikawa *et al.* (2002) conducted interesting research on the green note compounds in malt whisky using a multidimensional GC coupled with mass spectrometry and olfactometry (GCxGC-MS-O). Three alcohols and two aldehydes probably related to lipid oxidation and degradation were found to contribute to the green scent, namely 1-octen-3-ol, 4-hepten-1-ol and nonan-2-ol, and E,Z-2,6-nonenal and E-2-nonenal respectively. In the other study, diacetyl was identified as a potent contributor to the sweet caramel descriptor often associated with the aroma of port wine (Rogerson *et al.*, 2001). On the other hand, de Freitas *et al.* (1999) assigned 2,2,6-trimethylcyclohexanone and ethyl dihydrocinnamate to the balsamic sweet, rock-rose (*Cistus ladaniferus*) aroma described frequently in young port wines; and da Silva Ferreira and collaborators found the *trans*-5-hydroxy-2-methyl-1,3-dioxane mainly responsible for the typical sweet old port wine aroma (da Silva Ferreira *et al.*, 2002).

Another related objective of GC-O analysis is to discover new significant odour-active components present in a beverage at trace levels, which cannot be easily detected using conventional detectors due to their insufficient sensitivity. In this case, a preliminary olfactometric screening can help determine the retention

Table 5.1 Wine aroma profiles and odorant compounds

Wine type	Key odorants	Odour description	Reference
Chardonnay wine	Ethyl butanoate	Fruity, strawberry	Lorrain *et al.*, 2006
	Octanoic acid	Animal, spicy	
	Phenylacetaldehyde	Floral, spicy	
	4-Vinylphenol	Spicy, pharmaceutical	
	2-Methyltetrahydrothiophen-3-one	Gas, diesel oil	
	δ-Decalactone	Coconut, floral	
	3-Methylbutyl acetate	Banana	
	Decanoic acid	Vinegar, animal	
	4-Vinyl-2-methoxyphenol	Spicy, pharmaceutical	
	Linalool	Floral, burnt	
Cabernet Sauvignon wine	Acetic acid	Vinegar, solvent	Falcao *et al.*, 2008
	Butyric acid	Cheese, rancid	
	Isovalerianic acid	Dirty socks, old cheese	
	2-Phenylethanol	Rose, flowery	
	Methional	Cooked potato, soup	
	2-Methoxy-3-isobutylpyrazine	Vegetative, bell pepper	
	β-Damascenone	Peach, canned apple	
	β-Ionone	Violet, flowery	
	Furaneol	Jam, cooked red fruits, caramel	

(Continued overleaf)

Table 5.1 Continued

Wine type	Key odorants	Odour description	Reference
Riesling wine	2-Phenylethanol	Flowery, rose	Komes *et al.*, 2006
	3-Methyl-1-butanol	Chemical, harsh, stale	
	3-(Methylthio)-1-propanol	Potato	
	Ethyl propanoate	Fruity	
	Ethyl butanoate	Fruity, aromatic	
	Ethyl 3-methylbutanoate	Fruity, sweet-like	
	3-Methyl-1-butanol acetate	Fruity, sweet-like, banana	
	Ethyl hexanoate	Fruity, candy	
	Ethyl octanoate	Fruity, fresh, sweet-like	
	Ethyl 3-hydroxybutanoate	Fresh, fruity	
	2-Phenylethyl acetate + β-Damascenone	Flowery, fruity, cooked apple, marmalade	
	Hexanoic acid	Unpleasant, rancid, metallic	
	3-Methylbutanoic acid	Cheese	
	Butanoic acid,	Cheese	
	γ-Undecalactone	Creamy, fruity, sweet-like, nutty	
	4-Vinylguaiacol	Smoky, spicy, smoked meat	
Bordeaux wines (Cabernet Sauvignon and Merlot grapes)	2-/3-Methylbutanols	Nail polish	Kotseridis and Baumes, 2000
	2-Phenylethanol	Rose	
	2-Methyl-3-sulfanylfuran	Sunflower seeds	

	Acetic acid	Vinegar	
	3-(Methylsulfanyl)propanal	Baked potato	
	2-/3-Methylbutanoic acids	Parmesan cheese	
	β-Damascenone	Canned apple	
	3-Sulfanylhexan-1-ol	Grapefruit	
	Furaneol	Caramel	
	Homofuraneol	Caramel	
Palm wine	3-Isobutyl-2-methoxypyrazine	Earthy	Lasekan *et al.*, 2007
	Acetoin	Buttery	
	Ethyl hexanoate	Fruity	
	3-Methylbutyl acetate	Banana	
	2-Acetyl-1-pyrroline	Popcorn	
Orange wine	Ethyl butanoate	Fruity, sweet	Selli *et al.*, 2008
	3-Methyl-1-pentanol	Roasty	
	Linalool	Floral, citrus	
	γ-Butyrolactone	Cheese, burnt sugar	
	3-(Methylthio)-propanol	Boiled potato, rubber	
	Geraniol	Floral, citrus	
	2-Phenylethanol	Floral, rose	

time of a given analyte for the purpose of its selective enrichment and identification. One method is cryo-trapping the eluate in a selected range of retention times in an empty deactivated capillary using liquid nitrogen as the cooling agent (Callemien *et al.*, 2006). A stream splitter allowing for the electronic switching of the eluate flow in the trap allows for selective enrichment of the unidentified components of beer aroma responsible for the undesirable phenolic–tobacco odour created during the ageing of beer. Interestingly, analyte identification by mass spectrometry became possible only after capturing the fraction from thirty consecutive chromatographic runs, while the olfactometric detection did not require any additional pre-concentration of the extract.

A second possibility is to look at the aroma profile of another food product, plant or any material whose odour properties associate with the smell of the unknown compound (Fretz *et al.*, 2005). Ong and Acree (1999) identified the key aroma compound of Gewürztraminer wine recognizable by its specific bouquet reminiscent of the tropical lychee fruit. GC-O analysis of the fruit extract revealed that *cis*-rose oxide is a common potent odorant in both matrices. What is more, the absence of the oxide in other wine samples confirmed the possibility of using this compound as a marker of the intensity of the Gewürztraminer character of wine in viticultural and enological practice.

Each product has a characteristic and unique pattern of aroma components, a so-called aroma profile, aromagram or olfactogram. As presented above, the application of the GC-O technique allows a detailed characterization of the aroma profile of an alcoholic beverage. Furthermore, a comparison of aromagrams obtained from different samples can indicate differences between products of different origins or differing in quality. In such an approach, the information is provided not only by single compounds, but also by mutually qualitative and quantitative relations taking place between odour-active components (Plutowska and Wardencki, 2007).

5.3.2 GC-O in reconstitution studies

An important feature of GC-O analysis is that the odorants are chromatographically separated and detected individually. This approach does not take into account the interactions between the compounds such as the additive, synergistic or suppressive effects that normally occur in the mixture. For this reason GC-O is often considered as an initial stage of the broader studies used for the identification and selection of the most potentially active components for further investigations. Dilution to threshold methods are usually involved in reconstitution studies where the compounds revealed by the GC-O methodology are first quantified, hierarchically arranged and then used to mimic the original product. Subsequently, classical sensory analysis methods are utilized in order to evaluate the degree of compliance between the original sample and the reconstituted one.

The most common approach relies on determining the individual odour activity values (OAV) or flavour dilution factor (FD) values for each identified compound and reconstructing the fragrance using substances with the highest OAV or FD values (Ferreira *et al.*, 2002b; Lorrain *et al.*, 2006). Quite often, a satisfactory simulation of the real flavour comes from a few compounds with the highest sensory activity. However, oversimplification can be disadvantageous. For example, Ferreira *et al.* (2002b) found that the greatest similarity to the original Spanish Grenache wine showed models based on mixtures of odour extracts of OAV values above 0.5, while the addition of compounds with OAV values below 0.5 had practically no influence on the obtained aroma. On the other hand, mixtures made exclusively from compounds with OAV values over 10 had completely different sensory properties from the original samples. Other studies indicated that the specificity of the aroma of a given alcoholic beverage can be determined by individual components with relatively low OAVs, but characterized by a very different smell from the rest of compounds (Escudero *et al.*, 2004).

To identify key aroma compounds, various tests are used, such as omission tests or addition tests (Ferreira *et al.*, 2002b; Guth, 1997b), which assess the changes in odour that occur as a result of the omission of individual compounds in a reproduced composition, or changes due to the addition of individual odour-active substances (Escudero *et al.*, 2004). Only tests aiming at the reconstruction of the actual aroma composition can yield information on the effects of the summation or mutual buffering of compounds included in the aroma of alcoholic beverages (Lorrain *et al.*, 2006).

5.3.3 Application in the production of alcoholic beverages

Investigation of the influence of raw materials on the development of aroma

A low quality of alcoholic beverages often results from low-quality raw materials used in their production. Fermentation of rotting or mouldy fruits or grains results in low organoleptic quality as a result of the presence of degradation products in the raw materials, as well as undesirable microbial metabolites. Mould development is particularly undesirable because of the earthy, fungi or musty smells that appear in the aroma. These smells are caused by compounds that usually have very low detection thresholds, such as geosmine, 2-methyl isoborneol, 1-octen-3-one or chloroanisoles (Darriet *et al.*, 2000; Plutowska and Wardencki, 2009). The possibility of discovering and identifying these compounds by GC-O during the early stages of raw material rot, as well as during fermentation or in a finished product is very important. Current studies aim not only at the identification of sensory-active metabolites of undesirable microorganisms appearing in alcoholic beverages, but also at linking lower organoleptic quality with the composition of the microflora, which could help avoid the development of fungi and other microorganisms during raw material storage and during fermentation (Darriet *et al.*, 2002; Guerche *et al.*, 2006).

The flavour of wines strongly depends on the specific kind of grapes used for fermentation (Sarrazin *et al.*, 2007b; Serot *et al.*, 2001; Souid *et al.*, 2007). Using GC-O, Genovese *et al.* (2007) investigated the influence of over-ripeness of grapes on the volatile composition and the quality of wine. Takoi *et al.* (2009)

discovered that several volatile thiols contributed to the cultivar-specific flavour of Sauvignon Blanc wines. Studies of grape and wine were recently paid wild attention by experts in China (Fan *et al.*, 2010).

In beers, typical odour-active compounds may be derived from hops (*Humulus luplus* L.), indispensable components used in brewing. Various hop cultivars have different aroma nuances, and beers brewed with certain hops have cultivar-specific flavours (Takoi *et al.*, 2009). Characteristic compounds such as linalool, γ-nonalactone, humuladienone, ethyl cinnamate and sulfides were formed during the boiling and fermentation portion of the brewing process (Lermusieau *et al.*, 2001). It was found that 4-mercapto-4-methyl pentane-2-one was the main contributor to the fruity aroma in beer hopped with some US cultivars (Kishimoto *et al.*, 2008).

Application in the production process

The quantitative and qualitative composition of active-odour compounds formed during the production of alcoholic beverages depends strictly on the conditions of the process, and is the result of enzymatic, microbiological and/or thermal processes. These processes may imply a large change in the composition of the final product, including the appearance of new odorants, which may impact the aroma of different alcoholic beverages.

GC-O can be used to study the influence of individual factors, such as the type of yeast used for fermentation. Studies indicate that different yeast varieties have different abilities to produce and transform various odour compounds in wines, and that, other than the type and maturity of grapes (Ebeler *et al.*, 2000), they determine the final odour of the wine (Comuzzo *et al.*, 2007; Delfini *et al.*, 2001; Gomez-Miguez *et al.*, 2007).

Gas chromatography with olfactometric detection was also used to study the formation of the aroma profile of alcoholic beverages providing important information for the improvement and optimization of the operational conditions (harvest parameters, fermentation techniques, use of yeasts, bacteria and enzymes, etc.) (Campo *et al.*, 2008; Falcao *et al.*, 2008; Gomez-Miguez *et al.*, 2007; Kechagia *et al.*, 2008). Such studies can be especially useful during the pre-production stages of new beverages made of previously unused raw materials, such as apple cashew, which thus far has been considered waste a by-product of the popular nut (Garruti *et al.*, 2006). Poisson and Schieberle characterized the most odour-active compounds in an American Bourbon whisky to reveal the influence of each technological step on the aroma compound formation during whisky processing (Poisson and Schieberle, 2008). The influence of pre-fermentative treatments (effects of skin contact and the must clarification process) on the major aroma compounds of Assyrtico wines have been investigated by Kechagia *et al.* (2008).

Determination of volatile odour substances using GC-O can be used to control and optimize the technological processes. For example, the determination of an increasing content of aldehydes, such as methional or phenylacetaldehyde, formed as a result of the degradation of amino acids according to Strecker's mechanism during the maturation and storage of beer, can be an indicator of ageing and the decreasing of its organoleptic quality (Soares da Costa *et al.*, 2004). Gijs *et al.* (2002) found that the determination of the level of odour compounds such as *β*-damascenone or dimethyl trisulfide can be a helpful indicator in controlling and optimizing pH during beer brewing, as the dependence between the pH of beer and the production of these compounds during brewing has been established. Studies of a wider scope were conducted by Guyot-Declerck *et al.* (2005) regarding the influence of the maturation time and the pH of beer on its organoleptic properties. These authors stressed the dependence of the concentrations of other compounds, such as *trans*-2-nonenal and methional, on the pH of maturing beer. They also claimed that the intensity of dimethyl trisulfide odour increases during maturation regardless of pH, and that the *β*-damascenone odour does not depend on the pH or maturation time. Oliveira e Silva *et al.* (2008) investigated the impact of a forced-ageing process on Madeira wine to determine the optimal temperature and baking time. It was found that sotolon is the compound with the highest impact on typical Madeira wine flavour.

Another example is the determination of volatile products of Maillard non-enzymatic browning reactions in oak wood extracts coming from barrels used for the fermentation and storage of high-quality wines and spirits (Cutzach *et al.*, 1997; Oliveira e Silva *et al.*, 2008). Odour compounds formed during the thermal processing of wood can influence the aroma both positively and negatively; therefore, barrel production conditions can have a significant influence on the quality of the finished product.

Monitoring of the ageing process and storage conditions

A significant problem, especially in southern countries, is premature degradation of wine aromas because of oxidation, which in these regions proceeds faster as a result of higher temperatures, pH and enzymatic activity in fruits during grape collection and the production of wine. From the point of view of organoleptic properties, one observes in these cases lowered impressions of freshness and fruitiness, accompanied by the appearance of odours resembling paper, cooked, rancid or rotten food, as well as the smell of fodder, hay or wood. Such phenomena might shorten the expiry date, which is why the technique discussed was used to analyze changes in odour profiles of young, white wines, related to the processes of oxidation (Escudero *et al.*, 2000a). The emergence of 19 compounds likely to have a significant influence on sensory changes in wine aromas, as well as the disappearance or lowering of the concentration of 15 desirable aroma components, were observed in the study. The effect of individual odour components was determined using AEDA. This method was also used to identify undesirable odour compounds related to oxidation changes in white Portuguese wines (Silva Ferreira *et al.*, 2003). The authors, basing their analysis on an odour comparison of spoiled wine and samples of wine spiked with the analyzed compounds, identified three compounds, amongst which 3-methylthiopropionic aldehyde was selected as the most characteristic indicator of wines spoiled as a result of oxidation.

5.3.4 Quality evaluation of alcoholic beverages

For alcoholic beverages, especially high-quality wines and beers, an important prerequisite for marketing and commercial consideration is that the products are free of imperfections such as taints and off-odours. Undesirable flavours in alcoholic beverages can be derived from the cultivar, formed during fermentation and by microbiological spoilage, during maturation in barrels or bottles. The problem with the identification of undesirable odorants in alcoholic beverages is the fact that off-flavours are often present at trace levels or even below the detection limits of chemical detector concentrations. GC-O is a powerful tool in quality control to identify the compounds responsible for foreign taints or off-flavours and to determine their cause. The analytical approach for determining the odorants responsible for an off-flavour involves analyzing the compounds isolated from the complex aroma mixture by GC and smelling the column effluent. Ideally, aroma isolation procedures should yield an isolate sufficiently concentrated for GC-MS identification of the tainting compound. If there is no peak on the chromatogram but the offending aroma is smelled, then the isolation procedure needs to be changed, or the sample increased.

The feasibility of GC-O was demonstrated for quality control to identify compounds that caused external taints or off-flavours in wine and beer (Grosch, 1994; Murakami *et al.*, 2003). With the aid of this technique, components causing undesirable flavours can be found in the complex mixture of many hundreds of aroma compounds. One of the most significant organoleptic defects in wines is reported as 'cork taint' and is related to a moldy/musty note (Cullere *et al.*, 2009; Simpson *et al.*, 2004). The list of possible compounds producing unpleasant aromas in cork is relatively large, but many reports showed that the highest incidence of musty, earthy odour are due to 2,4,6-trichloroanisole (TCA), 2-methoxy-3,5-dimethylpirazine (MDMP) responsible for fungal must and 2,4,6-tribromoanisole (TBA). Other compounds, such as 1-octen-3-ol, 1-octen-3-one (mushroom aroma), guaiacol (smoky, phenolic aroma) and geosmin (trans-1,10-dimethyl-trans-9-decanol) and 2-methylisoborneol (earthy aroma) have also been reported (Prat *et al.*, 2009). Disinfectant-like or swimming pool-like taints have been recognized in some wines, and chlorophenols have been associated with this particular taint (Capone *et al.*, 2010).

Sulphur compounds are another group of compounds responsible for an unpleasant aroma from alcoholic beverages. Thiols are most likely involved in the aroma of fresh beer (Vermeulen *et al.*, 2006). Polyfunctional thiols have been mentioned in the brewing literature, e.g. 3-methyl-2-buten-1-thiol (MBT) inducing a repulsive skunky odour (Lermusieau and Collin, 2003).

5.3.5 Investigation of the organoleptic properties of alcoholic beverages

An important feature of alcoholic beverages is their sensory perception during drinking. GC-O was applied for the investigation of the intensity and duration of odour compounds during and after the testing and consumption of alcoholic beverages. Immediate aroma perception is received when a beverage is present in the oral cavity or has just been swallowed. After swallowing, a prolonged retronasal aroma perception is perceived; often called the after-taste, after-odour or after-smell.

Tamura *et al.* (2009) has evaluated the fishy aftertaste that is sometimes generated in wine and seafood pairing. A formation mechanism for this taste was proposed. Ferrous ion in wine can instantly promote the formation of hexanal, heptanal, 1-octen-3-one, (E,Z)-heptadienal, nonanal and decanal by the breakdown of preformed lipid hydroperoxides derived from unsaturated fatty acids in fish and seafood.

A novel approach for the identification of potent odorants intra-orally and the determination of their dwell time within the oral cavity, called the buccal odour screening system (BOSS), based on modified stir-bar sorptive extraction (SBSE) was developed (Buettner and Welle, 2004). A comparison of the results of olfactometric analysis of the extracts of two Chardonnay wines, obtained using BOSS analysis upon extracts obtained by dichloromethane extraction and vacuum distillation, indicated that most of the compounds comprising wine aromas, except the most polar ones which interact with saliva, were responsible for the emergence of the after-taste. However, the holding times of the individual compounds in the oral cavities of the evaluators differed significantly (Buettner, 2004).

In similar studies on the aroma of a white wine made with Devín grapes, a specially designed extractor imitating the oral cavity, based on the purge-and-trap principle, was used for the isolation of compounds taking part in retronasal perception (Petka *et al.*, 2006). The sorption trap was placed at the top of a flask containing a mixture of wine and artificial saliva, through which a stream of nitrogen was bubbled.

5.4 Conclusions

The introduction of the combination of olfactometry and gas chromatography into food analysis was a breakthrough in aroma research, enabling the differentiation of a multitude of volatiles in odour-active and non-odour active compounds, related to their existing concentrations in the matrix under investigation. Over the years, many attempts have been made to identify beverage odorant profiles. Knowledge concerning an aroma profile, especially the odorant compounds responsible for the characteristic aroma of the alcoholic beverages, is extremely important for better understanding of the differences between them. Therefore, many papers, presenting different techniques for designing GC-O experiments, recording the data and estimating the sensory combination of aroma compounds, have been published. The general aspects of food aroma analysis with using GC-O techniques may be found in several books (Blank, 1996; Deibler *et al.*, 1999; Ebeler, 1999; Grosch, 2007; Leland *et al.*, 2001) and in the reviews published in journals (d'Acampora Zellner *et al.*, 2008; Delahunty *et al.*, 2006; Plutowska and Wardencki, 2008; van Ruth, 2001).

Many factors influence the correct detection and assessment of odour active compounds using GC-O technique, i.e. the extraction procedure, the method of data collection and the separation capability of the gas chromatographic column. The choice of an appropriate sample preparation method becomes crucial because the aroma profile is closely related to the isolation procedure, which should yield a product that is representative of the sample. Many extraction techniques have been developed to assure representative extracts, but the best way to attain an optimum recovery of the flavour chemicals and getting a complete reproduction of a flavour's profile is the employment of more than one extraction technique. The choice of a GC-O data measurement method is of extreme importance for the correct characterization of a matrix, since the application of different methods to an identical real sample can be distinctively selective and rank the odour-active compounds according to their potency and/or intensity.

A limitation of GC-O is the co-elution of odour-active compounds, which results in the perception of odour clusters. Using a different stationary phase may allow the intensity and quality of the odorants to be evaluated individually without co-elution. Another possible solution for identifying active odours when co-elution occurs is to use comprehensive multidimensional gas chromatography (GCxGC). The application of this technique in conjunction with the identification capabilities of mass spectrometry facilitates the identification of character-impact odorants in very complex samples.

In spite of the fact that GC-O is quite a mature technique, used for almost 50 years, further investigations are still being conducted in order to improve its capability, i.e. to achieve a higher sensitivity and better repeatability of results. These aims may be principally realized by applying a broader range of sample preparation techniques, particularly so-called solventless ones. These techniques are preferred because they combine extraction and pre-concentration into one step, do not require initial sample preparation and, unlike solvent techniques, require only small amounts of the sample. An important advantage of such an approach is that they are preservative techniques, allowing preservation of natural composition and sample characteristics.

Another perceptible tendency in the development of GC-O is aimed at development of the highest repeatability and reliability of the obtained results by unifying, simplifying and shortening the applied procedure. It can be done by improving instrument use in olfactometric detection, e.g. by the application of a specially cooled interface between two chromatographic columns, increasing chromatographic selectivity (Begnaud *et al.*, 2006). Better repeatability and reliability can also be obtained by using simultaneous detection by several parallel connected olfactometers, performed at the same time by several investigators.

5.5 References

ACREE T E and BARNARD J (1984), 'The analysis of odor-active volatiles in gas chromatographic effluents', in Schreier P (ed), *Analysis of Volatiles. Methods and Applications*, Walter de Gruyter, Berlin, 251–267.

ACREE T E, BARNARD J and CUNNINGHAM D G (1984), 'A procedure for the sensory analysis of gas chromatographic effluents', *Food Chem*, **14**, 273–286.

AZNAR M, LOPEZ R, CACHO J F and FERREIRA V (2001), 'Identification and quantification of impact odorants of aged red wines from Rioja. GC–olfactometry, quantitative GC–MS, and odor evaluation of HPLC fractions', *J Agric Food Chem*, **49**, 2924–2929.

BAILLY S, JERKOVIC V, MARCHAND-BRYNAERT J and COLLIN S (2006), 'Aroma extraction dilution analysis of sauternes wines. Key role of polyfunctional thiols', *J Agric Food Chem*, **54**, 7227–7234.

BARBE J-C, PINEAUB B and SILVA FERREIRA A C (2008), 'Instrumental and sensory approaches for the characterization of compounds responsible for wine aroma', *Chem Biodivers*, **5**, 1170–1183.

BEGNAUD F, STARKENMANN C, WAAL M and CHAINTREAU A (2006), 'Chiral multidimensional gas chromatography (MDGC) and chiral GC-olfactometry with a double-cool-strand interface: application to malodors', *Chem Biodiv*, **3**, 150–160.

BENN S M and PEPPARD T L (1996), 'Characterization of tequila flavor by instrumental and sensory analysis', *J Agric Food Chem*, **44**, 557–566.

BERDAGUE J L, TOURNAYRE P and CAMBOU S (2007), 'Novel multi-gas chromatography–olfactometry device and software for the identification of odour-active compounds', *J Chromatogr A*, **1146**, 85–92.

BERNET C, DIRNINGER N, CLAUDEL P, ETIEVANT P and SCHAEFFER A (2002), 'Application of finger span cross modality matching method (FSCM) by naïve assessors for olfactometric discrimination of Gewurztraminer wines', *Lebensm -Wiss u -Technol*, **35**, 244–253.

BLANK I (1996), 'Gas chromatography-olfactometry in food aroma analysis', in Marsili R (ed), *Techniques for Analyzing Food Aroma*, Marcel Dekker, New York, 293–329.

BOIDO E, LLORET A, MEDINA K, FARINA L, CARRAU F *et al.* (2003), 'Aroma composition of *Vitis vinifera* Cv. Tannat: the typical red wine from Uruguay', *J Agric Food Chem*, **51**, 5408–5413.

BONINO M, SCHELLINO R, RIZZI C, AIGOTTI R, DELFINI C *et al.* (2003), 'Aroma compounds of an Italian wine (Ruché) by HS-SPME analysis coupled with GC-ITMS', *Food Chem*, **80**, 125–133.

BOTELHO G, CALDEIRA I, MENDES-FAIA A and CLIMACO M C (2007), 'Evaluation of two quantitative gas chromatography–olfactometry methods for clonal red wines differentiation', *Flavour Fragr J*, **22**, 414–420.

BUETTNER A (2004), 'Investigation of potent odorants and afterodor development in two Chardonnay wines using the buccal odor screening system (BOSS)', *J Agric Food Chem*, **52**, 2339–2346.

BUETTNER A and WELLE F (2004), 'Intra-oral detection of potent odorants using a modified stir-bar sorptive extraction system in combination with HRGC-O, known as the buccal odour screening system (BOSS)', *Flavour Fragr J*, **19**, 505–514.

CALLEMENT G, BOCHET M, LANGLOIS D, ETIEVANT P X and SALLES C (2001), 'Odor intensity measurements in gas chromatography–olfactometry using cross modality matching: evaluation of training effects', in Leland J V, Schieberle P, Buettner A and Acree T E (eds), *Gas Chromatography–Olfactometry: The State of the Art*, ACS Symposium Series, 782, Washington, DC: American Chemical Society, 172–186.

CALLEMIEN D, DASNOY S and COLLIN S (2006), 'Identification of a stale-beer-like odorant in extracts of naturally aged beer', *J Agric Food Chem*, **54**, 1409–1413.

CAMPO E, CACHO J and FERREIRA V (2008), 'The chemical characterization of the aroma of dessert and sparkling white wines (Pedro Ximenez, Fino, Sauternes, and Cava) by gas chromatography-olfactometry and chemical quantitative analysis', *J Agric Food Chem*, **56**, 2477–2484.

CAMPO E, FERREIRA V, ESCUDERO A and CACHO J (2005), 'Prediction of the wine sensory properties related to grape variety from dynamic–headspace gas chromatography–olfactometry data', *J Agric Food Chem*, **53**, 5682–5690.

CAPONE D L, VAN LEEUWEN K A, PARDON K H, DANIEL M A, ELSEY G M *et al.* (2010), 'Identification and analysis of 2-chloro-6-methylphenol, 2,6-dichlorophenol and indole: causes of taints and off-flavours in wines', *Australian J Grape Wine Res*, **16**, 210–217.

CASIMIR D J and WHITFIELD F B (1978), 'Flavour impact values: a new concept for assigning numerical values for the potency of individual flavour components', in *Proceedings of the Symposium 'Aromastoffe in Fruchten und Fruchtsaften'*, Switzerland, 325–347.

CHISHOLM M G, GUIHER L A, VONAH T M and BEAUMONT J L (1994), 'Comparison of some French-American hybrid wines with white Riesling using gas chromatography-olfactometry', *Am J Enol Viticult*, **45**, 201–212.

COMUZZO P, TAT L, TONIZZO A and BATTISTUTTA F (2006), 'Yeast derivatives (extracts and autolysates) in winemaking: Release of volatile compounds and effects on wine aroma volatility', *Food Chem*, **99**, 217–230.

CULLERE L, AZNAR M, CACHO J and FERREIRA V (2003), 'Fast fractionation of complex organic extracts by normal-phase chromatography on a solid-phase extraction polymeric sorbent. Optimization of a method to fractionate wine flavour extracts', *J Chromatogr A*, **1017**, 17–26.

CULLERE L, CACHO J and FERREIRA V (2009), 'Comparative study of the aromatic profile of different kinds of wine cork stoppers', *Food Chem*, **112**, 381–387.

CULLERE L, ESCUDERO A, CACHO J and FERREIRA V (2004), 'Gas chromatography–olfactometry and chemical quantitative study of the aroma of six premium quality Spanish aged red wines', *J Agric Food Chem*, **52**, 1653–1660.

CULLERE L, ESCUDERO A, PEREZ-TRUJILLO J P, CACHO J and FERREIRA V (2008), '2-Methyl-3-(methyldithio)furan: A new odorant identified in different monovarietal red wines from the Canary Islands and aromatic profile of these wines', *J Food Compos Anal*, **21**, 708–715.

CUTZACH I, CHATONNET P, HENRY R and DUBORDIEU D (1997), 'Identification of volatile compounds with a "toasty" aroma in heated oak used in barrelmaking', *J Agric Food Chem*, **45**, 2217–2224.

D'ACAMPORA ZELLNER B, DUGOB P, DUGOA G and MONDELLO L (2008), 'Gas chromatography–olfactometry in food flavour analysis', *J Chromatogr A*, **1186**, 123–143.

DA SILVA FERREIRA A C, BARBE J -C and BERTRAND A (2002), 'Heterocyclic acetals from glycerol and acetaldehyde in Port wines: evolution with aging', *J Agric Food Chem*, **50**, 2560–2564.

DARRIET P, PONS M, HENRY R, DUMONT O, FINDELING V *et al.* (2002), 'Impact odorants contributing to the fungus type aroma from grape berries contaminated by powdery mildew (*Uncinula necator*); incidence of enzymatic activities of the yeast *Saccharomyces cerevisiae*', *J Agric Food Chem*, **50**, 3277–3282.

DARRIET P, PONS M, LAMY S and DUBORDIEAU D (2000), 'Identification and quantification of geosmin, an earthy odorant contaminating wines', *J Agric Food Chem*, **48**, 4835–4838.

DE FREITAS V A P, RAMALHO P S, AZEVEDO Z and MACEDO A (1999), 'Identification of some volatile descriptors of the rock-rose-like aroma of fortified red wines from Douro Demarcated region', *J Agric Food Chem*, **47**, 4327–4331.

DEBONNEVILLE C, ORSIER B, FLAMENT I and CHAINTREAU A (2002), 'Improved hardware and software for quick gas chromatography–olfactometry using CHARM and GC-'SNIF' analysis', *Anal Chem*, **74**, 2345–2351.

DEIBLER K D, ACREE T E and LAVIN E H (1999), 'Gas chromatography–olfactometry (GC/O) of vapor phases', in: Teranishi R, Wick E L and Hornstein I (eds), *Flavor Chemistry:30 Years of Progress*, Kluwer Academic/Plenum Publishers, New York, 387–396.

DELAHUNTY C M, EYRES G and DUFOUR J-P (2006), 'Gas chromatography-olfactometry', *J Sep Sci*, **29**, 2107–2125.

DELFINI C, COCITO C, BONINO M, SCHELLINO R, GAIA P and BAIOCCHI C (2001), 'Definitive evidence for the actual contribution of yeast in the transformation of neutral precursors of grape aromas', *J Agric Food Chem*, **49**, 5397–5408.

DRAVNIEKS A (1985), *Atlas of Odor Character Profiles*, ASTM, Philadelphia, 354–370.

EBELER S E (1999), 'Linking flavor chemistry to sensory analysis in wine', in Teranishi R, Wick E L and Hornstein I (eds), *Flavor Chemistry:30 Years of Progress*, Kluwer Academic/Plenum Publishers, New York, 409–422.

EBELER S, TERRIEN M B and BUTZKE C E (2000), 'Analysis of brandy aroma by solid-phase microextraction and liquid–liquid extraction', *J Sci Food Agric*, **80**, 625–630.
ESCUDERO A and ETIEVANT P (1999), 'Effect of antioxidants on the flavor characteristics and the gas chromatography/olfactometry profiles of Champagne extracts', *J Agric Food Chem*, **47**, 3303–3308.
ESCUDERO A, CACHO J and FERREIRA V (2000a), 'Isolation and identification of odorants generated in wine during its oxidation: a gas chromatography-olfactometric study', *Eur Food Res Technol*, **211**, 105–110.
ESCUDERO A, CAMPO E, FARINA L, CACHO J and FERREIRA V (2007), 'Analytical characterization of the aroma of five premium red wines. Insights into the role of odor families and the concept of fruitiness of wines', *J Agric Food Chem*, **55**, 4501–4510.
ESCUDERO A, CHARPENTIER M and ETIEVANT P X (2000), 'Characterisation of aged champagne wine aroma by GC-O and descriptive profile analysis', *Sci Aliments*, **20**, 331–346.
ESCUDERO A, GOGORZA B, MELUS M A, ORTIN N, CACHO J *et al.* (2004), 'Characterization of the aroma of a wine from Maccabeo. Key role played by compounds with low odor activity values', *J Agric Food Chem*, **52**, 3516–3524.
ESCUDERO A, HERNANDEZ-ORTE P, CACHO J and FERREIRA V (2000b), 'Clues about the role of methional as character impact odorant of some oxidized wines', *J Agric Food Chem*, **48**, 4268–4272.
ETIEVANT P X, CALLEMENT G, LANGLOIS D, ISSANCHOU S and COQUIBUS N (1999), 'Odor intensity evaluation in gas chromatography-olfactometry by finger-span method', *J Agric Food Chem*, **47**, 1673–1680.
EYRES G T, MARRIOTT P J and DUFOUR J–P (2007), 'Comparison of odor-active compounds in the spicy fraction of hop (*Humulus lupulus* L.). Essential oil from four different varieties', *J Agric Food Chem*, **55**, 6252–6261.
FALCAO L D, DE REVEL G, ROSIER J P and BORDIGNON-LUIZ M T (2008), 'Aroma impact components of Brazilian Cabernet Sauvignon wines using detection frequency analysis (GC–olfactometry)', *Food Chem*, **107**, 497–505.
FALQUE LOPEZ E and FERNANDEZ GOMEZ E (2000), 'Comparison of solvents for determination of monoterpenes in wine using liquid–liquid extraction', *Chromatogr*, **52**, 798–802.
FAN W and QIAN M C (2005), 'Headspace solid phase microextraction and gas chromatography-olfactometry dilution analysis of young and aged Chinese "Yanghe Daqu" liquors', *J Agric Food Chem*, **53**, 7931–7938.
FAN W and QIAN M C (2006a), 'Characterization of aroma compounds of Chinese "Wuliangye" and "Jiannanchun" liquors by aroma extract dilution analysis', *J Agric Food Chem*, **54**, 2695–2704.
FAN W and QIAN M C (2006b), 'Identification of aroma compounds in Chinese 'Yanghe Daqu' liquor by normal phase chromatography fractionation followed by gas chromatography/olfactometry', *Flavour Fragr J*, **21**, 333–342.
FAN W, XU Y, JIANG W and LI J (2010), 'Identification and quantification of impact aroma compounds in 4 nonfloral *Vitis vinifera* varieties grapes', *J Food Sci*, **75**, 81–88.
FANG Y and QIAN, M (2005), 'Aroma compounds in Oregon Pinot Noir wine determined by aroma extract dilution analysis (AEDA)', *Flavour Fragr J*, **20**, 22–29.
FERRARI G, LABLANQUIE O, CANTAGREL R, LEDAUPHIN J, PAYOT T *et al.* (2004), 'Determination of key odorant compounds in freshly distilled Cognac using GC-O, GC-MS, and sensory evaluation', *J Agric Food Chem*, **52**, 5670–5676.
FERREIRA V, AZNAR M, LOPEZ R and CACHO J (2001), 'Quantitative gas chromatography–olfactometry carried out at different dilutions of an extract. Key differences in the odor profiles of four high-quality Spanish aged red wines', *J Agric Food Chem*, **49**, 4818–4824.
FERREIRA V, HERNANDEZ-ORTE P, ESCUDERO A, LOPEZ R and CACHO J (1999), 'Semipreparative reversed-phase liquid chromatographic fractionation of aroma extracts from wine and other alcoholic beverages', *J Chromatogr A*, **864**, 77–88.
FERREIRA V, LOPEZ R, ESCUDERO A and CACHO J F (1998), 'The aroma of Grenache red wine: hierarchy and nature of its main odorants', *J Sci Food Agric*, **77**, 259–267.

FERREIRA V, ORTIN N, ESCUDERO A, LOPEZ R and CACHO J (2002a), 'Chemical characterization of the aroma of Grenache Rose wines: aroma extract dilution analysis, quantitative determination, and sensory reconstitution studies', *J Agric Food Chem*, **50**, 4048–4054.

FERREIRA V, PETKA J and AZNAR M (2002b), 'Aroma extract dilution analysis. Precision and optimal experimental design', *J Agric Food Chem*, **50**, 1508–1514.

FERREIRA V, SAN JUAN F, ESCUDERO A, CULLERE L, FERNANDEZ-ZURBANO P *et al.* (2009), 'Modeling quality of premium Spanish red wines from gas chromatography-olfactometry data', *J Agric Food Chem*, **57**, 7490–7498.

FRETZ C, KANEL S, LUISIER J -L and AMADO R (2005), 'Analysis of volatile components of Petite Arvine wine', *Eur Food Res Technol*, **221**, 504–510.

FRITSCH H T and SCHIEBERLE (2005), 'Identification based on quantitative measurements and aroma recombination of the character impact odorants in a Bavarian Pilsner-type beer', *J Agric Food Chem*, **53**, 7544–7551.

FULLER G H, STELTENKAMP R and TISSERAND G A (1964), 'The gas chromatograph with human sensor: perfumer model', *Ann NY Acad Sci*, **116**, 711–724.

GARRUTI D S, FRANCO M R B, DA SILVA M A A P, JANZANTTI N S and ALVES G L (2006), 'Assessment of aroma impact compounds in a cashew apple-based alcoholic beverage by GC-MS and GC-olfactometry', *Lebensm-Wiss u-Technol*, **39**, 372–377.

GENOVESE A, GAMBUTI A, PIOMBINO P and MOIO L (2007), 'Sensory properties and aroma compounds of sweet Fiano wine', *Food Chem*, **103**, 1228–1236.

GIJS L, CHEVANCE F, JERKOVIC V and COLLIN S (2002), 'How low pH can intensify β-damascenone and dimethyl trisulfide production through beer aging', *J Agric Food Chem*, **50**, 5612–5616.

GOMEZ-MIGUEZ M J, CACHO J F, FERREIRA V, VICARIO I M and HEREDIA F J (2007), 'Volatile components of Zalema white wines', *Food Chem*, **100**, 1464–1473.

GROSCH W (1994), 'Determination of potent odourants in foods by aroma extract dilution analysis (AEDA) and calculation of odour activity values (OAVs)', *Flavour Fragr J*, **9**, 147–158.

GROSCH W (2007), 'Gas chromatography-olfactometry of aroma compounds', in Berger R G (ed), *Flavors and Fragrances. Chemistry, Bioprocessing and Sustainability*, Springer-Verlag, Berlin Heidelberg, 363–378.

GUERCHE S, DAUPHIN B, PONS M, BLANCARD D and DARRIET P (2006), 'Characterization of some mushroom and earthy off-odors microbially induced by the development of rot on grapes', *J Agric Food Chem*, **54**, 9193–9200.

GUICHARD H, GUICHARD E, LANGLOIS D, ISSANCHOU S and ABBOT N (1995), 'GC sniffing analysis: olfactive intensity measurement by two methods', *Z Lebensm Unters Forsch*, **201**, 344–350.

GURBUZ O, ROUSEFF J M and ROUSEFF R L (2006), 'Comparison of aroma volatiles in commercial Merlot and Cabernet Sauvignon wines using gas chromatography-olfactometry and gas chromatography-mass spectrometry', *J Agric Food Chem*, **54**, 3990–3996.

GUTH H (1997a), 'Identification of character impact odorants of different white wine varieties', *J Agric Food Chem*, **45**, 3022–3026.

GUTH H (1997b), 'Quantitation and sensory studies of character impact odorants of different white wine varieties', *J Agric Food Chem*, **45**, 3027–3032.

GUYOT-DECLERCK C, FRANCOIS N, RITTER C, GOVAERTS B and COLLIN S (2005), 'Influence of pH and ageing on beer organoleptic properties. A sensory analysis based on AEDA data', *Food Qual Prefer*, **16**, 157–162.

HANAOKA K, VALLET N, GIAMPAOLI P, HEYD B and MAC LEOD P (2001) 'Possible influence of breathing on detection frequency and intensity rating in gas chromatography-olfactometry', *Food Chem*, **72**, 97–103.

HOCHEREAU C and BRUCHET A (2004), 'Design and application of a GCSNIFF/ MS system for solving taste and odour episodes in drinking water', *Water Sci Technol*, **49**, 81–87.

JANACOVA A, SADECKA J, KOHAJDOVA Z and SPANIK I (2008), 'The identification of aroma-active compounds in Slovak brandies using GC-sniffing, GC–MS and sensory evaluation', *Chromatogr*, **67**, S113–S121.

KECHAGIA D, PARASKEVOPOUOS Y, SYMEOU E, GALIOTOU-PANAYOTOU M and KOTSERIDIS Y (2008), 'Influence of prefermentative treatments to the major volatile compounds of Assyrtiko wines', *J Agric Food Chem*, **56**, 4555–4563.

KISHIMOTO T, KOBAYASHI M, YAKO N, IIDA A and WANIKAWA A (2008), 'Comparison of 4-mercapto-4-methylpentan-2-one contents in hop cultivars from different growing regions', *J Agric Food Chem*, **56**, 1051–1057.

KISHIMOTO T, WANIKAWA A, KONO K and SHIBATA K (2006), 'Comparison of the odor-active compounds in unhopped beer and beers hopped with different hop varieties', *J Agric Food Chem*, **54**, 8855–8861.

KOMES D, ULRICH D and LOVRIC T (2006), 'Characterization of odor-active compounds in Croatian Rhine Riesling wine, subregion Zagorje', *Eur Food Res Technol*, **222**, 1–7.

KOTSERIDIS Y and BAUMES R (2000), 'Identification of impact odorants in Bordeaux Red grape juice, in the commercial yeast used for its fermentation, and in the produced wine', *J Agric Food Chem*, **48**, 400–406.

LABLANQUIE O, SNAKKERS G, CANTAGREL R and FERRARI G (2002), 'Characterization of young Cognac spirit aromatic quality', *Anal Chim Acta*, **458**, 191–196.

LASEKAN O, BUETTNER A and CHRISTLBAUER M (2007), 'Investigation of important odorants of palm wine (*Elaeis guineensis*)', *Food Chem*, **105**, 15–23.

LE FUR Y, MERCURIO V, MOIO L, BLANQUET J and MEUNIER J M (2003), 'A new approach to examine the relationships between sensory and gas chromatography-olfactometry data using generalized procrustes analysis applied to six French Chardonnay wines', *J Agric Food Chem*, **51**, 443–452.

LEDAUPHIN J, GUICHARD H, SAINT-CLAIR J-F, PICOCHE B and BARILLIER D (2003), 'Chemical and sensorial aroma characterization of freshly distilled Calvados. 2. Identification of volatile compounds and key odorants', *J Agric Food Chem*, **51**, 433–442.

LEE S-J and NOBLE A C (2003), 'Characterization of odor-active compounds in Californian Chardonnay wines using GC-olfactometry and GC-mass spectrometry', *J Agric Food Chem*, **51**, 8036–8044.

LELAND J V, SCHEIBERLE P, BUETTNER A and ACREE T E (2001), *Gas Chromatography-Olfactometry: the State of the Art*, ACS Symposium Series 782, American Chemical Society.

LERMUSIEAU G and COLLIN S (2003), 'Volatile sulfur compounds in hops and residual concentrations in beers – A review', *J Am Soc Brew Chem*, **61**, 109–113.

LERMUSIEAU G, BULENS M and COLLIN S (2001), 'Use of GC–olfactometry to identify the hop aromatic compounds in beer', *J Agric Food Chem*, **49**, 3867–3874.

LINSSEN J P H, JANSSENS J L G M, ROOZEN J P and POSTHUMUS M A (1993), 'Combined gas chromatography and sniffing port analysis of volatile compounds of mineral water packed in polyethylene laminated packages', *Food Chem*, **46**, 367–371.

LOPEZ R, FERREIRA V, HERNANDEZ P and CACHO J F (1999), 'Identification of impact odorants of young red wines made with Merlot, Cabernet Sauvignon and Grenache grape varieties: a comparative study', *J Sci Food Agric*, **79**, 1461–1467.

LOPEZ R, ORTIN N, PEREZ-TRUJILLO J P, CACHO J and FERREIRA V (2003), 'Impact odorants of different young white wines from the Canary Islands', *J Agric Food Chem*, **51**, 3419–3425.

LORRAIN B, BALLESTER J, THOMAS-DANGUIN T, BLANQUET J, MEUNIER JM *et al.* (2006), 'Selection of potential impact odorants and sensory validation of their importance in typical Chardonnay wines', *J Agric Food Chem*, **54**, 3973–3981.

MARTI M P, MESTRES M, SALA C, BUSTO O and GUASCH J (2003), 'Solid-Phase microextraction and gas chromatography olfactometry analysis of successively diluted samples. A new approach of the aroma extract dilution analysis applied to the characterization of wine aroma', *J Agric Food Chem*, **51**, 7861–7865.

MCDANIEL M R, MIRANDA-LOPEZ R, WALSON B T, MICHEALS N J and LIBBEY L M (1990), 'Pinot Noir aroma: A sensory/gas chromatographic approach', in Charalambous G (ed), *Flavors and Off-Flavors*, Elsevier, Amsterdam, 23–36.

MIRANDA-LOPEZ R, LIBBEY L M, WATSON B T and MCDANIEL M R (1992), 'Odor analysis of Pinot-Noir wines from grapes of different maturities by a gas-chromatography olfactometry technique (Osme)', *J Food Sci*, **57**, 985.

MOIO L, SCHLICH P and ETIEVANT P (1994), 'Acquisition and analysis of aromagrams of Chardonnay Burgundy wines', *Sci Aliments*, **14**, 601–608.

MOYANO L, ZEA L, MORENO J A and MEDINA M (2010), 'Evaluation of the active odorants in Amontillado sherry wines during the aging process', *J Agric Food Chem*, **58**, 6900–6904.

MURAKAMI A A, GOLDSTEIN H, NAVARRO A, SEABROOKS J R and RYDER D S (2003), 'Investigation of beer flavor by gas chromatography-olfactometry', *J Am Soc Brew Chem*, **61**, 23–32.

NONATO E A, CARAZZA F, SILVA F C, CARVALHO C R and DE L CARDEAL Z (2001), 'A headspace solid-phase microextraction method for the determination of some secondary compounds of Brazilian sugar cane spirits by gas chromatography', *J Agric Food Chem*, **49**, 3533–3539.

OLIVEIRA E SILVA H, GUEDES DE PINHO P, MACHADO B P, HOGG T, MARQUES J C *et al.* (2008), 'Impact of forced-aging process on Madeira wine flavor', *J Agric Food Chem*, **56**, 11989–11996.

ONG P K C and ACREE T E (1999), 'Similarities in the aroma chemistry of Gewurztraminer variety wines and lychee (*Litchi chinesis* Sonn.) fruit', *J Agric Food Chem*, **47**, 665–670.

PETKA J, FERREIRA V, GONZALES-VINAS M A and CACHO J (2006), 'Sensory and chemical characterization of the aroma of a white wine made with Devin grapes', *J Agric Food Chem*, **54**, 909–915.

PLUTOWSKA B and WARDENCKI W (2007), 'Aromagrams – Aromatic profiles in the appreciation of food quality', *Food Chem*, **101**, 845–872.

PLUTOWSKA B and WARDENCKI W (2008), 'Application of gas chromatography–olfactometry (GC–O) in analysis and quality assessment of alcoholic beverages – A review', *Food Chem*, **107**, 449–463.

PLUTOWSKA B and WARDENCKI W (2009), 'Headspace solid-phase microextraction and gas chromatography–olfactometry analysis of raw spirits of different organoleptic quality', *Flavour Fragr J*, **24**, 177–185.

POISSON L and SCHIEBERLE P (2008), 'Characterization of the most odor-active compounds in an American Bourbon whisky by application of the aroma extract dilution analysis', *J Agric Food Chem*, **56**, 5813–5819.

POLLIEN P, OTT A, MONTIGON F, BAUMGARTNER M, MUNOZ-BOX R *et al.* (1997), 'Hyphenated headspace-gas chromatography-sniffing technique: screening of impact odorants and quantitative aromagram comparisons', *J Agric Food Chem*, **45**, 2630–2637.

PONS A, LAVIGNE V, ERIC F, DARRIET P and DUBOURDIEU D (2008), 'Identification of volatile compounds responsible for prune aroma in prematurely aged red wines', *J Agric Food Chem*, **56**, 5285–5290.

PRAT C, TRIAS R, CULLERE L, ESCUDERO A, ANTICO E and BANERAS L (2009), 'Off-odor compounds produced in cork by isolated bacteria and fungi: A Gas chromatography-mass spectrometry and gas chromatography-olfactometry study', *J Agric Food Chem*, **57**, 7473–7479.

PRISER C, ETIEVANT P X, NICKLAUS S and BRUN O (1997), 'Representative Champagne wine extracts for gas chromatography olfactometry analysis', *J Agric Food Chem*, **45**, 3511–3514.

REINECCIUS G (2006), 'Choosing the correct analytical technique in aroma analysis', in Voilley A and Etievant P (eds), *Flavour in Food*, Woodhead Publishing Limited and CRC Press LLC Cambridge, 81–97.

ROGERSON F S S, CASTRO H, FORTUNATO N, AZEVEDO Z, MACEDO A *et al.* (2001), 'Chemicals with sweet aroma descriptors found in Portuguese wines from the Douro region: 2,6,6-trimethylcyclohex-2-ene-1,4-dione and diacetyl', *J Agric Food Chem*, **49**, 263–269.

SAENZ-NAVAJAS M-P, CAMPO E, CULLERE L, FERNANDEZ-ZURBANO P, VALENTIN D *et al.* (2010), 'Effects of the Nonvolatile matrix on the aroma perception of wine', *J Agric Food Chem*, **58**, 5574–5585.

SANCHEZ N B, LEDERER C L, NICKERSON G B, LIBBEY L M and MCDANIEL M R (1992), 'Sensory and analytical evaluation of beers brewed with three varieties of hops and an unhopped beer', in Charalambous G (ed), *Proceedings of the 6th International Flavor Conference*, Rethymnon, Crete, Amsterdam, Elsevier, 403–426.

SARRAZIN E, DUBOURDIEU D and DARRIET P (2007a), 'Characterization of key-aroma compounds of botrytized wines, influence of grape botrytization', *Food Chem*, **103**, 536–545.

SARRAZIN E, SHINKARUK S, TOMINAGA T, BENNETAU B, FREAROT E *et al.* (2007b), 'Odorous impact of volatile thiols on the aroma of young botrytized sweet wines: identification and quantification of new sulfanyl alcohols', *J Agric Food Chem*, **55**, 1437–1444.

SCHIEBERLE P and GROSCH W (1987), 'Evaluation of the flavour of wheat and rye bread crusts by aroma extract dilution analysis', *Z Lebensm Unters Forsch*, **185**, 111–113.

SCHNEIDER R, BAUMES R, BAYONOVE C and RAZUNGLES A (1998), 'Volatile compounds involved in the aroma of sweet fortified wines (vins doux naturels) from Grenache Noir', *J Agric Food Chem*, **46**, 3230–3237.

SELLI S, CABAROGLU T, CANBAS A, ERTEN H, NURGEL C *et al.* (2004), 'Volatile composition of red wine from cv. Kalecik Karasu grown in central Anatolia', *Food Chem*, **85**, 207–213.

SELLI S, CANBAS A, VARLET V, KELEBEK H, PROST C *et al.* (2008), 'Characterization of the most odor-active volatiles of orange wine made from a Turkish cv. Kozan (*Citrus sinensis* L. Osbeck)', *J Agric Food Chem*, **56**, 227–234.

SEROT T, PROST C, VISAN L and BURCEA M (2001), 'Identification of the main odor-active compounds in musts from French and Romanian hybrids by three olfactometric methods', *J Agric Food Chem*, **49**, 1909–1914.

SILVA FERREIRA A C, HOGG T and GUEDES DE PINHO P (2003), 'Identification of key odorants related to the typical aroma of oxidation-spoiled white wines', *J Agric Food Chem*, **51**, 1377–1381.

SIMPSON R F, CAPONE D and SEFTON M A (2004), 'Isolation and identification of 2-methoxy-3,5-dimethylpyrazine, a potent musty compound from wine corks', *J Agric Food Chem*, **52**, 5425–5430.

SOARES DA COSTA M, GONCALVES A, FERREIRA A, IBSEN C, GUEDES DE PINHO P *et al.* (2004), 'Further insights into the role of methional and phenylacetaldehyde in lager beer flavor stability', *J Agric Food Chem*, **52**, 7911–7917.

SOUID I, HASSENE Z, SANCHEZ PALOMO E, PEREZ-COELLO M S and GHORBEL A (2007), 'Varietal aroma compounds of *Vitis vinifera* cv. Khamri grown in Tunisia', *J Food Qual*, **30**, 718–730.

STEVENS S S (1971), 'Issues in psychophysical measurement', *Psychol Rev*, **78**, 426–50.

TAKOI K, DEGUEIL M, SHINKARUK S, THIBON C, MAEDA K *et al.* (2009), 'Identification and characteristics of new volatile thiols derived from the hop (*Humulus luplus* L.) cultivar Nelson Sauvin', *J Agric Food Chem*, **57**, 2493–2502.

TAMURA T, TANIGUCHI K, SUZUKI Y, OKUBO T, TAKATA R *et al.* (2009), 'Iron is an essential cause of fishy aftertaste formation in wine and seafood pairing', *J Agric Food Chem*, **57**, 8550–8556.

TORRENS J, RIU-AUMATELL M, VICHI S, LOPEZ-TAMAMES E and BUXADERAS S (2010), 'Assessment of volatile and sensory profiles between base and sparkling wines', *J Agric Food Chem*, **58**, 2455–2461.

ULLRICH F and GROSCH W (1987), 'Identification of the most intense volatile flavour compounds formed during autoxidation of linoleic acid', *Z Lebensm Unters Forsch*, **184**, 277–282.

VAN RUTH S M (2001), 'Methods for gas chromatography-olfactometry: a review', *Biomol Eng*, **17**, 121–128.

VAN RUTH S M, ROOZEN J P and COZIJNSEN J L (1996), 'Gas chromatograpy/sniffing port analysis evaluated for aroma release from rehydrated French beans (*Phaseolus vulgaris*)', *Food Chem*, 56, 343–346.

VERMEULEN C, LEJEUNE I, TRAN T T H and COLLIN S (2006), 'Occurrence of polyfunctional thiols in fresh lager beers', *J Agric Food Chem*, **54**, 5061–5068.

WANIKAWA A, HOSOI K, KATO T and NAKAGAWA K (2002), 'Identification of green note compounds in malt whisky using multidimensional gas chromatography', *Flavour Fragr J*, **17**, 207–211.

XU Y, FAN W and QIAN M C (2007), 'Characterization of aroma compounds in apple cider using solvent-assisted flavor evaporation and headspace solid–phase microextraction', *J Agric Food Chem*, **55**, 3051–3057.

Part II

Fermented products

6

Beer: production, sensory characteristics and sensory analysis

D. K. Parker, Campden BRI, UK

Abstract: Beer is a very complex beverage, and the objective evaluation of beer flavour is a skill that many breweries can ill afford to be without. This chapter provides insights into the origins, chemistry and sensory character of the most common beer components. Details of the main ingredients used in brewing are described, with the processes used to produce them. The influence of water composition, the chemistry behind the formation of malt flavour and colour and the key flavours derived from hops and yeast are described. This section also outlines the brewing process and how different brewing practices impact on the flavour and style of the beer. The most common off-flavours encountered in beer are also described, and guidelines on sensory evaluation practices in breweries are offered.

Key words: beer flavour evaluation, malt flavour, hop aroma compounds, flavour wheel, sensory evaluation of beer.

6.1 Introduction

> 'There is more to life than beer alone, but beer makes those other things even better.'
>
> Stephen Morris

Beer is a wonderful invention. It has evoked emotion and passion from the earliest days of the ancient Egyptians and Babylonians to the present day. It is a liquid that has sustained the lives of those when water was unfit to drink, it has fed armies, comforted and nourished sailors on long voyages, seen kings and queens, presidents and prohibition come and go. It is used to help people celebrate, commiserate and contemplate, and is woven into the human psyche.

On the surface, beer is a simple beverage composed of four main ingredients: water, barley malt, hops and yeast. The composition of each raw material and the way in which the flavours derived from these raw materials combine during the brewing process has a huge impact on the final character of the beer, and allows the large range of beer styles available today. The interaction of the flavour compounds in beer is very complex and this chapter hopes to provide some insight into the origins, chemistry and sensory character of the most common beer components. Thus, the origins of beer flavour and details of the main ingredients used in brewing are described. The effect of the mineral composition of water, the different malt types and the processes used to produce them, together with the chemistry behind the formation of malt flavour and colour are discussed. The flavours derived from hops, including bitterness, and flavours derived from yeast during fermentation are outlined. This section also gives an overview of the brewing process and how different brewing practices impact on the flavour of the beer, together with the methods for producing different beer styles. Following sections cover the most common off-flavours encountered in beer and their origins, and offers guidelines on the sensory evaluation practices in breweries.

6.2 The origins of beer flavour

> 'Not all chemicals are bad. Without chemicals such as hydrogen and oxygen, for example, there would be no way to make water, a vital ingredient in beer.'
>
> Dave Barry

Beer flavour is very complex. The main brewing ingredients are water, barley malt, hops and yeast. For a 4.0% abv beer, the composition would comprise about 94% water, 1–2% residual sugars, 4% ethanol and 0.1% flavour compounds. Although the amount of flavour compounds present in beer is very small in comparison to other components, it is these compounds that are responsible for the unique flavour of each individual beer. So far over 1000 different flavour compounds have been identified in beer originating directly from the raw materials and being formed during processing or packaging. To complicate matters further, beer flavour is not static, and changes occur from packaging throughout the life of the beer during storage. The contribution of the raw materials to beer flavour and the effect of current brewing practices on final beer flavour are summarised below.

6.2.1 Raw materials

Water

Water is the main constituent in beer comprising normally 92–95% (Briggs *et al.*, 1986). The brewing process uses substantially more water than that which ends up in beer, as large amounts are required for cleaning, cooling and steam generation for heating as well as that used in the brewing process itself. As water is an expensive commodity, much emphasis has been placed in recent years on the development of technologies to reduce water consumption in breweries and to recycle as much water as possible and reduce effluent. Therefore, a modern brewery will have various streams of water servicing different needs and there will be various treatments including charcoal filtration, ultra-filtration and deionisation to ensure the desired balance of ions in the water relevant for the process involved (Bamforth, 1998).

Traditionally, breweries were established where there was a plentiful supply of water, and the composition of water influenced the style of beer that was produced. Thus the hard waters of Burton-on-Trent, which have high levels of calcium and magnesium sulphates, produced bitter pale-ale style beers. The soft waters of Pilsen contain small quantities of ions except carbonates and produced lagers that have a smooth, rounded palate. In many breweries today, the salt composition of the water is adjusted to match that of Burton (a practice called Burtonisation), or salts are removed if a Pilsen-style beer is being brewed. It is known that certain ions influence the brewing process itself and others contribute towards the flavour and mouth-feel of the beer. Calcium is important in processing, such as affecting wort pH, as a co-factor for enzyme activity, oxalate precipitation (beer stone) and yeast flocculation, but has little direct effect on flavour. Chloride and sulphate ions, however, are considered to be important for the palate of the beer and many brewers specify the chloride–sulphate ratio for brewing liquor. It is considered that sulphates enhance the dryness of a beer and chlorides afford body and fullness, although there is a lack of research publications to support this idea.

Malt

Barley (*Hordeum vulgare*) is the principal cereal used to make malt, and barley malt comprises 90% of global beer production. As with other cereals, sugar is held within the barley grain in the form of starch, which is unfermentable and cannot be used by yeast as a food source during fermentation. Before barley is used in the brewing process, it must be malted in order to generate the enzymes required to break down (or modify) the protein–glucan matrix surrounding the starch.

In barley, the starch is held within cells in the endosperm of the grain surrounded by a wall of protein, pentosan and glucan material. Enzymes naturally present in the barley must degrade this material to release the starch, and this occurs during the malting of the grain. It is then during the mashing stage in the brewing process that the starch is broken down by the malt enzymes to sugars that yeast can ferment. Barley malt contains more than enough enzymes to convert the starch and surrounding proteins and glucans. Barley also has an outer coat, or husk, which acts as a filter bed during the brewing process. In contrast, wheat is a 'naked' grain (the husk is removed during threshing) and hence cannot be used as the only cereal in brewing as it results in set mashes.

The main varieties of barley used for brewing fall into two categories, Spring and Winter. Winter barleys are sown in the autumn and are traditionally grown in the UK and parts of Europe where ale brewing is favoured. Varieties of Winter barley include Pearl, Flagon and Cassata and the traditional UK ale variety, Maris Otter. Spring varieties are sown from February through to April and varieties include Tipple, Quench and Concerto, and are used traditionally in lager brewing. Australian and New Zealand varieties include Schooner, Gairdner, Baudin, Cellar and Fairview. The barley plant itself also has two forms of structure, relating to the number of rows of grains attached to the central stem. Two-row barleys are commonly used across Europe and Australia for malting. Six-row barleys tend to have higher enzyme potential and are used in the US where other unmalted cereals, such as maize and rice, are used as alternative sources of starch to supplement the barley malt. North American and Canadian barley varieties include the six-row Tradition and Lacey and the two-row Metcalf and Moravian varieties.

Malt is produced by the controlled steeping, germination and kilning of barley to provide an available source of carbohydrate (starch) and enzymes for brewing.

The barley grain is first hydrated by a regime of alternate steeping and air resting which allows water and oxygen to activate the embryo to produce hormones and enzymes. As the grain begins to germinate, enzymes are produced that break down the cell walls around the starch and the grain starts to grow, producing roots and shoots. Before too much of the starch is broken down, germination is stopped by kilning. Kilning is the controlled drying of the grain in order to preserve the enzymes and starch for the brewing process.

Kilning is the stage of the malting process that primarily allows different types of malt to be produced, depending on the time taken to dry the malt and the final temperature reached. As kilning temperature increases, more flavour and colour develop in the malt. The majority of malt produced comprises lager and ale malts, called white or enzymic malts as they contain the enzymes necessary to break down the starch. These white malts comprise the majority of the brewers' grist and produce a golden-coloured beer with typically cereal, sweet and nutty flavours. Speciality malts, such as caramalt, crystal, amber, chocolate and black malts, are produced by varying degrees of roasting in specialised roasting drums and as such do not have active enzymes. These malts are used in small percentages in the grist to provide additional colour and flavour. Altering kilning or roasting conditions can produce a huge range of different malt types. Speciality malts produced with low levels of heating, such as crystal malt, have sweet, fruitcake and caramel flavours and provide an orange/brown colour, whereas those malts experiencing intense heat, such as chocolate and black malt, are very dark and have a burnt, smoky character. Barley can be roasted without malting and is typically used in stouts to give a distinctive burnt, roasted character. The range of colours that be can be achieved by malting is also enormous, ranging from 3 European Brewing Convention (EBC) units for lager malts and up to 1000 EBC units for chocolate malt. Some of the common malt types together with key descriptions of flavour and colour range can be found in Table 6.1.

Malt provides the principal colour for beer, and different malt types and malting conditions have been shown to be responsible for both positive and negative flavours in beer. During the malting process, modification of the endosperm and development of growing areas produce and release amino acids, sugars, lipids,

Table 6.1 Typical malt flavour and colour specifications

Malt		Flavour	Colour
Green malt		Green, grassy, green pea	
Lager malt		Green, malty, nutty, sweet	2–4
Ale malt		Malty, nutty, sweet, toasted	4–7
Caramelised			
	Caramalt	Sweet, caramel	10–30
	Crystal	Sweet, caramel, stewed fruit, toffee, black treacle, burnt	75–750
Roasted (dry)			
	Amber malt	Biscuity, baked, malty, bitter	40–60
	Chocolate malt	Dark chocolate, burnt, sharp	800–1200
	Black malt	Burnt, black coffee, sharp, some acrid	1200–1400
	Roasted barley	Burnt, sharp, acrid, dry, black coffee	1200–1500

sulphur compounds and phenols. All of these compounds are precursors for flavour compounds.

The most important control point for flavour production is the kilning stage where control of temperature, time and humidity are critical. Only small changes in relative humidity (RH) profiles, achieved by careful control of air-recirculation, can change the flavour of malts with similar chemical specification (Chandra *et al.*, 1997). Increasing kilning time and using greater temperatures yields a wide range of malt styles with increasing colour and flavour intensities. As temperature increases, non-enzymic browning reactions occur such as Maillard reactions, caramelisation and pyrolysis (Pengelly, 2001).

Maillard reactions are very complex and can occur in quite mild conditions, (e.g. temperatures above 50 °C) and do occur within the malt during kilning. Maillard reactions begin with the condensation of a reducing sugar (such as maltose or glucose) with an amino acid to form intermediates, which can either polymerise to give colour (melanoidins) or react in a different way to give flavour-active materials (Bamforth, 2001). The spectrum of Maillard products alters depending on the level of malt modification and ratios of nitrogen, carbohydrate and sulphur. The key flavour products from the kilning of white malts (lager and ale malts) to more roasted, dark malts such as chocolate malts (speciality malts) are furans (sweet, nutty, bready) and pyrans (malty, toasted, bread crust). The incorporation of nitrogen in the Maillard product allows the formation of pyrroles (sweet, nutty, liquorice), pyridines (fishy, popcorn) and pyrazines (cocoa, roasted nuts, peanut butter, caramel, coffee). Pyrazines are key flavour notes of dark beers. The incorporation of sulphur (e.g. methionine, cysteine) forms products such as thiophenes and thiazoles (nutty, green, bean sprout). Maillard reactions also contribute to flavour formation during the mashing and wort boiling stages of the brewing process.

Caramelisation of the grain can occur during the production of speciality malts, such as crystal malts, in which the grain is 'stewed' at temperatures above 129 °C. The endosperm of the malt is effectively a concentrated sugar solution, and as the grain is heated, caramelisation leads to the formation of furans, furanones and pyranones (maltol and isomaltol), which have sweet, fruity and caramel flavours. As temperatures rise, (above 250 °C), pyrolysis results in the scorching or burning of the sugar molecules, which leads to very burnt, ashy and astringent flavours which are associated with dark roasted malts, such as chocolate and black malts and roasted barley. Intense heating can also produce phenols, which have harsh, smoky and medicinal flavours. Dark malts also have higher levels of aldehydes formed by the Strecker degradation of amino acids during kilning and which are associated with staling flavours in beer.

Other less favourable flavours associated with malt are astringent flavours linked to glycosides present in the acrospires (Kageyama *et al.*, 2007) and lipid oxidation products. Carbonyl compounds are purported to be the source of staling character in beer (Bamforth, 2001a) with aldehydes such as *trans*-2-nonenal (T2N) giving cardboard flavours. Lipid oxidation and T2N potential are considerably influenced by the temperature and moisture content of partially kilned malt (Guido *et al.*, 2005). Three indicators for staling potential include lipoxygenase activity, hydroperoxide activity and antioxidant activity. Although dark malts have higher antioxidant activity (from melanoidins and reductones) the T2N potential of worts produced from grists containing substantial amounts of coloured and dark malts (as found in English ales and other dark beers) is significantly increased (Liegeois and Collin, 2003). Lipoxygenase activity and hydroperoxide potential develops during germination and increases during the early stages of kilning, then decreases once temperatures rise above 60–70 °C (Dumoulin and Bolvin, 2001).

Studies have shown that increasing the duration of the initial drying phase and/or the final high-temperature phase of the malt kilning improves the flavour stability of beer (Preuss *et al.*, 2001) and that single-floor industrial kilns favour malts with higher antioxidant activity and better flavour stability compared to two-floored kilns (Dumoulin and Bolvin, 2001). Single-floor kilns also produce malts with higher dimethyl sulphide (DMS) and dimethyl sulphoxide (DMSO) levels. DMS is a flavour compound originating from S-methyl methionine (a modified amino acid) found in the growing areas in malt. DMS imparts a sweetcorn note and is an important feature of European lager flavour.

Hops

Hops (*Humulus lupulus*) were classified as a member of the Cannabinaceae family until replaced in 1972 as members of the Cannabaceae family (which also includes the genera *Cannabis* (hemp) and *Celtis* (blackberries)). The hop plant is a perennial climbing vine, which has separate male and female plants. The cones from the female plant are used in brewing. Before hops were introduced in the fifteenth century, ales were flavoured with herbs and spices.

Hops contribute to the bitterness and hoppy flavour of beer, and also enhance foam formation and foam stability. In addition, hops have antibacterial properties and provide protection against spoilage by certain microorganisms.

Varieties of hops include those high in bittering components, such as Target, Admiral, Nugget, Pride of Ringwood and Super Pride to those low in bittering components but with a good hoppy aroma, such as Fuggles, Goldings, Saaz, Willamette, Cascade and Cluster. Aroma hops grown in Germany, such as Hallertau and Hersbrucker, are called Noble hops. Recently, dual-purpose varieties such as First Gold and Northern Brewer have been bred which have reasonable levels of bittering compounds and good hoppy flavours.

Hops provide the bitterness to beer and the hoppy aroma and taste. Hops contain a range of chemical species, including cellulose and lignins, proteins, lipids and resins, oils and tannins (Bamforth, 1998). Hops chemistry is very complex and is beyond the scope of this chapter but a summary follows.

The key compounds important in hops for flavour are the resins and essential oils. The resin fraction consists of several components, the most significant being the α-acids or humulones, which are the source of bittering compounds. There are three variants of the acids, cohumulone, adhumulone and humulone, making up 2–15% dry weight. The acids undergo isomerisation when heated during wort boiling to more soluble and more bitter compounds called iso-α-acids, which also exist as isomers. All six iso-α-acids differ in the quality and intensity of their bitterness (Table 6.2). The perceived bitterness in beer depends on the overall

Table 6.2 Summary table of key groups of beer flavours and their origin

Origin and compound	Flavour descriptors
Water	
Chlorides	Palate fullness, body
Sulphates	Dryness
Sodium	Low levels – sweet
Iron	Metallic
Malt	
Furan	
Furfural	Sweet, woody, almond, baked bread, fragrant
Furfuryl alcohol	Cooked sugar
5-Methylfurfural	Sweet, spicy, caramel, bitter almond
2-Acetylfuran	Sweet, balsamic, cocoa. coffee
Hydroxydimethylfurananone	Sweet, fruity, strawberry, hop sugar
Isomaltol	Fresh bread, malt
Hydroxymethylfurfural	Stale, fatty, musty, waxy, vegetable oil
Pyran	
Maltol	Malt, bread crust
5-Hydroxy-5,6-dihydromaltol	Malt, toasted
Pyrrole	Nutty, sweet, ethereal
2-Acetpyrrole	Bread, walnut, liquorice

(Continued overleaf)

Table 6.2 Continued

Origin and compound	Flavour descriptors
Thiazol	Nutty
4-Methylthiazole	Nutty, green
Pyridine	Sour, putrid, fishy, amine
2/3-acetypryidine	Sweet, nutty, popcorn
Pyrazine	Pungent, sweet
2-Methylpyrazine	Green, nutty, cocoa, roasted chocolate
2,3/5,6-dimethylpyrazine	Green, nutty, coffee, caramel, cocoa, roasted nuts, peanut butter, walnut
2-ethyl-3,5(6)-dimethylpyrazine	Coffee, roasted nuts, chocolate, burnt almond
Dimethylsulphude (DMS)	Sweetcorn, tinned tomatoes, blackcurrant, seaweed
3-Methylthiopropionaldehyde (Methional)	Worty, mashed potato
Glycosides	Astringent, harsh
Trans-2-nonenal	Papery, cardboard, oxidised, stale
Hops	
Hop essential oil	
Myrcene	Sweet, balsamic, plastic
Humulene	Isomer of caryophylene
Caryophylene	Cloves spicy, terpentine
Farnescene	
ß-Pinene	Woody, spicy, pine
Selinene	
Geraniol	Sweet, floral, rose, fruity
Linalool	Floral, citrus, lemon, orange, fragrant, sweet
Nerolidol	Rose, apple, green, citrus, sl woody, waxy
Citral	Lemon
Limonene	Sweet, orange, lemon, citrus
Nerol	Sweet rose
Hop resins (% in beer)	Ranking bitterness
trans-Isocohumulone 7%	4
cis-Isocohumulone 30%	2=
trans-Isohumulone 10%	2=
cis-Isohumulone 40%	1
trans-Isoadhumulone 3%	?
cis-Isoadhumulone 10%	?
3-Methyl-butene-1-thiol	Lightstruck, skunky, pungent, leek-like
Yeast	
Alcohols	
Ethanol	Vodka, antiseptic wipe
n-Propanol	Alcohol, sweet
iso-Butanol	Wine-like
iso-Amyl alcohol	Fusel oil, whisky
Phenylethanol	Roses
Esters	
Ethyl acetate	Solvent, fruity
Butyl acetate	Fruity, banana

iso-Amyl acetate	Fruity, banana, sweet, peardrop
iso-Butyl acetate	Sweet, fruity, apple-banana, ethereal
Ethyl valerate	Papaya
iso-Amyl propionate	Pineapple, aniseed
Ethyl hexanoate	Fruity, wine-like, apple, banana, brandy, pineapple
Phenylethyl acetate	Roses, honey
Methyl caprate	Coconut
Octyl caproate	Orange peel
iso-Amyl caprate	Tropical fruits
Ethyl nicotinate	Medicinal
Miscellaneous	
Acetaldehyde	Green apples, emulsion paint
Acetic acid	Vinegar, sour, acidic
Diacetyl	Butterscotch, buttery, toffee
Sulphurs	
Hydrogen sulphide	Eggy
Sulphur dioxide	Struck match, burnt hair
Dimethyl disulphide	Cooked vegetable, cabbage
Ethyl mercaptan	Onion, garlic
Diethyl sulphide	Garlic, burnt rubber
Methional	Mashed potato, worty
Methanethiol	Rotting vegetable
Ethanethiol	Leek, onion, garlic

level of iso-α-acids and their relative proportions. These proportions depend on the hop variety, the manner in which the hop has been processed (whole leaf hop, pellets, carbon dioxide (CO_2) extract, isomerised hop extract) and the final bitterness determined by level of losses during boiling and fermentation (utilisation).

The hop oil fraction makes up only 0.5–2% of the whole hop but contains over 200 compounds. The most abundant is myrcene (up to 60%), which has a pleasant aroma but a disagreeable taste reminiscent of engine oil (Pengelly, 2002). Other volatile hydrocarbons include humulene, caryophylene, farnescene, pinene and selinene. These compounds are primarily responsible for the fresh hop aroma. It is estimated that 85% of the hop oil is evaporated in the kettle, so to achieve a hoppy aroma and taste, aroma hops are added to the wort kettle about 10 minutes before the end of the boil, a practice called late-hopping, or whole hops can be added to the fermentation vessels or directly into casks after fermentation, called dry hopping.

Hop oil also contains small amounts of more pleasant, oxygen-containing terpenoids, which result from the oxidation of the major volatile hydrocarbons. Oxidation of caryophylene and humulene yield epoxides which hydrolyse to give terpene alcohols, important for herbal, fruity and spicy flavours. Geraniol and linalool give floral flavours while citral, limonene and nerol add citrus flavours. Therefore, some oxidation of hops is necessary for the proper development of pleasant hop flavour and aroma. Excessive oxidation, however, results in the degradation of α- and β- acids and the production of compounds such as isovaleric and butyric acids, which give cheesy and rancid flavours, as well as harsh bitterness from oxidised β-acids.

Hop oil extracts are now available for adding hop flavours to beer. These extracts can be purified fractions of hop aroma, e.g. floral, spicy, fruity, herbal, or extracts of a single hop variety, e.g. Golding, Saaz and Cascade. These products can be added to the wort kettle or most often post-fermentation or even into the beer stream prior to packaging as these compounds can be lost through venting, absorption and even transformed by yeast metabolism during fermentation (Bamforth, 2001b).

Another important flavour originating from hops but which is less desirable in beer is the light-struck or 'skunky' flavour. This is produced through the degradation of the iso-a-acid bitter substances to form 3-methyl-2-butene-1-thiol (MBT) and this degradation is brought on by light, whether natural sunlight or artificial (fluorescent). Brown glass affords some protection from the wavelength of light that causes damage, but green and clear glass does not. An option for brewers is to use chemically modified (reduced) iso-α-acids, such as tetra or rho-iso-α-acids (Weiss *et al.*, 2002). These compounds achieve better light stability, and as an added advantage better foam stability too, but the quality of the bitterness decreases as the levels of modified hops increases.

Yeast

Yeast (*Saccharomyces cerevisiae*) plays a vital role in the production of all alcoholic beverages. The final character of the beer will depend on the composition of the wort and the yeast strain used. Brewers began to understand that yeast was central to the specific flavour characteristics of the beer. However, before Emil Hansen developed a technique for purifying yeast in 1883, beers were often contaminated by off-flavours from unwanted bacteria and wild yeast strains. Hansen worked for the Carlsberg Laboratory in Copenhagen and his method to purify yeast is still used today to allow brewers to brew beers with consistent flavours.

Brewers split yeast into two groups. Those that work at warmer temperatures (16–22 °C) are traditionally used for ale brewing. Once fermentation is complete, the yeasts flocculate, or aggregate, at the top of the vessel and are skimmed off for re-pitching into new brews. These top-fermenting yeasts are called *Saccharomyces cerevisiae*. Those yeast strains that ferment at lower temperatures (6–16 °C) are traditionally used for lager-style beers and will settle to the bottom of the fermentation vessel at the end of fermentation. These bottom-fermenting yeasts are known as *Saccharomyces carlsbergensis*.

The final flavour of a beer will depend very much on the strain of yeast used. Many different strains of brewing yeast exist, and some beers are brewed with a single yeast strain and others are fermented using mixed strains. It is essential to use a supply of good healthy yeast, and yeast is usually collected at the end of a brew for subsequent re-use in the following brew. In most modern breweries a fresh batch of yeast is propagated or grown every 8–12 generations. This regular propagation helps to maintain the yeast quality and performance. The yeasts used in each brewery are different and provide the unique flavour fingerprint for each beer. It is no surprise, then, that brewers guard their yeasts very carefully and many keep samples of their yeasts for safekeeping in research establishments.

Of the 1000 flavour compounds described so far in beer, over 600 are yeast-derived. As with most flavour compounds, some are present at mg/l, μg/l and ng/l levels. Flavour-active products from yeast metabolism include ethanol and higher alcohols, glycerol, CO_2, esters, acetaldehyde, fatty acids, organic acids, phenols and a range of sulphur compounds. Factors affecting the spectrum of flavour compounds derived from yeast metabolism are the yeast strain and its health, the composition of the wort and physical factors that can stress the yeast. A schematic diagram of the flavours produced by yeast during fermentation is shown in Fig. 6.1.

Ethanol content for beer usually ranges between 3–7% for the majority of beers and is a product of anaerobic sugar metabolism. The higher (fusel) alcohols are produced as by-products of biosynthetic amino acid synthesis or are produced from amino acids taken up from the wort via the Ehrlich pathway (Bamforth, 2001b). Esters are compounds formed between an acid and an alcohol. Many acids are produced from yeast of which acetic acid is the simplest; but longer chain fatty acids are also important and are derived from lipid metabolism. The range of alcohols to make esters is diverse. Ethanol which has largely a subtle (some say no) flavour, is present in the highest quantity but higher alcohols such as iso-butanol, iso-amyl alcohol, propanol and 2-phenylethanol are all very important for flavour and ester production. Overall, esters make a more significant contribution to beer flavour than do the alcohols or acids from which they are derived. Ester flavours include fruity notes, banana (iso-amyl acetate), pineapple/red apple/aniseed (ethyl hexanoate) and solvent esters (ethyl acetate). Yeast strain

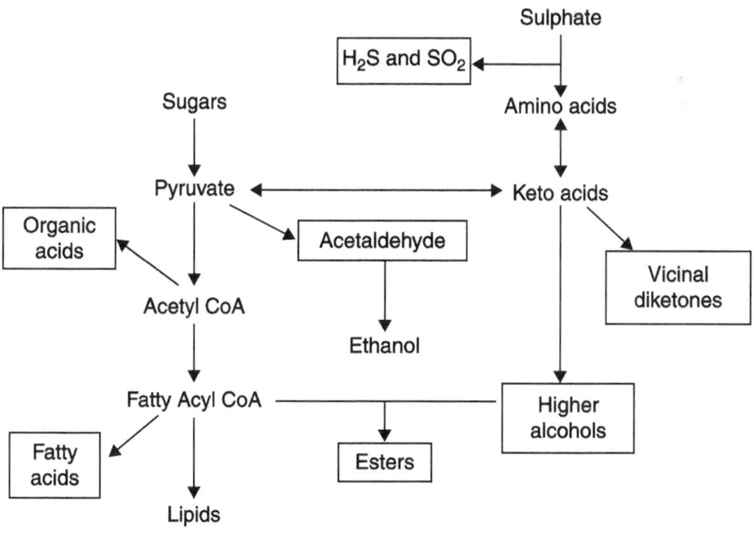

Fig. 6.1 Schematic diagram of yeast metabolism and flavour production.

is very important for higher alcohol and ester formation, sometimes more important than environmental factors, but for a given yeast strain, increased temperature leads to elevated higher alcohol and ester formation.

Another important flavour group, which is a by-product of amino acid metabolism, are the vicinal diketones. The most important of these is diacetyl, which has an intense butterscotch flavour. Diacetyl is produced in large amounts by yeast during fermentation but is then re-absorbed by the yeast and converted to less flavour-active substances. To allow time for this to happen, brewers include a 'diacetyl stand' at the end of fermentation to take levels below the flavour threshold of 0.15 mg/l (Meilgaard, 1975).

Yeast also produces lots of different sulphur-containing compounds during fermentation. Most are driven off with fermentation gases and during maturation and conditioning, and so are present in beer generally at low levels. Low levels of sulphur compounds in beer can contribute to the overall flavour, and are a characteristic of some beers (e.g. English ales brewed in Burton-on-Trent) but at higher levels can become undesirable. Common sulphur-containing compounds produced by yeast include hydrogen sulphide (eggy), sulphur dioxide (SO_2), DMS (sweetcorn) and dimethyl disulphide (cooked vegetable, cabbage). These sulphur compounds tend to have very low flavour thresholds (μg/l/).

One other compound worthy of mention is 3-methylthiopropionaldehyde (methional). This compound is formed during malt kilning, mashing and wort boiling and is principally identified as being responsible for the worty flavour associated with alcohol-free beers (Perpete and Collin, 1999a). Methional has a higher boiling point than other similar flavoured wort constituents, so less is lost during boiling. Methional is not found in normal beer because it is reduced to a flavour-neutral alcohol by yeast during fermentation. Alcohol-free beers produced by dealcoholising normal beer also do not have this flavour. However, alcohol-free beers can also be produced by the 'cold contact' method, whereby pitched wort is too cold for alcoholic fermentation but allows some yeast metabolic processes to take place to form flavour compounds. Producing alcohol-free beer using the cold contact method does not result in the reduction of methional and levels remain in the beer above threshold level (Perpete and Collin, 1999b).

6.2.2 The brewing process

> 'Give a man a pint of beer and he will drink it. But teach a man to brew, and he will rule the world.'
>
> *Anon.* (Travelling philosopher from the last century)

Brewhouse: mashing

Malt is mixed according to the recipe required for the beer and then is ground using roller mills to preserve the integrity of the husk. The ground malt (called grist) is mixed with hot water (liquor) at a specific temperature to allow enzymes to convert the insoluble starch into fermentable sugars.

In traditional ale brews, the striking temperature of the liquor to which the malt is added is around 65 °C, which is the gelatinisation temperature of starch. In lager brewing, malts are generally less well modified than ale malts and have more protein and glucan still present from the malting stage. It is usual in lager brewing to mash-in at lower temperatures (around 45 °C) to allow proteases and glucanases to break down these materials before the temperature is ramped up. It normally takes about an hour to convert all of the starch into sugar. In ale brewing, conversion takes place in a single vessel called a mashtun, from which the sugar extract, or wort, is separated from the mash solids by filtration through the mash bed, which is formed from the malt husks, For lager brewing, the mash is pumped into another vessel called a lauter-tun for separation. It is during the mashing stage that the flavour and colour compounds present in the malt are extracted from the malt grist and remain in the malt extract, called wort.

It is at the mashing stage also, that other cereals, called adjuncts, can be added as alternative sources of starch. Common adjuncts include unmalted rice, maize, wheat, sorghum and brewing syrups. The use of adjuncts results in a different style of beer to that of beers brewed using an all-malt grist. The use of adjuncts tends to 'dilute' the flavours and colour provided by malt, therefore adjunct beers tend to be pale-coloured and light-flavoured compared to all-malt brews. In the US, beers can contain as much as 40% adjunct, mainly maize and rice in the form of flakes or grits. In the UK, ales contain more than 80% of malt. Wheat flour is used in small proportions to enhance beer foam. The style of beer known as Wheat beer uses around 40% malted wheat in the grist. Sugar syrups are also commonly used as adjuncts and these are added at the boiling stage of the brewing process. This is typically done in Australian breweries.

Brewhouse: wort boiling

The sweet wort passes to the copper or kettle where the wort is boiled for about an hour. Wort boiling has several aims: the first is to sterilise the wort and provide microbiological stability; the second is to help precipitate protein–polyphenol complexes that would otherwise form haze material further down the process, and a third is to add the hops. The bitter resins within hops are insoluble and not very bitter. During the boiling stage, the α-acids undergo isomerisation to the more bitter iso-α-acids. Bittering hops are added at the beginning of the boil to allow this chemical change to occur. The disadvantage of adding hops at the beginning of the boil is that any hoppy aroma and flavours derived from the essential oils within the hop are boiled away. Therefore, to add hoppy flavours to beer, other varieties of hops called Aroma or Noble hops are added about ten minutes from the end of the boil to provide hop flavour. This late addition of hops is termed 'late-hopping' and gives subtle floral, spicy and fruity notes from the hops. Alternatively, leaf or cone hops can be added post-fermentation direct to beer in cask to provide the resinous, fresh hop flavours typical of English ales. More recently, hop extracts and hop oils are available which can be added to finished beer to provide different hop characteristics. Apart from adding desirable hop flavours, another objective of the wort boiling process is to drive off unwanted green flavours and sulphur notes from the malt. It is essential to get a good rolling boil in the wort kettle otherwise these green notes may remain in the wort, resulting in grainy and stewed flavours in the final beer.

Fermentation, maturation and conditioning
The hot, hopped wort is cooled before the yeast is pitched and a small amount of air or oxygen is added to the cooled wort to promote vigorous yeast growth. The speed of fermentation depends on temperature. Ales, which have fuller flavours generally, ferment at higher temperatures, 16–22 °C, and take 5–6 days. Lagers ferment at cooler temperatures, 6–16 °C, and take 2–3 weeks. During fermentation, the yeast converts the sugar in the wort to alcohol, carbon dioxide and a myriad of flavour compounds. After fermentation, the 'green' beer generally has sweet, fruity and raw alcohol and solvent notes, together with fresh yeast notes, butterscotch flavours and pungent sulphurs. These green beer flavours disappear as the beer is allowed to mature and condition and some other flavours develop, giving the beer a more rounded and mellow flavour.

At the end of fermentation the beer still contains live yeast and a small amount of fermentable sugar. Some brews undergo maturation, conditioning and filtering before final packaging in the brewery. This brewery-conditioned beer is held cold (below 0 °C) for a minimum of three days after fermentation to allow stabilisation and the natural sedimentation of the yeast. The beer is then filtered to remove the remaining yeast and the carbonation level adjusted ready for packaging into bottles, cans or kegs. Other beer such as cask beer undergoes final conditioning in package. At the end of fermentation the beer is sent to holding vessels before being racked into casks together with added sugar and finings to help sediment the yeast. Once in the cask, the beer undergoes a secondary fermentation that allows final development of flavour and builds up natural carbonation. Once air is allowed to enter the cask ready for dispense, cask beer has a short shelf life of five to seven days before oxidation and acetic flavours begin to develop. For beer packaged in bottles and cans, it is essential to keep any oxygen pick-up during filling to a minimum to reduce the formation of stale flavours (e.g. ribes) during the shelf life of the products. Following packaging it is usual to stabilise the beer microbiologically by pasteurisation or by sterile filter. Care must be exercised using either method as over-pasteurisation can lead to cooked flavours in the beer, and over-filtering can remove colour and bitterness.

A summary of the key groups of beer flavours and their origin discussed so far is given on Table 6.2.

6.3 Off-flavours and their origin

There are many flavours that have not yet been covered, mostly age-related and oxidised off-flavours, such as ribes, or contamination flavours, such as musty (trichloroanisole). There follows a brief description of the most commonly encountered off-flavours and their potential source.

6.3.1 Contaminants

Trichloroanisole

Water-borne contaminants include trichloroanisole (TCA), which has a musty taint typical of damp cellars and mould. TCA has a very low threshold and can be detected at the μg/l level.

Chlorophenols, bromophenols

Often described as medicinal, having antiseptic, hospital flavours, these compounds arise when chlorine from water reacts with phenolic compounds found naturally in beer. They have very low flavour thresholds and can readily be detected at the μg/l level. In draught beer, medicinal TCP taints can arise from reactions between the plastic tubing and hypochlorite-containing cleaning agents.

6.3.2 Microbiological spoilage

Beer pH lies typically in the range 3.5–4.5. Acidic and sour flavours can arise from infection by lactic acid bacteria and by acetobacteria. Acetic acid is described as vinegar and lactic acid as off-milk and yoghurt-like.

Sulphidic (hydrogen sulphide) is a by-product from yeast and bacterial growth and has an odour and taste of rotten eggs. In cask beer, a sulphidic flavour can be a feature of certain yeast strains and the flavour of specific brands of beer. In other cask products, a sulphidic flavour can indicate young, green beer that needs to mature further. If present in other products it can indicate possible bacterial spoilage. Phenolic is a flavour note described as spicy or clove-like. It is a definite character of certain beer styles such as wheat beers, but can also indicate a wild yeast contamination.

Dimethyl sulphide (DMS) has a flavour described mainly as sweetcorn, baked beans and tinned tomatoes. DMS is a characteristic flavour in many lager styles. Most of the DMS originates from malt but can also be produced by yeast and, if present in excess, can be an indication of bacterial infection.

Butyric acid is often referred to as 'rancid butter' and has a flavour described as baby-sick or vomit. It is produced by infection of wort, fermentations and sugar syrups.

6.3.3 Oxidation/staling off-flavours

The way in which a beer ages is very specific to the beer itself, and depends on many factors, such as the beer composition, the dissolved oxygen content, the level of antioxidants present, pasteurisation conditions and storage temperatures. Flavour changes that are generally known to occur as a beer ages include the development of staling flavours, such as cardboard, ribes, sweet caramel-toffee notes and wine-like flavours, and the loss of more positive flavour notes, such as bitterness and hoppy aroma (Fig. 6.2). Recently, the European Brewery Convention organised a collaborative study to further classify the way in which beers age and to develop the additional terms on the Flavour Stability Wheel (Hill, 2003). There are specific flavours regarded as oxidation or staling flavours in all beers and these are described below.

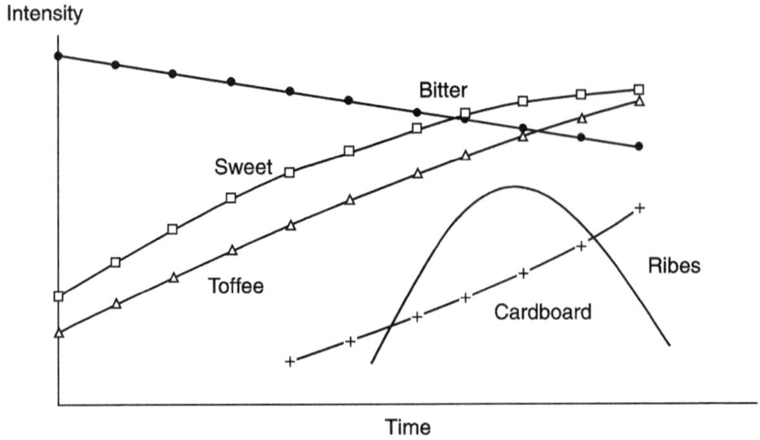

Fig. 6.2 Example of sensory changes in beer flavour during ageing.

Acetaldehyde is described as having the flavour of green apples or green leaves (Meilgaard, 1975). Acetaldehyde can be formed by the oxidation of ethanol and in excess is a typical character of aged beer and in particular cask-conditioned beer. Acetaldehyde is also produced during slow fermentations or the use of low viability yeasts.

Cardboard/papery flavour develops during the initial stages of beer staling due to the formation of unsaturated aldehydes from malt lipids. *trans*-2-nonenal is an example of one of these aldehydes and has a low flavour threshold of 0.1 μg/l (Taylor and Organ, 2009).

Ribes is an aged flavour usually associated with small pack beers and is described as 'catty' or 'tom-cat'. The flavour is transient and is reminiscent of blackcurrant leaves and tomato plants, and is a result of the formation of the sulphur compound p-menthane-8-thiol-3-one during ageing. This flavour is less apparent in beers that have low dissolved oxygen levels.

One of the final stages of staling is the formation of whisky-like and Madeira flavours.

6.4 Sensory analysis practice in the brewing industry

> 'A fine beer may be judged with only one sip, but it's better to be thoroughly sure.'
>
> Czech proverb

6.4.1 Sensory evaluation of beer: flavour terminology

Beer contains many flavour-active substances derived from the raw materials and from processing. The flavour thresholds for many of the volatile compounds have

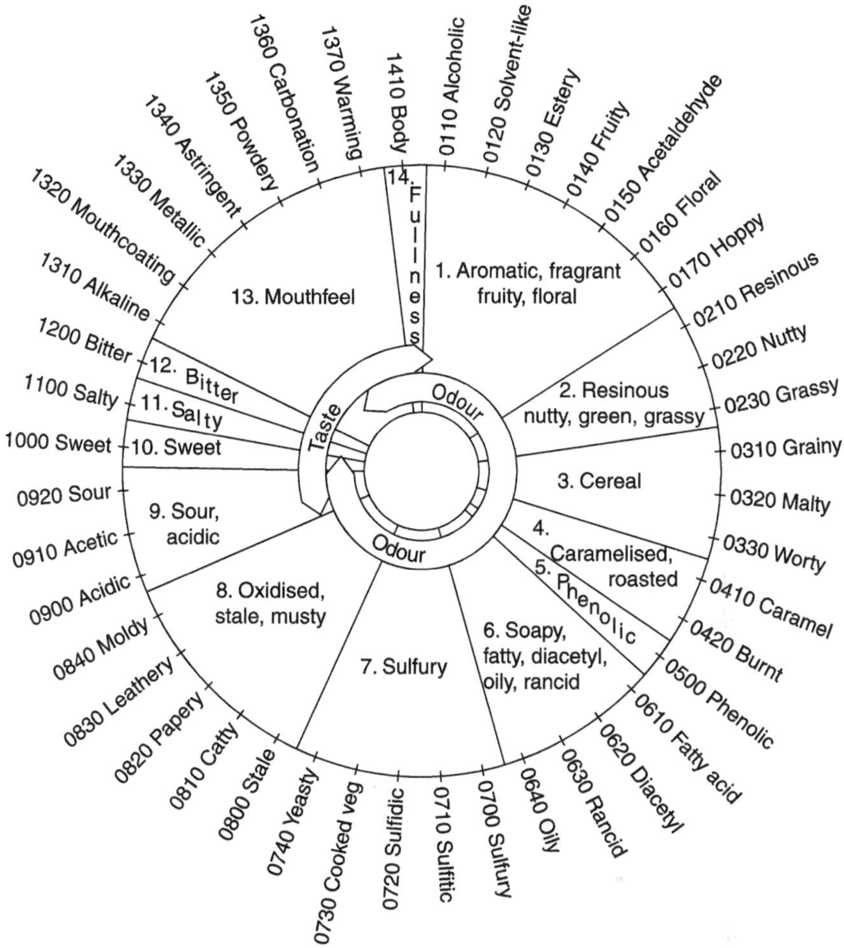

Fig. 6.3 The Flavour Wheel, reproduced with permission from the American Society of Brewing Chemists, St Paul, MN, USA.

been studied by Meilgaard (1975). The Beer Flavour Wheel (Fig. 6.3) developed by Meilgaard *et al.* (1979a, 1979b) describes beer in a technical language that brewers understand, and includes most groups of flavours encountered in beer. The basic principals of the Flavour Wheel are to provide a simple and easily understood terminology to enable brewers to communicate effectively about beer flavour and to provide a more comprehensive system to describe and define each separately identifiable flavour note in beer. The system contains three kinds of descriptors: class terms, first-tier terms and second-tier terms. The 14 class terms indicate the type of flavours that they contain, such as cereal, bitter (1200) or sulphury (0700). A four-digit number indicates that the term is intended for use as a descriptor. The first tiers contain familiar terms that represent flavours found in that class, for example,

0310 grainy and (0710) sulphitic. The second-tier terms name and define each separately identifiable flavour note in beer, and are used by the most expert and specially trained panels. These terms also have reference standards where available, such as (0732) DMS and (0620) diacetyl (Meilgaard *et al.*, 1979a, 1979b). In addition to similar flavours being grouped together, no terms are duplicated in the Flavour Wheel and subjective terms, such as good/bad, balanced/unbalanced are not included. This ensures that the Flavour Wheel terminology is a completely objective description of beer flavour and provides a technical and diagnostic language that can be related directly to flavours derived from raw materials or processing.

The Beer Flavour Wheel terminology is recognised and accepted throughout the brewing industry by bodies such as the European Brewing Convention (EBC), the American Society of Brewing Chemists (ASBC) and the Institute of Brewing (now the Institute of Brewing and Distilling (IBD)) and is recommended for use as the universal terminology to describe beer. The Sensory Analysis sections of the Analytica-EBC (1997) and the ASBC Methods of Analysis (1993) provide more details of reference standards and concentrations recommended for the training of these flavours in beer.

The brewing industry has the advantage over most other food products in that it already has an established vocabulary and set of training aids in order to build sensory panels within the brewery. In addition, there are published guidelines for the level of training recommended for brewery operatives depending on their degree of responsibility (Analytica-EBC, 1997; Taylor and Organ, 2009). Training panels for the objective evaluation of beer for quality control and measurement of production consistency can be tailored to reflect specific areas of production, thus increasing efficiency and building quality assurance.

6.4.2 Sensory evaluation of beer: protocol for flavour assessment

Beer dispense

As with other products, it is important that assessors evaluate beer using the same technique. Whether the beer is packaged in small-pack format or draught, the best temperature to assess beer flavour is around 8–12 °C. Serving beer at lower temperatures tends to suppress many of the positive (and negative) flavour volatiles, and serving the beer at higher temperatures results in many of the key flavour volatile components 'flashing off' during dispense. Whether the beers are presented in dark or clear glasses depends on the objective of the assessment, but ideally glasses should be covered with lids to trap the volatiles in the headspace for easy detection. The beers should be presented following good sensory practice, again depending on the objectives of the tasting. Beers should be presented with consideration of strength of flavour effects (alcohol strength and bitterness carry-over), and palate cleansers such as water and water biscuits should be provided.

Tasting technique

Beer has many volatile flavour components (e.g. esters, sulphur compounds and hop aroma compounds) as well as non-volatile tastants (e.g. hop bittering components) and compounds contributing to mouth-feel characteristics (e.g. alcohol, carbonation and polyphenols). It is the complexity of interaction of these compounds in the matrix of beer that determines the aroma and flavour of the final beer. Recent studies on the dynamics and mechanisms of aroma release from model systems show that aroma release in alcoholic beverages is more complex than other beverages and that further studies are required to provide insights into how aroma is perceived in a glass prior to consumption. It is essential, therefore, for each panellist to follow the same tasting procedure so that this complexity of beer flavour can be evaluated in the same way.

- Appearance: assess the beer foam for head depth, bubble size and colour and lacing (foam adhering to the glass during consumption). The colour and clarity of the beer is also considered.
- Aroma: swirl the beer in the glass with the lid on to release volatile components into the headspace. Remove the lid and assess the initial aroma, ensuring that a mental note is taken as some volatiles will flash off very quickly (e.g. sulphur compounds). Then take a few small sniffs to evaluate the intensities of the other volatile compounds.
- Taste: take a sample into the mouth and swirl to cover the inside of the mouth and tongue. Although there are taste buds all over the tongue, the bitterness of beer is detected more toward the back of the tongue and after swallowing. Assess the intensities of the different malt, hop and yeast-derived flavours. Also, the mouth-feel characteristics should be considered.
- After-taste: once swallowed, the after-taste of the beer is assessed, including the lasting flavours and length of linger.

6.4.3 Sensory evaluation of beer: raw materials and in-process tasting

During beer production it is important to ensure that any potential flavour problems are highlighted as early as possible. The evaluation of raw materials is essential to ensure that they give the correct sensory characters as expected and are free from any taints or defects that could affect the final beer. It is recommended that any raw materials and processing aids that may come into contact with the product at any stage be checked. Tasters should be familiar with the normal sensory characters of the raw materials, including visual appearance, smell and taste. In some cases where it may not be safe to taste samples due to potential microbiological hazards, aroma assessment will suffice, as a human's sense of smell is more sensitive than taste; but it must be noted that some taints, such as metallic and astringency, are not detected on aroma. In-process samples should be tasted routinely to ensure that the brew is progressing without deviation from the norm. Therefore, it is also vital that the people responsible for assessing the sensory quality have the correct training in order to be able to evaluate samples effectively. The number of tasters will depend on the resources available but it is recommended that no fewer than two trained assessors evaluate each sample.

Water
Brewing liquors, dilution waters and any steam condensates that are re-used in the main beer-stream should be assessed for aroma and taste for common water-borne contaminants such as TCA and sulphur compounds, and oily flavours from condenser leaks. Some phenolic taints can also be imparted from incorrect regeneration of de-ionisers. Untreated water should not be tasted on microbiological grounds but should be assessed on aroma. In addition, brewing salts should be assessed for off-flavours.

Malt
All malts, including speciality malts, should be checked visually and for aroma and taste. Common defects (apart from receiving the wrong malt) are green flavours from malts that have picked up moisture and have started to germinate, and musty flavours from mould growth. Suspect mouldy malts should not be tasted, as mycotoxins could potentially be present. It is advisable to microwave malts to destroy some of the microflora present on the malt (but not for too long as unusual flavours can be created in the malt).

If the result of a visual assessment is acceptable, some breweries will assess the flavour of the malt by chewing the malt or preparing a hot-water extract. Grinding the malt and mixing with hot water, and then passing through a filter, gives a hot-water extract. Hot-water extracts do not fully provide a true representation of the flavours imparted from the malt, as the amylases present in the malt and in the tasters' own mouth converts the starch in the malt to sugars, and therefore the flavour of these extracts is typically overwhelmingly sweet. Grinding the malt and mixing with cold water to give a paste can achieve a more representative flavour of the malt (Murray *et al.*, 1999). The aroma and taste of the malt can be assessed and in this way more subtle malt flavours can be detected, including any taints and astringent compounds.

Adjuncts such as rice, maize grits, wheat flour and brewing syrups should be assessed visually and by aroma to detect potential defects. Common taints in brewing syrups are butyric and lactic acid due to bacterial contamination. Brewing enzymes should also be assessed for off-notes.

Hops
Raw hops, leaf cones or pellets should be assessed visually and by rubbing for signs of deterioration; yellow/brown colouration and cheesy, catty notes on aroma are signs that the hops have been exposed to moisture or are aged. Hot-water extracts can be prepared for hop pellets but the extracts can be overpowering and very heady, leading quickly to sensory fatigue and even headaches, so caution must be exerted when preparing and evaluating hop teas.

Yeast
Pitching yeast should be assessed visually and for aroma. Typical fresh yeast flavours include pleasant fresh bready yeast with banana esters and sulphur notes. There should be no autolysed yeast (meaty) flavours.

Processing aids
Clarification and stabilisation agents such as isinglass, PVPP, PGA, antifoams, kieselguhr and other filter aids should be assessed visually and by aroma to ensure that they are taint-free. Many of these materials are hygroscopic and can pick up off-flavours easily. In addition, crowns and can lids should be checked for the presence of discolouration (rust indicates exposure to water) and musty, oily or phenolic taints picked up during transit. Any recovered carbon dioxide should also be checked.

In-process samples
The number of in-process samples evaluated depends on the size of the brewery and its resources. However, irrespective of the size of the brewery, there are key critical points during the brewing process at which it is pertinent to evaluate production samples to ensure that all is well with the brew.

These critical points include sweet wort and hopped wort production, fermentation tanks, maturation tanks and bright beer tanks. At each stage it is important for the assessor to be aware that the samples are exhibiting the correct flavour notes (even if that means that diacetyl and sulphur notes should be present) and the critical feature is that the beer is free from taints. Up to bright beer production, if the beer has a slightly atypical flavour profile, breweries have the option to blend the beer with other brews. The most critical stage is the evaluation of packaged beer for clearance for sale. At this point, the packaged beer can be evaluated using a number of different methods and these are covered in the next section.

6.4.4 Methods used for evaluating production beer

The following section outlines methods commonly used in breweries to evaluate the flavour quality of in-process and beer samples. Further details of each test can be found in Chapter 2.

Pass out
This is the simplest method to assess whether the sample can be released to the next stage of the brewing process for in-process samples or for release to trade for packaged beer. The procedure requires assessors to be familiar with the sensory characteristics of the samples they are evaluating and must have training in relevant taints that may be encountered. The result of the test is a pass/fail, go/no go.

Trueness-to-type
This descriptive test appears to be unique to the brewing industry for final packaged beer and does not feature in sensory textbooks (Meilgaard *et al.*, 2007; Stone and Sidel, 1993; Lawless and Heymann, 1999). One reference was found in the *Sensory Analysis Manual* (Institute of Brewing 1995). The method is being incorporated into Section 13 of the Analytica-EBC (European Brewing Convention, 1997). Assessors evaluate the beer based on whether the beer is true

to the brand. This involves determining whether the beer exhibits all the correct appearance and flavour characteristics and is free from off-flavours. The assessors are told the brand identity before the evaluation and must be familiar with the flavour of each brand that they assess. It is common to compare the intensities of key flavour characteristics against a reference profile for the brand and score up/equal/down accordingly. At the end of the assessment, the overall trueness-to-type score is calculated from individual responses and, usually, a mean is taken or a histogram generated for all the scores. It is also common to weigh the presence or absence of key flavour characteristics and certain off-flavours. This test requires a lot of training to conduct and is time-consuming for the panel leader to generate results but is also very common in breweries for flavour quality assessment and release to trade. It is suggested that a minimum of 10–20 assessors conduct this test but for many smaller breweries this number is impractical and as many trained panellists as possible are used.

Difference testing

Difference tests most commonly used in larger breweries include triangle tests, difference from control and pass/fail tests (go/no go). Small breweries have the disadvantage of numbers recommended to conduct some of these tests (18+ panellists for difference testing and difference from control and 42+ for similarity testing) and tend to conduct pass/fail tests, trueness-to-type tests or descriptive profiling. Ranking tests sometimes are used for blending or shelf-life studies but in the main are reserved for training panellists in the determination of flavour intensities or threshold testing.

Profiling

Full descriptive profiling of attributes in a beer gives the definitive flavour fingerprint for quality assessment. Panellists are trained in the individual flavour attributes as recommended by the European Brewing Convention (1997) and American Society of Brewing Chemists (1993) and a comprehensive guide is given by Taylor and Organ (2009). Training a profile panel is time-consuming and expensive but the rewards provided by a fully trained and consistently performing panel is considerable. Implementing an expertly trained profile panel can provide information on the types and intensities of flavour attributes of a beer, measure differences within own beers and competitor beers, contribute to quality assurance and brand consistency and, ultimately, brand success and consumer loyalty.

Consumer research for NPD: Preference testing including central location tests

Consumer loyalty: the very words that determine the success or failure of a product. Consumer research within breweries is typically the final stage of one or two trial samples to determine which is the most suitable for full-scale production (Taylor and Organ, 2009). In consumer tests, the question to be answered is normally two-fold: which is the preferred sample, and which sample has the greatest drinkability.

Drinkability is a term widely used to describe the characteristic of beer that results in the drinker wanting to have another glass. In research studies, drinkability has been assigned both as a specific attribute in beer and as a response (Mattos and Moretti, 2005) and is therefore itself not a widely understood or studied issue. To date there is no widely accepted methodology to evaluate the drinkability (or sessionability) of beer.

In order to explore the concept of drinkability, brewers often use extended drinkability testing. Consumer research is conducted using a central location test (CLT). At the start of the test the respondents are each presented with a sample of beer (usually blind and unbranded) and are asked to complete a short questionnaire upon presentation of the first sample. The questions asked in the test questionnaire at this stage normally relate to the visual aspects of the sample, and, perhaps, expectation of the product. The respondents are then asked to consume the sample in their own time and complete a second questionnaire. Questions at this stage relate to the aspects of the taste and liking of the sample. The test continues until the end of the session time or the required number of pints for each product is consumed. At the end of the session, the respondents complete an exit questionnaire usually asking about preference and purchase intent.

In general, the types of questions asked are unstructured, using both hedonic and just-about-right (JAR) scales. The preferred sample is the one with the highest preference score, and the most drinkable beer is deemed that for which the most volume was consumed. It must be emphasised that 'preference' and 'drinkability' do not mean the same thing. The beer that is preferred for the taste is not necessarily the same that will be drunk in larger volumes by the consumer (Mattos and Moretti, 2005).

6.5 Future trends

The challenges facing today's brewers is the globalisation of major brewers and amalgamation of those breweries having unique and household named brands, an increasingly competitive market and a worldwide recession hitting consumers' pockets.

Never before has consumer loyalty been so precious, and the assurance that the product delivers to the expectation of the consumer is one of paramount importance. Sensory evaluation has never had such an important role in the brewery, with micro, small, medium, large and global breweries investing in the assurance that a competent sensory capability can give. More schemes for proficiency testing of the performance of panels are increasing. Breweries increasingly have to be mindful, too, of the amount of beer that their panels are being asked to consume during daily testing, and potential panellists may also be asked to sign disclaimers and even to agree to a medical examination before becoming a member of the tasting panel.

Also, the industry is searching for more acceptable ways of testing beer in consumer tests, as traditional extended drinkability testing involved male respondents consuming relatively large quantities of beer. This method of testing has now led to concerns over the ethics of asking respondents to drink in excess of their daily recommended number of alcohol units, particularly in the current climate of the 'responsible drinking' campaign spearheaded by government.

International guidelines exist for recommended drinking levels, that is, the daily maximum recommended number of units of alcohol to be consumed. These definitions of moderate consumption vary country by country and are based according to human body size. For example, in the UK for men 4 units (32 g alcohol) per day is the recommended drinking level and less for women (24 g alcohol) whereas in France the level for men is 60 g alcohol per day.

Beer is one of the few products where preference for a particular beer can change with the consumption of that product, and the consumer may switch to another brand upon a second, third, etc., drink. Therefore, a degree of brand switching can occur within a session. New sensory methods are being developed using advances in consumer science to look behind the decision-making of consumers to see what is really controlling choice and satisfaction. Consumer and sensory science methods have been applied to investigate factors influencing liking and preference and the desire to continue drinking more of the same product, but using a small volume taste test to comply with industry compliance of moderate drinking (Parker, 2009).

6.6 Sources of further information

For more in-depth information on barley, malt, hops and the basics of brewing, the following textbooks are an excellent source:

- BRIGGS, D.E., BOULTON, C.A., BROOKES, P.A. and STEVENS, R. 2004 *Brewing: Science and Practice*. Woodhead Publishing Ltd, CRC Press LLC Cambridge.
- KUNZE, W. 2004 *Technology Brewing and Malting* (3rd edn). Verlagsabteilung der VLB Berlin, Berlin.
- TAYLOR, B. and ORGAN. G. 2009 *Sensory Evaluation*. Handbook of Brewing: Processes, Technology, Markets. 29.Wiley VCH Berlin.

For research centres in brewing:

- Campden BRI, Nutfield, UK, www.bri-advantage.com, foremost provider of scientific support and technical services to the brewing industry worldwide,
- Heriot Watt University, Edinburgh, UK www.icdb.hw.ac.uk,
- Siebels Institute of Technology, Chicago, USA www.siebelinstitute.com.

For brewing organisation websites:

- Institute of Brewing and Distilling www.Ibd.org.uk,
- British Beer and Pub Association www.beerandpub.com,
- European Brewing Convention www.europeanbreweryconvention.org,
- American Society of Brewing Chemists www.asbcnet.org,

- Master Brewers Association of the Americas www.mbaa.com,
- Maltsters Association of Great Britain www.ukmalt.com,
- The Brewers Association of Australia and New Zealand www.brewers-association.net.au,
- The Brewers Association of Japan www.brewers.or.jp,
- The Small Independent Brewers Association www.siba.co.uk.

6.7 References

AMERICAN SOCIETY OF BREWING CHEMISTS 1993, *ASBC Methods of Analysis, Sensory Analysis – 3 Choice of Method (International Method).*

AMERICAN SOCIETY OF BREWING CHEMISTS 1993, *ASBC Methods of Analysis, Sensory Analysis – 12. Flavour Terminology and Reference Standards (International Method).*

BAMFORTH, C.W. 1998 *Beer: Tap into the Art and Science of Brewing*, Plenum Press Berlin, 123.

BAMFORTH, C.W. 2001a, 'Beer flavour: malt and hop character', *Brewers' Guardian*, Nov. 6–18.

BAMFORTH, C.W. 2001b, 'Beer flavour: esters', *Brewers Guardian*, Sept. 32–43.

BRIGGS D.E., HOUGH, J.S. and YOUNG, T.W 1986, *Malting and Brewing Science*, Vol 1. 94–221.

CHANDRA, G.S. BOOER, C.D., DAIES, N.L. and PROUDLOVE, M.O. 1997, 'Effect of air-recirculation during kilning on malt and beer flavour', *6th Proceedings of the Institute of Brewing (C & SA Sect.), Durban*, 23–28.

DUMOULIN, M. and BOLVIN, P. 2001, 'Industrial kilning technologies and their influence on organoleptic quality of malt', *Proceedings of the European Brewing Convention*, Budapest, 10.

EUROPEAN BREWING CONVENTION (1997), *Analytica-EBC Section 13 Sensory Analysis. Method 13.12 Flavour Terminology and Reference Standards (IM)*, Fachverlag Hans Carl, Nurnberg.

EUROPEAN BREWING CONVENTION (1997), *Analytica-EBC Section 13 Sensory Analysis. Method 13.14. Routine Sensory Check of Raw Materials and In-process Samples*, Fachverlag Hans Carl, Nurnberg.

GUIDO, L.F., BOIVIN, P., BENISMAIL, N., GONCALVES, C.R. and BARROS, A.A. (2005), 'An early development of the nonenal potential in the malting process', *Proceedings of the European Brewing Convention*, Prague, 13.

HILL, P. (2003), 'Keeping the flavour wheel turning – The development of a flavour stability wheel', *Proceedings of the European Brewing Convention*, Dublin, Paper 75.

INSTITUTE OF BREWING (1995). *Sensory Analysis Manual*, Institute of Brewing, London.

KAGEYAMA, N., TADA, N., INUI, T., KOMURA, H., NAKAHARA, K. *et al.* (2007), 'Studies on astringent substances in malt', *Proceedings of the European Brewing Convention*, Venice, 21.

LAWLESS, H.T. and HEYMANN, H. (1999), *Sensory evaluation of food.* Gaithersburg, Maryland, Aspen Publishers, Inc.

LIEGEOIS, C. and COLLIN, S. (2003), 'Contribution of malt kilning to the cardboard flavour of aged beers', *Proceedings of the European Brewing Convention*, Dublin, 9.

MATTOS, R. and MORETTI, R.H. (2005), 'Beer drinkability: A review', *Technical Quarterly of the Master Brewers Association of the Americas,* **42**:13–15.

MEILGAARD, M.C. (1975), 'Flavour chemistry of beer: Part II: Flavour and threshold of 239 aroma volatiles', *Technical Quarterly of the Master Brewers Association of the Americas*, **12**(3), 151–168.

MEILGAARD, M.C., DALGEISH, C.E. and CLAPPERTON, J.F. (1979a), 'Beer flavour terminology', *Journal of the American Society of Brewing Chemists*, **37**, 47–52.

MEILGAARD, M.C., DALGEISH, C.E. and CLAPPERTON, J.F. (1979b), 'Beer flavour terminology', *Journal of the Institute of Brewing*, **85**(1), 38–42.

MEILGAARD, M.C., CIVILLE, G.V. and CARR, B.T. (2007), *Sensory evaluation techniques*. CRC Press Cambridge.

MURRAY, J.P., BENNETT, S.E., CHANDRA, G.S., DAVIES, N.L. and PICKLES, J.L. (1999), 'Sensory analysis of malt', *Technical Quarterly of the Master Brewers Association of the Americas*, **36**(1), 15–19.

PARKER, D. (2009), 'Stay of swap: The decision is yours', *European Brewing Convention*, Hamburg, L32.

PENGELLY, W. (2001), 'Malt flavour and aroma', *The New Brewer*, Nov/Dec. 15–17.

PENGELLY, W. (2002), 'Hop flavour and aroma', *The New Brewer*, March/April, 19–24.

PERPETE, P. and COLLIN, S. (1999a), 'Contribution of 3-methylthiopropionaldehyde to the worty flavour of alcohol-free beers', *Journal of Agriculture, Food and Chemistry*, **47**(6), 2374–2378.

PERPETE, P. and COLLIN, S. (1999b), 'Fate of worty flavours in a cold contact fermentation', *Food Chemistry*, **66**(3), 359–363.

PREUSS, T., FORSTER, C., THUM, B. and BACK, W. (2001), 'Investigations into flavour and flavour stability of dark beer and dependence on malting and kilning parameters', *Proceedings of the European Brewing Convention*, Budapest, 9.

STONE, H. and SIDEL, J.L. (1993), *Sensory Evaluation Practices*. Academic Press Inc USA.

TAYLOR, A.J., TSACHAKI, M., LOPEZ, R. and LINFORTH, R.S.T. (2007), 'Aroma release from beer: Dynamic mechanistic studies', *Proceedings of the European Brewing Convention*, Venice, 109.

TAYLOR, B. and ORGAN, G. (2009) *Sensory Evaluation. Handbook of Brewing: Processes, Technology, Markets. 29.* Wiley VCH Berlin.

WEISS, A., SCHONBERGER, C.H., MITTER, W., BIENDL, M., BACK, W., *et al.* (2002), 'Sensory and analytical characterisation of reduced, isomerised hop extracts and their influence and use in beer', *Journal of the Institute of Brewing*, **108**(2), 236–241.

7

Fortified wines: styles, production and flavour chemistry

A. G. J. Tredoux, University of Stellenbosch, South Africa and A. C. Silva Ferreira, University of Stellenbosch, South Africa and Universidade Católica Portuguesa, Portugal

Abstract: Fortified wines, traditionally from Europe but currently produced worldwide, are considered to be a specialty product in the wine industry. These wines are known to be of exceptional quality as a result of extensive and specific production processes followed, including exceptionally lengthy ageing steps, and have a significant economic impact in the global wine industry. The chapter gives an overview of the products' status in terms of the types produced, countries of origin, and the various grape cultivars utilized. Styles and types are discussed is terms of production processes; including sugar and spirit addition and ageing regimes. The second section deals with flavour chemistry of fortified wines and considers the main mechanisms for the evolution of flavour compounds and the role that specific, key aroma compounds play. Regulatory and health aspects are discussed.

Key words: fortified wine, production, aroma compounds, flavour evolution mechanisms, authentication.

7.1 Introduction to fortified wines

Fortified wines are wines to which a distilled beverage, usually brandy (grape spirits), has been added (Lichine, 1987). The production of fortified wine can broadly be divided into the following stages: selection of grape cultivar, treatment and crushing of grapes, alcoholic fermentation, addition of grape spirits, and maturation (ageing), either in oak barrels or in the bottle. Each of these steps, as well as the sequence in which they are carried out, has a decisive influence on the style and quality of the final product.

The historical reasons behind the introduction of the fortification stage in the winemaking process have been somewhat romanticized, principally for marketing purposes. The reasons usually cited, however, are largely based on anecdotal evidence and may arise from traditional folklore, mythology or ancient beliefs; as a result, we consider them to fall outside the scope of this chapter, and will not discuss them further here. Nevertheless, given that the introduction of grape spirits into the fermentation process does in fact have a significant effect on the shelf life of the product, it can be assumed that fortification was originally intended as a means of improving the sensory stability of wine.

The chemical compounds responsible for the aroma of wine arise from the variety of the grape and from alcoholic fermentation and maturation, and are usually categorized according to the production stage from which they originate. Those from the grapes, including the 'varietal aroma', are broadly described as primary compounds; fermentation-derived compounds constitute the secondary group; and compounds arising from maturation make up the tertiary 'aged bouquet'. There is much less diversity in the aged bouquet (aroma) of aged wines compared to that of young wines. This sensory convergence is a result of chemical changes in wine over a long period (20–50 years) and happens spontaneously according to thermodynamic restrictions. These changes occur irrespective of factors such as the cultivar or variety used, terroir, viticultural practices, vinification procedures and ageing regimes.

In the case of fortified wines, however, there is a fourth component, the addition of grape spirits, which adds to their distinct aroma and complexity. The effect of these spirits on the aroma of the final product is dependent upon the chemical composition of the spirits used, as well as the stage of the fermentation process at which they are introduced.

In fact, the addition of grape spirits can play a role in both principal wine spoilage processes, namely microbial action and chemical reactions such as oxidation.

Firstly, grape spirits can provide microbial stability by creating an unfavourable environment for microbial action, thus increasing the shelf life of the product. A long shelf life is extremely important for fortified wines, as it is generally accepted that the quality and value of the product increases with age. However, although grape spirits can provide microbial stability, they can also cause chemical reactions such as oxidation, which lead to a change in the sensory characteristics of the wine. Consumers usually judge oxidation of foodstuffs as unacceptable; however, in the exceptional case of fortified wines, oxidation actually provides the desirable 'aged' oxidative character that consumers expect from the product.

The chemical mechanisms responsible for oxidation and other chemical changes in fortified wines are far from completely understood; our understanding of most of the reactions involved remains limited, with the exception of the Maillard reaction and Strecker degradation. We are thus at a disadvantage when it comes to adequately managing the development of the wine over time. If we could acquire a more comprehensive understanding of the underlying chemical and kinetic mechanisms, it would then become possible to manage ageing processes more accurately and in turn to control and predict the shelf life of the product as well as its expected bouquet after many years.

7.1.1 Countries of origin, production and regional distribution

Fortified wine historically mainly originates from Europe but is now also extensively produced in the 'new world', most often employing the same traditional European production principles (Fig. 7.1).

The most recognized Portuguese fortified wines are the port wines, produced from grapes growing in Douro Valley in northern Portugal, the Madeira wines, made on the Madeira Island, and the Moscatel wines, growing in the Setúbal region of Portugal.

In Spain, the most famous fortified wines are the sherry wines produced in the regions of Jerez, Montilla-Moriles and Manzanilla in southern Spain, and Málaga and Montilla wines, from the province of Andalucía along the Mediterranean coast.

In France, 'vins doux naturels' (VDN) and 'vins de liqueur' are sweet fortified wines from the Rhone, Roussillon and Languedoc regions. Vermouth is also a French fortified wine but in addition it is aromatized as well.

In Italy, the most renowned fortified wine is the Marsala wine, from Italy's sunny southern region of Sicily. Vermouth is also produced in Italy and it is believed that this aromatized fortified wine originating from France, was produced initially as a 'tonic' for certain ailments.

In Greece, the Mavrodaphne fortified wine is a Greek dessert wine produced in the Achaia region in Peloponnese.

Regarding the new world fortified wines, a few countries of production are discussed below. This is, however, not intended to be an exhaustive discussion and since many of these wines are produced using grape varieties not corresponding to the traditional varieties used but rather following mainly the traditional production process, the cultivars commonly used will be included here, in contrast to the discussion on the 'traditional' cultivars used in the following section.

In Australia, new world port-style wines are produced extensively from Shiraz, Grenache, and Carignan grape varieties, while sherry-style wines are mainly produced

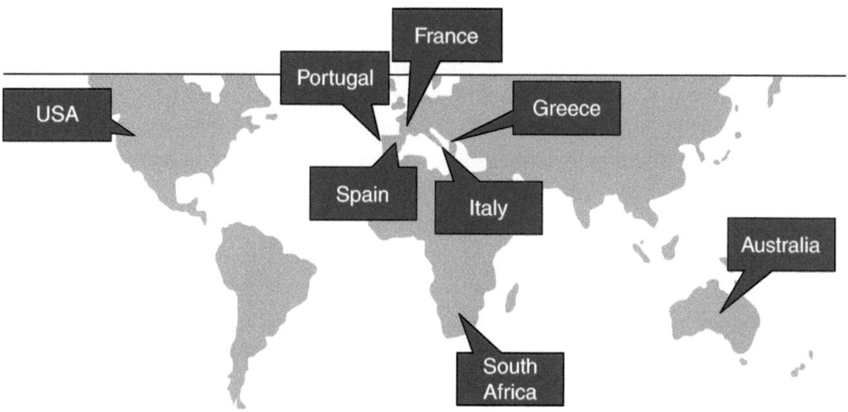

Fig. 7.1 Distribution of fortified wine production worldwide.

from Palomino grapes, and aged according to the 'Solera system'. Other popular fortified wines include Tawny (made from Shiraz, Grenache, Touriga, Mataro and Sauvignon Blanc) and Muscat (Winemakers' Federation of Australia, 2011).

In South Africa (SA), new world sherry-style wines are produced predominantly from Chenin Blanc (the most planted grape variety in SA) and Palomino grapes, grown in the Western Cape region, and are also matured according to the Solera system. Port-style wine is produced from Shiraz and Pinotage, as well as Portuguese varieties such as Tinta Barroca and Touriga Nacional. The styles produced closely resemble their Portuguese counterparts. Muscat (known locally as Muskadel) is produced extensively from Muscat grapes grown locally (*Wine Magazine*, 2005; Jackson, 2000).

In the United States (California), port-style wine is produced from Carignan, Petite Sirah, and Zinfandel grapes grown in the cooler regions, while Ruby-red and Sousão are the preferred varieties in the warmer regions. Concerning sherry production in the US and Canada, the preferred method appears to be the baking technique. Grapes used are Niagara in eastern North America and Palomino; and Thompson seedless and Tokay in California (Jackson, 2000).

Worthwhile mentioning here is that in California, Australia as well as Canada, a flor procedure significantly different from the Spanish model has been adopted. This submerged culture technique requires constant agitation to keep the yeast submerged in the wine and has the advantage of increasing the rate of chemical changes, notably the formation of acetaldehyde (Posson, 1981).

7.1.2 Cultivars

More than thirty grape varieties are grown in the port wine region but only five (red grape varieties) are considered essential components of fine port wine; these are Touriga Nacional, Touriga Francesa, Tinta Roriz, Tinto Cão, and Tinta Barocca. The varieties are extremely resilient as the climate in this region is very harsh. The winters are bitterly cold, although it rarely snows, whilst in summer the temperature often reaches 41 °C. Where the Tourigas provide berry aromas of a rich intensity, Tinto Cão contributes its own unique aromatic complexity, and its wines are also unusually long-lasting. Tinta Barocca is able to withstand cool conditions, and it is usually planted on north-facing slopes, where its grapes grow in large bunches (Taylor Fladgate, 2011). Port wine can be made with white grape varieties as well, and mostly Donzelinho Branco, Esgana-Cão, Folgasão, Gouveio, Malvasia Fina, Rabigato and Viosinho are used.

Madeira wines are made from the noble white grape varieties Malvazia, Boal, Verdelho, and Sercial. A red grape variety Tinta Negra Mole can also be used.

Moscatel wines from the Setúbal region of Portugal are produced from white or red grape Moscatel vine varieties. During the winemaking process, after the addition of spirits to stop fermentation, the grape seeds and the skins of the white Moscatel grape vine varieties stay in contact with the wine for several months.

The principal wine grape variety of sherry wines is Palomino (90%), although Pedro Ximénez, Perruno, Cañocazo, Albillo, Mantuos and Moscatel are also used.

Grape bunches are generally long and cylindrical in shape with a medium to high density of spherical, medium sized, thin-skinned berries that are yellowish-green in colour. The grapes are juicy, fragile, sweet and flavourful with colourless juice.

The two principal white grape varieties used for production of Málaga and Montilla wines are Pedro Ximénez and Moscatel.

VDN can also be made from white and red grapes. The white grapes are Muscat varieties, mainly Muscat of Alexandria and Muscat Blanc a Petits Grains. The red VDN are mostly made from red Grenache grapes and denote burnt notes of cocoa, chocolate, caramel and coffee, combined with aromas of dried fruits, prune, jam, cherries in brandy and crystallized fruits. Vins de liqueur are also made from Muscat grape varieties. The only difference between these wines and VDN wines is the time of spirit addition.

The four main white grape varieties of Marsala wines are Grillo, Catarratto, the highly aromatic Inzolia grapes, and Damaschino grapes. The ruby-coloured Marsalas hail from local red grape varieties like Pignatello, Calabrese, Nerello Mascalese, and Nero d'Avola.

Vermouth originated in Italy in the 1700s. Two distinct types are made; i.e. the sweet Italian Vermouth and the dry French Vermouth. Often the wine used is a neutral-flavoured white variety, although the very aroma-rich Muscato bianco variety from Piedmond is used for the world's most renowned Vermouths. A notable difference between this wine and other fortified wines is the addition of herbs and spices for aromatizing purposes.

7.2 Comparison of styles and types of fortified wines and their production methods

7.2.1 Sugar content

According to the sugar content, five types of port wines can be distinguished: very sweet (typically denotes a residual sugar content above 130 g/l), sweet (residual sugar typically between 90–130 g/l), medium dry (residual sugar typically between 65–90 g/l), dry (residual sugar content between 40–65 g/l), and extra dry (residual sugar content less than 40 g/l).

Regarding Madeira wines, four basic types can be distinguished according to the sugar content: sweet, medium sweet, medium dry and dry. They are named after the principal grape variety from which they were obtained. Malvazia is fortified early in order to produce a distinctly sweet wine with a residual sugar content of approximately 110 g/l. Boal is fortified after approximately one half of the sugar has been converted to ethanol. This yields a medium sweet wine (90 g/l). The Verdelho ferments even longer than than Boal to produce a medium dry wine (65 g/l of sugar). Sercial is allowed to ferment until a large fraction of the total sugar has been converted to ethanol, giving a dry wine with 25 g/l of sugar (Câmara *et al.*, 2004a).

Sherry wines are typically sweetened with PX or mistelle. PX is juice extracted from the sun-dried berries of the 'Pedro Ximénez' variety and has a remarkable raison aroma and fortified to 9% v/v ethanol content. Mistelle is mado of Palomina grape juice to which ethanol is added to about 20% v/v. (Franco *et al.*, 2004). As is the case with other fortified wines, sherry wines are also produced in various degrees of sweetness, ranging from extra dry to sweet (cream, 'dulce') (Reader and Dominguez, 2003).

Italian Vermouth contain between 40–160 g/l, depending on the sweetness, whereas French Vermouth contain about 40 g/l. For sweetening purposes mistelle is added.

7.2.2 Addition of spirits

Generally, fortification of port wines, which involves the addition of natural grape spirit to the fermenting juice, intentionally interrupts the fermentation process at a point when approximately half of the grapes' natural sugar has been converted into alcohol, resulting in wine containing 19–21% alcohol. The spirit added to port wine, in contrast to many other fortified wines, is not highly rectified (77% v/v) and therefore still contains several other potentially odour-active volatiles, notably higher alcohols. This accounts for port's characteristic rich, luscious style and also contributes to the wine's considerable ageing potential. Given the short fermentation cycle it is crucial to extract as much flavour, colour and tannins as possible from the grape skins (Pissarra *et al.*, 2005; Rogerson *et al.*, 2002).

Madeira wine can be characterized by typical vinification and ageing methods that include fortification with neutral rectified grape spirits (95% v/v), in order to obtain an ethanol content of about 18–20% (v/v), followed by a baking process known as 'estufagem' (Câmara *et al.*, 2004a).

During the winemaking process of Moscatel wine of Setúbal, after the addition of spirits to stop fermentation, the grape seeds and the skins of the white Moscatel grape vine varieties stay in contact with the wine for several months (Wines of Portugal, 2011).

Sherry wines are classified as fino, oloroso, amontillado and cream sherries, and are fractionally blended three or four times a year using a procedure named the Solera system (Martínez de la Ossa *et al.*, 1987a, 1987b).

Fino sherries result from biological ageing by a mixture of unique flor yeasts such as *Saccharomyces beticus, Saccharomyces cheresiensis, Saccharomyces montuliensis* and *Saccharomyces rouxii*, growing on the surface of the wine during production. The mainly aerobic character of this metabolism results in a typical flavour and endows the sherry with a light colour (Llaguno, 1968; Goswell, 1986). Manzanilla sherry presents a special kind of fino. This is as a result of maintaining a specific microclimate allowing the flor yeast to continue growing during the colder winter months, in contrast to the traditional sherries of the Jerez region. Sherry wines are fortified with a 50/50 blend of 95% neutral rectified spirits and aged sherry, called miteado. Oloroso sherry results from the addition of up to 18% ethanol to the wine, thus preventing the growth of flor yeasts. Under these physio-chemical oxidative ageing conditions olorosos develop a dark color and a typical flavour. Amontillado sherries are products of a combination of the above-mentioned methods. When production commences wines are treated similarly to fino sherries. After the period of biological ageing the product is laced with ethanol and subjected to oxidative ageing comparable with the oloroso method. Cream sherries, also named dulce or sweet sherries, are produced and fortified usually in accordance to the oloroso style and sweetened with concentrated grape must or blended with a sweet dessert wine (Patzold *et al.*, 2003). Addition of both grape juice concentrate as well as grape spirits to these wines is common, as is fractional blending employing the Solera system (Berlanga *et al.*, 2004; Margalit, 2004).

Regarding Málaga wine, regulations state that the grapes must be brought to Málaga to age in its warehouses. This rich, sweet, raisiny wine was traditionally made by leaving the grapes outdoors in the sun on grass mats for 7–20 days to concentrate the naturally occurring sugars. Nowadays, other methods such as the addition of boiled-down must (Arrope) and interrupting the fermentation with grape spirits are also employed to achieve the same effect (Reader and Dominguez, 2003).

Montilla wine has a high natural strength (14–16% alcohol). The wines are fermented quickly in tall open clay jars called 'tinajas'. The wines fall into the same classifications as sherry: fino, oloroso, and amontillado, but outside Spain must be sold as pale dry, medium dry, pale cream and cream (Reader and Dominguez, 2003).

In Vermouth, the spirit added is rectified ethanol (85% v/v); in the case of Italian Vermouth it is already infused with herbs in comparison with its French counterpart where the herbs are added to the wine after fortification.

7.2.3 Ageing

After vinification, port wines are blended and matured in the lodges of Vila Nova de Gaia in oak casks and can be divided into two major categories according to the method of ageing: Tawny style or barrel aged wine and Ruby style or bottle aged port. Interestingly, these 'bottle aged' wines are in fact also kept in the barrels, however only for a significantly shorter period of time. The Ruby style comprises the following categories, in increasing order of quality: Ruby, Reserve, Late Bottled Vintage (LBV) and Vintage. The Tawny style wine is obtained by blending several different wines that have been aged for different periods of time in the casks or in the vats. For these wines the age indicated on the bottle is an average age of all blended wines. The present categories in this style are: Tawny, Tawny Reserve, Tawny with an indication of age (10, 20, 30 and 40 years old), and 'Colheita'. These are blends of wines from several years, except for Colheita, which is a wine of a single year that is similar to an aged Tawny of the same age.

7.2.4 Comparative overview of fortified wine production processes

The process for producing port wine is illustrated graphically in Fig. 7.2. Since this process has many similarities with the production of other fortified wines, the

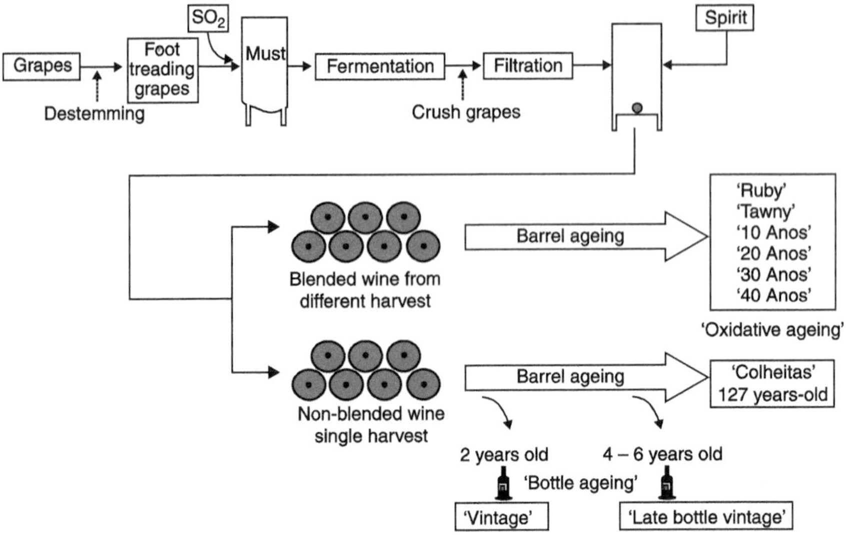

Fig. 7.2 Schematic representation of the port wine production process.

production processes for other fortified wines will be discussed in terms of similarities and differences from this production scheme. A number of these aspects have been discussed in previous sections and therefore this comparative overview will focus on key production steps rather than a detailed elaboration.

For the production of port, sherry and Madeira, grapes are harvested at optimum maturity. Of these three, port and Madeira are made of red and white varieties but these are always vinified and matured separately; in contrast to sherry for which only white grapes are used. After harvesting, white grapes (in the case of sherry and white Madeira) are crushed and pressed and undergo alcoholic fermentation. Red varieties (port and Madeira) are fermented with skin contact and pressing is only done after a relatively short fermentation period (48–72 hours) and subsequently filtered. As mentioned earlier, the time of spirit addition is highly dependent on the desired style to be achieved. For example, when dry base wines are required for the production of fino and oloroso sherries, fermentation is allowed to proceed to dryness, whereas for the production of, for example, Moscatel and Pedro Ximénez, spirit addition is performed once the desired sweetness is achieved. Considering blending and ageing, port wines are initially stored in oak barrels and in the case of bottle-aged port, transferred to bottles after a short period of time. Barrel-aged ports are kept in oak vats for about two years after which they are blended and aged further for up to thirty years or longer. Sherry wines are transferred to oak barrels and blended and aged using the fractional blending system (Solera system) for up to twelve years. Madeira wines are first subjected to a three-month heat treatment (Estufagem) in sealed tanks and then transferred to wooden vessels and aged for up to one and a half years. This is followed by blending and ageing for up to thirty years (Reader and Dominguez, 2003).

7.3 Flavour chemistry and sensory properties

7.3.1 Flavour evolution mechanisms

Chemical changes occurring during winemaking are results of both enzymatic and non-enzymatic reactions and are responsible for differences in the sensorial perception throughout the process. Several simple mechanisms are recognized as role-players in this phenomenon and, as a result of interactions and overlapping, these are responsible for the compositional complexity perceived, such as changes in colour, taste and flavour; and ultimately influencing the quality of the final product. The kinetic chemical parameters that affect the rate and the extent of each of these reactions can be calculated, and therefore, theoretically, it would be possible to modulate them, and ultimately, if the overall topography of the dynamic system created by the contribution of these kinetic parameters of each mechanism is known, control the overall process.

A common denominator in the production of most fortified wines is that they are produced under oxidative conditions. Therefore, oxidation reactions are one of the most important processes determining these wines' typical aroma and flavour profile. Other mechanisms occurring during grape juice heating or storage are caramelization and the Maillard reaction (Maillard, 1912), and are influenced by a combination of numerous factors including time, pH, temperature, dissolved oxygen and SO_2, but with temperature having the largest effect.

Sugar degradation: caramelization and the Maillard reaction

For caramelization to occur, high temperatures are required. This can occur during the production of some sweet wines in red and white wines vinified from raisined berries (Zoecklein *et al.*, 1995). In some Spanish regions a special product named 'grape syrup', obtained after boiling the grape must for a long time in a base wine, is added and increases sweetness and brown tonalities.

Melanins, melanoidins and caramels have several structural similarities. Melanins are brown pigments and are generally a result of enzymatic oxidation, in contrast to melanoidins and caramels that are formed via non-enzymatic oxidation under heating and final products of Maillard reactions. The Maillard reaction (Fig. 7.3), a reaction involving condensation of reducing sugars with amino acids and proteins, occurs in foods during processing, cooking, even during storage, and proceeds well at 50 °C, at a favoured pH of 4–7. Although the Maillard reaction has been found to occur in beer and other foods, there is little supportive evidence for the role it plays in wine browning (Main, 1992; Zoecklein *et al.*, 1995). Recent studies indicate that a number of volatile compounds responsible for typical aromas or ageing aromas of some sweet natural wines seem to be chemically linked with similar Maillard-like reactions between sugars and amino acids (Potman, 1989; Vernin, 1982; Marchand *et al.*, 2002; Moreno *et al.*, 2007; López de Lerma, 2009,

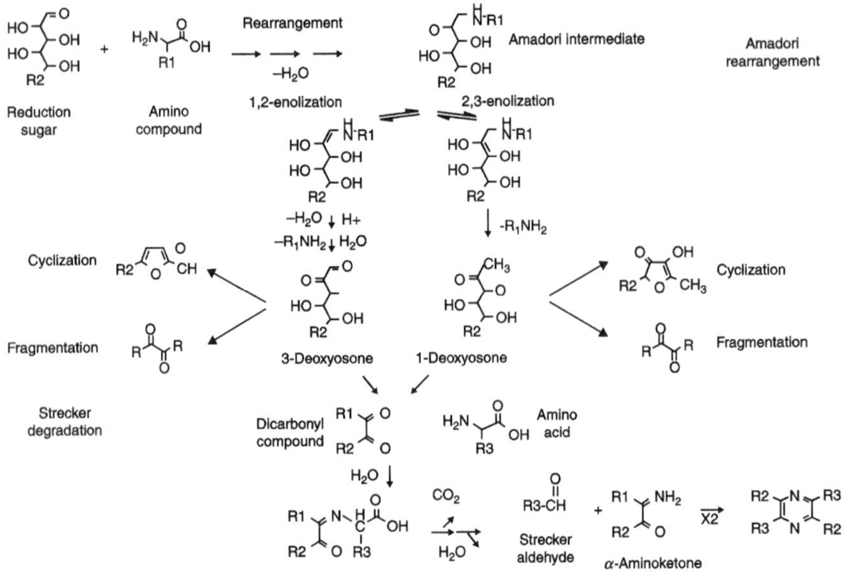

Fig. 7.3 The Maillard reaction.

2010). These reactions can also occur during off-vine drying or during exposure to direct sunlight, causing browning and anti oxidant activity (Franco *et al.*, 2004; Frank *et al.*, 2005; Moreno *et al.*, 2007).

The Strecker degradation is a minor pathway in the Maillard reaction; nevertheless, it is very important in the formation of many volatile organic compounds (Chan and Reineccius, 1994). This degradation reaction comprises a complex mechanism. An amino acid, in the presence of α-dicarbonyl compounds, is decarboxylated and deaminated, producing an aldehyde with one carbon atom less than the amino acid and is known as 'Strecker aldehyde' (Shonberg and Moubacher, 1952; Pripis-Nicolau *et al.*, 2000; Marchand *et al.*, 2000; Keim *et al.*, 2002).

Although the formation of baked potato aroma (methional) and floral notes (phenylacetaldehyde), occurring during storage, can be a result of direct oxidation of the respective alcohols, there is evidence that it can also be formed via Strecker degradation of the respective amino acid (Escudero *et al.*, 2000; Silva Ferreira *et al.*, 2002); however, the main pathway for their formation is not clearly established. Some authors have suggested the possible involvement of an intermediary keto-acid between the amino acid in question and the Strecker aldehyde formed (Silva Ferreira, 1998; Hofmann *et al.*, 2000; Granvogl *et al.*, 2006).

Other mechanisms, such as esterification, can also contribute to amino acid consumption in wine, and the respective products could also participate in wine aroma (Herraiz and Ough, 1992, 1993; Herraiz *et al.*, 1993).

Another chemical change involving the Maillard reaction is the conversion of L amino acids to D amino acids. This was demonstrated by heating a mixture of fructose and sucrose with L amino acids producing D amino acids (Bruckner *et al.*, 2001) as a result of non-enzymatic browning (Maillard reaction). This led to the assumption that the early stages of this multi-step reaction are favoured under production conditions of fortified wines (Patzold *et al.*, 2003). The mechanism proposed includes the (reversible) formation of Schiff bases from L amino acids, glucose and fructose. These Schiff bases may form azomethine carbanions by reversible deprotonation, producing the racemate.

The Amadori or Heyns rearrangement (Fig. 7.3), the subsequent reversible reactions of the Maillard reaction is accompanied with the formation of enaminols, release of amino acids and the generation of desoxyosones. This may be accompanied by racemization of amino acids. At the advanced stages of the Maillard reaction amino acids are destroyed during the development of aroma or flavour compounds and melanoidins. Of the different sherry wines, fino sherries had relatively the largest concentrations of D amino acids (Patzold *et al.*, 2003).

Oxidative reactions occurring during the ageing process of port wine stored in barrels were shown to lead to an increase in the concentration levels of several aldehydes and methyl ketones (Silva Ferreira *et al.*, 1996).

The chemistry of reactions involving oxygen (Fig. 7.4) presents several particularities which makes its presence, if not controlled, a serious risk to wine sensory stability. Despite being virtually innocuous towards organic molecules when existing in its ground state, the reactivity of oxygen species increases with their reduction state (i.e. superoxide anion < perhydroxyl radical < hydroxyl radical). The concentration of free radicals during wine ageing rises with increasing iron/copper ion concentrations, with increasing dissolved oxygen level and with storage temperature (Fig. 7.4).

Among these compounds, acetaldehyde is the major aliphatic aldehyde in wine and also the chemical compound that demonstrated the most significant concentration-increasing trend during barrel storage.

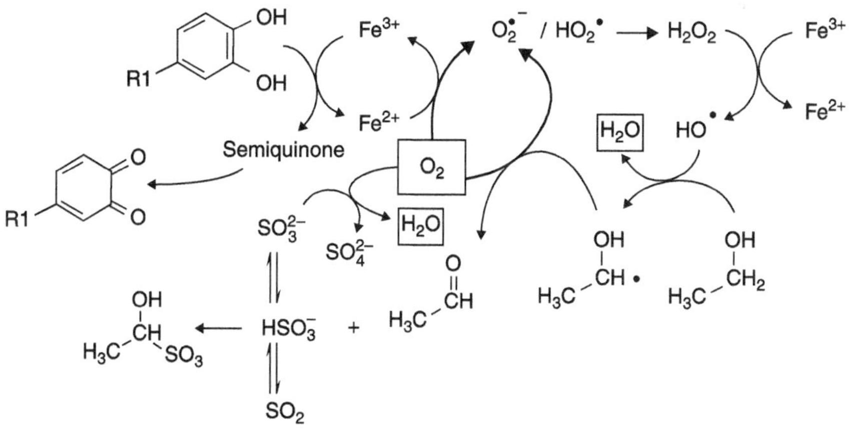

Fig. 7.4 Oxygen reduction mechanism.

7.3.2 Key aroma compounds in various fortified wines

Norisoprenoids and terpenoids

Norisoprenoids and terpenoids are known to have characteristic floral, fruity or sweet attributes. Norisoprenoid compounds (β-damascenone, β-ionone, 2,2,6-trimethylcyclohexanone (TCH), 1,1,6-trimethyl-1,2-dihydronaphthalene (TDN) and vitispirane) were determined in 14 young port wines and 45 old port wines (Silva Ferreira and Guedes de Pinho, 2004). In a high-oxygen environment, as is the case in port wine, it was found that the levels of all these compounds decreased while oxygen is being consumed. However, during barrel ageing, two opposite behaviours were observed: TDN, vitispirane and TCH increased significantly together with a decrease of levels of β-ionone and β-damascenone.

Carotenoids, present in port wines (Guedes de Pinho *et al.*, 2001) are known to be precursors of norisoprenoids. It was found that norisoprenoids could originate from direct degradation of carotenoid molecules such as β-carotene, lutein, neoxanthin and violaxanthin (Kanasawud and Crouzet, 1990; Kotseridis *et al.*, 1998; Marais *et al.*, 1989; Mordi *et al.*, 1991; Winterhalter, 1992), and also from the hydrolysis of glucoside molecules (Gunata *et al.*, 1995; di Stephano *et al.*, 1998; Skouroumounis *et al.*, 1992). The finding that young port wines contain significantly more carotenoids and chlorophyll-like compounds than old port wines (Mendes-Pinto *et al.*, 2005) indicated that an increase in norisoprenoids during ageing could be correlated with a decrease of the aforementioned precursors. This was subsequently proven in simulated experiments, monitoring the changes in the β-carotene/lutein concentration ratio (Silva Ferreira *et al.*, 2008).

The levels of norisoprenoids and monterpenols, among other volatile constituents, have been determined in the musts of Malvazia, Boal, Verdelho, and Sercial grapes, used for production of Madeira wines (Câmara *et al.*, 2004b; Câmara *et al.*, 2007). The results were subjected to principal component analysis (PCA) and grouped according to variety based on the terpenoid compounds. Boal and Verdelho were characterized by neral, linalool, citronellol and β-ionone, and Malvasia with neral and linalool. Sercial must were mostly correlated with α-terpineol, β-damascenone and geraniol (Câmara *et al.*, 2004b). Additionally 36 wine samples of the same varieties as above were analyzed for the free fraction of monoterpenols and C_{13}-norisoprenoids. This indicated that these wines have characteristic profiles of terpenoids. Malvazia had the highest levels of free monterpenols, in contrast with Verdelho which had the lowest levels of terpenoids but the highest concentration of farnesol (Câmara *et al.*, 2007).

Volatile sulphur compounds

Volatile sulphur compounds are commonly associated with off-odours; however, some thiols have contrarily demonstrated to be powerful impact odourants, adding pleasant fruity notes (Mateo-Vivaracho *et al.*, 2008).

Volatile sulfur compounds in 15 young port wines and 12 old port wines were determined (Silva Ferreira *et al.*, 2003b) and significant differences in the levels of these compounds between the two groups were observed. Subsequently, the influence of the presence of oxygen (together with pH, free SO_2 and temperature/time) on the behaviour of dimethyl sulfide, 2-mercaptoethanol, dimethyl sulfone, and methionol was performed. Dimethyl sulfide and dimethyl sulfone was formed during the experimental period in the presence of O_2. Methionol and 2-mercaptoethanol (considered to be an important off-flavour in dry wines) decreased significantly in the presence of O_2. These results corroborate the fact that old port wine (barrel aged) never develops off-flavours associated with the presence of methionol (cauliflower), 2-mercaptoethanol (rubber/burnt), or methional (cooked potato). In fact, temperature and oxygen are the major factors in the consumption of these molecules. However, some notes of 'quince' and 'metallic' can appear during port wine ageing, and these can be associated with the presence of dimethyl sulfide (Silva Ferreira *et al.*, 2003b).

Sotolon

Previous studies indicate that sotolon (3-hydroxy-4,5-dimethyl-2(5H)-furanone) presents a nutty, spicy or perceived 'aged' character and has been suggested as a key odourant in the aroma of aged port wines (Silva Ferreira *et al.*, 2003a). Sotolon was also found to contribute to the aroma of several other fortified wines, including sherry wines (Dubois *et al.*, 1976; Martin *et al.*, 1991, 1992; Moreno *et al.*, 2005), French fortified wines VDN (Schneider *et al.*, 1998), and Madeira wines (Câmara *et al.*, 2004a, 2006). Other wines with oxidative character have also reported this odour active molecule: Jura wines 'vin jaunes' (Pham *et al.*, 1995), Tokay wines (Martin *et al.*, 1991; Guichard *et al.*, 1993), and botrytized wines (Masuda *et al.*, 1994).

Several mechanisms for the presence of sotolon in foodstuff have been proposed. The first proposed explanation was that sotolon is formed as a result of aldol condensation between 2-ketobutyric acid and pyruvic acid (Dubois *et al.*, 1976), similar to the proposed mechanism for the formation of 5-ethyl-3-hydroxy-4-methyl-2(5H)-furanone (abexon) via protein hydroxylates (Sulser *et al.*, 1967). The same mechanism (aldol condensation) was also suggested to be responsible for the formation of sotolon, specifically in fortified wines (Martin *et al.*, 1991). Other mechanisms, excluding 2-ketobutyric acid have also been proposed, in particular:

- Sotolon formation from the chemical or enzymatic conversion of 4-hydroxisoleucine (Lerk and Ambuhl, 1995; Blank *et al.*, 1996),
- As a product of reaction from hexoses and pentoses in the presence of cysteine (Hofmann and Schieberle, 1997),
- From the aldol condensation of hydroxyacetaldehyde and buta-2,3-dione (Hofmann and Schieberle, 1996),
- And more recently it has been found that sotolon can be formed from ascorbic acid degradation in products, resulting in off-flavours (Konig *et al.*, 1999).

Regarding port wine, it was recently demonstrated that sotolon concentration levels correlate highly with the degree of barrel ageing. Additionally, in the

Colheita and Vintage categories, sotolon formation was found to be highly related to temperature and the amount of dissolved oxygen in the wine.

Fatty acid ethyl esters, acetates and fatty acids

Barrel ageing (1, 11 and 25 years) of Madeira varieties was studied and the results indicated that during ageing, the concentrations of the ethyl esters of fatty acids, acetates and fatty acids decrease significantly while ethyl esters of diprotic acids increased markedly (Câmara *et al.*, 2006). A strong correlation between sotolon, 2-furfural, 5-methyl-2-furfural, 5-hydroxymethyl-2-furfural and 5-ethoxymethyl-2-furfural were found with wine ageing. Therefore these compounds can be considered as age markers.

Heterocyclic acetals

Concerning heterocyclic acetals in Madeira wines, the formation of 1,3-dioxanes and 1,3-dioxolanes was investigated in order to determine levels and for possible use as indicators of wine age. The results show a linear correlation of the investigated acetals with wine age but also suggest that the acetalization reaction is not particularly affected by the drastic oxidative conditions observed during maturation (Câmara *et al.*, 2003). Moreover, there is a linear increase of 1,3-dioxanes and 1,3-dioxolanes isomers with time. The ratio between *cis*- and *trans*-forms remains constant independently of the wine age 2.35 ± 0.31 for *cis-/trans*-dioxane and 1.60 ± 0.18 for *cis-/trans*-dioxolane (Câmara *et al.*, 2006).

In order to determine the optimal temperature and baking time to obtain a quality Madeira wine, considered as typical of this style, several molecules were selected to be quantified for baked wines, on the basis of gas chromatography-olfactometry (GC-O) results and expected Maillard volatiles, such as sotolon, furfural, 5-methylfurfural, 5-ethoxymethylfurfural, methional, and phenylacetaldehyde. A positive correlation was also observed between baking time and sensory panel typicity scores. According to the present data the perceived quality increased for long periods of baking processes, for both Malvasia and Sercial wines. Baking time has a higher impact on the typicity of Malvasia wines than of Sercial wines (Oliveira e Silva *et al.*, 2008).

7.4 Legal aspects and health considerations

7.4.1 Regulatory aspects

As is the case for wine in general, fortified wines are also regulated in terms of ethanol content, production methods, allowable additions, labelling and conditions of use of specific terms, amongst others. These regulations normally are determined by a country's equivalent of a wine and spirits board (as, for example, determined in South Africa). The purpose of these regulations are two-fold: firstly to protect 'brand names' associated with a certain country or production area, thus guaranteeing that the product is made according to specific procedures ensuring consistent quality and distinctiveness and region of origin. Secondly, the wines have to comply with regulations guaranteeing the product safe for consumption and, when intended for export, compliance with the regulations of the country of destination.

In the EU, for instance, all products are required to comply with 'Council Regulation (EC) No 1493/1999 of 17 May 1999 on the common organisation of the market in wine'. In Australia a separate 'Fortified Wine Code of Practice' exists as part of an agreement between the EU and Australia.

7.4.2 Authentication

All fortified wines are produced using a multitude of steps and in order to be certified as a specific variety or style, proof must be provided that these steps and the sequence in which they were performed complies with regulatory requirements. The process for authentication of port wine is illustrated graphically in Fig. 7.5. As with the production processes of other fortified wines; this will be discussed in terms of similarities and differences from this process.

Methods of authentication are often required for certification purposes or when doubt exists regarding authenticity of the product. Studies addressing this issue include the authentication and identification of *Saccharomyces cerevisiae* 'flor' yeast traces involved in sherry ageing (Esteve-Zarzoso *et al.*, 2004), and authentication of type of yeast by using infrared (IR) or near infrared (NIR).

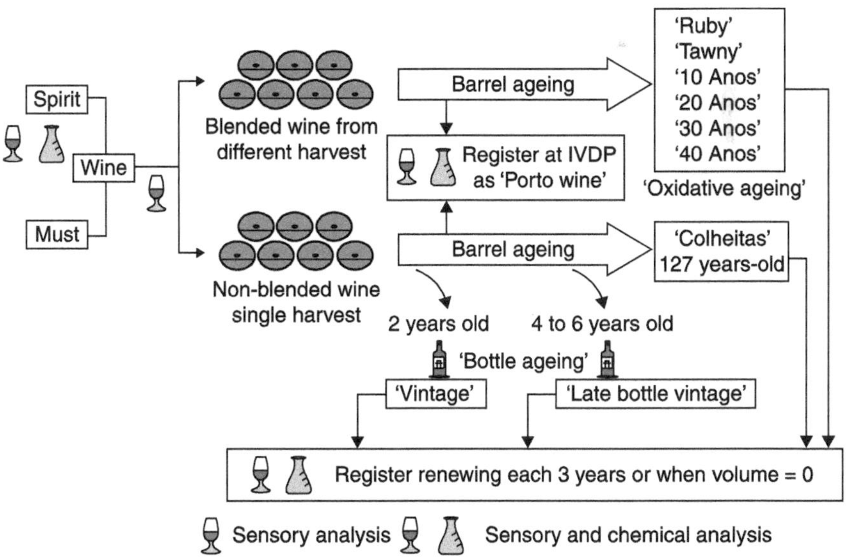

Fig. 7.5 Schematic representation of the authentication of port wine.

7.4.3 Health aspects

Wine in general, as well as fortified wine, is deemed a wholesome product and increasing attention is being focused on the effects that the composition of a product will have on human health.

Resveratrol is a chemical compound naturally occurring in grapes and wine. Several studies indicated that resveratrol, being a powerful anti oxidant, has a beneficial effect on human health. Resveratrol was furthermore found in fortified wines (Bravo *et al.*, 2008), aiding the belief in the health benefits of the product.

Conversely, the consumer also wants to be assured that the product is free from harmful chemicals, and for many of these compounds official limits were established that may not be exceeded. Ethyl carbamate, considered potentially toxic and reclassified in 2007 as a possible carcinogen, was found in Madeira wine (Perestrelo *et al.*, 2010). In 2004 a survey was performed in the UK concerning the prevalence of ethyl carbamate in fermented foods and the study included popular fortified wines such as port, sherry and Madeira (Hasnip *et al.* 2007).

Ochratoxin A, a mycotoxin and considered as possibly carcinogenic was found in fortified wines (Festas *et al.*, 2000).

Pesticides are worldwide strictly regulated, both in terms of use and presence in foodstuffs. The fungicide Fenhexamid was found in must and wine destined for production of fortified wines (Mercader and Abad-Fuentes, 2009).

Low-end fortified wines are commonly consumed as a means of accelerating intoxication and are popular as a result of the increased alcohol content.

All the above-mentioned aspects are of concern to the consumer and should be strictly monitored and well managed, both at the producers' level but also at a regulatory level to ensure that fortified wine is of a safe and wholesome nature, as is expected by the consumer.

7.5 Future trends

Although a large number of producers of fortified beverages prefer to still follow the 'traditional' production methods there is, as in the production of other beverages, a demand for faster production processes, leading to a more rapid turnover. However, the customer still expects the quality of the product to be consistent. Therefore some producers are moving towards methods of accelerated ageing and scaling up the process for faster production of larger volumes. In addition, if more insight is gained into the chemical processes responsible for the development of unique flavour and aroma attributes of certain types and styles, the possibility of modulating the overall process to produce a consistent high-quality fortified wine will become a more viable option.

7.6 References

BERLANGA, T.M., PEINADO, R., MILLÁN, C., ORTEGA, J.M. (2004), 'Discriminant analysis of sherry wine during biological aging', *Am. J. Enol. Vitic.*, **55**:4:407–411.

BLANK, I., LIN, J., FUMEAUX, R., WELTI, D.H., FAY, L.B. (1996), 'Formation of 3-hydroxy-4,5-dimethyl-2(5h)-furanone (sotolone) from 4-hydroxy-l-isoleucine and 3-amino-4,5-dimethyl-3,4-dihydro-2(5h)-furanone', *J. Agric. Food Chem.*, **44**, 1851–1856.

BRAVO, M.N., FELICIANO R., SILVA S., COELHO, A.V., VILAS BOAS L. *et al.* (2008), 'Analysis of trans-resveratrol: Comparison of methods and contents in Muscatel fortified wines from Setubal region in Portugal', *J. Food Comp. and Anal.*, **21**, 634–643.

BRUCKNER, H., JUSTUS, J., KIRSCHBAUM, (2001), 'Saccharide induced racemization of amino acids in the course of the Maillard reaction', *J. Amino Acids*, **21**, 429–433.

CÂMARA, J.S., MARQUES, J.C., ALVES, A., SILVA FERREIRA, A.C. (2003), 'Heterocyclic acetals in Madeira wines', *Anal. And Bioanal. Chem.*, **375**, 1221–1224.

CÂMARA, J.S., MARQUES, J.C., ALVES, M.A., and SILVA FERREIRA, A.C. (2004a), '3-Hydroxy-4,5-dimethyl-2(5H)-furanone levels in fortified Madeira wines: Relationship to sugar content', *J. Agric. Food Chem.*, **52**, 6765–6769.

CÂMARA, J.S., HERBERT, P., MARQUES, J.C., ALVES, M.A. (2004b), 'Varietal flavor compounds of four grape varieties producing Madeira wines', *Anal. Chim. Acta*, **513**, 203–207.

CÂMARA, J.S., ALVES, M.A., MARQUES, J.C. (2006), 'Development of headspace solid phase microextraction gas chromatography-mass spectrometry methodology for analysis of terpenoids in Madeira wines', *Anal. Chim. Acta*, **555**, 191–200.

CÂMARA, J.S., ALVES, M.A., MARQUES, J.C. (2007), 'Classification of Boal, Malvazia, Sercial and Verdelho wines based on terpenoid patterns', *Food Chem.*, **101**, 475–484.

CHAN, F., REINECCIUS, G.A. (1994) 'The reaction kinetics for the formation of isovaleraldehyde, 2-acetyl-1-pyrroline, di(H)di(OH)-6-methylpyranone, phenylacetaldehyde, 5-methyl-2-phenyl-2-hexenal and 2-acetylfuran in model systems' in Labuza, T.P., Reineccius, G.A., Monnier, V.M., O'Brien, J. and Baynes, J.W., (eds), *Maillard Reactions in Chemistry, Food and Health*, Cambridge, The Royal Society of Chemistry, pp. 131–139.

DI STEPHANO, R., BOTTERO, S., PIGELLO, R., BORSA, D., BAZZO, G. *et al.* (1998), 'Glycosylated aroma precursors in red winemaking grape cv.', *Enotecnico*, **34**, 63.

DUBOIS, P., RIGAUD, J., DEKIMPE, J. (1976), 'Identification of 4,5-dimethyltetrahydrofuran-2,3-dione in a Flor sherry wine', *Lebensm. Wiss. Technol.*, **9**, 366–368.

ESCUDERO, A., CACHO, J., FERREIRA, V. (2000), 'Isolation and identification of odorants generated in wine during its oxidation: a gas chromatography-olfactometry study', *Eur. Food Res. Technol.*, **211**, 105–110.

ESTEVE-ZARZOSO, B.M., FERNANDEZ-ESPINAR, M.T., QUEROL, A. (2004), 'Authentication and identification of *Saccharomyces cerevisiae* 'flor' yeast races involved in sherry ageing', *Antonie van Leeuwenhoek*, **85**, 151–158.

FESTAS, I., HERBERT, P., SANTOS, L., CABRAL, M., BARROS, P. *et al.* (2000), 'Ochratoxin A in some Portuguese wines: Method validation and screening in Port Wine and Vinho Verde', *American Journal of Enology and Viticulture*, **51**, 150–154.

FRANCO, M., PEINADO, R.A., MEDINA, M., MORENO, J. (2004), 'Off-vine grape drying effect on volatile compounds and aromatic series in must from Pedro Ximénez grape variety', *J. Agric. Food Chem.*, **52**, 3905–3910.

FRANK, D., GOULD, I., MILLIKAN, M. (2005), 'Browning reactions during storage of low-moisture Australian sultanas: effects of vine nitrogen nutrition on subsequent arginine-mediated Maillard reactions during storage of dried fruit', *Australian Journal of Grape and Wine Research*, **11**, 1523.

GOSWELL, R.W. (1986), 'Microbiology of fortified wines', in Robinson, R.K. (ed.), *Developments in Food Microbiology*, 2nd ed., London, Elsevier Applied Science Publishers.

GRANVOGL, M., BUGAN, S., SCHIEBERLE, P. (2006), 'Formation of amines and aldehydes from parent amino acids during thermal processing of cocoa and model systems: new insights into pathways of the Strecker reaction', *J. Agric. Food Chem.*, **54**(5), 1730–1739.

GUEDES DE PINHO, P., FERREIRA, A.C., MENDES PINTO, M., GOMEZ BENITEZ, J., HOGG, T. (2001), 'Determination of carotenoid profiles in grapes, musts and fortified wines from Douro varieties of *Vitis vinifera*', *J. Agric. Food Chem.*, **49**(11), 5484–5488.

GUICHARD, E., PHAM, T.T., ETIEVANT, P.X. (1993), 'Quantitative determination of sotolon in wines by high-performance liquid chromatography', *Chromatographia*, **37**, 539–542.

GUNATA, Y.Z., BAYONOVE, C.L., BAUMES, R.L., CORDONNIER, R.E. (1995), 'The aroma of grapes. I. Extraction and determination of free and glycosodically bound fractions of some grape aroma components', *J. Chromatogr.*, **331**, 83.

GUNTERT, M. (ed.), *Flavor Precursors*, Washington, D.C., American Chemical Society, pp. 98–115.

HASNIP, S., CREWS, C., POTTER, N., CHRITY, J., CHAN, D. *et al.* (2007), 'Survey of ethyl carbamate in fermented foods sold in the United Kingdom in 2004', *J. Agric. Food Chem.*, **55**, 2755–2759.

HERRAIZ, T., OUGH, C.S. (1992), 'Identification and determination of amino acid ethyl esters in wines by capillary gas chromatography-mass spectrometry', *J. Agric. Food Chem.*, **40**(6), 1015–1021.

HERRAIZ, T., OUGH, C.S. (1993), 'Formation of ethyl esters of amino acids by yeasts during the alcoholic fermentation of grape juice', *Am. J. Enol. Vitic.*, **44**(1), 41–48.

HERRAIZ, T., HUANG, Z., OUGH, C.S. (1993), 'Amino acids and ethyl esters of amino acids in sparkling and 'Sur Lie' wines', *Italian J. Food Sci.*, **5**(1), 11–20.

HOFMANN, T., SCHIEBERLE, P. (1996), 'Identification of the key odorants in processed ribose-cysteine Maillard mixtures by instrumental analysis and sensory studies', *Spec. Publ. – R. Soc. Chem., 197 (Flavour Science)*, 175–181.

HOFMANN, T., SCHIEBERLE, P. (1997), 'Identification of potent aroma compounds in thermally treated mixtures of glucose/cysteine and rhamnose/cysteine using aroma extract dilution techniques', *J. Agric. Food Chem.*, **45**, 898–906.

HOFMANN, T., MUNCH, P., SCHIEBERLE, P. (2000), 'Quantitative model studies on the formation of aroma-active aldehydes and acids by Strecker-type reactions', *J. Agric. Food Chem.*, **48**(2), 434–440.

JACKSON, R.S. (2000), *Wine Science: Principles, Practice, Perception*, 2nd edition, San Diego Academic Press, pp 559.

KANASAWUD, P., CROUZET, J.C. (1990), 'Mechanism of formation of volatile compounds by thermal degradation of b-carotene in aqueous medium. 1. b-carotene degradation', *J. Agric. Food Chem.*, **38**, 237–243.

KEIM, H., REVEL, G., MARCHAND, S., BERTRAND, A. (2002), 'Method for determining nitrogenous heterocycle compounds in wine', *J. Agric. Food Chem.*, **50**(21), 5803–5807.

KÖNIG, T., GUTSCHE, B., HARTL, M., HÜBSCHER, R., SCHREIER, P. *et al.*(1999), '3-Hydroxy-4, 5-dimethyl-2(5H)-furanone (sotolon) causing an off-flavor: Elucidation of its formation pathways during storage of citrus soft drinks', *J. Agric. Food Chem.*, **47**, 3288–3291.

KOTSERIDIS, Y., BAUMES, R., SKOUROUMOUNIS, G.K. (1998), 'Synthesis of labelled b-damascenone, 2-methoxy-3-isobutylpyrazine, a-ionone and b-ionone for quantification in grapes, juices and wines', *J. Chromatogr* A, **824**, 71–78.

LERK, K., AMBUHL, M. (1995), 'Biotechnological production of 4,5-diméthyl-3-hydroxy-2(5h)-furanone' in Etiévant, p., Schreier, p., (eds), *Bioflavor 95. Analysis-Precursor Studies-Biotechnology, Les Colloques no.75, INRA:* Paris, p.p. 381–384.

LICHINE, A. (1987). *Alexis Lichine's New Encyclopedia of Wines & Spirits*, 5th ed., New York, Alfred A. Knopf, Inc., pp 236.

LLAGUNO, C. (1968), 'Flor yeasts in sherry', *Process Biochem.*, **3**, 11–15.

LÓPEZ DE LERMA, N. (2009), 'Vinos dulces obtenidos por fermentación parcial de mostos de uvas Tempranillo pasificadas', University of Córdoba, degree thesis.

LÓPEZ DE LERMA, N. (2010), in press LWT – *Food Science and Technology*.

MAILLARD, L.C. (1912), 'Action des acides aminés sur les sucres: Formation des mélanoidines par voie méthodique', *Comptes Rendus Hebdomadaires des Seances de l'Academie des Science*, **154**, 66–68.

MAIN, G.L. (1992), 'Juice fining treatments to reduce browning and use of sulfur dioxide in white wine production', Ph.D. Thesis, University of Arkansas, United States.

MARAIS, J., VAN WYK, C., RAPP, A. (1989), 'Carotenoids in grapes. Flavors and off-flavors', in *Proceedings of the 6th International Flavor Conference*, Rethymnon, Crete, Greece.

MARCHAND, S., DE REVEL, G., BERTRAND, A. (2000), 'Approaches to wine aroma: Release of aroma compounds from reactions between cysteine and carbonyl compounds in wine', *J. Agric. Food Chem.*, **48**(10), 4890–4895.

MARCHAND, S., DE REVEL, G., BERTRAND, A. (2002), 'Possible mechanism for involvement of cysteine in aroma production in wine', *J. Agric. Food Chem.*, **50**(21), 6160–6164.

MARGALIT, Y. (2004), *Concepts in Wine Chemistry*, San Francisco, The Wine Appreciation Guild, pp 342.

MARTIN B., ETIEVANT, P.X., QUÉRÉ, J.-L.L. (1991), 'More clues of the occurrence and flavor impact of solerone in wine', *J. Agric. Food Chem.*, **39**, pp. 1501–1503.

MARTIN, B., ETIÉVANT, P.X., LE QUÉRÉ, J.L., SCHLICH, P. (1992), 'More clues about the sensory impact of Sotolon in some flor sherry wines', *J. Agric. Food Chem.* **40**, 475–478.

MARTÍNEZ DE LA OSSA, E., PÉREZ, L., CARO, I. (1987a), 'Variations of the major volatiles through aging of sherry', *Am. J. Enol. Vitic.*, **38**, 293–297.

MARTÍNEZ DE LA OSSA, E., CARO, I., BONAT, M., PÉREZ, L., DOMECQ, B. (1987b), 'Dry extract in sherry and its evolution in the aging process', *Am. J. Enol. Vitic.*, **38**, 321–325.

MASUDA, M., OKAWA, E., NISHIMURA, K., YUNOME, H. (1994), 'Identification of 4,5-dimethyl-3-Hydroxy-2(5H)-furanone (Sotolon) and ethyl 9-hydroxynonanoate in Botrytised wine and evaluation of the roles of compounds characteristic', *Agric. Biol. Chem.*, **48**, 2707–2710.

MATEO-VIVARACHO, L., CACHO, J., FERREIRA, V. (2008), 'Improved solid-phase extraction procedure for the isolation and in-sorbent pentafluorobenzyl alkylation of polyfunctional mercaptans. Optimized procedure and analytical applications', *J. Chromatogr.* A, **1185**, 1, 9–18.

MENDES-PINTO, M.M., SILVA FERREIRA, A.C., CARIS-VEYRAT, C., GUEDES DE PINHO, P. (2005), 'Carotenoid, chlorophyll, and chlorophyll-derived compounds in grapes and Port wines', *J. Agric. Food Chem.*, **53**, 10034–10041.

MERCADER, J.V., ABAD-FUENTES, A. (2009), 'Monoclonal antibody generation and direct competitive enzyme-linked immunosorbent assay evaluation for the analysis of the fungicide fenhexamid in must and wine', *J. Agric. Food Chem.*, **57**, 5129–5135.

MORDI, C.R., WALTON, J.C., BURTON, G.W., HUGHES, L., INGOLD, K.U. *et al.* (1991), 'Exploratory study of β-carotene auto-oxidation', *Tetrahedron Letters*, **32**, 4203–4206.

MORENO, J.A., ZEA, L., MOYANO, L., MEDINA, M. (2005), 'Aroma compounds as markers of the changes in sherry wines subjected to biological ageing', Food Control, **16**, 333–338.

MORENO, J., PEINADO, J., PEINADO, R.A. (2007), 'Antioxidant activity of musts from Pedro Ximénez grapes subjected to off-vine drying process', *Food Chem.*, **104**, 224–228.

OLIVEIRA, E., SLIVA, H., GUEDES DE PINHO, P., MACHADO, B., HOGG, T., MARQUES, J.C. *et al.* (2008), 'Impact of forced-aging process on Madeira wine flavor', *J. Agric. Food Chem.*, **56**, 11989–11996.

PATZOLD, R., NIETO-RODRIGUEZ, A., BRUCKNER, H. (2003), 'Chiral gas chromatographic analysis of amino acids in fortified wines', *Chromatographia Supplement*, **57**, Suppl., S-207–S-212.

PERESTRELO, R., PETRONILHO, S., CÂMARA, J.S., ROCHA, S.M. (2010), 'Comprehensive two-dimensional gas chromatography with time-of-flight mass spectrometry combined with solid phase microextraction as a powerful tool for quantification of ethyl carbamate in fortified wines. The case study of Madeira wine', *J. Chromatogr.* A, **1217**, 3441–3445.

PHAM, T.T., GUICHARD, E., PASCAL, S., CHARPENTIER, C. (1995), 'Optimal conditions for the formation of Sotolon from α-ketobutyric acid in the French 'Vin Jaune'', *J. Agric. Food Chem.*, **43**, 2616–2619.

PISSARRA, J.I., LOURENCO, S., MACHADO, J.M., MATEUS, N., GUIMARAENS, D. *et al.* (2005), 'Contribution and importance of wine spirit to the port wine final quality – initial approach', *J. Sci. Food Agric.*, **85**(7), 1091–1097.

POSSON, P. (1981) 'Production of baked and submerged culture sherry-type wines in California 1960–1980' in Amerine, M.A. (ed), *Wine Production Technology in the United States*, Washington, DC, American Chemical Society, pp 143.

POTMAN, R.P., VAN WIJK, T.A. (1989) 'Mechanistic studies of the Maillard reaction with emphasis on phosphate-mediated catalysis' in *Phosphate-Mediated Catalysis*; Washington, DC, American Chemical Society, 182–195.

PRIPIS-NICOLAU, L., REVEL, G., BERTRAND, A., MAUJEAN, A. (2000), 'Formation of flavor components by the reaction of amino acid and carbonyl compounds in mild conditions', *J. Agric. Food Chem.*, **48**(9), 3761–3766.

READER and DOMINGUEZ (2003), 'Fortified wines: sherry, port and Madeira' in Lea, A.G.H. and Piggott, J.R. (eds), *Fermented Beverage Production*, 2nd ed., New York, Kluwer Academic, pp 176.

ROGERSON, F.S.S., DE FREITAS, V.A.P. (2002), 'Fortification spirit, a contributor to the aroma complexity of port', *J. Food Sci.*, **67**(4), 1564–1569.

SCHNEIDER, R., BAUMES, R., BAYONOVE, C., RAZUNGLES, A. (1998), 'Volatiles compounds involved in the aroma of sweet fortified wines (vins doux naturels) from Grenache Noir', *J. Agric. Food Chem.*, **46**, 3230–3237.

SHONBERG, A., MOUBACHER, R. (1952), 'The Strecker degradation of α-amino acids', *Chem. Rev.*, **50**, 261–277.

SILVA FERREIRA, A.C. (1998), 'Caractérisation du Vieillissement du Vin de Porto. Approche Chimique et Statistique. Rôle Aromatique du Sotolon', Thèse de Doctorat de L'Université Victor Segalen Bordeaux II, France, p 593.

SILVA FERREIRA, A.C. and BERTRAND, A. (1996), 'Evolution de quelques constituants du vin de Porto au cours du vieillissement. Etude particulière des composés carbonylés', in *OEnologie 95, 5ème Symposium International d'OEnologie;* Lavoisier, Paris, France, 520–523.

SILVA FERREIRA, A.C., GUEDES DE PINHO, P., RODRIGUES, P., HOGG, T. (2002), 'Kinetics of oxidative degradation of white wines and how they are affected by selected technological parameters', *J. Agric. Food Chem.*, **50**, 5919–5924.

SILVA FERREIRA, A.C., BARBE, J.-C., BERTRAND, A.P. (2003a), '3-hydroxy-4,5-dimethyl-2(5H)-furanone: A key odorant of the typical aroma of oxidative aged port wine', *J. Agric. Food Chem.*, **51**, 4356–4363.

SILVA FERREIRA, A.C., RODRIGUES, P., HOGG, T., GUEDES DE PINHO, P. (2003b), 'Identification of key odorants related to the typical aroma of oxidation-spoiled white wines', *J. Agric. Food Chem.*, **51**, 1377–1381.

SILVA FERREIRA, A.C., GUEDES DE PINHO, P. (2004), 'Norisoprenoids profile during Port wine ageing-influence of some technological parameters', *Anal. Chim. Acta*, **513**, 169–176.

SILVA FERREIRA, A.C., MONTEIRO, J., OLIVEIRA, C., GUEDES DE PINHO, P. (2008), 'Study of major aromatic compounds in port wines from carotenoid degradation', *Food Chem.*, **110**(1), 83–87.

SKOUROUMOUNIS, G.K., MASSY-WESTROPP, R.A., SEFTON, M.A., WILLIAMS, P.J. (1992), 'Precursors of damascenone in fruit juices', *Tetrahedron Lett.*, **33**, 3533.

SULSER, H., DEPIZZOL, J., BUCHI, W. (1967), 'A probable flavoring principle in vegetable-protein hydrolysates', *T. Food Sci.*, **32**, 611–615.

TAYLOR FLADGATE, (2011), What is Port?, available from http://www.taylor.pt/pdf/Site/WhatIsPort.pdf.

VERNIN, G. (1982), 'Heterocyclic aroma compound in foods: occurrence and organoleptic properties' in Vernin, G. (ed), *Chemistry of Heterocyclic Compounds in Flavours and Aromas*, Chichester, Ellis Horwood, 72–140.

WINE MAGAZINE (2005), 'Fortified Wines', available from http://www.winemag.co.za/article/fortified-wines-2005-11-15.

WINEMAKERS' FEDERATION OF AUSTRALIA (2011), 'Australian fortified wines: the dawning of a new era', available from http://www.wineaustralia.com/australia/Portals/2/pdf/Australian%20Fortified%20Wines%20Strategy.pdf.

WINES OF PORTUGAL (2011), 'Wine styles', available from http://www.winesofportugal.info/pagina.php?codNode=18007.

WINTERHALTER, P. (1992), 'Oxygenated C_{13}-norisoprenoids – important flavor precursors. Flavour precursors – thermal and enzymatic conversion' in Teranishi, R., Takeoka, G.R., Guntert, M. (eds.), *Flavor Precursors*, Washington, D.C., American Chemical Society, pp. 98–115.

ZOECKLEIN, B.C., FUGELSANG, K.C., GUMP, B.H., NURY, F.S. (1995), *Wine Analysis and Production.* New York, Kluwer Academic/Plenum Publishers.

8

Sake: quality characteristics, flavour chemistry and sensory analysis

S. Furukawa, Kizakura Co. Ltd, Japan

Abstract: This chapter discusses Japan's traditional brew, sake. Although sake is a fermented beverage just like beer and wine, sake's brewing process is more complex. That is where the flavour and aroma of sake are generated. The chapter reviews quality of sake, flavour and aroma of sake, and sensory evaluation methods.

Key words: sake, koji, *Seishu-kobo*, multiple parallel fermentation, sake tasting.

8.1 Introduction

Japan's traditional brew, sake, is the national beverage. It has played a central role in Japanese life and culture for about two thousand years, and the knowledge and skills required for sake production have spread almost all over the country (Japan Sake and Shochu Makers Association, 2011). In the 2009 brewery year (1 July 2009–30 June 2010), 1302 breweries of all sizes were engaged in the production of sake (National Tax Agency Japan, 2011). Although it can be made almost anywhere in Japan (see Fig. 8.1), most famous sakes are produced in regions that have a supply of good water or good quality rice. Nada in Hyogo, Fushimi in Kyoto and Saijo in Hiroshima are some famous sake-producing regions (Japan Technical Information Services Corporation, 2006).

Although sake is a fermented beverage just like beer and wine, sake's brewing process is more complex. In sake brewing, polished rice grains are steamed after steeping in water. Steamed rice is fermented with water, yeast and koji. In this process of brewing sake, rice starch is saccharified by koji mold enzymes and the liberated glucose is fermented into ethanol by yeast. The processes of saccharification by koji and fermentation by yeast proceed simultaneously in the same vessel. This simultaneous fermentation is called 'multiple parallel fermentation'. It is unique among the methods used for making alcoholic beverages (see Fig. 8.2).

Fig. 8.1 Breweries exist in the unshaded areas, i.e. in all 46 administrative divisions of Japan except Kagoshima, where there is no sake brewery.

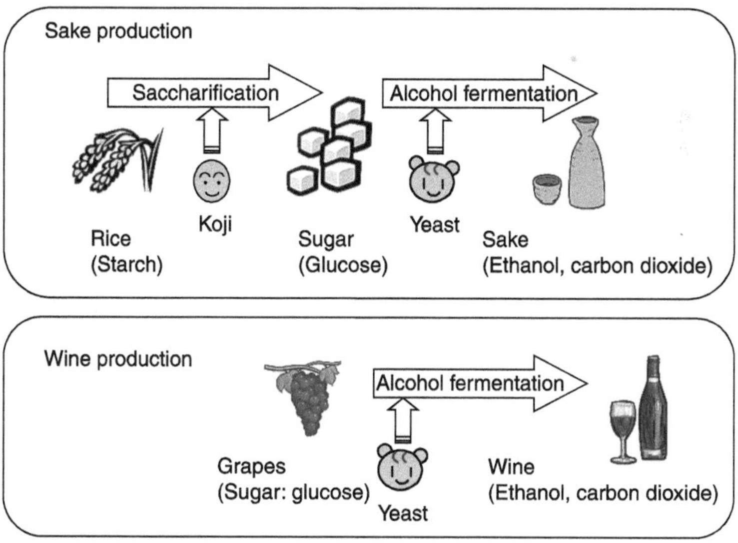

Fig. 8.2 Comparison of the sake brewing process with wine production. Source: Japan Sake and Shochu Makers Association (2011).

Compared with other alcoholic beverages, sake can be enjoyed at a much wider range of temperatures: approximately 5–55 °C. In this chapter, the quality, the components, the flavour and aroma of sake are described, and the sensory evaluation method and future trends are explained.

8.2 Quality of sake

The quality of sake is an important point, equal to all other characteristics in sake selection for the consumer. Consequentially, it is the most important factor determining market value. In this section, the ingredients and the components of sake, which most influence its quality, are described.

8.2.1 Ingredients

Sake is made mainly from rice, koji, and water. These materials greatly influence the quality of sake. A good many others such as yeast and lactic acid bacteria also play their roles in the brewing process. This exact, delicate collaboration generates a synergistic effect that enhances the flavour and aroma of sake.

Rice

Rice is an essential element in sake brewing. The characteristics of sake are controlled by the choice and combination of rice cultivars. Rice can be roughly classified into two types (see Fig. 8.3), ordinary table rice and rice suitable for sake brewing (*shuzo-kotekimai*). Ordinary table rice, commonly used for eating, encompasses a number of sub-varieties, depending on the manner of cultivation and the region in which it is grown. *Shuzo-kotekimai*, sake rice, is used exclusively

Fig. 8.3 Rice types. (a) Rice suitable for sake brewing of the cultivar *Oryza sativa* L. Hyogokitanishiki in which the white core, known as *shinpaku*, can be seen in the centre of the grain. (b) Ordinary table rice of the cultivar *Oryza sativa* L. Koshihikari. Bar, 5 mm. (Source: WeNARC, 2007).

for sake production. Ordinary table rice can also be used to make sake, but, in general, sake made from *shuzo-kotekimai* has a better reputation for quality (Anon., 2005).

Shuzo-kotekimai has a large grain, white-core structure (*shinpaku*, which is often observed in the centre of the grains), and contains less protein. These features are best suited to sake production. However, *shuzo-kotekimai* is taller and has a weaker resistance to lodging (falling down) requiring advanced cultivation techniques.

The typical cultivars of *shuzo-kotekimai* include *Oryza sativa* L. Yamadanishiki, *Oryza sativa* L. Gohyakumangoku, *Oryza sativa* L. Omachi, *Oryza sativa* L. Miyamanishiki and *Oryza sativa* L. Hyogokitanishiki. In Japan, the top ranked *shuzo-kotekimai* is Yamadanishiki, especially that grown in Hyogo.

Brewing rice can be also classified into two usage types. The first, koji rice, used for the growth media for koji mould, is called *kojimai*. The second, used as a fermentation substrate, is called *kakemai*.

The first major component of rice is starch, which approximately accounts for 78% (14% moisture) or 90% (dry) of milled rice by weight (Champagne, 2004). The second major component of rice is protein. It accounts for 4–11% of milled rice by dry weight (Champagne, 2004), which is affected by cultural conditions such as the amount of fertilizer used and changes in the temperature, as well as by rice variety (Champagne, 2004) and the degree of rice milling. Although the amount of protein is a small fraction compared with that of starch, protein is still an important factor in sake brewing. Koji enzymes break down the proteins into amino acids and peptides. These amino acids and peptides not only provide nutrients for yeast growth but also provide flavour and aroma components. Too much protein brings a *zatsumi* taste. *Zatsumi* is difficult to translate into English. Suffice to say that it represents a poor and 'lacking polish' taste. Therefore, techniques that appropriately control the dissolution of rice proteins during the sake brewing process are required (Furukawa, 2008).

Protein concentration is highest on the surface layers and decreases toward the centre of the grain (Champagne, 2004).Oil, minerals and vitamins, etc. are also most concentrated in the outer layer of the rice grain.

Oils in rice cause *zatsumi* in the same way as does having too much protein. However, because the outer portion is removed in the rice milling process these oils do not usually cause a problem. The degree of milling for eating rice is usually about 90%; however brewing rice is usually polished from 70% in regular quality sake to 35% in an excellent quality sake such as *Daiginjo-shu* (see Fig. 8.4). It is difficult but exquisite and delicate work to polish rice to a small, bead-like size with little but the starchy centre remaining. But polishing the rice this way produces superior taste. On the other hand, the more the rice is milled, the less perceptible its cultivar characteristics become; and because the minerals and vitamins which promote yeast health and good fermentation are also lost with high milling rates, advanced techniques to supplement the mash for a clean fermentation process are needed.

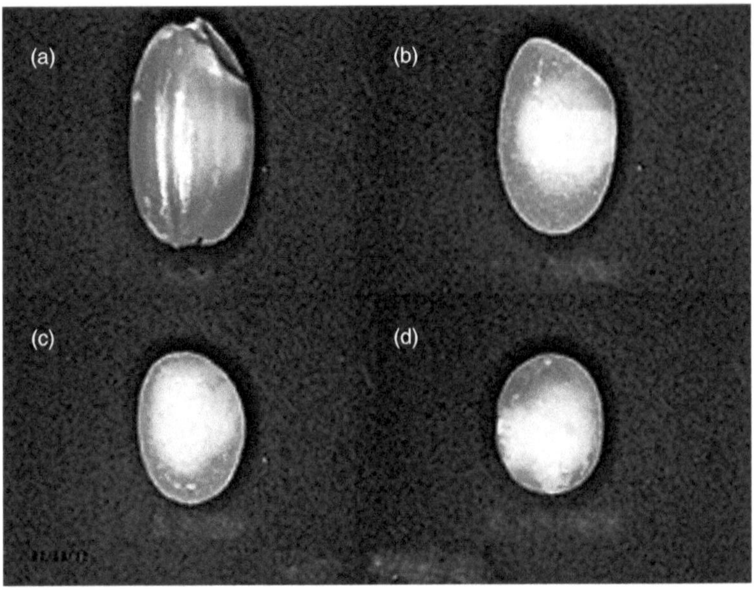

Fig. 8.4 Polishing of rice in sake brewing. (a) Brown rice. (b) Polished rice, degree of milling 70% (used for normal class sake). (c) Polished rice, degree of milling 50% (used for *ginjo-shu* and *daiginjo-shu*). (d) Polished rice, degree of milling 35% (used for special *daiginjo-shu*).

Koji

Koji (rice koji) is the mould (*Aspergillus oryzae*) grown on a steamed rice medium. Because rice, a cereal, is mainly composed of polysaccharide starch, yeast cannot use it as an energy source as it is. The glucoamylase and alpha amylase in koji break down rice starch into glucose, and make other resources in the rice available for yeast during the sake brewing process. This process is known as saccharification. Koji has proteases as mentioned in the rice section. Specifically acid protease and acid carboxypeptidase transform rice proteins into amino acids and peptides that have a great impact on yeast growth, sake flavour and aroma.

In the wine production process, saccharification is not necessary, because glucose is already a significant component of grape juice. The wine production process is a single fermentation process. In the East, koji is used not only in sake brewing but also in other alcoholic beverages such as the Japanese distilled spirit *shochu* and sweet cooking rice wine (*mirin*). In addition to its use in alcoholic beverages, koji is used in fermented food such as soybean paste (*miso*) and soy sauce (*shoyu*). This is why the food culture in the East is called a 'multi-fermentation culture' or 'moulds culture'. Far East climate characteristics make these methods possible. The moderately high-temperature humid zone from South-East Asia to East Asia enables the effective utilization of koji as a mould. Koji used for sake brewing is the same species (*kikoji-kin*) used for *miso* and *shoyu*.

Yeast

Yeast is indispensable for sake brewing though it is not considered a main ingredient. Many kinds of yeasts exist in the natural world. The yeast used to brew sake is called sake yeast (*seishu-kobo*), mostly *Saccharomyces cerevisiae*. Different characteristics of various yeasts are strong contributors in deciding the flavour, aroma and quality of sake. Strains of yeasts with especially high brewing quality are usually used. In many cases, a single strain is not necessarily used alone. The advantages of each strain are combined by mixing several strains or blending the brewed sake.

In the wine production process, the aroma of grapes largely contributes to the flavour of wine. On the other hand, rice has neither a characteristic flavour nor a profound effect on the flavour of sake. Instead, the complicated process of sake brewing gives flavour and aroma (Champagne, 2004). For example, the special yeast employed in *ginjo-shu* brewing (*ginjo-kobo*) happens to produce fruity flavours and aromas like banana or apple during fermentation. No artificial flavouring is added (Anon., 2005).

Recently in Japan, yeast taken from various flowers such as cherry blossoms, roses, and pinks (called flower yeast), were tried for sake brewing in many different locations. The sake brewed with flower yeast presents different characteristics according to their flower of origin.

8.2.2 Components of sake

Table 8.1 shows the composition of sake samples brewed in fiscal 2009 (National Tax Agency Japan, 2010).

There are two types of sake: special designation sake (*tokutei meisho-shu*) and normal sake (*ippan-shu* or *futsu-shu*). *Tokutei meisho-shu*, special designation sake, are premium sakes distinguished by the degree to which the rice is polished and the ingredients. There are eight varieties of special designation sake:

- *Ginjo-shu* (special brew),
- *Daiginjo-shu* (very special brew),
- *Junmai-shu* (pure rice),
- *Junmai Ginjo-shu* (pure rice, special brew),
- *Junmai Daiginjo-shu* (pure rice, very special brew),
- *Tokubetsu Junmai-shu* (special pure rice),
- *Honjozo-shu* (genuine brew),
- *Tokubetsu Honjozo-shu* (special genuine brew).

Ippan-shu, normal sake, accounts for all other varieties except for special designation sake. Table 8.1 shows the components of *ippan-shu, ginjo-shu, junmai-shu* and *honjozo-shu*.

Nihonshu-do is not a component of sake but an indicator for evaluation. *Nihonshu-do* means sake meter value (SMV) and is a rough indication of the level of sweetness/dryness. In fact, *nihonshu-do* indicates the specific gravity of sake relative to water at 4 °C. When sake contains a large amount of extracted

Table 8.1 Composition of sake brewed in fiscal 2009

Grade		*Ippan-shu*[a]	*Ginjo-shu*[b]	*Junmai-shu*[c]	*Honjozo-shu*[d]
Number of samples		543	489	462	462
Alcohol content (%, v/v)	Mean ± SD	15.41 ± 0.76	15.94 ± 0.89	15.52 ± 0.80	15.54 ± 0.74
Nihonshu-do	Mean ± SD	3.8 ± 3.0	4.6 ± 2.7	4.1 ± 4.3	5.0 ± 3.0
Acidity (ml/10 ml sake)	Mean ± SD	1.18 ± 0.20	1.30 ± 0.25	1.47 ± 0.24	1.25 ± 0.19
Amino acidity (ml/10 ml sake)	Mean ± SD	1.31 ± 0.43	1.28 ± 0.35	1.59 ± 0.41	1.41 ± 0.37

Notes: Sakes were classified into four types by the degree to which the rice is polished and whether brewer's alcohol has been added or not. A total of 1956 samples were analyzed in the study.
[a] *Ippan-shu*: normal sake with no special grade.
[b] *Ginjo-shu*: special grade sake brewed with the rice milled down to 60% or less of the original grain.
[c] *Junmai-shu*: sake made with only rice, koji, yeast and water. Specifically, no brewer's alcohol is added.
[d] *Honjozo-shu*: sake that is at least *Junmai-shu* grade (made with rice, milled until at least no more than 70% remains) and has a small amount of brewer's alcohol added to open up the flavour and aroma.

Source: National Tax Agency Japan (2010), Auld (2011).

components it is heavier than water, and this is indicated by a negative SMV. When sake has a small amount of extracted components it is lighter than water, which is indicated by a positive SMV. In general, the lower *nihonshu-do* (the larger the negative figure), the sweeter and stronger the taste. The higher the *nihonshu-do* (the larger the positive figure is), the drier and cleaner the taste. However, our taste sense of sweetness/dryness is actually influenced by acidity.

The acidity measurement shows the titration volume (ml) of 0.1 M NaOH necessary to neutralize 10 ml sake. It indicates the density of free organic acids such as malic acid, lactic acid, succinic acid, and so on. Organic acids bring sourness and *umami* (Gekkeikan, 1997–2010). We can sense a sharp and dry taste if we drink sake with high acidity. In contrast, we can sense a rich and sweet taste if we drink sake with low acidity (Gekkeikan, 1997–2010).

The balance of *nihonshu-do* and acidity controls our taste sense of sweetness/dryness. When comparing sake with the same level of *nihonshu-do*, the higher the acidity, the drier the sake seems because sweetness is effectively neutralized by sourness. The lower the acidity, the sweeter the sake seems. There may be differences in perception due to the drinking temperature and the physical conditions.

The amino acidity shows titration volume (ml) of 0.1 M NaOH necessary to determine the amino acids in 10 ml sake by formol titration. It mainly indicates the density of *umami* and *gokumi* constituents (these terms are difficult to translate into English, but may be seen as a rich and favourable balance of the basic five tastes (sour, sweet, bitter, dry and salty) in sake). Sake with high amino acidity, as long as it is not too high, has a rich and profound taste (*no-jun*). Sake with low amino acidity has a clean and light taste (*tan-rei*). Sake contains about 20 kinds of amino acids such as alanine and arginine. These amino acids are produced by the breakdown of rice proteins. The specific qualities of *umami* in sake are determined according to the compositions of amino acids. The various compositions of amino acids greatly contribute to the complexity and the richness of sake's taste. The long fermentation process, which can be as much as three to four times that of wine, produces a wide variety of amino acids, giving sake a balanced, rounded taste and fresh aroma.

The alcohol content indicates the percentage by weight of ethanol contained in 100 ml of sake. As mentioned above, sake production is a unique process in which saccharification by koji and fermentation by yeast progress simultaneously, a process called 'multiple parallel fermentation'. Multiple parallel fermentation is a key factor in producing sake with a higher alcohol content than other alcoholic beverages. Sake contains not only ethanol but also the higher alcohols, which greatly contribute to the flavour and aroma. The details of higher alcohols are described below.

8.3 Flavour and aroma of sake

The flavour and aroma of sake is created with complex and entangled elements of sweetness, sourness, dryness, bitterness and acerbity. In addition, *umami* and aroma all come together to make up the whole (Gekkeikan, 1997–2010). In this section, the major components of flavour and aroma, and the classification according to flavour and aroma, are described.

8.3.1 Components of flavour and aroma

Higher alcohols and esters are considered to be the most important component contributors to the flavour and aroma of sake.

Higher alcohols

The higher alcohols are formed in the main sake mash (*moromi*), accompanied by the generation of ethanol, and are not produced when fermentation becomes gentle. Three yeast generation pathways routes are known: the Ehrlich pathway, the pathway from amino acid biosynthesis and the pathway from acetic acid. The pathway through the condensation of acetic acid does not greatly contribute to the generation of higher alcohols (Brewing Society of Japan, 1999). Alcohols influence not only the flavour but also the aroma of sake (Yoshizawa *et al.*, 1997). Table 8.2 shows the flavour, the aroma and the threshold value of alcohols (Brewing Society of Japan, 1999). Ethanol and isoamyl alcohol, which are the

Table 8.2 Flavour, aroma and threshold value of alcohols

Component name	Flavour and aroma	Threshold value† (mg/l)	Threshold value (%)
Ethanol	Alcohol		1.4
n-Propanol	Alcohol	600–800	
n-Buthanol	Alcohol	50	
Isobuthanol	Alcohol	100–600	
Isoamyl alcohol	Alcohol, sweet aroma	50–70	
Active amyl alcohol	Alcohol, sweet aroma	50–65	
n-Hexanol	Green leaves	5	
β-phenethyl alcohol	Rose	40–75	
Methionol	Pickles	1.7–2.9	
2.3-Butanediol	Sweetness	4 500	
Tyrosol	Bitterness	200–300	
Glycerol	Sweetness	3 800	

Source: Brewing Society of Japan 1999. †Williams 1974.

aroma components with the lowest boiling point, form the top note (*uwadachi-ka*). Beta-phenethyl alcohol, which is the aroma component with the highest boiling point, forms the basic note. Tyrosol and glycerol influence the taste strength of sake as bitter and sweet constituents. The high concentration of ethanol can mask other flavours. Components of higher alcohols emphasize *uwadachi-ka* and *fukumi-ka* (the flavour which extends in the mouth when you drink sake) (Brewing Society of Japan, 1999).

Esters

The esters in sake are mainly composed of carboxylic acid esters. Table 8.3 shows the carboxylic acid esters and other esters in sake (Brewing Society of Japan, 1999). The production of fatty acid ethyl ester and esters, formed by the condensation of higher alcohols with acetic acid, increases as fermentation advances in sake mash, and it decreases a little during later stages of fermentation (Ikemi *et al.*, 1984). The esters in sake are generated by yeast.

A sweet fruity flavour *ginjo* bouquet (*ginjo-ka*) like a banana comes largely from isoamyl acetate. It is generated by alcohol acetyltransferase (AATase; EC 2.3.1.84.) of yeast with acetyl CoA and isoamyl alcohol as substrates (Yoshioka and Hashimoto, 1987). AATase is unstable at temperature (Yoshioka and Hashimoto, 1987); isoamyl acetate formation increases in the fermentation mash at low temperatures (Brewing Society of Japan, 1999). Therefore, sake which is brewed slowly at low temperatures, such as *daiginjo-shu* and *ginjo-shu*, has high levels of *ginjo-ka* contributed by isoamyl acetate.

Ethyl caproate is also a major constituent of *ginjo-ka* but has an apple-like aroma. Alcohol acyl transferase (AACTase; EC 2.3.1.) and esterase (EC 3.1.1.1.) seem to form ethyl esters of short-chain fatty acids such as ethyl caproate (Brewing Society of Japan, 1999). AACTase reacts with ethanol and acyl CoA as substrates,

Table 8.3 Flavour, aroma and threshold value of esters

Component name	Flavour and taste	Threshold value† (mg/l)
Ethyl acetate	Flesh, ether, fruit	25
n-Propyl acetate	Ethyl acetate-like, fruit	30
Isobutyl acetate	Fruit, pear	0.5
Ethyl butyrate	Pineapple	0.4
Isoamyl acetate	Banana, fruit	1.6
Ethyl caproate	Fruit aroma with sweetness	0.2
Ethyl caprylate	Fruit	0.2
Capric acid ethyl	Cognac, rose	1.5
Pelargonic acid ethyl	Quince, rose	1.2
Lauric acid ethyl	Strong fruit	3.5
Phenethyl acetate	Aroma, rose	3.8
Ethyl lactate	Weak	14
4-Ethyl hydroxybutyrate	Sweet aroma	40–50
Leucine acid ethyl	Aroma	1.0

Source: Brewing Society of Japan, 1999. †Hara, 1977; Akita, 1987, 1990.

and the esterase reacts with ethanol and short-chain fatty acids as substrates (Brewing Society of Japan, 1999).

Table 8.3 shows the flavour, the aroma and the threshold value of the major esters in sake (Hara, 1977, p. 17; Akita *et al.*, 1987). Ethyl acetate, isobutyl acetate, isoamyl acetate and ethyl caproate, which are the aroma components with low boiling points, form *uwadachi-ka*. Phenethyl acetate and ethyl caprylate, which are the aroma components with high boiling points, form the basic note (Takahashi, 1990).

Esters also influence the sake taste; isoamyl acetate and phenethyl acetate emphasize the sweetness and the sourness respectively (Yoshizawa *et al.*, 1997).

Organic acids

About 70% of the organic acids in sake are generated in the main mash by yeast, and most of these are composed of malic acid, lactic acid and succinic acid (Brewing Society of Japan, 1999).

Malic acid gives the fresh sourness, lactic acid gives the mild and tannic sourness, and succinic acid gives the rich and heavy sourness.

Amino acids

Most of the amino acids in sake are derived from rice proteins. The amino acids in sake are created from rice proteins by acid protease and acid carboxypeptidase from koji. In general, the sake main mash contains a large amount of arginine, alanine, threonine, glycine, proline, glutamic acid and leucine (Kondo *et al.*, 1983). The tastes of these are as follows:

- Arginine and threonine, gentle sweet,
- Alanine, very sweet,
- Glycine, fresh sweet,
- Proline, sweet and sour,
- Glutamic acid, strong *umami* and sour,
- Leucine, faintly sweet.

(Brewing Society of Japan, 1999).

Peptides
Most of the peptides in sake derive from rice proteins, as with amino acids. Glutathione is made by yeast and is rare in sake. Peptides bring a round and mellow taste to sake. However, the taste contribution given by individual peptides is unclear (Brewing Society of Japan, 1999).

8.3.2 Classification according to flavour and aroma

Although sake has various types, it is possible to generally classify sake into the following four types (see Fig. 8.5) in terms of flavour and aroma (Sake Service Institute, 1996–2009).

- Flavourful sake (*kun-shu*): sake with a glamorous aroma like flowers or fruits. This type has a fresh light taste and low acidity. Various types exist from dry to sweet. It is very popular outside Japan because of its fruity aroma.

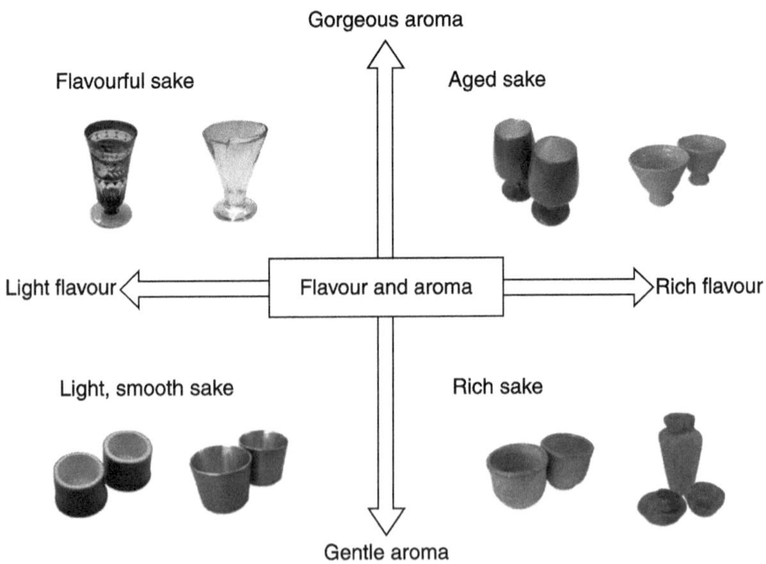

Fig. 8.5 Sake classification according to flavour and aroma. (Source: Sake Service Institute, 1996–2009).

- Light, smooth sake (*so-shu*): sake with a fresh dry taste with moderate acidity. Although the aroma in this type is mild overall, sake has fresh, cool *fukumi-ka*, and juicy flavour. This clean, light taste, *tanrei-karakuchi*, is liked by everybody.
- Rich sake (*jun-shu*): sake with a light, mild aroma and rich, smooth taste. This type has the flavour from the rice, and abundant *umami*. This is the essence of traditional and classic sake.
- Aged sake (*juku-shu*): sake with a light golden colour, a spicy nutty taste and an aroma reminiscent of sherry. This type has a mature complex aroma like dry fruits or spices, a rich sweetness, with a strong taste that is composed of sourness and abundant *umami*. Dedicated drinkers likes this type of sake.

This classification structure is able to offer information such as suitable drinking temperatures, glasses and bottles, and compatible cuisines (Sake Service Institute, 1996–2009).

8.4 Sensory evaluation method

Sake tasting (*kiki-shu*) is a technique that evaluates the quality of sake with the five human tastes. Because sake is a favourite drink, its quality evaluation is heavily dependent on individual sensibility. If we enjoyed drinking the sake we bought, it will generate repeat consumption because of its high quality. The sake with a value-added appeal for consumers will encourage a willingness to buy. These are key factors in a favourite drink. For this reason, evaluating the quality of sake by sensory analysis has great significance. However, in general, the accuracy and repeatability of sensory evaluations are inferior to instrumental analysis because of its subjectivity. Therefore, a sensory panel must be trained to evaluate scientifically and appropriately.

In sensory evaluation of sake it is important to first decide what information is required. The information to be obtained may include quality evaluation, process control, etc. The conditions must then be defined.

Special cups, called (*janome-*) *kiki-choko* or brown amber glasses are used (see Fig. 8.6). Sake samples are maintained at a constant temperature.

The order of sensory evaluation is as follows (Sato, 2007):

1. Look at the colour tone with the white part, and transparency with the blue part, of the *kiki-choko*. This step is only possible using *janome-kiki-choko*.
2. Aroma evaluations *uwadachi-ka, hikikomi-ka* (the aroma sensed when your nose gets close to the glass and you draw a deep breath), *kochu-ka* (the aroma that volatilizes from sake that is warmed in the mouth) and the aroma sensed when your breath goes through your nose after expectoration.
3. Taste evaluation of the initial five tastes (sour, sweet, bitter, dry and salty), the middle tastes (for example, rich, light, clean, smooth, etc.), and the aftertaste.

The sensory evaluation method includes rating (classifying the sample), scoring, identification (is there a difference between samples?), ranking, and so on. The

(a) (b)

Fig. 8.6 Vessels for sensory evaluation. (a) *Janome-kiki-choko*. (b) A brown amber glass. (Source: (a) Sake-no-awaya, 2010, (b) Miyakehonten, 2001–2011).

National Research Institute of Brewing (NRIB) provides quality evaluation terms and a standard sample for flavour of sake as a standard for an accurate sensory evaluation (Utsunomiya *et al.*, 2006).

8.5 Future trends

In Japan, the birthplace of sake, consumer preferences are changing, and alcoholic beverages are diversified into various kinds. This has led to reduced sake consumption. In 2009 brewery year, the value of sake production (equivalent units 20% alcohol) is 469 378 kilolitres (a reduction of 4.9% compared with the previous brewery year), and this has reduced by approximately 9% compared with 2005 brewery year (522 281 kilolitres) (National Tax Agency Japan, 2010).

On the other hand, in the USA and France, demand for sake is expanding briskly. In the UK, a sake section was established in the traditional international wine contest in 2007. Sake is becoming more familiar to people all over. In South Korea, although there has been a boom in sake for several years, sake is treated as a luxury liquor because of high tariffs.

Today, consumers emphasize not only inexpensive, delicious food but also safe and healthy food. In the USA, consumers are fuelling the continued growth of Asian cuisine, and their emphasis on a healthy lifestyle has also fuelled sake's move into the mainstream (Griffith and Gauntner, 2002).

Sake has had a healthy image in foreign countries, but this is not always the case in Japan. The harmful effects of drinking excessive amounts of alcohol tarnished the idea that sake could be healthy. However, sake has been consumed by people for two thousand years or more. Sake is a really safe alcoholic beverage if we keep to a responsible level of drinking. And with recent advances in research in the sake industry, it is now possible to expect that delicious sake can become a healthy drink.

Sake brewers are working hard to turn around the declining consumption of sake. For instance, rice cultivars with good eating quality such as *Oryza sativa* L. Koshihikari have recently been used in sake brewing. Traditionally these cultivars have not been used in sake brewing (Furukawa, 2010). The use of Koshihikari has extended to the entire industry. Rice cultivars with good eating quality are marketed, and they are more familiar than *shuzo-kotekimai* cultivars to consumers. Koshihikari is the best-known cultivar of delicious rice in Japan. Japanese consumers can easily visualize its palatability. Therefore, it is also showing great promise in attracting increased consumer demand for sake.

In addition, sake made using rice with mutated proteins such as *Oryza sativa* L. Shyunyo and *Oryza sativa* L. Mizuhonoka are being marketed. These sakes have a flavour similar to white wine, with clean, fresh sourness, and are gaining much attention for their innovation.

Sake must be made with the traditional charm and novelty and at the same time not be trapped by stereotypical views.

The author, being in the sake brewing industry, sincerely hopes that this wonderful alcoholic beverage, sake, is loved by people all over the world.

8.6 Sources of further information

To the reader who wants to know more details about sake the e-book by Griffith and Gauntner (2002), *Sake Pure + Simple* is recommended. A paperback edition (published by Stone Bridge Press in 1999, ISBN-10:188065637X, ISBN-13: 978-1880656372) is available. This is a nice book in which sake is introduced politely, plainly and in a fascinating way.

8.7 Acknowledgement

The author expresses sincere gratitude to Mr Will Auld, owner and operator of HomeBrewSake.com, for his appropriate, kind proofreading, and our continued and long-lasting friendship.

The author is grateful to Miyakehonten Co. Ltd and Sake-no-awaya Ltd for kindly providing the photographs of vessels for the sensory evaluation of sake.

8.8 References

AKITA O, HASUO T, TAKAHASHI K, MIYANO N and YOSHIZAWA K (1987), 'Generation of aroma component in post-fermentation with yeast', *Nihon-Jozo-kyokai-shi*, **82**, 703–708.

AKITA O, IDA S, OBATA T and HARA S (1990), 'Genetic background and use for sake brewing of high productivity yeast of Beta-phenethyl alcohol and phenethyl acetate', *Nihon-Jozo-kyokai-SHI*, **85**, 501–505.

ANON (2005), *Saké – The Making and Tradition of Japan's National Beverage*, Tokyo, Japan Saké Brewers Association.

AULD W (2011), *Glossary*, Portland, Home Brew Sake. Available from: http://homebrewsake.com/glossary (accessed 8 February 2011).

BREWING SOCIETY OF JAPAN (1999), *Components of Alcoholic Beverages Fermented Seasonings*, Tokyo, Brewing Society of Japan.

CHAMPAGNE E T (2004), *Rice: Chemistry and Technology*, 3rd ed. Minnesota, USA, AACC.

FURUKAWA S (2008), 'Delicious relation between eating quality and brewing aptitude that the rice proteins bring', *Nihon-Jozo-kyokai-shi*, **103**, 145–149.

FURUKAWA S (2010), 'Development of new sake that makes the best use of characteristic of rice – the accomplishment obtained from the industry-university-government cooperation to commercialization', *New Food Industry*, **52**, 25–33.

GEKKEIKAN (1997–2010), '*Nihonshu-do*, acidity and amino acidity (relation of element and taste of sake)', Kyoto, Gekkeikan. Available from: http://www.gekkeikan.co.jp/enjoy/encyclopedia/00145.html (accessed 20 December 2010).

GRIFFITH F and GAUNTNER J, (2002), '*Sake pure + simple Electric Edition 2002*', California, Stone Bridge Press. ISBN 1-880656-37. X.

HARA S (1977), *New Edition Brewing Element List*, Tokyo, Brewing Society of Japan.

IKEMI M, SAITO H, KOIZUMI T and NOZIRO K (1984), 'Trend of odor component in sake mash (1) Variation per day of higher alcohols and esters in *Ginjo-shu* mash', *Nihon-Jozo-kyokai-shi*, **79**, 119–123.

JAPAN SAKE AND SHOCHU MAKERS ASSOCIATION (2011), 'Welcome to the World of Japanese Sake', Tokyo, Japan Sake and Shochu Makers Association. Available from: http://www.japansake.or.jp/sake/english/index.html (accessed 8 February 2011).

JAPAN TECHNICAL INFORMATION SERVICES CORPORATION (2006), *JAPAN/NIPPON: The Land and Its People*, 8th ed. Tokyo, Gakuseisha.

KONDO K, NAKAZAWA E, KUSAMA T and TAKEMURA S (1983), 'Organic acid and amino acid composition of Namazake', *Nihon-Jozo-kyokai-shi*, **78**, 760–762.

MIYAKEHONTEN (2001–2011), *The Basics About Sake*, Hiroshima, Miyakehonten. Available from: http://www.sempuku.co.jp/kenkyu/kiso.html#kikisake (accessed 17 February 2011).

NATIONAL AGRICULTURAL RESEARCH CENTER FOR WESTERN (WENARC) (2007), 'Cultivation of new sake rice variety *Oryza sativa* L. 'Mizuhonoka' by which sake with clear taste is brewed', Hiroshima, WeNARC, press release. Available from: http://wenarc.naro.affrc.go.jp/pre/20071003/20071003_document_cgk.pdf (accessed 17 February 2011).

NATIONAL TAX AGENCY JAPAN (2010), 'Survey results sake on the market in fiscal 2009', Tokyo, National Tax Agency Japan. Available from: http://www.nta.go.jp/shiraberu/senmonjoho/sake/shiori-gaikyo/seizojokyo/2009/pdf/01.pdf (accessed 3 February 2011).

NATIONAL TAX AGENCY JAPAN (2011), *Sake production In 2009 Brewery Year*, Tokyo, National Tax Agency Japan. Available from: http://www.nta.go.jp/shiraberu/senmonjoho/sake/shiori-gaikyo/seibun/2010/pdf/01.pdf (accessed 3 February 2011).

SAKE-NO-AWAYA (2010), *Sake Vessel and Goods*, Fukuoka, Sake-no-awaya. Available from: http://www4.ocn.ne.jp/~awaya/kikityoko.htm (accessed 17 February 2011).

SAKE SERVICE INSTITUTE (1996–2009), *Type Classification of Sake*, Tokyo, Sake Service Institute. Available from: http://www.sakejapan.com/index.php?option=com_content&view=article&id=22&Itemid=43 (accessed 20 December 2010).

SATO S (2007), *Practice – Sake Tasting Seminar Sake Section*, Tokyo, Brewing Society of Japan.

TAKAHASHI K (1990), 'On desired physiological function of sake', *Nihon-Jozo-kyokai-shi*, **85**, 286–292.

UTSUNOMIYA H, ISOGAI A, IWATA H and NAKANO S (2006), *Quality Evaluation Term and Standard Sample Concerning Flavour in Sensory Rating of Sake*, Hirosima, National Research Institute of Brewing. Available from: http://www.nrib.go.jp/data/pdf/seikoumihou.pdf (accessed 20 December 2010).

YOSHIOKA K and HASHIMOTO N (1983), 'Cellular fatty acid and ester formation by brewer's yeast', *Agric. Biol. Chem.*, **47**, 2287–2294.

YOSHIOKA, K and HASHIMOTO N (1987), 'Ester formation by alcohol acetyltransferase from brewer's yeast', *Agric. Biol. Chem.*, **45**, 2183–2190.

YOSHIZAWA K, SUZUKI D, SHINDO H, KAKUTA K and KOIZUMI T (1997), 'Influence of sake aroma component addition on flavour', *Nihon-Jozo-kyokai-shi*, **92**, 217–223.

WILLIAMS A A (1974), 'Flavour research and the cider industry', *J. Inst. Brew.*, **80**, 455–470.

9

Table wines: sensory characteristics and sensory analysis

R. S. Jackson, Brock University, Canada

Abstract: Wine is traditionally viewed as the fermented by-product of grapes. Attributes that make grapes ideally suited for wine production are discussed, as well as the chemical origin of its distinctive sensory attributes. Table wines come in a range of styles, based as much on the grape cultivars used as on unique production procedures. The latter can markedly influence a wine's sensory characteristics. The protocol traditionally used to assess the varietal, stylistic, regional and artistic attributes of wine is discussed, as well as how this information is recorded for analysis. Aspects related to assessing consumer wine preferences are also noted.

Key words: table wine styles, varietal aroma, wine tasting procedure, tasting panel, consumer preference assessment.

9.1 Introduction

In its European usage, wine refers only to beverages derived from fermented grapes – notably *Vitis vinifera*. Varieties used in wine production differ in several aspects from table and raisin grapes. They tend to be higher in sugar, acid and phenolic contents, lower in pectin level, and possess relatively fragile skins. The primarily fermentable nature of their sugar content permits their rapid and near-complete fermentation to ethanol. The typical alcohol content generated (11–14%) has the potential, under anaerobic conditions, to donate considerable microbial stability. This conversion transforms a fragile, short-lived, sweet, juicy fruit into a long-lived, salubrious, and potentially inebriating beverage. The acid content of grapes further accentuates wine's microbial stability. This is further aided by the majority of the acidity being in the form of tartaric acid. This is a relatively rare acid in fruit and is metabolized by few microbes. The low pH of wine (typically in the range of 3.0–3.6) also enhances color stability and provides a fresh taste.

The phenolic content, predominantly in the form of flavonoid tannins, supplies additional preservative potential in the form of antioxidants and further promotes color stability. Tannins also generate most of a wine's bitter and astringent attributes, as well as many of its health benefits. In addition, for red and rosé wines, a subset of flavonoids (anthocyanins) supplies the wine's distinctive color. The limited pectin content of ripe wine grapes facilitates their rupture and juice release, simplifying pressing, clarification, fining and other winemaking practices. The low pectin content also results in wine possessing a low methanol content, methanol being primarily a by-product of pectin breakdown. The relatively thin skins of mature wine grapes further eases the rupture and release of the juice, but can limit the applicability of mechanical harvesting.

Although several wine-grape cultivars possess a distinctive aroma, most seemingly do not. When present, this feature may be evident in the mature fruit (Mateo and Jimenez, 2000). However, it is more common for the expression of the varietal aroma to be limited in the fruit. This results from their being primarily bound in non-volatile complexes (Strauss *et al.*, 1987). These complexes frequently involve sugars (glycosides) or amino acids such as cysteine (Murat, 2005). Their sensory impact becomes increasing evident as the complexes breakdown, permitting the aromatics to volatilize. This may occur following grape crushing (release and activation of hydrolytic enzymes), during fermentation (yeast hydrolytic enzymes), or during maturation (acid- or ethanol-induced hydrolysis). For other cultivars, impact varietal constituents are generated during aging from a variety of breakdown reactions. In most cases, the chemical nature of varietal aromas is still imperfectly understood.

Bottle aging results in extensive modification of a wine's aromatic characteristics. This involves not only degradation of existing volatile constituents, derived from grapes or yeast metabolism, but also the generation of new compounds. Some of these flavorants donate what is termed the wine's aged bouquet. It is this potential to change and improve aromatically, for up to decades, that has given cultivars such as Riesling, Chardonnay, Tempranillo, Sangiovese, Shiraz and Cabernet Sauvignon much of their cult status. This also has led to their being given preferential treatment during cultivation, as well as fermentation, maturation and post-bottling aging. That they are excellent is beyond doubt. What is in question is whether less well-known cultivars, given the same care, would express equivalent qualities. Regrettably, the highly conservative nature of the wine trade limits this likelihood by curtailing their widespread cultivation and distribution, and correspondingly adequate financial returns to the grape grower and winemaker. It is one of the many paradoxical situations in the wine trade.

As noted, the vast majority of wines are derived from *Vitis vinifera* cultivars. Nonetheless, hybrids between *V. vinifera* and other grape species, or direct selections from the latter, are used in some regions. Of these, the most extensively planted are French-American hybrids. Most are complex backcrosses between several North American *Vitis* spp. and *V. vinifera*. They were initially bred to incorporate the phylloxera tolerance of North American species with the wine-making properties of European cultivars. Their use sidestepped the problems and expense associated with grafting European cultivars on tolerant North American rootstocks. Grafting was the only effective means of controlling the devastation being caused by the accidental introduction of phylloxera (a root louse) into Europe in the late 1800s. Although successful in their intended role, some expressed novel aromatic attributes. To traditionalists, these new features assailed their sensibilities. In addition, their popularity with grape growers came to be viewed as a threat to the status of several established wine regions. They were supplanting existing cultivars due to their improved fruitfulness. This also contributed to the growing problem of overproduction. Eventually their cultivation was banned throughout Europe. In contrast, they became widely adopted and used for wine production in much of the north-eastern United States, Canada and Japan. Only in the past few decades have European *vinifera* cultivars come to dominate wine production in these regions, in a desire to simulate European wines. Applauded by most Eurocentric wine critics, it has lead to a loss in the pleasant distinctiveness that once characterized eastern North American wines. This also included wines from selections or crosses with *V. labrusca* parentage. In the south-eastern coastal regions of the United States, selections and cultivars bred from indigenous North American species (notably *V. aestivalis* and *V. rotundifolia*) are still the basis for most wines produced in these regions. Their continued use largely results from the sensitivity of most *V. vinifera* cultivars to endemic diseases (to which local *Vitis* spp. are comparatively resistant/tolerant). Particularly limiting in this regard is Pierce's disease. Wines produced from these indigenous species have interesting and novel flavors absent in *V. vinifera* and French-American hybrids.

A commonly expressed view in wine circles is that 'wine is made in the vineyard'. True to a degree, it is a major exaggeration. Certainly vineyard microclimate affects the attributes expressed by the grapes. Nonetheless, without a skilled winemaker, the inherent qualities of the grapes will never develop their full enologic potential. Despite this, winemakers are a less marketable consumer commodity. The mystique that surrounds the exotic expression *terroir*, the wine's provenance, has more appeal. This geographic feature is easily promoted as being unique and linked to regional history and cultural traditions. That these features rarely if ever consistently affect the sensory aspects of the wine seems immaterial. Beyond the macro- and meso-climatic features of a region, it is soil and atmospheric uniformity that bestow some vineyards with superior qualities. Microclimatic uniformity maximizes the potential for grapes to ripen evenly throughout the vineyard. The result is, assuming good vintage conditions, a crop that can be harvested at both optimal and uniform ripeness. Variability in ripeness is one of the principle limiting factors reducing wine quality. However, as with all generalizations there can be exceptions. In a poor vintage year, vineyard variability may provide some fruit of a quality to enhance the sensory attributes of the harvest or permit selective harvesting. Equally, planting several distinct clones may compensate for a poorer climate, or simply supply the complexity a single clone cannot provide. In all cases, though, the grape variety and wine production procedures set the final limits on the characteristics and quality potentially expressed by the wine.

One of the current major trends in vinification is termed 'precision viticulture'. It employs sophisticated microclimatic assessments with GPS to pinpoint areas of non-uniformity throughout the vineyard. This permits site-specific corrective action to improve vineyard uniformity, localize disease and pest control, and direct selective harvesting based on desired fruit characteristics. Assuming the economic incentives are appropriate, fruit uniformity can be considerably improved using precision viticulture (Bramley, 2005; Proffitt and Malcolm, 2005).

9.2 Categories of table wines

Table wines are traditionally grouped into red, white and rosé categories. This provides a quick and simple approximation of the sensory attributes reasonably expected from the wine. It also simplifies pairing a wine with a meal. Any pairing, of course, depends on consumer preferences, and whether they accept standard views on balancing flavor intensities.

As a general rule, red wines are the most intensely flavored of table wines. In addition, they frequently possess bitter and astringent attributes, unless very well aged. Well-aged red wines are best sampled alone, or at least paired with mildly flavored foods. This allows their more subtle and exquisite flavors to express themselves clearly in the absence of prominent food flavors. In contrast, consuming young red wines with a meal has the advantage that their less polymerized, but more bitter and astringent tannins will bind with food proteins (see Jackson, 2009). This reduces the amount available to react with bitter receptors on the tongue and astringency-sensing trigeminal nerve endings throughout the mouth. Food minimizes the least pleasant sensory aspects of the wine; cheese suppresses some of its less enjoyable attributes (Nygren *et al.,* 2003).

Young red wines are also typified by the presence of a diversity of berry-like to jammy flavors. Wines made from different cultivars can occasionally be identified by the combination of berry-like flavors they express. As the wine ages, these distinctive aromas fade. If the wine has good 'aging potential' the varietal aromas are replaced by an aged bouquet. Its fragrance is relatively independent of the wine's varietal origin (see Broadbent, 1980). Typically, red wines age longer than any other table wine category. This is thought to arise from their higher phenolic and alcohol contents. Alcohol provides antimicrobial activity, and tannins have antioxidant potential.

Rosé wines are ostensibly red wines, given short maceration to limit pigment uptake from the skins. This also results in an even more limited extraction of seed and skin tannins. They are released much more slowly than anthocyanins. This not only reduces the wine's astringency, but also limits its shelf life. Tannins not only delay the loss of fruity aromas (Roussis *et al.*, 2005), but are also essential for color stability (Cheynier *et al.*, 2000). Short maceration times also limit the extraction of varietal aromatics from the grapes. Although retarding the uptake of bitter/astringent flavonoid phenolics, the soluble, bitterish, non-flavonoid phenolics found in the juice are present. These are probably the source of the bitter aspect (Ong and Nagel, 1978; Hufnagel and Hofmann, 2008) typically detected in most dry rosés.

White wines are most commonly produced from white (yellowish) grapes, but can be derived from most purple-skinned cultivars if the grapes are pressed without or immediately after crushing. Anthocyanin presence is usually restricted to the outermost layers of skin cells. This not only limits pigment uptake, but also limits the extraction of antioxidant tannins. Correspondingly, the wine must be protected from exposure to oxygen much more than red wines. Nonetheless, important aroma compounds, such as terpenes and norisoprenoids, usually found in the juice are readily extracted. In addition, because white grapes are commonly harvested earlier than red grapes, most white wines are higher in acidity and lower in alcohol content than a corresponding red wine. This is part of the origin of the fresher, lighter character of white wines. The higher acidity also improves shelf life by slowing oxidative browning and through its antimicrobial activity.

Because of the limited uptake of grape aromatics, most young white wines owe much of their fruity fragrance to the production of esters during fermentation. To retard their breakdown, white wines benefit from storage at cooler temperatures than red wines. In addition, a cool temperature also provides the wines with added interest, in the same way as it enhances the character of sparkling wines.

9.2.1 Subcategories

Red wines come in a remarkable range of styles, from comparatively light and fruity to jammy and intense. Although the grape variety or varieties used in the wine's production markedly influence its sensory characteristics, the length of contact with the seeds and skins (maceration) during fermentation is equally important. For lighter reds, to be enjoyed without prolonged aging, maceration is often limited to a few days. The partially fermented juice is then separated from the seeds and skins (pomace) to complete its fermentation. Two to three days is often adequate to extract sufficient anthocyanins to produce a deep red wine. However, if a more intensely flavored wine is desired, the fermenting juice may be left in contact with the pomace for a week or more. This assures the extraction of considerable flavonoid tannins. Although providing a longer shelf life, the technique also demands prolonged aging before the wine develops it optimal characteristics. Prolonged aging also results in a wine of greater complexity, due to the combined presence of both a distinctive grape aroma and an aged bouquet.

Alternately, distinctive wine styles can be produced using techniques such as carbonic maceration, appassimento, or the governo processes. Carbonic maceration is used primarily in the production of wines such as Beaujolais. It involves a whole-grape fermentation prior to crushing and the commencement of yeast fermentation. The process donates a distinctive fruity aspect and permits drinkability within weeks of fermentation. This contrasts with about two years for traditionally produced red wines. Although permitting early consumption, carbonic maceration is usually associated with a short shelf life (little more than one year).

Appassimento wines involve a slow, partial dehydration of the grapes prior to crushing and fermentation at relatively cool temperatures. Occasionally, during the extended drying process (several months), nascent *Botrytis cinerea* infections reactivate. *Botrytis* produces glycerol and induces distinctive chemical transformations typical of other botrytized wines (see below). In addition, several phenolic compounds are oxidized. This probably explains the distinctive, oxidized-phenol fragrance often associated with wines like Amarone della Valpolicella.

The governo process, previously typical of Chianti wines, involves the separation of part of the harvest for partial drying under well-aerated, cool conditions. The rest of the harvest is fermented in a traditional manner. After the storage period, often lasting two months, the partially dried grapes are crushed. Once fermentation initiates, it is added to the wine derived from the main component of the harvest. The early onset of malolactic fermentation (induced by the bacterium *Oenococcus oeni*) may partially explain the early maturity (and drinkability) formerly associated with Chianti wines.

Rosé wines can be produced in a range of styles, from sweet and bubbly to dry and still. What is not permitted, except for rosé sparkling wines, is the addition of a small amount of red wine to a white wine. Typically, rosés are produced by crushing red grapes gently, and leaving the juice in contact with the seeds and skins at cool temperatures for between 12–24 h (or until sufficient color has been extracted). Cool maceration temperatures are used to retard the onset of fermentation. Alternatively, grapes may be slowly pressed, without prior crushing, to limit color extraction. In either case, once the juice is drawn off it is fermented as if it were a white wine. Occasionally, with grapes of limited anthocyanin content, the juice may be temporarily fermented with the pomace, prior to draining off the must (fermenting juice) for the completion of fermentation.

In comparison with red and rosé wines, white table wines come in an even wider range of styles. Most are dry, acidic, and from slightly floral to intensely fragrant. However, there also exists a wide diversity of sweet and sparkling versions. Dry versions are normally considered as a food beverage, for consumption with poultry or fish dishes. Intensely fragrant and mildly sweet styles are often consumed by themselves as a sipping wine. The sweetest versions are viewed as dessert wines. This typically means they become the dessert, rather than being an accompaniment for a dessert.

Most dry white wines are made from grapes pressed shortly, if not immediately, after crushing. Thus, they tend to have mild or minimal varietal fragrances. There is little opportunity for the extraction of flavorants from the skins. The wines are normally released for sale and consumption within a few months of production. This is preferable as their fragrance is often based on fruit esters (produced by yeast metabolism). These tend to revert to their alcohol and acid constituents relatively quickly. The low phenolic content of most white wines also makes them susceptible to early oxidation. Where wines with more varietal character and aging potential are desired, the juice is left in contact with the pomace for several hours after crushing at cool temperatures. As with all white wines, fermentation typically occurs after the juice has been separated from the pomace.

The most famous of slightly to markedly sweet white wines are made from *Botrytis*-infected grapes. Under special environmental conditions, just before harvest, the grapes may undergo a 'noble rot'. The effect is to concentrate the juice and modify its aromatic character – the greater the dehydration, the sweeter the resulting wine and the more noticeable the botrytized aspect. Classic examples are the auslesen wines of Germany, Sauternes (France) and Tokaji Eszencia (Hungary).

Another form of natural juice concentration involves leaving the grapes on the vine until frozen (<7 °C). Slow pressing of the frozen grapes releases the concentrated juice which, once fermented, is called icewine (*eiswein*).

Another, and ancient, technique of juice concentration involves either partial raisining on mats in the sun, or a slower dehydration under well-ventilated conditions. Classic versions of wine made from partially raisined *vs.* slowly dehydrated grapes are Samos from Greece and *vin santo* from Italy, respectively.

Where a natural, unmodified grape aroma is desired, a portion of the extracted juice is set aside and stored very cold (with or without sulfiting). This juice is sterifiltered and added to the dry wine, made previously, from the majority of the harvest.

9.3 Flavor chemistry and sensory aspects

The wine's principal organic constituents generate most of its gustatory (sweetness, sourness, bitterness) and sapid (astringency, viscosity, pain, prickle) sensations. These include ethanol, organic acids, glycerol, sugars, and tannins. In contrast, the minor constituents (often occurring at concentrations of about a few mg/l to ng/l) are those primarily involved with generating its varietal aroma.

Except for very sweet wines, ethanol is the most common organic constituent (typically in the range of 110–130 g/l). With the exception of glycerol, and occasionally sugars in sweet wines, most other major constituents occur at concentrations of less than 1 g/l. The most common of these are fusel alcohols, volatile acids, and fatty acid esters. Of these, fusel alcohols often constitute 50% of all volatile compounds. Although present in much smaller amounts, carbonyls, volatile phenolics, lactones, terpenes, hydrocarbons (and their derivatives), and volatile sulfur and nitrogen compounds are the most significant contributors to the varietal and unique sensory features that permit distinguishing among wines.

9.3.1 Alcohols

Ethanol is the principal alcohol in table wine, typically occurring at concentrations between 11–13%. Nonetheless, it may be as low as 7.5% in some German botrytized wines, and rise to 15% and above in some red and dessert white wines. As the concentration of ethanol increases, so does the perception of its burning sensation and its distinctive odor. The ethanol content also influences the relative volatility of other wine aromatics (D'Angelo *et al.*, 1994; Guth, 1998; Williams and Rosser, 1981), and the taste intensity of its acidic and phenolic contents (Lea and Arnold, 1978). The prime factors controlling ethanol production are the grape's sugar content, the yeast strain, and the fermentation temperature.

Despite the significant physiological, psychological and social consequences of the presence of alcohol in wine, ethanol is critical to a wine's development, stability, and sensory attributes. The accumulation of ethanol during fermentation tends to soon restrict yeast activity to *Saccharomyces cerevisiae*. The antimicrobial action of alcohol also contributes to the wine's shelf life. Furthermore, ethanol facilitates the extraction of tannins and non-polar aromatics from the pomace and oak cooperage, is a reactant in the synthesis of fruit esters, and is directly and indirectly involved in the development of the aged character of older wines.

Although wine does contain methanol, its content ($\leq$ 0.1–0.2 g/l) is lower than most alcoholic beverages. This results from the limited pectin content of mature grape juice.

The other alcohols found in wine are generally referred to as higher (fusel) alcohols, and possess more than two carbons per molecule. They are generally the by-products of yeast amino acid metabolism. Straight-chain alcohols, such as isoamyl and isobutyl alcohols, possess strong, pungent odors. However, their relatively high odor thresholds mean that they typically contribute only to the general fermentation odor of wines. The amounts present vary considerably among wines, and are markedly affected by wine-making conditions (Sponholz, 1988). Notable in this regard are the fermentation temperature, extent of pre-fermentation clarification, and yeast strain. Occasionally, sensorially significant additional sources of higher alcohols may be derived from grape or leaf hexanals and diseased fruit. At sufficient concentrations, hexanols can donate a vegetal/herbaceous odor, and the fungal metabolite 1-octen-3-ol can donate a mushroomy odor. Phenol alcohols, such as 2-phenylethanol, may come either from grapes or yeast metabolism. Depending on its concentration, 2-phenylethanol may add rose- to pepper-like notes.

Of polyols, the most significant is glycerol. In concentration it is often second only to ethanol as the most common organic component. It often reaches concentrations in the range of 10 mg/l in red wines and 7 mg/l in white wines. Nonetheless, at these concentrations it has little sensory impact. To affect perceived wine viscosity it needs to occur at concentrations above 26 mg/l (Noble and Bursick, 1984). This level is reached only occasionally in markedly botrytized wines. Glycerol has a sweetish taste, but it is too mild to be of marked significance in table wines.

9.3.2 Phenolics

Of phenolic compounds, the most sensorially significant are the flavonoids. The most common are anthocyanins and condensed tannins (polymers of catechins). Both are critical to the sensory characteristics of red wines. Non-flavonoid (hydrolyzable) tannins, largely derived from oak cooperage, are less significant (Pocock *et al.*, 1994). Most non-flavonoid phenolics are derivatives of hydroxycinnamic and hydroxybenzoic acids found dissolved in grape cell cytoplasm. Because of their water solubility they occur equally in white and red wines. In contrast, most grape tannins occur as poorly soluble polymers in grape seeds and skins. Because the juice of white grapes rarely remains in contact with the seeds and skins following crushing, grape tannins are not found in significant concentrations in white wines. However, if the wines are given extended maturation in oak cooperage, hydrolyzable tannins may be temporarily present before breaking down to gallic or ellagic acids.

As noted, the major flavonoids in wine are anthocyanins, catechins (flavan-3-ols), and their polymers (tannins). Anthocyanins are located almost exclusively in the outer skin cells of red grapes, whereas catechins and distinctive condensed tannins are found in both seed coats and the skin. Typically, only catechins and their smaller dimers and oligomers are extracted during maceration. During wine maturation, catechins and their smallest polymers (procyanidins) polymerize with themselves and anthocyanins. As catechins polymerize, their bitter attributes eventually become less marked, but their astringent characteristics increase (Brossaud *et al.*, 2001). Above a certain size, however, the ability of these polymers to bind effectively with proteins declines and, correspondingly, their ability to induce astringency. The binding of catechins and procyanidins with anthocyanins is important to the color stability of red wines (Cheynier *et al.*, 2000; Waterhouse and Kennedy, 2004). As anthocyanins are extracted, the protective, self-associated complexes in grapes separate. Because free anthocyanins are comparatively susceptible to oxidative browning, it is important that they quickly bind with catechins in relatively stable polymers. This is facilitated by the formation of acetaldehyde, generated indirectly as a consequence of phenol oxidation (Wildenradt and Singleton, 1974). Acetaldehyde is an essential reactant in polymerization reactions between anthocyanins to tannins (Timberlake and Bridle, 1976). Although achieving color stability, the rich purple color of free anthocyanins shifts to a less-intense, more reddish-brown shade. In addition to generating the typical colors of red and rosé wines, anthocyanins also play a temporary but significant antioxidant role in preserving one of the important flavorants in many rosé wines, 3-mercaptohexanol (Murat *et al.*, 2003). The antioxidant potential of flavonoid and non-flavonoid phenolics also protects several other important wine aromatics and limits the formation of undesirable oxidized taints and flavors. These are all important in prolonging wine shelf life.

Volatile phenols can influence the fragrance of wines but, with the major exception of methyl anthranilate in some *Vitis labrusca* cultivars, rarely contribute to the generation of varietal aromas. In contrast, volatile phenolics derived from oak cooperage, such as vanillin and eugenol, have considerable sensory impact. In addition, hydroxycinnamic acid derivatives may be metabolized by spoilage yeasts to ethylphenols, generating a variety of unpleasant animal, barnyardy, off-odors.

Halogenated phenols, found in cork or wood products, can occasionally be a source of important off-odors. Several fungi can metabolize these compounds to highly volatile halogenated anisoles. These may be absorbed by wine, in which they can be detectable in nanogram quantities. 2,4,6-Trichloroanisole, derived from chlorophenols, is the principal agent generating moldy (corked) odors in wines (Pollnitz *et al.*, 1996). An equivalently malodorous contaminant is 2,4,6-tribromoanisole (Chatonnet *et al.*, 2004). It also is a fungal metabolite derived from a commonly used wood preservative, 2,4,6-tribromophenol. It is often a component in the preparation of fire-retardant epoxy resins.

9.3.3 Sugars

The principal sugars in wine are glucose and fructose, often in near-equal quantities. In dry wines, they generally occur at less than 1.5 g/l. In this range they do not generate a detectable sweet sensation. This generally begins at about 2 g/l and above. The sweetness occasionally perceived in dry wines is usually an olfactory illusion, generated in the brain's orbitofrontal cortex. It arises from memories derived from former associations between fruity smells and sweetness. The brain cortex appears to invoke a hierarchy of significance in interpreting sensory input, with visual clues being dominant over olfactory and gustatory sensations, and olfactory over gustatory (see Prescott, 2004; Rolls, 2006).

Typically, sweet wines have sugar contents reaching or exceeding 10%. During ageing, structural transformation can lead to the formation of melanoidin compounds, generating a golden color. To avoid microbial spoilage, a constant risk with sweet wines, the wines are typically given special treatment prior to bottling.

Minor quantities of other sugars are commonly found in wines, notably pentose sugars such as arabinose, rhamnose and xylose. They neither affect taste nor are a major factor in microbial spoilage.

Wine may contain small quantities of complex sugars or their derivatives, notably gums and other complex polysaccharides. Gums are typically produced in botrytized grapes. While not directly affecting sensory attributes, their release into the juice on pressing can cause problems during wine clarification. However, the common polysaccharides found in wines are mannoproteins, polymers composed of mannans and protein. These escape into the wine as yeast cells die and begin to autolyze at the end of fermentation. They accumulate particularly in wines left in contact with the lees (primarily the precipitated remains of dead yeast cells) for extended periods. Mannoproteins can associate weakly with aromatic compounds in non-volatile complexes. As such, their dissociation constants can significantly affect the dynamics of how wine aromatics are liberated after pouring.

9.3.4 Acids

The principal grape acids are tartaric and malic acids. In warmer climates, the grapes respire much of the malic acid as they ripen. In cooler climates this usually does not occur and the wine is encouraged to undergo malolactic fermentation. In the process, residual malic acid is converted to lactic acid – a less sour-tasting acid.

Grape acids (plus any lactic acid derived from malolactic fermentation) give wine their 'fresh' flavor, donate balance to sweet wine (reducing perceived sweetness), augment a wine's antimicrobial properties, favor desirable aging reactions, and enhance color stability and intensity. Although most organic acids are relatively odorless, those derived from yeast and bacterial activity are often volatile. The most important of these is acetic acid. At normal concentrations (<300 mg/l), acetic acid may contribute to the complexity of the wine's character. However, at values greater than 300 mg/l, it increasingly donates an undesirable sour taste and begins to generate a vinegary odor. It can also react spontaneously with ethanol to form ethyl acetate.

9.3.5 Esters

Wine esters are normally grouped into three categories: acetate esters (based on acetic acid and higher alcohols) and two ethyl ester subgroups (based on ethanol and volatile fatty acids and non-volatile organic acids, respectively). Of these groups, the most sensorially significant are those based on acetic acid. Examples such as isoamyl, isobutyl and benzyl acetates play a significant role in the fruity bouquet of young white wines. Some ethyl esters, notably those based on shorter-chain fatty acids, such as ethyl butyrate and ethyl hexanoate, may also contribute to the fruity character of young wines. Their production is favored by fermentation at cool temperatures and by the choice of yeast strain. Because these esters often accumulate at above-equilibrium concentrations, they tend to break down slowly upon the completion of fermentation. This can be retarded by cool storage (Marais, 1986) and the presence of wine antioxidants (Roussis *et al.*, 2005).

In contrast, ethyl esters of non-volatile organic acids tend to accumulate slowly, and non-enzymatically, post-fermentation. Because of their low volatility and weak odors, they seldom play an important role in wine flavor development.

Although some ethyl esters enhance the pleasurable attributes of wines, the nail-polish remover aspect of ethyl acetate is a fault at all but threshold levels. At or just below detection thresholds (50 mg/l), ethyl acetate may contribute to an overall complex flavor. Its occurrence at concentrations of 150 mg/l and above is usually an indicator of contamination with acetic acid bacteria.

9.3.6 Carbonyls

The most frequently occurring and important carbonyl in wine is acetaldehyde. It can generate an oxidized odor, but typically accumulates to detectable levels only in sherries. Its primary sensory significance in table wines is favoring color stability. By reacting with anthocyanins and procyanidins, acetaldehyde favors their polymerization. This results in the red anthocyanin chromophore becoming more resistant to oxidative browning.

Hexanals and hexenals, metabolically derived from their corresponding alcohols, contribute to grassy/vegetal/herbaceous odors in some wines. Other aldehydes of aromatic significance are derived from contact with oak cooperage.

Examples are the vanilla odors derived from phenolic aldehydes, such as vanillin and syringaldehyde, the caramel-like fragrance of furfural aldehydes, and the almond attribute donated by benzaldehyde.

Of ketones, the most significant example is diacetyl (2,3-butanedione). At low concentrations, it can donate a buttery, nutty, or toasty note to wines. At higher concentration it can produce a lactic off-odor. Its presence at detectable levels is frequently associated with wines having undergone malolactic fermentation. In infected grapes, the fungal ketone, 1-octen-3-one, may contribute to the presence of a mushroomy note.

9.3.7 Terpenes

Terpenes are an important and diverse group of compounds involved in the fragrance of many plants, including several important white grape varieties, notably the Muscats, Riesling, and Gewürztraminer. Terpenes tend to accumulate in the pulp and are, thus, released with the juice. Because most terpenes are associated with sugars as non-volatile glycosides, glycosidic terpenes do not directly contribute to the fragrance of young wine. Nonetheless, they tend to break down during aging. The liberation of bound terpenes can partially compensate for loss of free terpenes oxidized during ageing. Oxidation not only changes their odor quality but also increases the sensory threshold.

More complex terpenes, such as sesquiterpenes, may be involved both in the generation of musty, earthy, off-odors, and important varietal odorants. An important example of the latter is rotundone. It generates the pepper aroma of Shiraz and some other Rhône varieties (Wood *et al.*, 2008).

9.3.8 Lactones

Except for a few cultivars, such as Riesling and the Muscats, lactones do not contribute to the varietal character of most wines. They can, however, contribute to the raisined character of sunburned grapes. Lactones such as sotolon can also be involved in the distinctive fragrance of botrytized wines and sherries. For most table wines, the most common source of sensory significant concentrations of lactones comes from maturation in oak cooperage. The most important of these are oak lactones (isomers of β-methyl-γ-octalactone) (Chatonnet *et al.*, 1990) and several γ-nonalactones. These can respectively donate coconut and oaky flavors.

9.3.9 Hydrocarbons

Several aromatic hydrocarbons, primarily degradation products of grape carotenoids, have sensory significance in some wines. Major examples are β-damascenone, α- and β-ionone, vitispirane, and 1,1,6-trimethyl-1,2-dihydronaphthalene (TDN). The latter can reach concentrations sufficient to contribute to the aged bouquet of some white wines.

9.3.10 Pyrazines

Pyrazines (cyclic nitrogen-containing compounds) contribute significantly to the flavor of many natural and baked foods. They also contribute to the varietal aroma of several grape cultivars. Particularly notable in this regard is 2-methoxy-3-isobutylpyrazine and several related pyrazines. They play a major role in the bell pepper odor of cultivars such as Cabernet Sauvignon and Merlot (Boison and Tomlinson, 1990). At concentrations of about 8–20 ng/l, it can be desirable, but above this it generates an overpowering vegetative/herbaceous aroma. A related molecule, 2-methoxy-3,5-dimethylpyrazine, has also been associated with some moldy off-odors (Simpson *et al.*, 2004).

9.3.11 Organosulfur compounds

Most reduced organosulfur compounds are considered undesirable. They typically induce foul odors at even trace amounts. Well-known examples are ethyl and methyl mercaptans and dimethyl sulfide. Their formation in wine depends on the development of strongly reducing conditions. Recently, though, a number of thiol compounds have been identified as critical constituents of several grape varietal aromas. Examples are 4-mercapto-4-methylpentan-2-ol and 3-mercaptohexan-1-ol. They produce odors reminiscent of citrus zest/cat urine and grapefruit, respectively. Both compounds are important in the varietal character of Sauvignon Blanc (Tominaga *et al.*, 1998), whereas the former is an important odorant in Scheurebe (Guth, 1997). In addition, 4-mercapto-4-methylpentan-2-one and 3-mercaptohexyl acetate (Tominaga *et al.*, 1996), and especially benzenemethanethiol (benzyl mercaptan) (Tominaga *et al.*, 2003) contribute to the varietal box-tree (smoky) aroma of some Sauvignon Blanc wines. Although seemingly not a varietally distinctive thiol, 3-mercaptohexanol is an important aromatic constituent of several rosé wines (Murat, 2005).

9.3.12 Gases

Carbon dioxide is a constituent in all wines, typically occurring at supersaturated conditions in young wine. This usually dissipates before bottling. However, if the wine is bottled early, the wine may still possess sufficient gas to form a mild petillance on pouring, or form bubbles at the bottom or sides of the glass. At these concentrations, it seldom generates the prickling sensation typical of sparkling wines.

More important is the presence of sulfur dioxide. It is usually added to wine (notably white wines) prior to bottling to augment its antioxidant and antimicrobial attributes. Sulfur dioxide is no longer added at concentrations sufficient to give poured wine a slight burnt-match note.

9.4 Sensory analysis applied to table wines

The application of sensory analysis to table wines differs considerably from its use with other alcoholic beverages, including sparkling and fortified wines. In the latter, sensory analysis is typically directed toward the formulation of distinctive brand styles, and the assessment of consumer preferences. This rarely applies to table wine, except for some large-volume, brand-named wines. Intriguingly, Singleton and Ough (1962) showed that blending two wines rarely averages their relative quality. Instead, the perceived quality of the blend was always greater than the mean, and occasionally superior to that of either of the samples blended. Thus, blending appears not as negative as normally portrayed in the popular media. Even the most prestigious champagnes, ports, brandies, etc. are blends. Nonetheless, even large firms tend to promote the local distinctiveness of their table wines. Producers realize the marketing value of provenance. Yearly variations in character (vintages) are accepted and often emphasized in vintage charts, rather than being harmonized through blending. These subtle distinctions are very important, probably essential, to the commercial survival of small wineries.

For small producers, the winemaker is the 'sensory panel'. Wines are often made on their palate. Commercial viability depends on consumers accepting, and often searching out, this possibly imaginary singularity. It often involves a longing to identify with a particular philosophic conviction. Much of the appeal of organic and biodynamic wine appears to be of this nature. The thought of blending out site distinctiveness is anathema. The same mindset also applies to those who consider they are not only purchasing wine, but also identifying with the region's or site's context, culture, or history. It is participation by association—the equivalent of achieving exclusivity by possessing a rare piece of art.

Other than those aficionados who are truly interested in the sensory attributes of wine, not its pedigree or provenance, detailed sensory analysis of wine is almost the exclusive domain of universities and research centers. Even the best wine competitions fail to rigorously apply satisfactory standards or apply adequate statistical analyses. They are frequently vehicles designed to promote a region's wines and/or collect medals to impress visitors at participant wineries.

Sensory analysis at most of the larger wineries usually has one of two objectives: detection of faults before they become irreversible, or to formulate blends to maintain brand consistency. Occasionally, it may be used to separate samples of different quality for independent processing and bottling. Tastings are usually conducted by a relatively small panel of winery employees with years of experience. The intentions are very pragmatic and rarely subjected to critical or statistical analysis. Thus, the techniques employed can vary widely and be conducted irregularly. Records, if kept, are used only as a guide for subsequent sessions. Nonetheless, Lesschaeve (2007) argues for the inclusion of sensory analysis as a complement in traditional, in-house, winery tastings to facilitate consumer-oriented wine production.

Another form of sensory analysis is used by large wholesale and retail firms. Tasting is intended to select wines for sale. For this, expression of varietal, regional and/or stylistic features, as well as price and availability, are the primary concerns. Training is often limited to sampling with the reigning authority of the firm or a trade school. Because no verifiable objective standards are available to compare experts, the quality of the education may be questionable. Nonetheless, without objective standards for wine norms, the tremendous variability in human sensory acuity, and the marked influence of psychological factors on perception, the lack of scientific rigor in the commercial sector may be of relatively little importance. In addition, with so much variability among wines (in contrast to the comparatively narrow range with commercial food products), it is very difficult for consumers to become effective critical tasters.

Within the scientific community, sensory analysis primarily has a research focus. This may be directed at determining if wines from separate and supposedly distinct subregions possess detectable and definable differences (Fig. 9.1). Regrettably, these studies typically investigate only the wines from a single or a few vintages. Thus, whether the results are consistently valid over time remains questionable. Alternatively, the study may be intended to ascertain the effects of variations in meso- and microclimate, grape-growing techniques, or wine-making procedures on the sensory attributes of the wine, and to what degree (Elliott-Fisk and Noble, 1992; Tomasi *et al.*, 2005). These influences may have significant implications on detectable wine quality and commercial success. Other investigations may be aimed at shedding light on features such as the chemical origins of a wine's varietal aroma (Ferreira *et al.*, 1998; Guth 1998; Kotseridis *et al.*, 2000; Lacey *et al.*, 1991), changes involved during aging, or the nature of wine faults (Mazzoleni and Maggi, 2007; Soleas *et al.*, 2002). Sensory analysis has also been used to study the nature of human sensory perception, especially as it applies to wine (Aznar *et al.*, 2004; Hemingway *et al.*, 1999; Guinard *et al.*, 1986), or how tasting conditions or prior knowledge can affect (bias) perception (Fig. 9.2). Because of the variety of uses to which sensory analysis may be applied, their experimental design often differs considerably. Despite this, some aspects

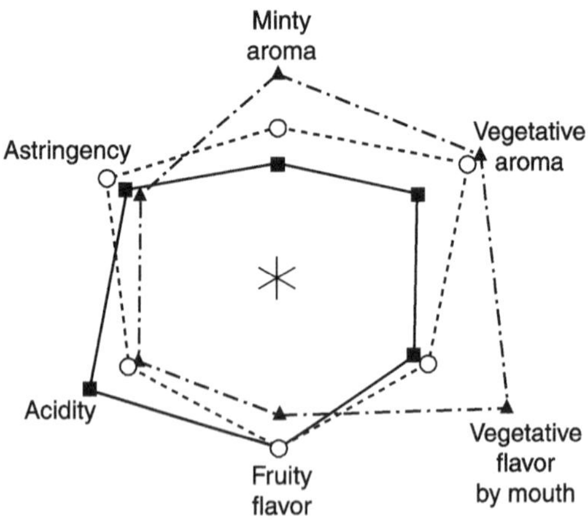

Fig. 9.1 Comparison of descriptive profiles of Cabernet Sauvignon wines from Napa Valley, Monterey, and Mendocino (Source: Schmidt, 1981, reproduced in Noble, 1988, © Springer-Verlag).

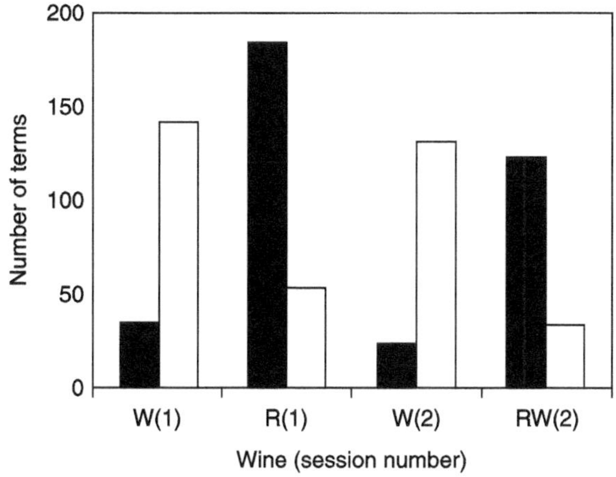

Fig. 9.2 Frequency distribution of olfactory terms representative of red/dark objects (solid bars) and yellow/light objects (open bars) used to describe two paired tastings of a white wine (W, Sémillon/Sauvignon Blanc), a red wine (R, Cabernet Sauvignon) and a white wine sample colored red with anthocyanins (RW, Sémillon/Sauvignon Blanc) (Source: Morrot *et al.*, 2001, with permission from Elsevier).

are relatively consistent, such as taster training and selection, tasting conditions, data collection, and statistical analysis. What almost consistently varies are the precise attributes for which tasters are trained to detect and differentiate, and the type of data collected.

Occasionally research centers may be employed to conduct consumer analysis studies. In these situations, a markedly different *modus operandi* is required (see below).

9.4.1 Training and selection of panels

Tasting panels are used as substitute analytic instruments, in instances where objective devices are insufficiently sensitive, or the researchers are more interested in subjective or perceptive data than chemical analysis. Panels of ten or more members are preferred, due to variability in human sensory acuity – both temporal and individual. Statistical measures are used to ascertain the relative level of confidence that can be placed on the interpretations drawn.

To limit the degree of sensory variability among panel members (Fig. 9.3), and to improve term usage consistency, prospective panelists undergo lexicon training suitable to the task. This may also involve employing candidates to develop an appropriate vocabulary, in whose precise use they are subsequently trained and tested. This has the advantage of producing a set of terms particularly relevant to the participants. More commonly, though, training involves experience in the use of terms chosen by the researchers. The descriptors selected are typically concrete (for example, specific fruits, flowers, vegetables, nuts, fungi, etc.) rather than conceptual (masculine, perfumed, intense, smooth). In most instances specific

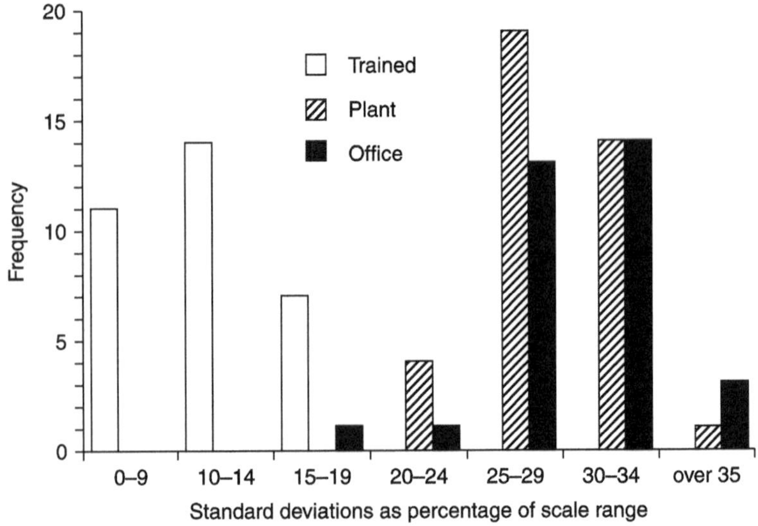

Fig. 9.3 Standard deviations as a percentage of scale range in fragrance descriptive analysis. Data from a trained panel and two consumer model panels (untrained). Ratings were made across ten commercial air freshener fragrances on nine-attribute scales (Source: Lawless, 1999, with permission from Elsevier).

chemical names are avoided. Panelists tend to be far better at using terms resembling already familiar objects than chemical nomenclature (Wise *et al.*, 2000). In addition, even individuals with extensive sensory experience have difficulty identifying specific compounds when combined in mixtures (Fig. 9.4). Chemical terminology may be invoked when they are industry standards, familiar to the panelists, or when domestic descriptors are nonexistent. Examples are sulfur dioxide, hydrogen sulfide, quaiacol, diacetyl, and mercaptan.

Incorporated into the training process is selection. Trainees who do not achieve a preset level of consistent term use are eliminated. Due to this selection (removal of natural variation), the data derived from such panels is not applicable to consumer preferences. To maintain panel consistency, periodic retraining is often employed, and sensory reference standards are available during analysis sessions.

Besides lexicon training, potentially used in sensory profiling, or measuring the intensity of particular attributes, training and selection may also involve general sensory skills. This is particularly relevant when the panel assesses subjective aspects such as wine quality. In this instance, it is important to determine sensory acuity to a wide range of specific wine attributes. These may also include abilities to recognize and identify wines derived from a range of grape varieties, coming from separate wine regions, and/or produced by different techniques. It may also be important to determine the relative abilities of panelists to differentiate among similar wines, as well as recognize and identify a range of wine faults. Different training programs are required for each of these aspects (Jackson, 2009).

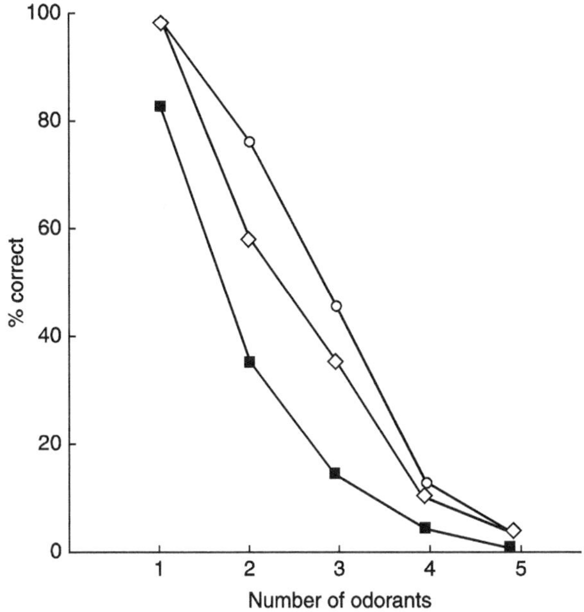

Fig. 9.4 Percent correct identification of the constituents of mixtures containing up to five odorous chemicals by untrained (■), trained (◊), and expert subjects (○). (Source: Laing, 1995).

The amount of training and the degree of proficiency required by the panelists will depend on the task(s) and the commercial significance of the results.

Exceptional sensory acuity and existing identification skill may be less important than the ability to learn and rapidly develop an adequate odor memory. It is the accurate retention of, and immediate access to, odor patterns that may be most critical to success as a sensory panelist. Identification of holistic aspects, such as a wine's varietal or geographic origin (if possible) is frequently a slower, more indirect, cognitive process, combining inductive and deductive reasoning skills. This often involves memories of a wide range of specific sensory attributes and their combination. Ideally, identification would be based on the presence of distinctive features that characterize a particular cultivar, provenance, or style. However, more often than not, identification also involves recognizing specific attributes that are absent. Lesschaeve (2007) distinguished two categories of wine tasters: wine experts and wine assessors. Wine experts were defined as those possessing greater product knowledge, whereas wine assessors were those possessing greater sensory acuity. However, both aspects are usually required in most routine sensory analysis tasks.

Because most panels consist of volunteers, for which tasting is only an accessory to their primary jobs, membership is often in a state of flux. New members need to be trained to fill vacant spots and become substitutes to replace members who may be unavailable. It is important to keep the number of members at any session relatively constant.

Because of the concentration involved in sensory evaluation, and the repetitive and prolonged duration of some tasks, clear recognition of the importance of their work is necessary to maintain panel moral and enthusiasm. It also requires panels to have dedication and an analytic and creative frame of mind. If panel members are drawn from within a large organization, department heads should also be made fully aware of the importance of the time that panelists spend away from their normal activities. It must not be viewed as a fun time away from work. It rarely is.

Under research conditions, the purpose of a panel may often change. This usually requires retraining for the new task. An appropriate proficiency level in one task is no guarantee that an equivalent level can be assumed for even a seemingly related task.

9.4.2 Tasting procedure

No specific tasting technique or sequence has been adopted as an industry standard. This probably reflects the diverse tasks panelists may be called upon to perform. In addition, in most circumstances, sensory analysis is not their primary job. In other words, sensory analysis is not their profession. Thus, development of an industry-wide standard has not been viewed as a necessity. Those professionally involved in sensory analysis typically organize tasting and do the analysis. They are not participants in the actual tasting. In addition, since the goals of assessments can be so diverse and specific, assessing all aspects of a wine's character is usually unnecessary. Only a few select sensory attributes are usually necessary to distinguish among similar wines. The data are usually presented in spider plots (Fig. 9.1). They are particularly useful in visually representing the sensory differences between sets of samples. They illustrate not only that differences exist, but also their nature and degree. However, if confidence bars are absent, the significance of the differences represented is in doubt. In addition, only a few variables can be presented simultaneously, otherwise the diagram becomes too cluttered. Nor do spider plots represent the holistic or subjective nature of the differences, any more than do spectrographs or gas chromatograms provide a sensory feeling for the wine's color or fragrance. For the purposes of its use, this is not a serious limitation. Most sensory analyses are designed to provide relatively objective data on the nature of differences generated by the variables studied. It is not intended to present a detailed sensory profile of the samples, or analyze subjective, holistic impressions. Assessing the comparative quality and typicity of wines is more the prerogative of winery or comparative tastings. For this, the wines are usually ranked within designated mark ranges for a series of attributes. Ranking criteria and procedures vary, depending on the precise intentions of the experimenter. By giving numerical expression to subjective impressions such data could, but seemingly rarely are, presented in spider plot form. This differs markedly from the popular, consumer-related, rankings based only upon the total accumulated point averages of the various samples. These often include a fudge factor permitting the judge to adjust the sum to reflect their overall ranking of the samples. They appeal to the modern, and almost insatiable desire of consumers to have quality reduced to a numerical expression.

Tasting techniques, such as those illustrated in Fig. 9.5, separate the assessment into four components. The first deals with the wine's countenance. In most instances, this is simply for pleasure. Although appearance can be used to predict certain sapid and olfactory attributes, these impressions can, too easily, subconsciously bias perception. The second phase involves smelling the wine in the glass. After an initial whiff, the wine is swirled in the glass and sniffed again. The first assessment detects the most volatile wine constituents. Subsequent assessment permits the detection of compounds released more slowly. Swirling facilitates the liberation of less volatile components. Cover glasses (or tightly fitting, small, plastic Petri dish tops) may be placed over the mouth of the glass. This has several advantages. If the wines possess subtle fragrances, holding the cover permits vigorous swirling, enhancing aroma release and capture in the headspace of the glass. Where the wines are very aromatic, covers help avoid contamination of the immediate vicinity of the tasting booth. Finally, if the wines are poured in advance of sampling, it guards against fragrance loss during the interval between pouring and tasting. Orthonasal assessment is conducted repeatedly throughout the sampling period. It permits detection of how a wine's fragrance changes in character (termed development) and intensity with time (termed duration). Repeated assessment can be important as the equilibrium among volatile compounds, in the gaseous phase above the wine, and its dissolved and weakly bound forms in the wine, may change considerably over the course of the tasting. 'Normal' smelling seems fully adequate (Laing, 1983), requiring no special procedure other than keeping the sniffs relatively short and separated by about 30 seconds. This minimizes the likelihood of odor adaptation. Nonetheless, variation in smelling pattern can affect the perception of fragrance intensity (and thereby potentially odor quality) (Fig. 9.6). Subsequently, in the third phase of tasting, a small sample of the wine is taken into the mouth. If considered important, the various gustatory perceptions may be assessed in sequence, for example sweetness, sourness, bitterness, astringency, and heat. To enhance the perception of the retronasal aspect of olfaction, the wine is often aspirated in the mouth (pursing the lips and slowly and methodically drawing air through the wine). It is the oral equivalent of swirling wine in the glass. Subsequently, the wine is expectorated. This helps avoid an increase in blood alcohol content that can affect sensory perception. In the fourth stage, attention shifts from concentration on individual attributes to holistic sensations, such as complexity, balance, and body. After expectoration, attention is directed to the wine's finish (the lingering sensations in the mouth). Although the latter is considered to be a major quality indicator for connoisseurs, the fourth stage is rarely incorporated into research-based wine assessments. Finally, for connoisseurs, attention shifts to quality interpretation, either artistic, stylistic, varietal, historic, or regional.

The intention of the sensory analysis will determine how and what wine attributes are recorded.

Each sample should be poured into identical, clear, tulip-shaped, wine glasses. They should each be filled (1/4 to 1/3 full) with the same volume of wine.

I. Appearance

1 – View each sample at a 30° to 45° angle against a bright, white background.
2 – Record separately the wine's:
clarity (absence of haze)
color hue (shade or tint) and depth (intensity or amount of pigment)
viscosity (resistance to flow)
effervescence (notably sparkling wines)

II. Odor "in-glass"

1 – Sniff each sample at the mouth of the glass before swirling.
2 – Study and record the nature and intensity of the fragrance[a].
3 – Swirl the glass to promote the release of aromatic constituents from the wine.
4 – Smell the wine, initially at the mouth and then deeper in the bowl.
5 – Study and record the nature at intensity of the fragrance.
6 – Proceed to other samples.
7 – Progress to tasting the wines (III)

III. "In-mouth" sensations

(a) Taste and mouth-feel
1 – Take a small (6 to 10 ml) sample into the mouth.
2 – Move the wine in the mouth to coat all surfaces of the tongue, cheeks and palate.
3 – For the various taste sensations (sweet, acid, bitter) note where they are perceived, when first detected, how long they last, and how they change in perception and intensity.
4 – Concentrate on the tactile (mouth-feel) sensations of astringency, prickling, body, temperature, and "heat".
5 – Record these perceptions and how they combine with one another.

(b) Odor
1 – Note the fragrance of the wine at the warmer temperatures of the mouth.
2 – Aspirate the wine by drawing air through the wine to enhance the release of its aromatic constituents.
3 – Concentrate on the nature, development and duration of the fragrance. Note and record any differences between the "in-mouth" and "in-glass" aspects of the fragrance.

(c) Aftersmell
1 – Draw air into the lungs that has been aspirated through the wine for 15 to 30 s.
2 – Swallow the wine (or spit it into a cuspidor).
3 – Breathe out the warmed vapors through the nose.
4 – Any odor detected in this manner is termed aftersmell; it is usually found only in the finest or most aromatic wines.

[a] Although fragrance is technically divided into the *aroma* (derived from the grapes) and *bouquet* (derived from fermentation, processing and aging), descriptive terms are more informative.

IV. Finish

1 – Concentrate on the olfactory and gustatory sensations that linger in the mouth.
2 – Compare these sensations with those previously detected.
3 – Note their character and duration.

V. Repetition of assessment

1 – Re-evaluate the aromatic and sapid sensations of the wines, beginning at II.3 – ideally several times over a period of 30 min.
2 – Study the duration and development (change in intensity and quality) of each sample.

Finally, make an overall assessment of the pleasurableness, complexity, subtlety, elegance, power, balance, and memorableness of the wine. With experience, you can begin to make evaluations of its *potential* – the likelihood of the wine improving in its character with additional aging.

Fig. 9.5 Wine tasting sequence. (Source: Jackson, 2009, reproduced with permission from Elsevier).

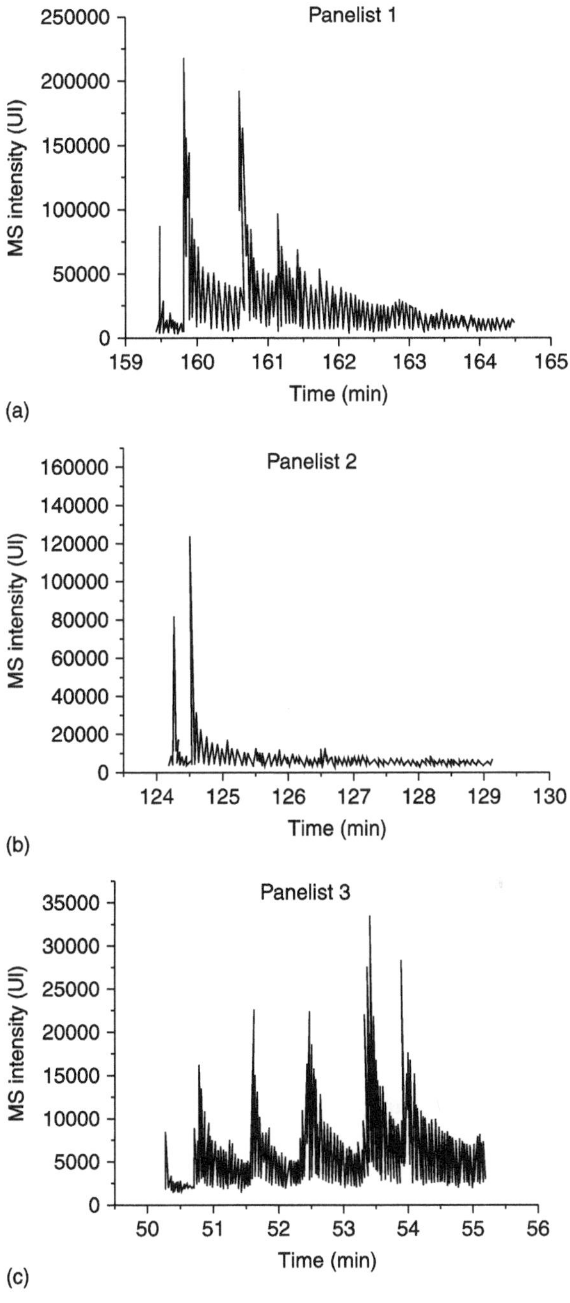

Fig. 9.6 Flavor release from a solution of menthol sampled by three different subjects. Each large peak corresponds to a swallow, smaller peaks to breathing. Panelist 3 is a flavorist who makes many small swallows. This behavior clearly prolongs flavor release (Source: Normand *et al.*, 2004).

9.4.3 Tasting room and conditions

It is preferable to have sensory evaluation conduced in a room dedicated to the purpose. Not only are dedicated rooms always available when needed, but also can contain permanent, individual, isolated, tasting booths. Such rooms should be properly ventilated (usually under positive air pressure to avoid the influx of odors from the exterior), air-conditioned, soundproofed, and provided with adequate and variable lighting. For example, low-intensity red light may be used to neutralize color bias. Color can have a major influence on the sensory perception of wine (Pangborn *et al.*, 1963; Tromp and van Wyk, 1977). Proper and uniform glassware should also be provided. Particularly useful are 8 oz ISO wine tasting glasses (Cliff, 2001). Black versions are available as an alternative to red light to prevent visual attributes from influencing gustatory and olfactory perception.

To assure presentation uniformity, all glasses should be identical and filled to the same level. Thirty ml samples are usually fully adequate. Care needs to be taken that the glasses are perfectly clean and assessed for odorlessness prior to usage.

As is now typical, every member tastes individual samples in isolation. This avoids visual or auditory reactions by one panelist biasing the perception of other members.

A palate cleanser should be available to avoid sensations from previous samples affecting the perception of subsequent wines (sequence effects). Water and/or unsalted crackers, or cubes of white bread have been traditional palate cleansers. However, several studies have demonstrated that solutions such as 1% pectin are more effective (Colonna *et al.*, 2004; Ross *et al.*, 2007). These are especially efficient at diminishing the lingering influence of tannins in red wines on the perception of subsequent samples. Presenting samples to individual members in different and random order is another precaution that can be taken to diminish sequence-induced errors.

The number of samples in, and duration of, each session will depend on the design and purpose of the sensory analysis. The more detailed each analysis, the fewer the number of wines that can be assessed adequately. Sensory fatigue is an important feature that must be taken into account to improve the validity of the data obtained.

9.4.4 Data collection and analysis

Optimally, data collection should include only that which is critical to the task involved, unless it is a 'fishing trip', looking for data that might prove important. As noted previously, only a few attributes are usually necessary to distinguish among similar wines. It saves not only on analysis time, but also is less fatiguing for the panel. However, where overall assessment is desired, one of the many published score sheets (Anonymous, 1994; Jackson, 2009) may prove useful. For example, the Davis score sheet (Amerine and Roessler, 1983) has had extensive use. However, like any score sheet, it has limitations. For example, attributes designated may not apply to the wines in question, whereas important attributes are not noted.

To facilitate both data collection and analysis, it is now standard to compile results directly into a central computer. This usually involves forms presented on monitors located at each tasting station. This is most applicable for aspects that can either be checked off, or intensities noted on a sliding scale, in which case a touchscreen monitor is preferable. For data that must be recorded verbally, not simply marked off from a set of terms or boxes, a keyboard should be available. Electronic data collection is clearly preferable. It not only permits almost immediate collation, but also avoids data-entry errors resulting from interpretation of panelists' handwriting. Where results can be entered or quickly converted to numerical form, the data can be directly subjected to analysis by statistical software packages. This has particular value in quickly indicating if additional data needs to be collected, or further assessment sessions scheduled while samples are available or fresh.

9.4.5 Consumer-related studies

These studies generally fall into two categories: those designed to ascertain the likes and dislikes of specific subsets of consumers, and in-house formulation of blends to meet the needs and desires of a particular market. Assuming that valid, detailed, and precise data can and have been obtained, formulation depends on an adequate supply of readily available and competitively priced wines, and the skill of the blenders at combining those wines to meet the preferences of the intended market group.

Consumer studies have a long history in the food industry, but have seen comparatively little use in the wine trade, except in the formulation of major, brand-named table wines. Part of the problem has been in accurately detecting definable wine consumer groups. Because of the tremendous diversity in commercially available wines, most consumers show limited brand awareness or loyalty. The confusion so generated produces the vinous expression of Barry Schwartz's 'Paradox of Choice' (Schwartz, 2004). In addition, consumer groups are often splintered by preferences based on features such as cultural biases and price consciousness. Even where clearly definable groups can be identified, there is concern about how best to assess and enunciate preferences in terms useful in formulating wine blends. Consumer term usage is different and more variable than that of trained judges (Lawless, 1984). In addition, simple preference choices in a wine shop or shopping center are unlikely to provide useful data. Most external tasting environments are too divorced from real-life situations to be relevant. Expert panels typically differ from consumers in their preferences, and thus are poor indicators of purchaser opinion (McDermott, 1990; Williams, 1982). This partially results from the tendency for exposure and training to modify preferences.

To offset these problems, Köster (2003) has advocated an approach where images of natural consumption conditions are evoked. Photos or videos, as well as auditory cues, may facilitate the generation of appropriate images, but may be too specific to evoke applicable domestic images. Alternatively, recounting a story has been suggested to be more effective in simulating realistic images in the mind of the consumer. Such an approach is more likely to have the desired effect than simply asking the consumer to imagine appropriate conditions. The latter demands more effort than most consumers are usually willing to contribute.

In evaluating samples, consumers are typically only asked to rank them in appropriateness to the situation imagined. If desired, ranking could be requested relative to different tasting conditions, such as consumption with friends at home, with business associates in a restaurant, or relaxing on the patio. However, suggesting too many conditions would probably become counterproductive, just as providing too many samples reduces testing accuracy. It is better to perform separate analyses for each condition rather than attempt to cram too much into any particular study.

Consumer testing should be highly focused, and as uncomplicated as possible, to obtain the most reliable data. Questions as to why the samples were ranked the way they were might provide useful data. However, employing a trained panel to assess the same samples is more likely to define, in terms useful to blenders, the underlying reasons behind the rankings.

Where questionnaires are used, extreme care must be taken in their preparation. It is too easy to influence responses by the way in which questions are phrased or presented (Köster, 2003). Although time-consuming and demanding to prepare, a well-designed and conducted consumer study can be very profitable. The data provided can help formulate a winning blend. The economic returns can be extraordinary.

9.5 Future trends

A current trend in sensory assessment involves investigations directed at detecting the origin of varietal characteristics. Although wines may contain hundreds of aromatic compounds, few occur at concentrations above their sensory threshold. Of those that do, most contribute only to a general, wine-like fragrance. Techniques such as aroma extract dilution analysis (AEDA) (Guth, 1998; Grosch, 2001), atmospheric pressure chemical ionization mass spectroscopy (APCI-MS) (Aznar *et al.*, 2004), and buccal odor screening system (BOSS) (Buettner, 2004) are being used to determine those few constituents that may be critical to expression of a distinctive varietal aroma. With some cultivars, the aroma appears to arise from the presence of one or a few compounds, often occurring at nanogram concentrations (Boison and Tomlinson, 1990; Lacey *et al.*, 1991; Tominaga *et al.*, 2000; Wood *et al.*, 2008). For most cultivars, though, the varietal aroma, if one exists, appears to depend on a unique combination of relatively common wine aromatics (Ferreira *et al.*, 1998; Le Fur *et al.*, 1996; Rapp, 1998). With such knowledge, it may be possible for viticultural practices to be adjusted to optimize the occurrence of these compounds.

Another distinctive trend is the investigation of the interaction between wine and food. Thus far, the studies have demonstrated that most interactions result in a mutual suppression in the flavor intensities of both elements, not an enhancement (Nygren *et al.*, 2002; Madrigal-Galan and Heymann (2006). Examples are a reduction in the bitter/astringent sensations of red wines, by the reaction between wine tannins and food proteins (see Haslam *et al.*, 1992); the action of food gums, gels and oils in coating taste receptors, making them less responsive to gustatory sensations (Kinsella, 1990; Hollowood *et al.*, 2002); influences of food chemicals and dilution in the mouth on the volatility of wine aromatics (Hodgson *et al.*, 2005; Bayarri *et al.*, 2007); and the action of metal ions in wine in promoting fatty acid oxidation (Tamura *et al.*, 2009).

Finally, there is increasing interest in the use of electronic noses and tongues in routine wine analysis (Martí *et al.*, 2005; Röck *et al.*, 2008). They are both more objective and always ready for use as needed. What is still lacking is diversity and, in many cases, sensitivity.

9.6 Sources of further information

Several texts on the technical aspects of wine production are available. Jackson (2008) is international in scope and covers both the viticultural and enologic aspects; Boulton *et al.* (1996) discuss winemaking primarily from the Californian perspective, as does Ribéreau-Gayon *et al.* (2006a, 2006b) from the Bordeaux viewpoint. The most recent text on wine sensory analysis is Jackson (2009). It discusses critical sensory analyses, along with some consumer aspects that can be research supported. Amerine and Roessler (1983) and Williams *et al.* (1979) are older, but still provide excellent material on sensory evaluation, providing examples of simpler statistical analysis (prior to the widespread availability of computer statistical packages). ISO manuals (2006, 2007, 2008) provide recommendations on procedures suggested by the International Standards Organization, while papers in Beech (1978) and Rantz (2000) provide various views on wine sensory analysis.

9.7 References

AMERINE, M. A. and ROESSLER, E. B. (1983). *Wines – Their Sensory Evaluation.* 2nd ed, Freeman, San Francisco, CA.

ANONYMOUS (1994). 'Resolution Eno 2/94. OIV standard for international wine competitions'. *Bull. O.I.V.* **67**, 558–597.

AZNAR, M., TSACHAKI, M., LINFORTH, R. S. T., FERREIRA, V., and TAYLOR, A. J. (2004). 'Headspace analysis of volatile organic compounds from ethanolic systems by direct APCI-MS'. *Int. J. Mass Spectrometry* **239**, 17–25.

BAYARRI, S., TAYLOR, A., and HORT, J. (2007). 'The role of fat in flavor perception: effect of partition and viscosity in model emulsions'. *J. Agric. Food Chem.* **54**, 8862–8868.

BEECH, F. W. (Coordinator). (1978). *The Fourth Wine Subject Day – Sensory Evaluation.* Long Ashton Research Station, University of Bristol.

BOISON, J. O. K., and TOMLINSON, R. H. (1990). 'New sensitive method for the examination of the volatile flavor fraction of Cabernet Sauvignon wines'. *J. Chromatogr.* **522**, 315–328.

BOULTON, R. B., SINGLETON, V. L., BISSON, L. F., and KUNKEE, R. E. (1996). *Principles and Practices of Winemaking*. Springer, New York.

BRAMLEY, R. G. V. (2005) 'Understanding variability in winegrape production systems. 2. Within vineyard variation in quality over several vintages'. *Aust. J. Grape Wine Res.* **11**, 33–42.

BROADBENT, M. (1980). *The Great Vintage Wine Book*. Alfred Knopf, New York.

BROSSAUD, F., CHEYNIER, V., and NOBLE, A. C. (2001). 'Bitterness and astringency of grape and wine polyphenols'. *J. Grape Wine Res.* **7**, 33–39.

BUETTNER, A. (2004). 'Investigation of potent odorants and afterodor development in two Chardonnay wines using the buccal odor screening system (BOSS)'. *J. Agric. Food Chem.* **52**, 2339–2346.

CHATONNET, P., BOIDRON, J. N., and PONS, M. (1990). 'Élevage des vins rouges en fûts de chêne. Évolution de certains composés volatils et leur impact arômatique'. *Sci. Aliments.* **10**, 565–587.

CHATONNET, P., BONNET, S., BOUTOU, S., and LABADIE, M.-D. (2004). 'Identification and responsibility of 2,4,6-tribromoanisole in musty, corked odors in wine'. *J. Agric. Food Chem.* **52**, 1255–1262.

CHEYNIER, V., REMY, S., and FULCRAND, H. (2000). 'Mechanisms of anthocyanin and tannin changes during winemaking and aging'. In *Proceedings of the ASEV 50th Anniversary Annual Meeting*, Seattle, Wash. (J. M. Rantz, ed.). pp. 337–344. American Society of Enology and Viticulture, Davis, CA.

CLIFF, M. A. (2001). 'Impact of wine glass shape on intensity of perception of color and aroma in wines'. *J. Wine Res.* **12**, 39–46.

COLONNA, A. E., ADAMS, D. O., and NOBLE, A. C. (2004). 'Comparison of procedures for reducing astringency carry-over effects in evaluation of red wines'. *Aust J. Grape Wine Res.* **10**, 26–31.

D'ANGELO, M., ONORI, G., and SANTUCCI, A. (1994) 'Self-association of monohydric alcohols in water: compressibility and infrared absorption measurements'. *J. Chem. Phys.* **100**, 3107–3113.

ELLIOTT-FISK, D., and NOBLE, A. (1992). 'The diversity of soils and environments in Napa Valley, California, and their influence on Cabernet Sauvignon flavors'. In *Viticulture in Geographic Perspective: Proceeding of the 1991 Miami AAG Symposium, Miami, Florida* (H. de Blij, ed.), pp. 45–71. Miami Geographical Society.

FERREIRA, V., LÓPEZ, R., ESCUDERO, A., and CACHO, J. F. (1998). 'The aroma of Grenache red wine: Hierarchy and nature of its main odorants'. *J. Sci. Food Agric.* **77**, 259–267.

GUINARD, J.-X., PANGBORN, R. M., and LEWIS, M. J. (1986). 'The time-course of astringency in wine upon repeated ingestion'. *Am. J. Enol. Vitic.* **37**, 184–189.

GROSCH, W. (2001). 'Evaluation of the key odorants of foods by dilution experiments, aroma models and omission'. *Chem. Senses* **26**, 533–545.

GUTH, H. (1997). 'Quantitation and sensory studies of character impact odorants of different white wine varieties'. *J. Agric. Food Chem.* **45**, 3027–3032.

GUTH, H. (1998). 'Comparison of different white wine varieties by instrumental and analyses and sensory studies'. In *Chemistry of Wine Flavor*. (L. A. Waterhouse and S. E. Ebeler, eds.), pp. 39–52. ACS Symposium Series #714., American Chemical Society, Washington, DC.

HASLAM, E., LILLEY, T. H., WARMINSKI, E., LIAO, H., CAI, Y. *et al.* (1992). 'Polyphenol complexation, a study in molecular recognition'. In *Phenolic Compounds in Food and Their Effects on Health, 1, Analysis, Occurrence, and Chemistry*. (C.-T. Ho *et al.*, eds.), pp. 8–50. ACS Symposium Series No. 506. American Chemical Society, Washington, DC.

HEMINGWAY, K. M., ALSTON, M. J., CHAPPELL, C. G., and TAYLOR, A. J. (1999). 'Carbohydrate-flavour conjugates in wine'. *Carbohydrate Polymers* **38**, 283–286.

HODGSON, M. D., LANGRIDGE, J. P., LINFORTH, R. S. T., and TAYLOR, A. J. (2005). 'Aroma release and delivery following the consumption of beverages'. *J. Agric. Food Chem.* **53**, 1700–1706.

HOLLOWOOD, T. A., LINFORTH, R. S. T., and TAYLOR, A. J. (2002). 'The effect of viscosity on the perception of flavour'. *Chem. Senses* **27**, 583–591.

HUFNAGEL, J. C., and HOFMANN, T. (2008). 'Orosensory-directed identification of astringent mouthfeel and bitter-tasting compounds in red wine'. *J. Agric Food Chem.* **56**, 9190–9199.

ISO. (2006). *Sensory analysis – Methodology – Initiation and training of assessors in the detection and recognition of odours.* #5496. International Standards Organization. Geneva, Switzerland.

ISO. (2007). *Sensory analysis – General guidance for the design of test rooms.* #8589. International Standards Organization. Geneva, Switzerland.

ISO. (2008). *Sensory analysis – General guidance for the selection, training and monitoring of assessors – Part 2: Expert sensory assessors.* # 8586. International Standards Organization. Geneva, Switzerland.

JACKSON, R. S. (2008). *Wine Science. Principles and Applications.* 3rd ed. Academic Press, San Diego, CA.

JACKSON, R. S. (2009). *Wine Tasting: A Professional Handbook.* 2nd ed. Academic Press, London, UK.

KINSELLA, J. E. (1990). 'Flavour perception and binding'. *Intern. News Fats Oils Related Materials: Inform.* **1**, 215–226.

KÖSTER, E. P. (2003). 'The psychology of food choice: some often encountered fallacies'. *Food Qual. Pref.* **14**, 359–373.

KOTSERIDIS, Y., RAZUNGLES, A., BERTRAND, A., and BAUMES, R. (2000). 'Differentiation of the aromas of Merlot and Cabernet Sauvignon wines using sensory and instrumental analysis'. *J. Agric. Food Chem.* **48**, 5383–5388.

LACEY, M. J., ALLEN, M. S., HARRIS, A. Q. N., and BROWN, W. V. (1991). 'Methoxypyrazines in Sauvignon blanc grapes and wines'. *Am. J. Enol. Vitic.* **42**, 103–108.

LAING, D. G. (1983). 'Natural sniffing gives optimum odour perception for humans'. *Perception* **12**, 99–117.

LAING, D. G. (1995). 'Perception of odor mixtures'. In *Handbook of Olfaction and Gustation* (R. L. Doty, ed.), pp. 283–297. Marcel Dekker, New York.

LAWLESS, H. T. (1984). 'Flavor description of white wine by expert and nonexpert wine consumers'. *J. Food Sci.* **49**, 120–123.

LAWLESS, H. T. (1999). 'Descriptive analysis of complex odors: Reality, model or illusion?' *Food Qual. Pref.* **10**, 325–332.

LE FUR, Y., LESSCHAEVE, I., and ETIÉVANT, P. (1996). 'Analysis of four potent odorants in Burgundy Chardonnay wines: Partial quantitative descriptive sensory analysis and optimization of simultaneous extraction method'. In *Proc. 4th Int. Symp. Cool Climate Vitic. Enol., Rochester, NY, July 16–20, 1996.* (T. Henick-Kling, T. E. Wolf, and E. M. Harkness, eds.), pp. VII-53–56. NY State Agricultural Experimental Station, Geneva, New York.

LEA, A. G. H., and ARNOLD, G. M. (1978). 'The phenolics of ciders, bitterness and astringency'. *J. Sci. Food Agric.* **29**, 478–483.

LESSCHAEVE, I. (2007). 'Sensory evaluation of wine and commercial realities: Review of current practices and perspectives'. *Am. J. Enol. Vitic.* **58**, 252–258.

MADRIGAL-GALAN, B. and HEYMANN, H. (2006). 'Sensory effects of consuming cheese prior to evaluating red wine flavor'. *Am. J. Enol. Vitic.* **57**, 12–22.

MARAIS, J. (1986). 'Effect of storage time and temperature of the volatile composition and quality of South African *Vitis vinifera* L. cv. Colombar wines'. In *Shelf Life of Foods and Beverages* (G. Charalambous, ed.), pp. 169–185. Elsevier, Amsterdam.

MARTÍ, M. P., BOQUÉ, R., BUSTO, O., and GUASCH, J. (2005). 'Electronic noses in the quality control of alcoholic beverages'. *TRAC – Trends Anal. Chem.* **24**, 57–66.

MATEO, J. J. and JIMENEZ, M. (2000). 'Monoterpenes in grape juice and wines'. *J. Chromatography A* **881**, 557–567.

MAZZOLENI, V., and MAGGI, L. (2007). 'Effect of wine style on the perception of 2,4,6-trichloroanisole, a compound related to cork taint in wine'. *Food Res. Int.* **40**, 694–699.

MCDERMOTT, B. J. (1990). 'Identifying consumers and consumer test subjects'. *Food Technol.* **44**, 154–158.

MORROT, G., BROCHET, F., and DUBOURDIEU, D. (2001). 'The color of odors'. *Brain Lang.* **79**, 309–320.

MURAT, M.-L. (2005). 'Recent findings on rosé wine aromas. Part I: identifying aromas studying the aromatic potential of grapes and juice'. *Aust. NZ Grapegrower Winemaker* **497a**, 64–65, 69, 71, 73–74, 76.

MURAT, M. L., TOMINAGA, T., SAUCIER, C., GLORIES, Y., and DUBOURDIEU, D. (2003). 'Effect of anthocyanins on stability of a key-odorous compound, 3-mercaptohexan-1-ol, in Bordeaux rosé wines'. *Am. J. Enol. Vitic.* **54**, 135–138.

NOBLE, A. C. (1988). 'Analysis of wine sensory properties'. In *Wine Analysis*. (H. F. Linskens and J. F. Jackson, eds.), pp. 9–28. Springer-Verlag, Berlin.

NOBLE, A. C., and BURSICK, G. F. (1984). 'The contribution of glycerol to perceived viscosity and sweetness in white wine'. *Am. J. Enol. Vitic.* **35**, 110–112.

NORMAND, V., AVISON, S., and PARKER, A. (2004). 'Modeling the kinetics of flavour release during drinking'. *Chem. Senses* **29**, 235–245.

NYGREN, I. T., GUSTAFSSON, I.-B., and JOHANSSON, L. (2002). 'Perceived flavour changes in white wine after tasting blue mould cheese'. *Food Service Technol.* **2**, 163–171.

NYGREN, I. T., GUSTAFSSON, I.-B., and JOHANSSON, L. (2003). 'Perceived flavour changes in blue mould cheese after tasting white wine'. *Food Service Technol.* **3**, 143–150.

ONG, B.Y., and NAGEL, C. W. (1978). 'Hydroxycinnamic acid-tartaric acid ester content in mature grapes and during the maturation of white Riesling grapes'. *Am. J. Enol. Vitic.* 29, 277–281.

PANGBORN, R. M., BERG, H. W., and HANSEN, B. 1963. 'The influence of colour on discrimination of sweetness in dry table wines'. *Am. J. Psyc.* **76**, 492–495.

POCOCK, K. F., SEFTON, M. A., and WILLIAMS, P. J. (1994). 'Taste thresholds of phenolic extracts of French and American oakwood: The influence of oak phenols on wine flavor'. *Am. J. Enol. Vitic.* **45**, 429–434.

POLLNITZ, A. P., PARDON, K. H., LIACOPOULOS, D., SKOUROUMOUNIS, G. K., and SEFTON, M. A. (1996). 'The analysis of 2,4,6–trichloranisole and other chloroanisoles in tainted wines and corks'. *Aust. J. Grape Wine Res.* **2**, 184–190.

PRESCOTT, J. (2004). 'Psychological processes in flavour perception'. In *Flavour Perception* (A. J. Taylor and D. Roberts, eds.), pp. 256–278. Blackwell Publishing, London.

PROFFITT, T., and MALCOLM, A. (2005) 'Zonal vineyard management through airborne remote sensing'. *Aust. NZ Grapegrower Winemaker* **502**, 22–24, 25–26.

RANTZ, J. M. (ed.) (2000). 'Sensory symposium'. In *Proc. ASEV 50th Anniv. Ann. Meeting*, Seattle, WA., June 19–23, 2000. pp. 1–31. American Society for Enology and Viticulture, Davis, CA.

RAPP, A. (1998). 'Volatile flavour of wine: correlation between instrumental analysis and sensory perception'. *Nahrung* **42**, 351–363.

RIBÉREAU-GAYON, P., DONÈCHE, B., DUBOURDIEU, D., and LONVAUD, A. (2006a). *Handbook of Enology – Volume I: The Microbiology of Wine and Vinifications*. 2nd ed., John Wiley and Sons, Chichester, UK.

RIBÉREAU-GAYON, P., DUBOURDIEU, D., GLORIES, Y., and MAUJEAN, A. (2006b). *Handbook of Enology – Volume II: The Chemistry of Wine Stabilization and Treatments*. 2nd ed., John Wiley and Sons, Chichester, UK.

RÖCK, F., BARSAN, N., and WEIMAR, U. (2008). 'Electronic nose: Current status and future trends'. *Chem. Rev.* **108**, 705–725.

ROLLS, E. T. (2006). 'Brain mechanisms underlying flavour and appetite'. *Phil. Trans. R. Soc. B* **361**, 1123–1136.

ROSS, C. F., HINKEN, C., and WELLER, K. (2007). 'Efficacy of palate cleansers for reduction of astringency carryover during repeated ingestions of red wine'. *J. Sens. Stud.* **22**, 293–312.

ROUSSIS, I. G., LAMBROPOULOS, I., and PAPADOPOULOU, D. (2005). 'Inhibition of the decline of volatile esters and terpenols during oxidative storage of Muscat-white and Xinomavro-red wine by caffeic acid and N-acetyl-cysteine'. *Food Chem.* **93**, 485–492.

SCHMIDT, J. O. (1981). 'Comparison of methods and rating scales used for sensory evaluation of wine'. Thesis, University of California, Davis, CA.

SCHWARTZ, B. (2004). *The Paradox of Choice: Why More is Less*. HarperCollins, New York.

SIMPSON, R. F., CAPONE, D. L., and SEFTON, M. A. (2004). 'Isolation and identification of 2-methoxy-3,5-dimethylpyrazine, a potent musty compound from wine corks'. *J. Agric. Food Chem.* **52**, 5425–5430.

SINGLETON, V. L., and OUGH, C. S. (1962). 'Complexity of flavor and blending of wines'. *J. Food Sci.* **12**, 189–196.

SOLEAS, G. J., YAN, J., SEAVER, T. and GOLDBERG, D. M. (2002). 'Method for the gas chromatographic assay with mass selective detection of trichloro compounds in corks and wines applied to elucidate the potential cause of cork taint'. *J. Agric. Food Chem.* **50**, 1032–1039.

SPONHOLZ, W. R. (1988). 'Alcohols derived from sugars and other sources and fullbodiedness of wines'. In *Wine Analysis*. (H. F. Linskens and J. F. Jackson, eds.), pp. 147–172. Springer-Verlag, Berlin.

STRAUSS, C. R., GOOLEY, P. R., WILSON, B., and WILLIAMS, P. J. (1987). 'Application of droplet countercurrent chromatography to the analysis of conjugated forms of terpenoids, phenols, and other constituents of grape juice'. *J. Agric. Food Chem.* **35**, 519–524.

TAMURA, T., TANIGUCHI, K., SUZUKI, Y., OKUBO, T., TAKATA, R. *et al.* (2009). 'Iron is an essential cause of fishy aftertaste formation in wine and seafood pairing'. *J. Agric. Food Chem.* **57**, 8550–8556.

TIMBERLAKE, C. F., and BRIDLE, P. (1976). 'Interactions between anthocyanins, phenolic compounds, and acetaldehyde and their significance in red wines'. *Am. J. Enol. Vitic.* **27**, 97–105.

TOMASI, D., SIVILOTTI, P., PASCARELLA, G., RASERA, M., CALÓ, A. (2005). 'Soil moisture: an indicator of zoning and wine quality also in small farm surfaces'. *Bull. O.I.V.* **78**, 447–459.

TOMINAGA, T., BALTENWECK-GUYOT, R., PEYROT DES GACHONS, C. and DUBOURDIEU, D. (2000). 'Contribution of volatile thiols to the aromas of white wines made from several *Vitis vinifera* grape varieties'. *Am. J. Enol. Vitic.* **51**, 178–181.

TOMINAGA, T., DARRIET, P., and DUBOURDIEU, D. (1996). 'Identification de l'acétate de 3–mercaptohexanol, composé à forte odeur de buis, intervenant dans l'arôme des vins de Sauvignon'. *Vitis* **35**, 207–210.

TOMINAGA, T., FURRER, A., HENRY, R., and DUBOURDIEU, D. (1998). 'Identification of new volatile thiols in the aroma of *Vitis vinifera* L. var. Sauvignon blanc wines'. *Flavour Fragr. J.* **13**, 159–162.

TOMINAGA, T., GUIMBERTAU, G., and DUBOURDIEU, D. (2003). 'Contribution of benzenemethanethiol to smoky aroma of certain *Vitis vinifera* L. wines'. *J. Agric. Food Chem.* **51**, 1373–1376.

TROMP, A., and VAN WYK, C. J. (1977). 'The influence of colour on the assessment of red wine quality'. *Proc. S. Afr. Soc. Enol. Vitic.* pp. 107–117.

WATERHOUSE, A. L. and KENNEDY, J. A. (eds.) (2004). *Red Wine Color. Revealing the Mysteries. ACS Symposium Series, No. 886*. American Chemical Society Publication, Washington D.C.

WILDENRADT, H. L., and SINGLETON, V. L. (1974). 'The production of aldehydes as a result of oxidation of polyphenolic compounds and its relation to wine aging'. *Am. J. Enol. Vitic.* **25**, 119–126.

WILLIAMS, A. A. (1982). 'Recent developments in the field of wine flavour research'. *J. Inst. Brew.* **88**, 43–53.

WILLIAMS, A. A. and ROSSER, P. R. (1981). 'Aroma enhancing effects of ethanol'. *Chem. Senses* **6**, 149–153.

WILLIAMS, A. A., SPENCER, H. W., CARTER, C. S., ARNOLD, G. M., and LANGRON, S. (1979). *Manual of Sensory Difference Testing of Wines*. Long Ashton Research Station, Bristol, UK.

WISE, P. M., OLSSON, M. J., and CAIN, W. S. (2000). 'Quantification of odor quality'. *Chem. Senses* **25**, 429–443.

WOOD, C., SIEBERT, T. E., PARKER, M., CAPONE, D. L., ELSEY *et al.* (2008). 'From wine to pepper: Rotundone, an obscure sesquiterpene, is a potent spicy aroma compound'. *J. Agric. Food Chem.* **56**, 3738–3744.

Part III

Distilled products

10

Anise spirits: types, sensory properties and sensory analysis

I. Zabetakis, University of Athens, Greece

Abstract: In this chapter, the history, legislation and production of anise spirits are described with particular focus on the three main representatives of this class of spirits, namely pastis from France, and ouzo and tsipouro from Greece. Production procedures and the application of sensory analysis in the characterization of the final products are thoroughly described. Finally, future trends are given and some sources of further information are provided.

Key words: anise, ouzo, pastis, tsipouro, anethole.

10.1 Overview of anise spirits (pastis, ouzo, tsipouro)

The history of distillates extends way back into the mists of history. One of the first biotechnological applications known was wine and, thanks to the existence of primitive distillation techniques, its distillates. Distillation was studied by Alexandrian sages and later by Arab alchemists and constituted a significant chapter in the science of chemistry. What we do not know is precisely when distillation using copper alembic stills began. The name indicates that it was probably a Greek rather than an Arab discovery. The *ambix*, from which the word *alembic* is derived, was a vessel with a broad base and thin neck, very similar to the boiler of a still. The Greek word seems to have passed into Arabic (*al-ambiq*), and through Arabic to other European languages. The need to process metals with various reagents at high temperatures and the desire to extract the 'spirits' of substances led to improvements of the distillation technique.

Recommendations by Arab scholars that additives be used, to absorb one substance in the effort to isolate another, may have led to the addition of salts to wine in the container of a still. The salts absorbed part of the water and produced a 'water' that burned. An early description of how to produce such a water comes from Salerno, Italy via an alchemical manuscript, known as the *Mappae Clavicula* (the little key to painting). The most recent version of the manuscript, which has survived in two versions, one from the tenth century and another from the twelfth century, but was probably written around AD 800, contains a partially encoded (perhaps out of fear of the Catholic Church) recipe, 'mixing pure and very strong wine with three parts of salt and heating it in a vessel suitable for this purpose produces a water that, when ignited, burns but leaves material unburned'. This was clearly a solution of alcohol that burned at such a low temperature that an underlying substance did not burn. This solution became known as *aqua ardens* (burning water). This concept probably originated from the ancient Greek belief in a 'spirit' that could act on anything and could possibly be transmuted into anything. From the thirteenth century on, further recipes for alcohol production materialized, although the word itself was first used by Theophrastus Paracelsus in the sixteenth century. It came from the Arabic *al-kuhl*, originally the name of very finely powdered antimony sulphide.

The history, formulation and evolution of anise spirits is closely related to the skills of coppersmiths, since without a copper still it is simply impossible to produce such distillates and therefore, as progress was made in working copper, the production of anise spirits evolved too. Later improvements in distilling equipment contributed to the large-scale distillation of marc (a by-product of wine-making) in Asia Minor and in the grape-producing regions of Greece. Until the mid-fifteenth century, Alexandria, Constantinople (today Istanbul) and Smyrna (today Izmir) were the most important centres of distillation skills. Monks in Mount Athos also knew about distillation well before the fifteenth century, and it was a customary practice of monasteries to distil the marc that remained after winemaking. Distillation, however, was affected by politics and the unchecked production of spirits was a matter of some concern to the authorities as the monk Alexandros Lavriotis wrote, '. . . it has been of great concern to Mount Athos, to the Turkish government and to the Ecumenical Patriarchate itself that alembics operate freely on the Holy Mountain (Athos) and produce untaxed spirits. More than twenty firmans on the matter have been issued since 1590 . . .'

10.2 History of anise spirits

Anise spirits are related to two parameters: distillation techniques and grape growing. During the Ottoman period, the distillation of marc abated as grape growing declined, yet despite the Muslim prohibition of drinking wine, viniculture recovered and soon *rakitzides* (i.e. producers of tsipouro or raki) began to appear, many of whom received the special privilege of supplying the upper classes of the empire. In Europe, Chaptale in 1780 and Adam in 1805 improved the alembic still, giving it the form in which it has more or less survived to this day. In 1863, the Greek state promulgated the first law for the taxation of alcohol and in 1896 the first permits were issued for the production of a beverage distilled from marc.

It is not known precisely when production of ouzo began, nor when this name began to be used but we know for sure that it was made on Greek territory and was developed from tsipouro production. This is the reason for its initial name, *metavrasmen* (distilled again) raki. The word ouzo means nothing in Greek and appears not to have been widely used until the second half of the nineteenth century. The etymology of the word has alternative theories: according to one, ouzo comes from the word *uso* (for use), a word that, along with the destination, was stamped on the crates with the select goods that were being exported to Europe. More likely answers are that the root of the word ouzo comes from either the ancient Greek *ozo* (to smell) because of its very strong aroma or the Turkish *űzűm* (bunch of grapes and infusion of currants), although ouzo is not related to grapes or marc.

Ouzo, initially a distillation of marc, owed its individuality to the Mediterranean herbs used to give it flavour. Anise, fennel, mastic, coriander, angelica root, linden blossoms, cardamom and other aromatic plants and herbs went into the alembics to be used in the recipes of many ouzo-producing family businesses.

10.2.1 Pastis

The First World War signalled the end for absinthe, the most popular drink at the end of the nineteenth century. In 1914, all drinks with more than 16% abv alcohol were rigorously outlawed. They were destined not to be re-authorized until the end of the war except for absinthe (forbidden by law in 1915). Despite the regulations in France (similar in some extent to the ones mentioned above regarding ouzo and the Ottomans), the habit of enjoying anise spirits (fresh or mixed with water) has never stopped. The French government eventually authorized individual types of anise spirits without, however, authorizing them as a group, perhaps fearing the return of absinthe.

Toward the end of the 1920s, a young 23-year-old salesman, son of a wine merchant, put into practice his inspirational idea to commercialize an anise spirit under his own name. Flying in the face of the authorities, Paul Ricard widely distributed this forbidden drink and was constantly fined by police and customs-agents. After a lot of intense lobbying, he eventually succeeded in obtaining in 1932, the 'legal' right to make and sell the drink. The fines were transformed into taxes. Paul Ricard was obliged to find a name for his drink and to define its composition. His recipe entered the public domain and 'pastis' was officially born. Its name was derived from the Provencal 'pastisson' and the Italian 'pasticcio', a synonym for blend or mixture. Pernod profited from the authorization of the drink by bringing out 'pernod' but they never labelled the bottles as pastis, despite the drink being an anise spirit of the same family. In 1936, the first French paid holidays popularized pastis, and it soon became the most popular aperitif in France; in 1938, Ricard obtained the right to sell it at 45% abv instead of merely 40% abv. During World War Two, alcoholic drinks of over 16% were again forbidden. Only in 1951 was pastis re-allowed. The Pernod family then counter-attacked and brought out his '51', which he called 'pastis de Marseille'. Product wars ensued between the two producers until 1974, when the directors of the two companies agreed to merge. The Pernod Ricard group has gone on to well-known success and growth. In 1996, the group bought 95% of the shares of the Lesvos Union of Distillers (EPOM) and the routes of pastis and ouzo joined.

10.2.2 Ouzo

The modern origins of ouzo began on the island of Lesvos, in a small town called Plomari, by the Barbayannis family who five generations later are still making ouzo today. Its founder, Efstathios Barbayannis, moved to Lesvos from Odessa (Ukraine) where he had learnt the secrets of distillation. Visiting Constantinople (today Istanbul), he bought a bronze alembic still, which was the first piece of machinery used to found the family ouzo distillery as a cottage industry, in 1860 (Fig. 10.1). The unique aspects of the place and time were that:

- Lesvos had endemic aniseed (in fact the village of Lisvori, where still today green anise (*Pimpinella anisum*) grows is only a few miles north-west of Plomari),
- The water of the island had superior organoleptic properties,
- Distillation skills were also there.

Fig. 10.1 Old copper alembic stills resting in the Barbayannis museum.

With a combination of these three factors, it holds no surprise that Lesvos became the birthplace of ouzo. Exactly the same fortunate combination of these parameters occurred during the birth of pastis in Provence. Starting from the island of Lesvos, ouzo's reputation soon became recognized all around the world. Today, there are more than a dozen established ouzo distilleries on the island, most of which are still independent family businesses. According to EC 110/2008 (Annex III), in the 'distilled anis' category, there are six geographical indications of ouzo originating from Greece.

10.2.3 Tsipouro

The history of tsipouro is closely related not only to distillation skills but also to wine-making experience. Given that one of the raw materials for tsipouro is the remnants of the vinification procedure (i.e. marc), it is essential to have access to marc in order to make tsipouro. According to EC 110/2008 (Annex III) (European Union 2008), in the 'grape marc spirits' category, there are four geographical indications of tsipouro originating from Greece.

10.3 Raw materials and production process

According to EU regulations (1576/1989 and 110/2008) (European Union 1989, 2008), spirits flavoured with aniseed constitute a special category. They are made with neutral ethyl alcohol of agricultural origin to which aromatic plants or products of them (seeds, roots, blossoms, buds, leaves, fruit, etc.) or the natural extracts thereof are added either before and/or after distillation. The plants used for this purpose are common anise, fennel and any other plant that contains the basic aromatic ingredient anethole. Almost all Mediterranean countries produce beverages that belong to this category. Following this pattern, it is not surprising to meet variations of anise spirits with different names in various countries of the Mediterranean basin (e.g. pastis in France, ouzo and tsipouro in Greece, raki in Turkey, arak/arack/arraki in Jordan, Lebanon and Syria, zivania in Cyprus or ojén in Spain).

10.3.1 Pastis

By legal definition (European Union, 1989), pastis is an anise-flavoured spirit that contains the additional flavour of licorice root, and is bottled with sugar (but no more than 100 g/l). Today, pastis is prepared by simply mixing ethyl alcohol with flavourings (essences and/or extracts). Pastis is often associated with its historical predecessor (absinthe), yet the two are in fact very different, as pastis does not contain grand wormwood (*Artemisia absinthium*), the herb from which absinthe derives its name. Also, pastis obtains its anise flavour from the distillation (or industrially prepared distillates) of star anise (*Illicium verum*), a herb of Asian origin, whereas absinthe traditionally obtains its base flavour from green anise

(*Pimpinella anisum*), a European herb. Furthermore, pastis traditionally exhibits the distinct flavour of licorice root (*Glycyrrhiza glabra*) (another herb of Asian origin), which is not a part of a traditional absinthe. Traditional absinthes were bottled at 45–74% abv, while pastis is typically bottled at 45–50% abv. Unlike a traditional absinthe, pastis is a liqueur, which means that it is always bottled with sugar. Today, like ouzo and tsipouro, each distillery has its own secret recipe and variation. Pastis becomes milky white and cloudy when mixed with water (1:5 ratio). This phenomenon is known as the pastis effect (Grillo, 2003) or ouzo effect (Vitale and Katz, 2003). There is a sharp, sweet smell of aniseed and it is called 'the milk of Provence'. The production process of pastis is the simplest of the three anise spirits in this chapter. Raw materials are mixed; water, sugar, distilled anise extract and ethyl alcohol are added; the mixture is stored in vats, filtered and bottled.

10.3.2 Ouzo

According to EU regulations, for an anise spirit to be called ouzo, it must:

- Have been produced exclusively in Greece.
- Have been produced by blending alcohols flavoured by means of distillation or maceration using aniseed and possibly fennel seed, mastic from the lentiscus indigenous to the island of Chios (*Pistacia lentiscus Chia* or *latifolia*) and other aromatic seeds, plants and fruits.
- The alcohol flavoured by distillation must represent at least 20% of the alcoholic strength of the final product.

Furthermore, that distillate must:

- Have been produced by distillation in traditional discontinuous copper stills with a capacity of 1000 litres or less,
- Have an alcoholic strength of not less than 55% abv and not more than 80% abv,
- The ouzo, as the end product, must be colourless and must have a sugar content of 50 g/l (or less) and a minimum alcoholic strength of 37.5% abv.

The label may bear the word 'aperitif', but not the wording 'from distillation', 'doubly distilled' or 're-distilled'. Only if the alcohol content of the ouzo is exclusively the product of distillation is it permitted to declare '100% distilled'. Nowadays, most quality ouzo labels are 100% distilled.

The production process for ouzo is as follows. Ethyl alcohol of plant origin, anise, other aromatic plants and water are distilled for 15 h. The distillate (79–80% abv) is then left to 'relax' for 20–25 days. Water is added to reduce the alcohol content to 37.5–45% abv. The mixture is then bottled. It is worth mentioning the practice of today's most popular brand in Lesvos (Pitsiladis) where they use a unique and unusual technique to maximize the extraction of flavour from aniseed: the anise seeds are left to soak in the sea for 24 h prior to distillation! According to the quality manager of this distillery, the role of the sea water is two-fold: firstly, the seeds are soaked and the yield of the flavour extraction during the subsequent distillation is maximized; secondly, the seeds absorb sea salt and this improves the overall flavour of the distillate.

10.3.3 Tsipouro

The production process for tsipouro is similar to that of ouzo with the only exception being that the initial raw material is not aniseed but marc. Marc (either from white or red grapes) is distilled and following the first distillation, a second distillation takes place when aniseed is usually added. The first few litres of the second distillation (heads) are discarded. This second distillate is usually 60–70% abv and it is diluted to 40–45% abv with water. In practice, from 100 kg of white grapes, 15 kg of marc is derived and this marc will eventually (after the second distillation) give two 750 ml bottles of final product (tsipouro of 40–45% abv).

10.4 Description of major variants

The major variants in anise spirits are mainly the quality of aromatic herbs and the water. The process (distillation, mixing, addition of water) is rather simple without adding differentiating parameters affecting the quality (flavour) of the final products. Variation is mainly derived from differences in the quality (taste, maturity, overall flavour) of the aromatic herbs and anise used. The fact that most anise spirits are made by small independent family businesses means there is a higher degree of versatility in terms of adopting the 'recipe' to the likes/dislikes of the owners. In other words, the most major variant could be the individual taste of the producer that affects the ratio of raw ingredients and therefore the flavour of the final product.

Another variant is that pastis is made from star anise (*Illicium verum*) whereas ouzo and tsipouro are made from green anise (*Pimpinella anisum*). The cultivation area and climate are also parameters affecting the flavour produced by these two plants, i.e. the microclimate of Provence, France or Lisvori, Lesvos should give a different fingerprint on the spectrum and levels of flavour molecules biosynthesized by the plants. In general, in Greece, due to the dryer and warmer weather, the plant secondary metabolism (i.e. biosynthesis of flavour) is favoured and this leads to plant extracts (in this case seeds) with richer and more complex flavour.

In the case of ouzo and tsipouro, anise is distilled with the mixture of alcohol and other ingredients, while in pastis, anise distillate is only mixed with the rest of the ingredients. This is another technical detail that should, thus, be considered as a major variant when the overall organoleptic properties of anise spirits are studied.

10.5 Overview of flavour chemistry and sensory properties

Anise spirits are produced by flavouring ethyl alcohol with anise extracts/distillates or by distilling ethyl alcohol with extracts or seeds of star anise, green anise, fennel (*Foeniculum vulgare*) or any other plant with similar principal aromatic constituents (European Union 1989).

The aromatic component that links all three major anise spirits (pastis, ouzo and tsipouro) is *trans*-anethole, which can be obtained by steam distillation from different plants (such as green anise, star anise and fennel). Thus, the main character of the flavour of the anise spirits is closely dependant on the flavour of the plant ingredients that are used for distillation. Chemical analysis of anise spirits has been performed in a number of studies (Jurado *et al.*, 2007; Kokkinofta and Theocharis, 2005; Gerogiannaki-Christopoulou *et al.*, 2006), giving an insight into the compounds responsible for the aroma character of these products. The main flavour component is *trans*-anethole (Jurado *et al.*, 2007; Tsachaki *et al.*, 2010); isoamyl alcohol, limonene, linalool, estragole and terpene hydrocarbons have also been identified as important flavour constituents in anise spirits (Kontominas, 1986; Jurado *et al.*, 2007; Tsachaki *et al.*, 2010).

In chronological order, the first study on the flavour of ouzo was carried out by Kontominas (1986). Almost twenty years later, Apostolopoulou *et al.* (2005) studied some differences in the concentration of the principal volatile constituents in commercially produced and homemade tsipouro. Some further insight on the impact of grape variety on the flavour of tsipouro was given by the work of Gerogiannaki-Christopoulou *et al.* (2006); in this work, white and red grape cultivars were used to produce tsipouro and the final distillates were analyzed by gas chromatography (GC). The volatile composition of anise spirits was studied by headspace solid-phase microextraction (HS-SPME) coupled with GS-MS (Jurado *et al.*, 2007). In this work, the flavour of pastis and tsipouro was further studied. The main flavour components in anise spirits are given in Table 10.1.

Table 10.1 The main flavour components in anise spirits

Compound	Pastis	Tsipouro		Ouzo		Reference
	(% peak area)	(% peak area)	(μg/ml)	(% peak area)	(μg/ml)	
α-Pinene				0.49		(Kontominas, 1986)
β-Pinene				Trace		(Kontominas, 1986)
cis-Anethole				0.07–0.27		(Kontominas, 1986)
	0.77–10.36	7.50–11.39				(Jurado *et al.*, 2007)
			0.0–4.0		1.2–8.0	(Tsachaki *et al.*, 2010)
trans-Anethole				93.18–95.90		(Kontominas, 1986)
	79.22–97.52	56.78–89.95				(Jurado *et al.*, 2007)
			0.0–154.0		120.0–876.0	(Tsachaki *et al.*, 2010)

p-Anisaldehyde				0.10–0.73		(Kontominas, 1986)
			1.5–8.0		0.0–26.0	(Tsachaki *et al.*, 2010)
Anisaldehyde diethyl acetal			1.5–4.5		0.0–16.0	(Tsachaki *et al.*, 2010)
Estragole	0.10–1.37	0.97–8.34				(Jurado *et al.*, 2007)
			3.0–8.0		0.0–13.0	(Tsachaki *et al.*, 2010)
Isoamyl alcohol			57.0–65.0		0.0	(Tsachaki *et al.*, 2010)
Limonene				0.29–0.72		(Kontominas, 1986)
					3.0–5.0	(Tsachaki *et al.*, 2010)
Linalool				0.19–0.36		(Kontominas, 1986)
			0.0–2.0		0.0–15.0	(Tsachaki *et al.*, 2010)

Finally, Tsachaki *et al.* (2010) developed a suitable lexicon for sensory studies of ouzo and tsipouro (Table 10.2). In this work, ouzo and tsipouro samples were analyzed by instrumental analytical methods (GC) and sensory analytical methods. Statistical analysis (principal component analysis (PCA), factor analysis and cluster analysis) was used with the chemical and sensory data to discriminate ouzo from tsipouro. PCA and factor analysis discriminated the products mainly

Table 10.2 Attributes (odour, taste and aftertaste) and their descriptors used during sensory analysis for testing of the ouzo and tsipouro products based on the lexicon developed by the panelists

Attribute	Descriptor	*p*-Values
Anise (odour)	Aniseed	1.39E-16
Mastic (odour)	Chios mastic gum	9.30E-02
Sweet (odour)	Icing sugar	4.63E-05
Alcoholic (odour)	Ethyl alcohol	8.58E-05
Herbs (odour)	Dill or fennel	3.69E-02
Vanilla (odour)	Vanilla rods	1.44E-03
Menthol (odour)	Menthol	3.20E-04
Strong (odour)	Alcoholic drink (e.g. vodka)	4.79E-03
Sweet (taste)	Sugar	1.51E-04
Alcoholic (taste)	Ethyl alcohol	1.69E-02
Rich (taste)	Full mouth-feel	5.12E-01
Spicy (taste)	Savoury spices (e.g. cardamom)	8.31E-01
Artificial (taste)	Rubber or dye	3.55E-01
Aromatic (taste)	Sweet aroma (e.g. vanilla-like)	3.00E-03

(Continued overleaf)

Table 10.2 Continued

Attribute	Descriptor	*p*-Values
Menthol (taste)	Cooling sensation (menthol chewing gum)	1.53E-02
Caustic (taste)	Pungent (e.g. nutmeg)	4.07E-01
Sweet (aftertaste)	Sugar	1.31E-02
Alcoholic (aftertaste)	Ethyl alcohol	2.13E-01
Artificial (aftertaste)	Rubber or dye	5.64E-01
Spicy (aftertaste)	Savoury spices (e.g. cardamom)	9.14E-01
Bitter (aftertaste)	Oregano	2.38E-01

Notes: Each descriptor gives information about the standard compound (or the implied characteristic) used in the training of the panelists.

p-Values of each attribute produced from ANOVA analysis are for discrimination of how different panelists comprehend each descriptor.

into two groups, one group of ouzo products and another of tsipouro products. Cluster analysis showed that the data from sensory analysis were able to discriminate the samples better than the chemical analysis data. Sensory analysis has also showed that, when a suitable lexicon is used, it is possible to discriminate firstly the two different groups of spirits (ouzo and tsipouro) and secondly the samples of tsipouro into further groups, containing anise or not (Fig. 10.2).

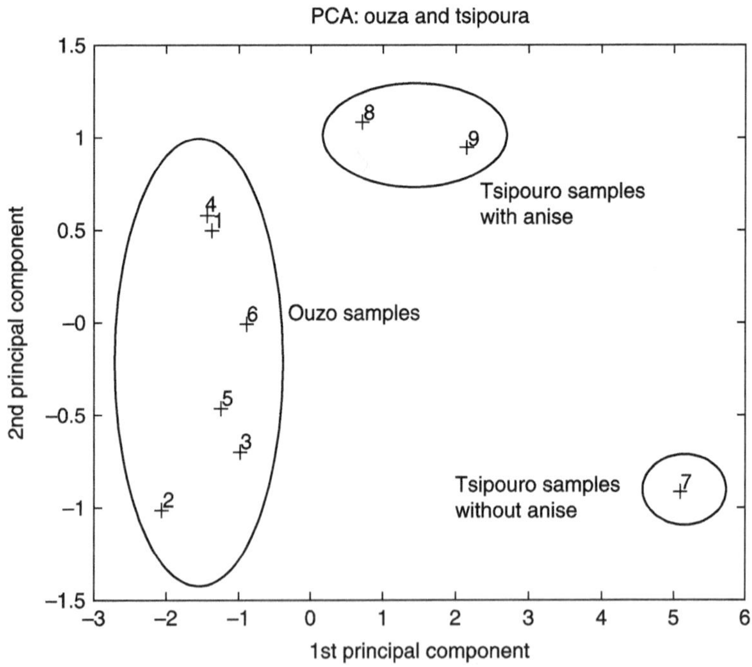

Fig. 10.2 PCA diagram demonstrating the grouping of samples from sensory analysis data. 1–6 different ouzo brands, 7–9 different tsipouro brands.

10.6 Description of sensory analysis practice in the anise spirit industry

Sensory analysis is carried out by using trained panelists (groups of 5–10 people of both sexes and various ages) on anise spirits. Each distillery uses its own lexicon adapted to the particular attributes of their own products. Panelists are asked to mark (on a scale of 1–10) different attributes using specific descriptors (as shown in Table 10.2). After marking, spider web plots are drawn and the quality differences of the samples are depicted. At this point, the link between distillers and marketing people is important in order to amalgamate the market needs with the feasible distillation improvements so that the product can satisfy changing consumer needs and tastes. Major distillers are producing at a pilot scale new products (based on new recipes with a slightly differentiated mixture of herbal plants) which are compared organoleptically with the existing marketed ones (as described above) before scaling up production and launching new brands. In Greece, however, this process is not very vigorous. Even the leading distillers follow more or less the same recipe that has been handed down by their predecessors.

10.7 How consumer research is managed and executed

Consumer research is executed in various ways. First, the final selling points of anise spirits (wine shops, tavernas, restaurants) receive feedback from consumers and, in most cases, (depending on the relationship of the owner of the catering unit with the distillery) this feedback is passed on to the relevant distillers. Secondly, the distiller's marketing team collects direct feedback from the customers by visiting the points (bars, tavernas, etc.) where their products are consumed. Unfortunately, there are no well-defined procedures for managing and executing consumer research within this industry in Greece and this is a significant problem in the development of new anise spirits. Despite the lack of such procedures, various people from the anise spirits market have confirmed that the current trend in Greece is to have anise spirits with less alcohol. Consumers are asking for more 'gentle', more 'sweet' and less 'sharp' taste and aroma.

10.8 Future trends

The future trends are towards two directions:

- Lessen the alcoholic content of anise spirits and enrich their flavour by using more natural raw ingredients and processes,
- Development of ready-to-drink (RTD) anise spirits.

Regarding the first direction, research carried out in the author's laboratory in collaboration with one of the leading distillers in Lesvos has shown that the addition of honey and thyme to the mixture of raw ingredients before distillation can have a distinct positive impact on 'smoothing' the flavour character of the final products.

Regarding the second direction, three RTD ouzo samples were made in the author's laboratory, by mixing ouzo with fruit juices in such a way so the final alcoholic content would be about 5% abv. Three juices were used (orange, lemon and peach). This (unpublished) research has shown that the lemon RTD ouzo could have a high acceptance level with young consumers. Until now, especially in Greece, Italy, Spain and Turkey and to a lesser degree in France, anise spirits are linked with the consumption of a certain type of food (*meze* in Greek) and so their consumption is mainly located in tapas bars, tavernas and restaurants. In order to widen the market, the challenge is to persuade other consumer groups (in bars, pubs and clubs) to drink anise spirits by offering them RTD anise spirits. Thus, the next challenging task is how to manage the mass production of such a drink.

10.9 Sources of further information and advice

Pastis

- http://www.foodreference.com/html/artpastis.html

Ouzo

- http://www.greecefoods.com/ouzo/
- http://www.barbayanni-ouzo.com/ouzo-museum-lesvos.html
- GIANNAKOPOULOS GA (2010), *Distillation of life ouzo Barbayanni*, Plomari Lesvos, Hellenic Facsimile Edition.
- http://www.wiw.gr/index.php?language=english&section=directory&category=ouzo-raki-tsipouro-tsikoudia

Tsipouro

- http://www.wiw.gr/index.php?language=english&section=directory&category=ouzo-raki-tsipouro-tsikoudia

10.10 References

APOSTOLOPOULOU AA, FLOUROS AI, DEMERTZIS PG and AKRIDA-DEMERTZI K (2005) 'Differences in concentration of principal volatile constituents in traditional Greek distillates', *Food Control*, **16**, 157–164.

BERBEROGLU H. 'PASTIS – the French national, versatile and much-loved drink', available from http://www.foodreference.com/html/artpastis.html. date accessed 11 oct 2011.

DELAHAYE M.-C., SALLÉ J and MARTINE NOUET M. 'Pastis . . . the History', available from http://www.lamaisondupastis.com/images/pastis/pastisgb.htm date accessed 11 oct 2011.

EUROPEAN UNION (1989) 'Council Regulation (EEC) No 1576/89 of 29 May 1989 laying down general rules on the definition, description and presentation of spirit drinks', available from http://eur-lex.europa.eu/LexUriServ/LexUriServ.do?uri=OJ: 2008: 030: 0016: 0054: EN: PDF date accessed 11 oct 2011.

EUROPEAN UNION (2008) 'Regulation (EC) No 110/2008 of the European Parliament and of the Council of 15 January 2008 on the definition, description, presentation, labelling and the protection of geographical indications of spirit drinks and repealing Council Regulation (EEC) No 1576/89', available from http://www.mee.government.bg/ind/doc/LexUriServ.pdf

GEROGIANNAKI-CHRISTOPOULOU M, KYRIAKIDIS NV and ATHANASOPOULOS PE (2006) 'The evaluation of grape pomace distillates from selected red grape varieties', *International Journal of Food Science and Technology*, **41**, 854–860.

GIANNAKOPOULOS GA (2010), *Distillation of life ouzo Barbayanni*, Plomari Lesvos, Hellenic Facsimile Edition.

GRILLO I (2003) 'Small-angle neutron scattering study of a world-wide known emulsion: Le Pastis', *Colloids Surface A*, **225**, 153–160.

JURADO JM, BALLESTEROS O, ALCÁZAR A, PABLOS F, MARTÍN MJ, ET AL. (2007) 'Characterization of aniseed-flavoured spirit drinks by headspace solid-phase microextraction gas chromatography-mass spectrometry and chemometrics', *Talanta*, **72**, 506–511.

KOKKINOFTA RI and THEOCHARIS CR (2005) 'Chemometric characterization of the Cypriot spirit 'zivania'', *Journal of Agricultural and Food Chemistry*, **53**, 5067–5073.

KONTOMINAS MG (1986) 'Volatile constituents of Greek ouzo', *Journal of Agricultural and Food Chemistry*, **34**, 847–849.

TSACHAKI M, ARNAOUTOPOULOU AP, MARGOMENOU L, ROUBEDAKIS S and ZABETAKIS I (2010) 'Development of a suitable lexicon for sensory studies of the anise-flavoured spirits ouzo and tsipouro', *Flavour and Fragnance Journal*, **25**, 468–474.

TSAKIRIS AN (1995), *Potografia*, Athens, Psychalos.

VITALE SA and KATZ JL (2003) 'Liquid droplet dispersions formed by homogeneous liquid-liquid nucleation: 'The ouzo effect', *Langmuir*, **19**, 4105–4110.

11

Cognac: production and aromatic characteristics

L. Lurton, G. Ferrari and G. Snakkers, Bureau National Interprofessionel du Cognac, France

Abstract: This chapter presents the production process and aromatic characteristics of Cognac, the world-famous wine spirit (*eau-de-vie de vin*) produced in the Charentes region of Southwest France. Coming from a delimited geographical area, Cognac has AOC (Appellation of Controlled Origin) certification. Its quality is largely determined by the raw material (grapes) used and the production process, both of which must comply with strict regulations. An olfactometric map of freshly distilled Cognac has been drawn up, bringing together the main aroma compounds generated at succeeding stages of the production process. A Cognac aroma wheel, recently created to raise consumers' awareness of the aromatic richness of Cognac, is also presented here.

Key words: wine spirit, Cognac, production process, aroma, olfactometry.

11.1 Raw materials, production process and major variants

11.1.1 A product with a long history

Grapevines have been grown in Charentes since the end of the first century AD. In the Middle Ages, thanks to the River Charente, the town of Cognac was already famous for its wine trade, a welcome addition to the salt trade, which had been a major source of income since the eleventh century. Wines from the Poitou vineyards, transported in Dutch ships coming to load up with Atlantic coast salt, were much appreciated in the lands lying alongside the North Sea.

In the sixteenth century, the Dutch began distilling the region's wines in order to preserve them better. They came to Cognac to seek out the celebrated wines produced in the Champagne and Borderies vineyards, distilling them back at home in order to preserve them. They named the product obtained 'brandwijn' ('burnt wine') – the origin of the term 'brandy'.

Double distillation made its appearance in the early seventeenth century, enabling the product to travel in the form of a stable wine spirit a great deal more concentrated than wine itself. The first stills were set up in Charente by the Dutch.

Numerous trading companies sprang up which, in the mid-nineteenth century, began dispatching wine spirit in bottles rather than in casks. This new form of trade gave rise to complementary industries such as glassmaking (which went on to develop local expertise in mechanisation of bottle-manufacturing processes), crate and cork manufacture and printing.

Around 1875 *Phylloxera vastatrix*, an insect of the hemiptera order, that attacks vines by sucking the sap from their roots, destroyed most of the Cognac vineyards which, by 1893, only covered some 40 600 hectares, as against 280 000 hectares before the phylloxera plague. As elsewhere in Europe, the Charentes vineyards were reconstituted by the use of American rootstocks. This episode led to the creation of the Comité de la Viticulture in 1888, which went on to become the Station Viticole in 1892, an inter-professional qualitative research body devoted to Cognac.

In the first half of the twentieth century, legislation was drawn up in order to conserve longstanding and unchanging local practices:

- 1909: delimitation of the geographical production area,
- 1936: certification of Cognac as an Appellation d'origine contrôlée (AOC),
- 1938: delimitation of regional appellations.

Historically a product for export, nowadays more than 95% of Cognac is consumed outside France, in over 160 different countries. Currently, around 150 million bottles a year are dispatched to waiting markets. For fans from the Far East, via Europe, to the continent of America, Cognac means top-quality wine spirit emblematic of the French art of living.

11.1.2 Current regulations relating to Cognac

With reference to European Regulation N° 110/2008 (European Union, 2008) on the definition, description, presentation and labelling of spirit drinks, Cognac is a wine spirit. The geographical indication 'Cognac', along with the complementary geographical indications associated with it, are in Annex III of the same regulation. Under French law, Cognac is an AOC, the rules for production of which are more restrictive than those provided for wine spirits by the aforementioned Regulation, and are laid down in Order no. 2009-1146 of 21 September 2009 by the French Ministry of Agriculture Food and Fisheries (2009).

11.1.3 A delimited geographical production area

Temperate maritime climate

The Cognac winegrowing region enjoys a temperate maritime climate, differing little except in coastal areas, which get more sunshine and experience a narrower range of temperatures. Because the ocean is so close by, rain, although heavier in winter, may fall at any time of the year. Periods of drought are few, ensuring the vine a regular water supply. The average annual temperature stands at around

13 °C, with fairly mild winters. Temperatures are high enough to ensure good maturation of grapes to be used for wine spirit production, but not high enough to burn them. The Cognac region's climate was described at the beginning of the twentieth century by Ravaz (1900), and later on by Lafon *et al.* (1964).

A recent study carried out by the BNIC Station Viticole sought to pinpoint recent changes in the Cognac region's climate and assess their impact on vine growing (Boitaud *et al.*, 2010).

Appellation area delimited into 'crus'

In the mid-nineteenth century, Henri Coquand (1811–1881) made a study of the Cognac region's geology (Coquand, 1858). With the help of a taster, he classified the different areas according to the quality of the wine spirits that their soils could produce. Around 1860, their work resulted in the demarcation of different 'crus' (growing areas) and served as a basis for the Order of 13 January 1938 delimiting the crus. Geographical denominations, complementary to the Cognac appellation, still make use of their historical names: Grande Champagne, Petite Champagne, Fine Champagne, Borderies, Fins Bois and Bons Bois, to which must be added Bois Ordinaires or Bois à Terroir (Fig. 11.1).

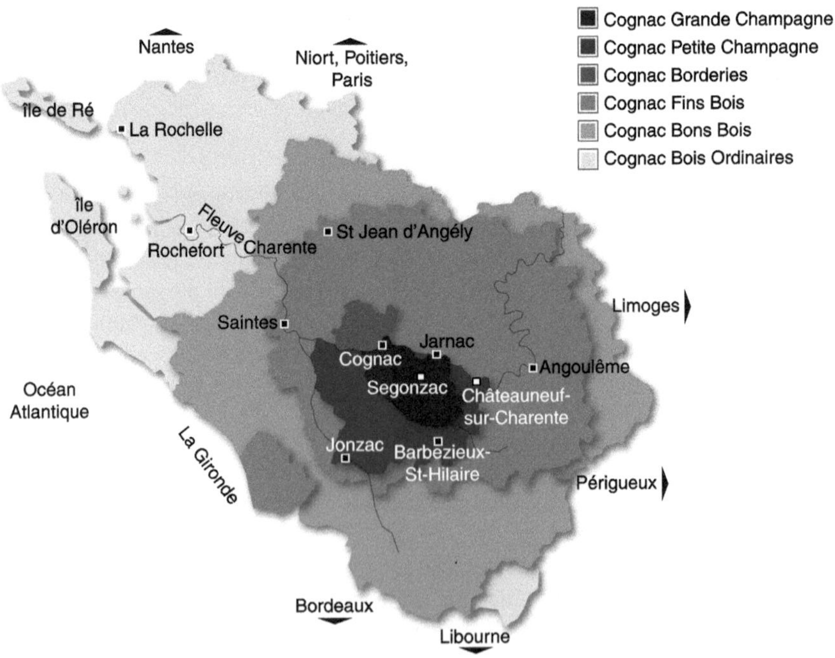

Fig. 11.1 A delimited production area divided into six crus (© BNIC).

According to the work carried out at the time, the dominant characteristics of soils typical of each denomination are as follows:

- Grande and Petite Champagne: thin argilo-calcareous soils over soft chalky Cretaceous limestone,
- Borderies: sandy clay soils containing flint nodules resulting from limestone decarbonation,
- Fins Bois: largely covered by 'groies', thin, stony red argilo-calcareous soils, from Jurassic limestone,
- The Bois (Bons Bois, Bois Ordinaires and Bois à Terroirs): sandy soils in coastal areas, certain valleys and over the entire southern part of the winegrowing area. The sand comes from erosion of the Central Massif.

Dumot *et al.* (1993) give a detailed description of the main types of soil to be found in the Cognac area, which has become a reference in the field.

Wine spirits obtained from the stills are marked by very considerable analytic and organoleptic diversity, largely due to their origin. Such diversity necessitates implementation of different ageing techniques of variable duration.

Grande Champagne produces wine spirits of remarkable finesse, characterised by great distinction and a long finish, and with a predominantly floral bouquet. Its wine spirits are slow to mature, and require long ageing in oak casks in order to reach full maturity.

Petite Champagne wine spirits have mostly the same characteristics as those from Grande Champagne, without, however, the latter's extreme finesse.

The Borderies produce smooth, rounded wine spirits with a strong bouquet and characterised by an aroma of violets. They have a reputation for reaching optimal quality after shorter maturation than those from the Champagne areas.

Fins Bois produce supple, rounded wine spirits that age quite rapidly and whose fruity bouquet is reminiscent of pressed grapes. Bons Bois produce wine spirits with fruity aromas that age rapidly.

Regulations provide for the indication of the cru (Grande Champagne, Petite Champagne or Borderies, for example) on the label. In such cases, the blend must be 'single cru' – in other words, all the products in the blend must come only from the area mentioned on the label. The term Fine Champagne designates a blend of Grande Champagne and Petite Champagne, with at least 50% from Grande Champagne.

11.1.4 Encepagement and vineyard management

Wines destined for elaboration of wine spirits come exclusively from the following grape varieties:

- Colombard B, Folle Blanche B, Montils B, Ugni Blanc B and Semillon B,
- Folignan B, up to 10% of the regional grape varieties.

There are around 75 000 hectares of vineyards devoted to Cognac production. Encepagement is the term used to describe the grape varieties planted within a vineyard. Ugni Blanc is the most commonly planted variety and accounts for almost 98% of vines in the Cognac area. Its rise in Cognac vineyards began following the phylloxera crisis, about a century ago (Ravaz, 1900). A variety originating in Italy, where it is known as 'Trebbiano Toscano', the Cognac region is its northern limit for ripening. In the mid-twentieth century, it became the first variety for Cognac production, due to its qualities with regard to productivity (its average yield in the Cognac region is between 120 and 130 hectolitres per hectare), late budding, and production of an acidic wine, low in alcohol content, that is particularly suited to production of quality wine spirits. In 2003, the Bureau National Interprofessionnel du Cognac undertook a major research programme in order to preserve the variety's genetic diversity (Dumot *et al.*, 2010).

Rules concerning vineyard management also define planting density, distance between rows, pruning, number of buds per hectare, percentage of missing plants, and minimum age of vines.

The management method used by the vine grower enables reinforcement of the required characteristics of high acidity and low alcohol content. Such maturation objectives are specific to wines meant for distillation. High acidity helps wine preservation through the winter months up to distillation, and low alcohol content enables greater concentration of the wines' aromas in the spirits.

Maximum annual production expressed in pure alcohol per hectare is fixed each year by inter-ministerial order.

11.1.5 An adapted vinification technique

Harvesting of grapes is currently largely mechanised in the Cognac area. Ensuring the grapes' integrity during the transfer and must-extraction stages is very important for the final quality of the spirits obtained. For this reason, the use of centrifugal pumps with pallets is prohibited for grape transfer. Grapes are pressed immediately after harvesting in traditional horizontal-plate presses or pneumatic presses. Continuous presses with Archimedes screws are prohibited. The musts are immediately transferred into vats for fermentation.

Alcoholic fermentation management is a key factor in ensuring the success of winemaking. The process produces ethanol as well as a large part of the wine's – and therefore spirit's – volatile compounds.

Studies on Cognac region yeasts were carried out from the 1970s (Park, 1974; Ribes, 1986) in order to provide producers with strains specifically adapted to distillation wines. In the 1990s, a programme bearing on characterisation and selection of yeast strains naturally present in the Cognac vineyards (Roulland *et al.*, 1995; Versavaud *et al.*, 1995) was carried out. The BNIC defined specifications for selecting the best adapted yeast strains. Six strains currently meet these specifications and are mostly used in the region (Fig. 11.2). In order to increase diversity in production of aromatic compounds, use of several different strains in the same winery is recommended (Ferrari *et al.*, 2010).

Use of sulphur dioxide is not authorised during the fermentation of distillation wines. Addition of sulphur dioxide leads to increased production of aldehydes,

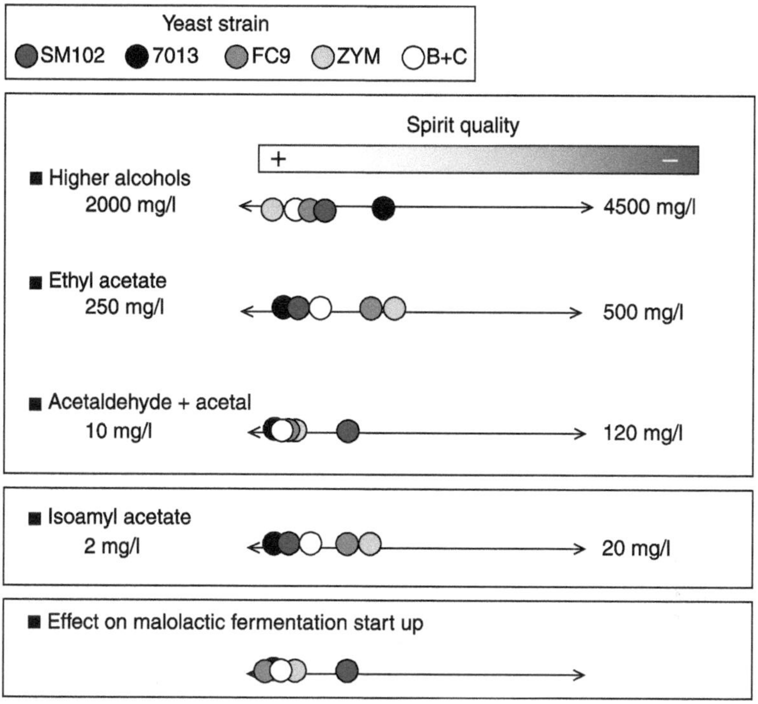

Fig. 11.2 Characteristics of the main yeast strains used in the Cognac area.

which would affect the quality of the spirits (Cantagrel *et al.*, 1998). Because of this, so as to minimise wine-preservation risks, the date limit for distillation of wines for Cognac production is set at 31 March of the year following harvesting.

11.1.6 A traditional method of distillation

Distillation of Cognac is determined by:

- The distillation principle implemented: discontinuous distillation or double distillation, also known as '*à repasse*',
- The type of still used: the so-called 'Charentais' still, whose shape, material (copper), capacity and heating method have been defined since 1936 and are primordial determining factors for wine spirit quality (Fig. 11.3).

This is because:

- The shape of the apparatus contributes to selection of volatile substances,
- Heating by a naked flame generates synthesis of complementary aromas, due to the contact between the wine and the surface of the boiler (cooking phenomenon),

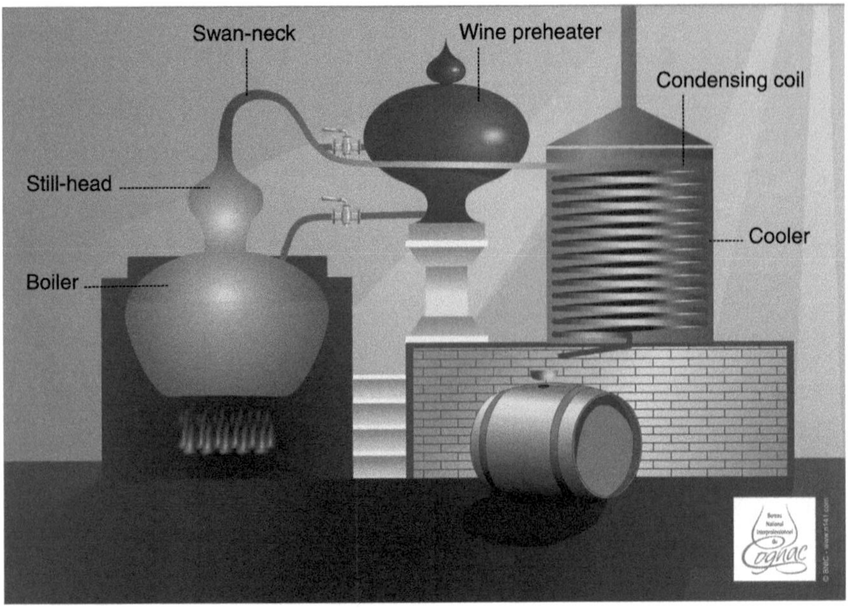

Fig. 11.3 The Charentais still: main components (© BNIC).

- Still components in contact with the wine, vapours and distillates are made entirely of copper, because of the metal's physical properties (malleability and good heat conductivity) and its chemical reactivity with different wine constituents.

For each distillation, this type of still requires the carrying out of a highly delicate operation: the 'coupe' (cutting). This consists of dividing up the distillate according to its proof and composition in terms of volatile substances, into the 'heart' (which will become Cognac) and the parts to be recycled in following distillations ('heads', 'seconds' and 'tails'). Depending on wine quality and their qualitative objectives, distillers judge the right moment for cutting along with methods of recycling. Distillation of Cognac therefore remains a manual operation whose success very much depends upon the distiller's skills.

There are two possible variations to the distillation process for Cognac wine spirits: one concerns the type of wine used in the still, and the other concerns methods of recycling during distillation.

- The distiller can distil either a clear decanted wine or a wine with its lees, i.e. containing the yeasts resulting from alcoholic fermentation. In this case, the distillate acquires a specific aromatic profile, due to higher content of long- and medium-chain fatty acid esters. This type of product is generally matured in casks over longer periods.
- The other variant concerns the distillation method itself, specifically the recycling of the different parts. During the second distillation, the distiller has

the choice of recycling the seconds either in a wine or in a 'brouillis' (first distillate). The choice calls for different parameters, such as the proof of the wines, seconds and brouillis, and the profile sought. The distillation method is usually included in the specifications laid down with the purchaser.

11.1.7 Strictly controlled ageing

Following distillation, wine spirits to be used for elaboration of Cognac are aged in oak barrels, in the appellation area, for at least two years.

After leaving the still, the new wine spirit stays in oak casks for several years (sometimes even several decades). Ageing of Cognac is a process helped along by the climatic conditions prevalent in the appellation area and by local knowledge developed over the course of the centuries. A range of physicochemical phenomena occur: evaporation of water and alcohol, changes in the concentration of various substances, extraction of compounds produced by wood, oxidation, etc. These phenomena are directed by the spirit's initial characteristics (proof and acidity), by the type of barrel in which ageing occurs, and by the hygrometric and temperature conditions in the cellars in which the casks are kept (Cantagrel *et al.*, 1993).

During the ageing process, the '*eaux-de-vie*' are exposed to moderately damp conditions and not-too-extreme changes of season. Localisation and construction of cellars are adapted to ensure harmonious ageing. Oaks of Troncais or Limousin types (mainly sessile and pedunculate oak) are used because they enable exchanges to take place over long periods between the wine spirit, the external environment and the wood. In collaboration with Cognac companies, the many cooperages active in the geographical area have developed expertise in the creation of storage conditions best adapted to the ageing of Cognac.

It is up to the Master Blender to select the most appropriate storage conditions depending on the spirits' initial characteristics, the stage of ageing, and his or her qualitative objectives. The Master Blender also decides when to end the ageing process. Once maturity is reached, he or she transfers the wine spirits into glass demijohns known as 'Dame-Jeanne', in which they may remain sheltered from the outside air for decades on end without further development. The oldest *eaux-de-vie* are stored in a dark storehouse, referred to as 'Paradise'.

A Cognac wine spirit usually reaches its peak after around 50–60 years of ageing. However, certain *eaux-de-vie* that have been stored in oak casks a good deal longer (sometimes up to 100 years) can be used in blends in very small quantities, to bring a final touch to the most prestigious Cognacs.

11.1.8 The art of blending

Ageing of Cognac is inseparable from the art of blending. Depending on its elaboration and ageing, every wine spirit has its own aromatic profile, which will be put to full use through its blending with other wine spirits having different characteristics. This crucial and highly complex stage cannot be accomplished through the application of simple technical recipes. The Master Blender depends on empirical knowledge, gained over the course of time and necessitating constant monitoring through tasting, along with a perfect sensory memory of *eaux-de-vie* at their different stages of elaboration. Such knowledge, which requires many years of apprenticeship under the elders of the profession, has been differentiated, maintained and transmitted through exchanges in the Cognac region's fabric of companies – between Master Blenders, vine growers, Cognac companies and brokers. On the scientific side, Cantagrel *et al.* (1991a) have described the development of chemical balances occurring during the blending and reduction of wine spirits.

11.1.9 Ageing designations

As explained above, Cognac must be aged in oak casks for at least two years before being put on the market for direct human consumption. Inventory and age control of all Cognac producers are performed by the Bureau National Interprofessionnel du Cognac (BNIC) (Executive order of July 2003).

The age of an *eau-de-vie* corresponds to the period during which it has matured in oak casks. In contrast to wine, *eau-de-vie* virtually ceases to age as soon as it is transferred to a glass container. A Cognac will always be the same age it was when bottled. Ageing designations refer to the age of the youngest component in the blend (Fig. 11.4). It is not the age of the Cognac in the bottle. Cognac companies generally use products much older than the minimum requirement in their blends. Those bearing the most prestigious designations may have aged for decades. Compte '0' begins on 1 April of the year following the harvest. Ageing designations are optional indications, regulated in application of European Community legislation. A decision of the French Government Commissioner to the BNIC

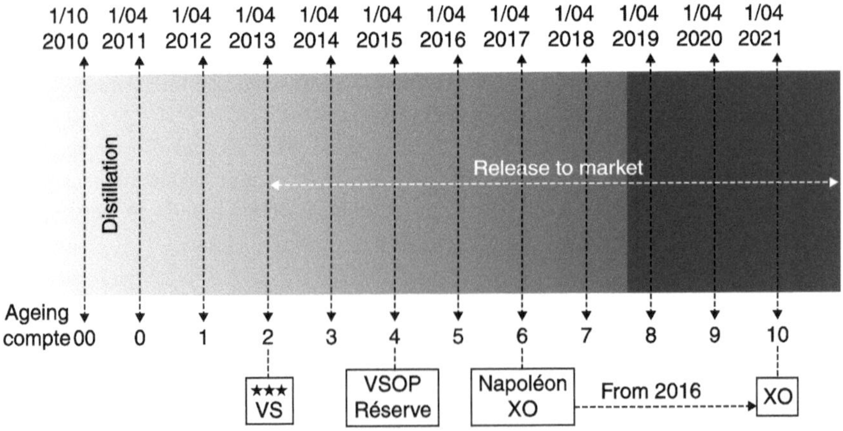

Fig. 11.4 Cognac ageing counts and designations (© BNIC).

(1983 decision) codified the designations to be used, based on the age of the spirits making up the blends:

- *** (3-star) or VS (Very Special): The youngest *eau-de-vie* in the blend is at least two years old (compte 2),
- VSOP (Very Superior Old Pale): The youngest *eau-de-vie* in the blend is at least four years old (compte 4),
- Napoléon, XO (Extra Old), Extra, Hors d'âge: The youngest *eau-de-vie* in the blend is at least six years old (compte 6). As of 2016, XO (Extra Old), Extra and Hors d'âge are expected to require compte 10 for the youngest *eau-de-vie*.

A producer may also create a 'vintage' Cognac, although this practice is not very common. In order to do so, he or she must make a request to the BNIC, and a specific traceability of the batch involved right up to its sale has to be set up. The wine spirit so declared must be stored in a separate warehouse and the BNIC carries out regular monitoring until it is marketed. In such vintage Cognacs, the wine spirits used are all produced in the same year, and it is therefore the harvest year that appears on the label (Fig. 11.4).

11.1.10 A well-established typicity

Compliance with the strict set of rules concerning origin and production conditions confers very specific organoleptic and analytic characteristics, a fact that does much to help combat counterfeiting (Mazerolles *et al.*, 1991, 1993), which is one of the BNIC's main tasks.

11.2 Overview of flavour chemistry and sensory properties

11.2.1 Aromatic complexity

A Cognac wine spirit's aroma results from the great variety of volatile substances formed and modified throughout the elaboration process, from vine to bottle. Production of quality grapes, winemaking, distillation and ageing – mastery of each stage in the production chain is essential to the aromatic quality of *eaux-de-vie* used for creation of Cognac (Cantagrel *et al.*, 1995; Ferrari *et al.*, 2004).

In 1900, Ravaz, Director of the Station Viticole de Cognac, expressed his approach to the aromatic composition of Cognac wine spirits and its connection with sensory perception as follows:

> 'Its qualities are due to all the other fixed (. . .) and above all volatile substances that accompany the ethyl alcohol during distillation; it is these that give the mellowness, aroma and bouquet (. . .).
>
> Separately, none of them, even when diluted, is any pleasanter than the ethyl alcohol. It is only when brought together in the right mixture of water and alcohol that they confer upon the latter those well known gustative qualities.

> And as most of them exist in imponderable quantities in the mixture, one can see that the slightest modification in their proportions may deeply affect the quality of the wine spirit. It only needs one of them to disappear or simply decrease in quantity for the taste and flavour of the spirit to change completely.
>
> Hence, the influence that soil, climate, variety of grape, distillation etc. has on the quality of the product is not hard to explain; a mere nothing can bring about radical changes in a product as complex as a wine spirit.'

Up until now, several hundred volatile compounds have been identified in Cognac (De Rijke and Ter Heide, 1983; Ter Heide, 1986; TNO, 1989; Ledauphin *et al.* 2010), with many chemical families represented: alcohols, acids, esters, ethers, aldehydes, acetyls, ketones, hydrocarbons, terpenes, C_{13}-norisoprenoids, lactones, sulphur and nitrogenous compounds, phenols, etc.

Such chemical diversity goes alongside a wide range of concentrations, from several hundred mg/l (esters and alcohols) to a few ng/l (pyrazines and lactones). Perception thresholds for these molecules are also extremely varied. Figure 11.5 illustrates the great diversity of volatile substances of which a wine spirit is

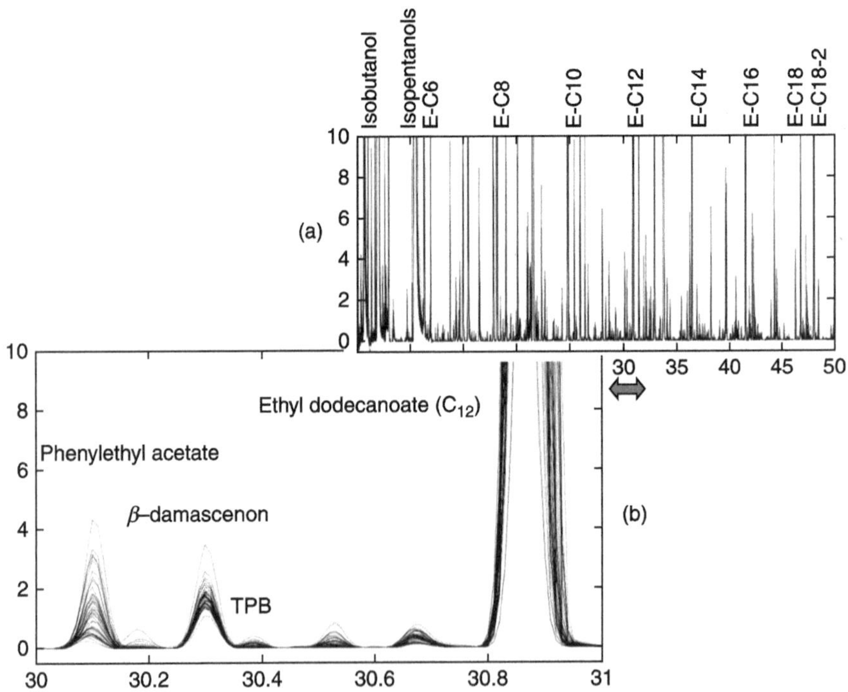

Fig. 11.5 Superposition of 40 FID chromatograms of Cognac wine spirits. (a) Global. (b) Focus upon a small zone. E-C6: ethyl hexanoate, E-C8: ethyl octanoate, E-C10: ethyl decanoate, E-C12: ethyl dodecanoate, E-C14: ethyl hexadecanoate, E-C16: ethyl hexadecanoate, E-C18: ethyl octadecanoate, E-C18-2: ethyl linoleate (© BNIC).

composed, as well as the variability of different products that makes no two Cognac wine spirits identical.

11.2.2 Compounds from the grape

A number of aroma substances come from the grapes themselves. Extracted during the early stages of winemaking and revealed by complex chemical and biochemical transformations (Cantagrel *et al.*, 1998), they are present in both new and aged wine spirits, to which they give floral, fruity and vegetal notes in particular.

The grape, whose composition is conditioned by terroir, climate vineyard, management and harvest date, contains numerous precursors whose transformation during winemaking and distillation generates aroma substances. Grape compounds found in wine spirits include six-carbon-atom alcohols (hexanol and hexenols), terpenes (linalool and derivatives, α-terpineol), products of carotenoid degradation (vitispiranes, α-ionone, β-damascenone and E)-1-(2,3,6-trimethylphenyl)buta-1,3-diène (TPB)), and aldehydes (Lurton *et al.*, 1991).

Some of these play a major part in the wine spirit's aroma. Analysis by gas chromatography coupled with olfactometry (GC-O) enables the linking of certain volatile compounds coming from the grape to aromatic notes perceptible in wine spirits (Fig. 11.6). Using the aroma extract dilution analysis (AEDA) olfactometric method, it has therefore been possible to demonstrate that β-damascenone (zone 41) and TPB (zone 42), stemming from odourless precursors present in grapes, are compounds essential to the aroma of a freshly distilled spirit, and are still perceptible at 1/100 dilution.

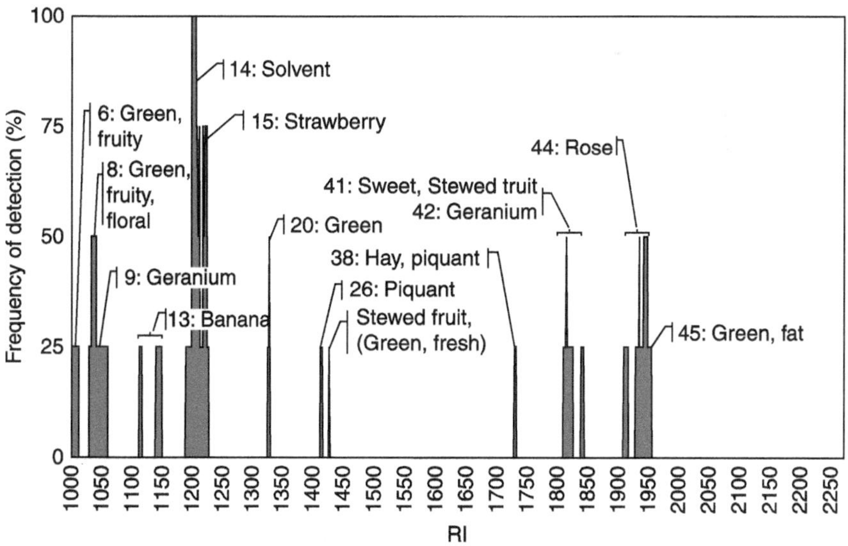

Fig. 11.6 Olfactogram of a 100-times diluted extract from a freshly distilled wine spirit (© BNIC).

The ripening stage at harvesting affects grapes' composition in aromatic compounds. As an example, Fig. 11.7 shows the development of the average *cis*-3-hexenol and β-damascenone content of maturing grapes.

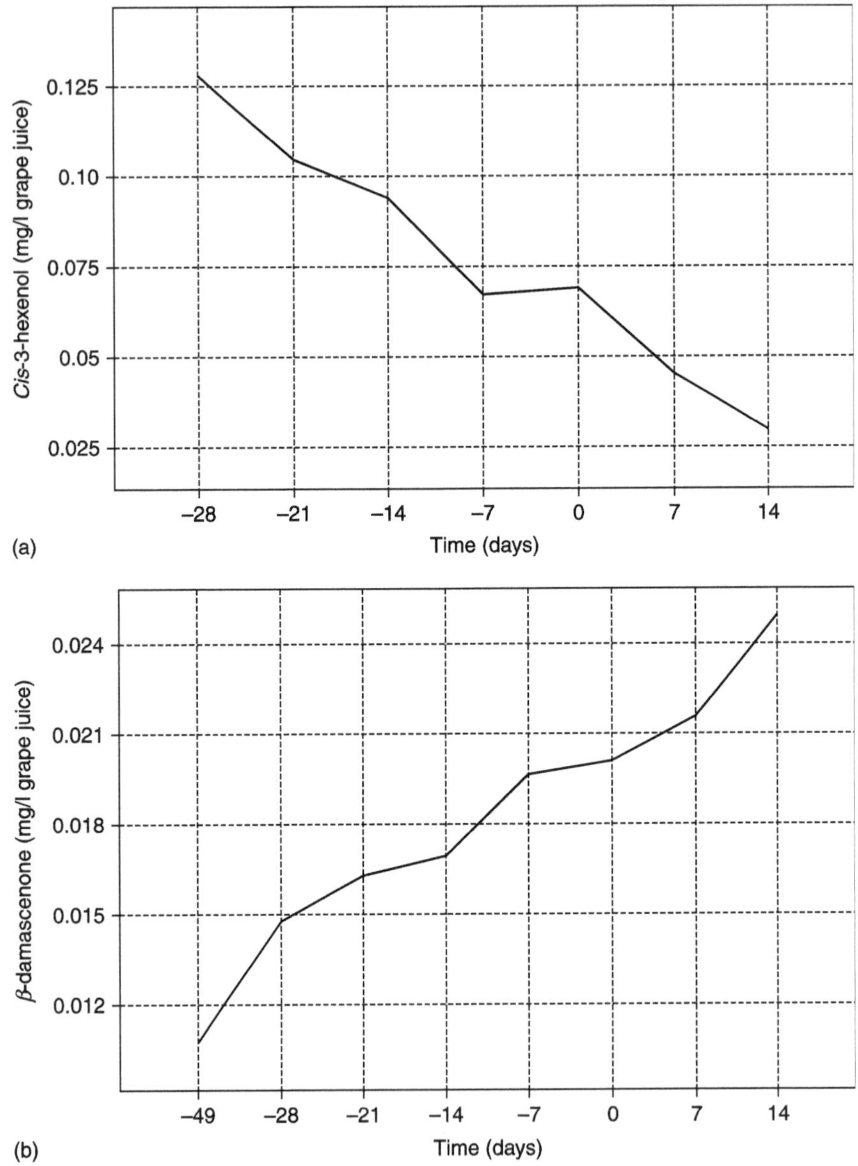

Fig. 11.7 Evolution of *cis*-3-hexenol and β-damascenone in the juice during grape ripening, 2008. (a) *cis*-3-hexenol. (b) β-damascenone. GC is performed on the micro-distillate of an average regional must sample from 55 different parcels. Date 0 = beginning of harvest (© BNIC).

Some unwanted notes may also find their origin in the grapes. It is the case with the 'kerosene-like' notes due to trimethyl-1-1-6-dihydro-1,2-naphthalene (TDN) (Vidal *et al.*, 1991) and connected to the conditions of grape mechanical processing before fermentation (Lurton *et al.*, 1991). It is also the case with certain markers linked to the presence of grey mould (*Botrytis cinerea*) in the crop, a study of which was made by Urruty *et al.* (2008).

11.2.3 Compounds of fermentative origin

Numerous volatile compounds (higher alcohols, esters, aldehydes, acetyls, fatty acids, etc.) are formed by yeasts during the alcoholic fermentation stage (Lurton *et al.*, 1995). Composition of grape musts (nitrogen, matter in suspension, etc.), along with conditions of fermentation (temperature, oxygen, yeast strain, etc.) and wine storage before distillation, determine their presence in a wine spirit.

Nitrogenous composition of the musts is especially important as it influences the nutrition of yeasts and hence the development of alcoholic fermentation (Nedjma, 1997; Ferrari, 2002).

Although they have high perception thresholds, ethyl esters, whose concentrations may vary in a wide range, contribute to the fruity perception (Léauté, 1990).

Distilling wine in the presence of yeasts contained in fermentation lees leads to enrichment in fatty acid esters, such as ethyl octanoate (fruity, floral, pineapple, apple and pear), and ethyl decanoate (fruity, pear, wine, etc.). Aromatic synergy between those esters strengthens their olfactory impact. Fermentation lees also supply the wine spirit with fatty acids (C_8–C_{18}, saturated and unsaturated) which are at the origin of aromatic derivatives formed by their oxidation into aldehydes or ketones during the ageing process.

Acetates floral and fruity notes decrease over time because of their gradual hydrolysis during preservation of wines and spirits.

11.2.4 The role of distillation

Distillation has a triple role of concentration of aromas, selection of certain fractions, and formation of new compounds (Cantagrel *et al.*, 1991b). Certain volatile substances, favourable or not to quality, may be formed or eliminated during the process. Its effect on the quality of Cognac wine spirits has been frequently described (Lafon *et al.*, 1964; Gay-Bellile, 1981; Léauté, 1990; Cantagrel *et al.*, 1991b).

Newly-formed compounds

Distillation performed by naked flame heating is accompanied by cooking phenomena that result in a wide range of chemical reactions dependent upon temperature conditions, duration, acidity of the wine, and also the role that copper plays as a catalyst. Gay-Bellile (1981) and Léauté (1990) describe such reactions as hydrolysis of esters and terpenes, as well as the production of new compounds: furans, pyridines and pyrazines (cocoa and grilled nuts notes) through Maillard reaction, and aldehydes and acetals through Strecker degradation. A typical example of such reactions is the production of furfural and its derivatives from sugars. Different compounds produced during Charentaise distillation have been identified (Gay-Bellile, 1981; Galy *et al.*, 2008).

The role of rectification

The alcohol content of freshly distilled Cognac spirits is limited to 72.4% vol. The low level of rectification during the Charentaise distillation process results in a high concentration of wine volatile substances in the distillate. This implies specific constraints in the choice of wines to be distilled, which must be free of defects likely to be concentrated in the wine spirit. Such constraints apply less in the case of spirits distilled to a higher degree, where greater rectification leads to lower content of aromatic substances coming from the wine.

11.2.5 The olfactometric profile of a freshly distilled spirit

Olfactometric techniques have been adapted to study the aromatic structures of freshly distilled spirits (Leclaire *et al.*, 1999; Lablanquie *et al.*, 2002). High-quality new spirits were selected in partnership with different Cognac companies, and a panel of olfactors was trained and assessed. The nasal impact frequencies (NIF) method was applied to the *eau-de-vie* selected, whose sensory profiles had previously been described. This approach allowed the identification of some forty major olfactory zones characteristic of the new high-quality wine spirits. A number of these zones are always present, while others are met with less systematically. This ensemble of zones may be regarded as the aromatic 'skeleton' of a newly distilled wine spirit (Ferrari *et al.*, 2004; Leyrat *et al.*, 2005). Major work bearing on identification of volatile compounds present in these zones was undertaken (Ledauphin *et al.*, 2004) with a view to pinpointing their olfactory roles and also to determining their origins (raw material, vinification or distillation). Once the compounds were identified, and when pure molecules were available, their aromatic impact was checked by measured-out addition. Recent extension to that work has been carried out by identification of new aroma-marker compounds in different zones, along with characterisation of the aromatic impact of some that were already known, in relation to their concentrations in wine spirits (Ferrari *et al.*, 2008). Figure 11.8 presents a typical olfactogram of a wine spirit of the 'fruity-floral' type. Table 11.1 presents the current state of knowledge with regard to these olfactory zones and the marker compounds identified.

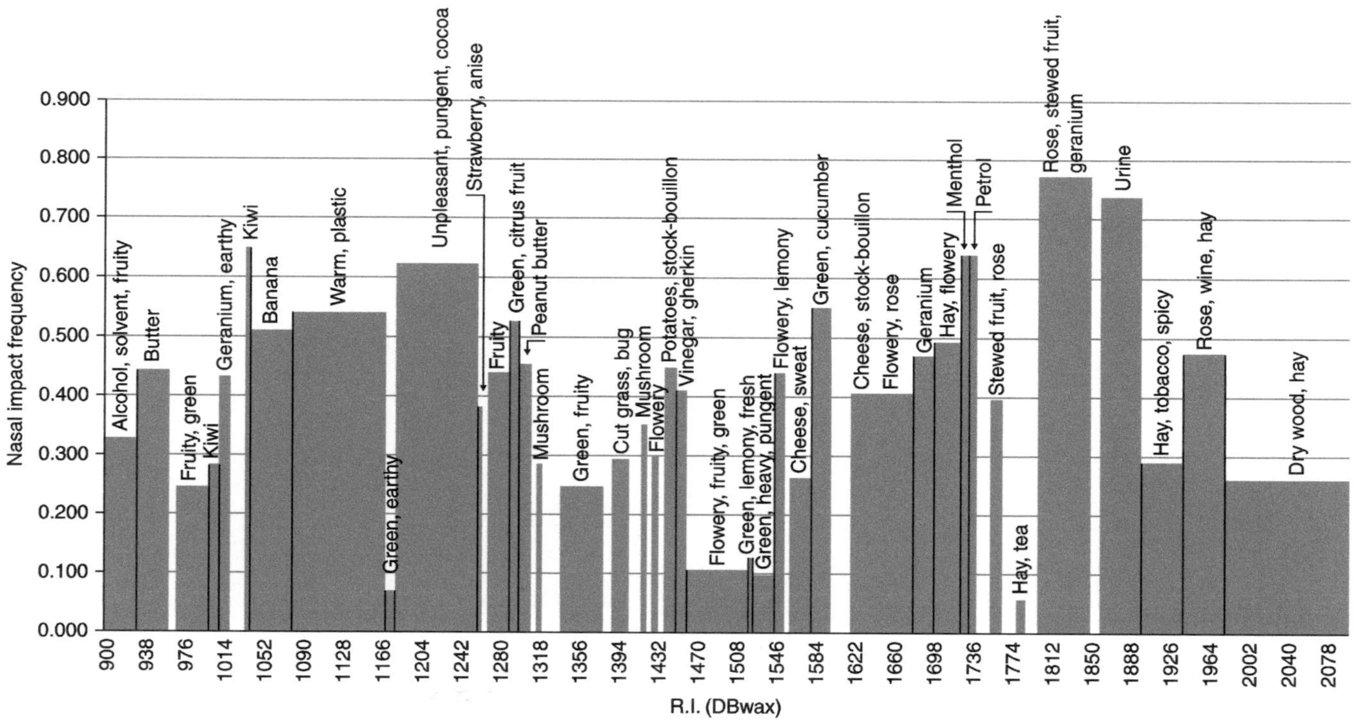

Fig. 11.8 Average olfactogram of a 'fruity-floral' type Cognac wine spirit (Leyrat, 2005).

Table 11.1 Gas chromatography-olfactometry profile of freshly distilled Cognac spirit: main zones, odour and markers

Zone	Retention index on DB-wax column	Origin	Odour	Compound responsible
1	<972		Solvent, methylated spirits	Acetaldehyde, isobutyraldehyde, ethyl acetate, 1,1-diethoxyethane
2	972	Fermentation	Butter	2,3-Butanedione
3	975		Methylated spirits	
4	989		Green, earthy, geranium	
5	991		Solvent, plastic	
6	996		Green, fruity	
7	1008		Solvent, earthy, geranium	
8	1032		Kiwi, fruity, slightly acid	Ethyl butyrate
[9]	1051		Geranium, earthy	
10	1053	Fermentation	Kiwi, fruity, slightly acid	Ethyl 2-methylbutyrate
11	1065	Fermentation	Kiwi, fruity, slightly acid	Ethyl 3-methylbutyrate
12	1072		Solvent, plastic	
13	1121	Fermentation	Banana	Isoamyl acetate
14	1181	Fermentation	Solvent, chocolate	Isoamyl alcohol
15	1238	Fermentation	Strawberry, anise	Ethyl hexanoate
[16]	1274		Sweet peanut	
17	1295	Grape	Mushroom	1-Octen-3-one
18	1335	Fermentation	Green, fat, pungent	Ethyl lactate
19	1345	Grape	Green fresh	Hexan-1-ol
20	1350		Green floral	
21	1365		Green, fat	
22	1370	Grape	Green, leaves, grass	(Z)-3-hexen-1-ol
23	1385		Green floral	
[24]	1390	Grape	Mushroom	1-Nonen-3-one
25	1420		Earthy	
26	1430		Pungent, acidic	Ethyl octanoate
[27]	1435	Fermentation	Potato	Methional
[28]	1450		Earthy	
[29]	1470		Mushroom	
[30]	1485		Green pepper	

[31]	1510		Green pepper	(Z)-2-nonenal
32	1520		Green, fat	(E)-2-nonenal
33	1530	Grape	Bergamot, orange blossom	Linalool
34	1565		Cucumber	(E,Z)-2,6-nonadienal
35	1620	Fermentation	Rose	Phenylacetaldehyde
36	1645		Toasty	
[37]	1680		Geranium, metallic	
38	1700		Hay	
[39]	1708	Grape	Hydrocarbon	1,1,6-Trimethyl-1,2-dihydro naphthalene (TDN)
[40]	1781		Rose	2-Phenylethyl acetate
[41]	1791	Grape	Stewed fruit	β-Damascenone
[42]	1824	Grape	Geranium	(E)-1-(2,3,6-trimethylphenyl) buta-1,3-Diene (TPB)
43	1845		Phenol	
44	1895	Fermentation	Rose	2-Phenylethanol
45	1938		Green, fat	

Note: Zones in square brackets [ ] are not systematically present in wine spirits. Compounds were subjected to validation based on their GC-O detection in hydroalcoholic solutions with a content estimated or similar to that found in wine spirits.

11.2.6 Ageing compounds

Extraction of wood compounds

New compounds are extracted from the oak during the early stages of the ageing process. They bring the *eau-de-vie* new, woody, spicy and empyreumatic notes, characteristic of spirits aged in wooden barrels. How this stage develops depends on a variety of factors. The status of the cask (new or previously used) influences levels of extraction as well as the nature of the compounds extracted (Calvo *et al.*, 1993; Snakkers *et al.*, 2008). The wood's botanical and geographical origin (Prida and Puech, 2006), its structure, which is linked to the way the forest was managed (Snakkers *et al.*, 2000), and cooperage techniques – the barrel toasting level in particular (Snakkers *et al.*, 2003) – also affect the contribution brought by the wood.

The main compounds extracted include derivatives of lignin (vanillin being the best known), lactones (mainly methyl octalactones), furanic derivates (furfural, etc.)

Oxidative reactions

Ageing in oak barrels also calls for a carefully monitored oxidation process. After the extraction stage, changes intervene in the pool of compounds extracted from the wood, e.g. coniferaldehyde is transformed into vanillin and vanillic acid (Calvo *et al.*, 1993).

Compounds coming from the new spirit also evolve: increase in methyl ketones by β-oxidation of fatty acids (Marché and Joseph, 1975; Vidal *et al.*, 1993; Watts *et al.*, 2003), and formation of acetaldehyde and ethyl acetate through oxidation of ethanol (Reazin, 1981).

Evaporation, concentration and modification of chemical balances

During ageing under wood, ethanol content diminishes through evaporation (Puech *et al.*, 1984; Cantagrel *et al.*, 1993). In contrast, concentrations of components less volatile than ethanol, such as higher alcohols, increase.

With the lowering of alcohol content, chemical balances regulated by ethanol are subjected to change. Higher alcohol esters produced during fermentation see their content decrease by transesterification. Their impact diminishes while the wine spirits' aromatic power increases.

Interactions between wine spirit and wood compounds

New molecules are formed by reactions between the wine spirit's constituents and compounds coming from the wood. Such reactions illustrate the notion of a 'marriage' between the wine spirit and the wood, which is regarded as a key element in the qualitative development of wine spirits during the ageing process. Among the compounds formed are esters (ethyl vanillate, ethyl syringate, etc.) and ethers (vanillin ethyl ether, etc.) produced by the reaction of ethanol with compounds extracted from the wood.

The 'Cognac rancio'

Different stages may be identified in this complex and continuous process, which takes place over many years. Léauté *et al.* (1998) cite over 50 terms used to describe the development of Cognac wine spirits during ageing (Fig. 11.9). Of these descriptors 19 correspond to fruity characters, 14 to floral characters, 10 to spicy characters, and 10 to woody characters. Aged wine spirits present specific characteristics that are grouped together under the term 'rancio charentais' or 'Cognac rancio'. 'The mixture of highly-smelling methyl-ketones with compounds produced by oxidative degradation of lignins, lignans and various oak resins smelling of vanilla brings old Cognac a balsamic scent that is specific to the spirit and which is known as the rancio charentais' (Marché and Joseph, 1975). Léauté *et al.* (1998) gave a highly detailed description of the various stages in the development of the Cognac rancio, distinguishing four phases: from 12–15 years of ageing (phase one), up to 50 or 60 years (phase four), to which they attached the main descriptors encountered.

11.3 Sensory analysis practice in the industry

Cognac is a wine spirit generally created by blending, a process carried out by the Master Blender and which is a continuous search for harmony and consistency. Just

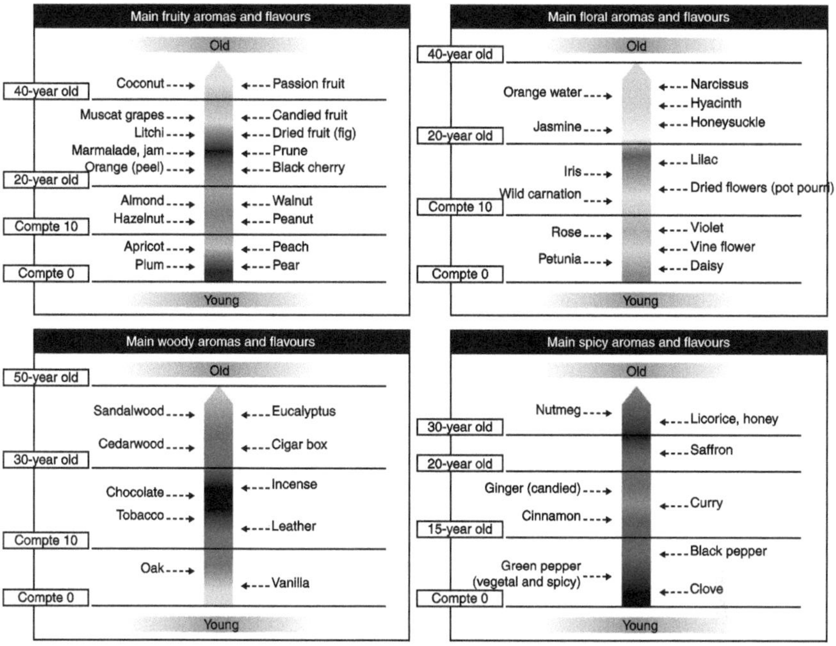

Fig. 11.9 The main aromatic characters of Cognac. Main descriptors for Cognac wine spirits belonging to the fruity, floral, spicy and woody families, for ageing periods of 40 or more years (Source: Léauté *et al.* 1998).

like the perfumer's 'nose', he or she creates subtle blends of wine spirits of different ages and from different crus to ensure that Cognac will not only retain its distinctive personality over the course of the years, but also the fidelity of its consumers.

It is he or she who, along with his or her team, selects and purchases the wine spirits from vine growers and watches over them throughout the ageing process. Once distilled, he or she selects those to be used for elaboration of a young Cognac and those to be aged for longer periods. Finally, it is he or she who creates their own blends, which constitute the real signature of each brand.

In this approach, each Cognac house has its own know-how, combining sensory and physicochemical analysis of the products. This calls upon increasingly sophisticated techniques, such as gas chromatography.

11.4 Consumer research: creation of the Cognac aroma wheel

In 2009, in order to better acquaint consumers with Cognac's wide-ranging aromatic palette, the BNIC undertook the creation of a 'Cognac aroma wheel'. Some 30 Cognac samples corresponding to the different stages in the ageing process were assessed by 44 professional tasters (sommeliers, cellar masters, wines-and-spirits journalists and writers), brought together at the BNIC in the course of the

International Cognac Summit (Nouet, 2010). Over 60 descriptors were used by the tasters in order to meaningfully describe the products involved. The results, processed by principal component analysis (PCA) are presented in Fig. 11.10.

Section I of the chart groups together the notes characteristic of young wine spirits: grape, vine flower, acacia, honeysuckle, litchi, lilac, etc. Such notes largely correspond to aromatic substances coming from the grape, such as terpenes and C_{13} norisoprenoids.

The lower part of the chart (sections II and III) features fruity notes (peach, banana, apricot and passion fruit), as well as notes typical of contact with oak (coconut, vanilla pods, coffee, toasted bread, cloves, etc.)

Finally, in section IV, come the notes typical of Cognac rancio and corresponding to long ageing of wine spirits, including leather, walnuts, hazelnuts, tobacco, cigar boxes, underwood, oak moss, truffles, cedarwood and sandalwood.

Organisation of descriptors used in the comparison chart and defined by components 1 and 2 of the PCA, clearly illustrates the diversity of wine spirits' aromatic profiles which increases during the ageing process as symbolised by arrows on the graph.

With a view to reaching the widest possible audience in acquainting the public with such aromatic diversity, the range of notes was grouped into four families, corresponding to the four seasons: spring, summer, autumn and winter. The choice of this presentation reached a consensus among International Cognac Summit participants in charge of the creation of the aroma wheel and representatives of Cognac companies gathered at the BNIC.

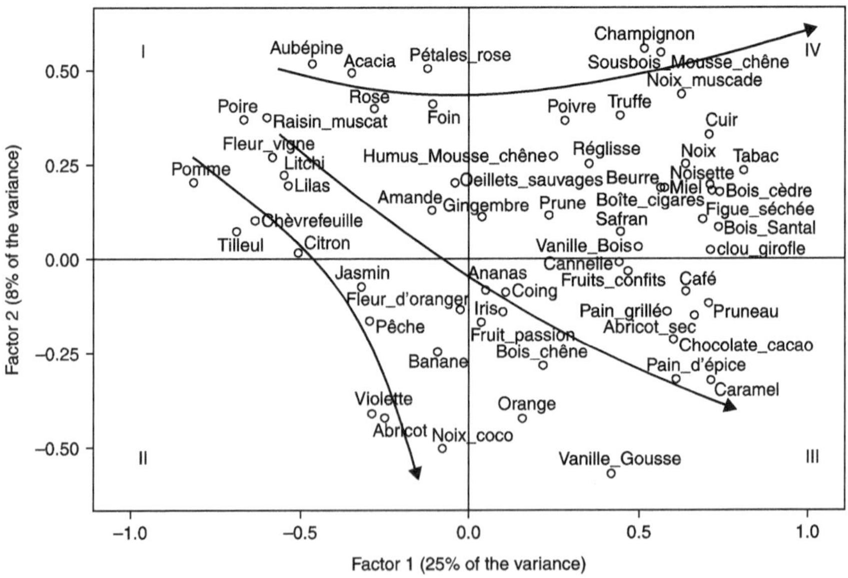

Fig. 11.10 Principal component analysis of the results of tastings carried out during the International Cognac Summit 2009, for the creation of a Cognac aroma wheel (© BNIC).

This classification led to the creation of an original and very complete aroma wheel (Fig. 11.11) that provides a dynamic image of the diversity of Cognac and its development during the ageing process, while leaving the consumer's imagination to run free. The wheel has been widely distributed by the BNIC and is used by Cognac companies and consumer motivators to raise consumer awareness with regard to Cognac's remarkable aromatic palette.

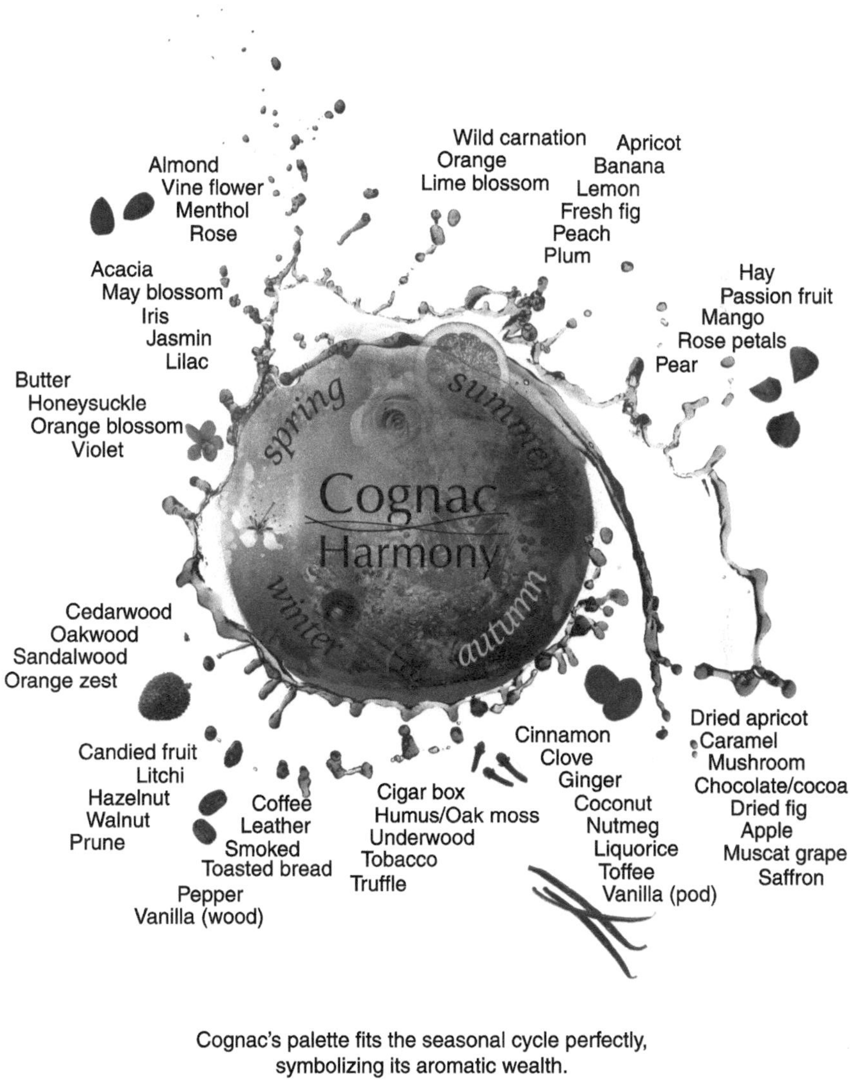

Fig. 11.11 The Cognac Aroma Wheel. Principal aromatic notes. (© BNIC/Gérard Martron).

11.5 References

BOITAUD L, DUMOT V, FERRARI G, LURTON L (2010), 'Constat et incidences du Changement climatique dans la région de Cognac'. Poster. XXXIIIème Congrès Mondial de la Vigne et du Vin, 20–27 Juin 2010, Tbilisi (Géorgie).

CALVO A, CAUMEIL M, PINEAU J (1993), 'Extraction des polyphénols et des aldéhydes aromatiques pendant le vieillissement du Cognac, en fonction du titre alcoolique et du « degré d'épuisement » des fûts', In *Elaboration et connaissance des spiritueux*, Édition BNIC, diffusion Lavoisier Tec et Doc, 562–566.

CANTAGREL R, MAZEROLLES G, VIDAL JP, LABLANQUIE O, BOULESTEIX JM (1991a), 'L'assemblage, une étape importante dans le processus d'élaboration des Cognac', In *1er Symposium International sur les Eaux-de-vie Traditionnelles D'Origine Viticole, Bordeaux 26-30/06/1990*, A. Bertrand, (ed), Lavoisier Tec & Doc, 243–253.

CANTAGREL R, LURTON L, VIDAL J P, GALY B (1991b), 'La distillation charentaise pour l'obtention des eaux-de-vie de Cognac', In *1er Symposium International sur les Eaux-de-vie Traditionnelles D'Origine Viticole, Bordeaux 26–30/06/1990*, A. Bertrand, (ed), Lavoisier Tec & Doc, 60–69.

CANTAGREL R, MAZEROLLES G, VIDAL JP, GALY B, BOULESTEIX JM *et al.* (1993), 'Evolution analytique et organoleptique des eaux-de-vie de Cognac au cours de leur vieillissement, 2ème partie: Incidence de la température et de l'hygrométrie des lieux de stockage', In *Elaboration et Connaissance des Spiritueux*, Édition BNIC, Diffusion Lavoisier Tec et Doc, 573–576.

CANTAGREL R, LURTON L, VIDAL JP, GALY B (1995), 'From vine to Cognac', In *Fermented Beverage Production*, A. G. H., Lea, Piggott, J. R., (Eds); Blackie Academic and Professional, London, 208–228.

CANTAGREL R, GALY B, JOURET C (1998), 'Eaux-de-vie d'origine viticole', In *Oenologie, Fondements Scientifiques et Technologiques*, Flanzy C (coordonateur). Lavoisier, Tec & Doc. Paris, France. 1084–1107.

COQUAND H (1858), 'Description physique, géologique, paléontologique du département de la Charente', T1, Besançon, 1858, T2, Marseille, 1860.

DE RIJKE D, TER HEIDE R (1983), 'Flavour compounds in Rum, Cognac and Whisky', In *Flavour of Distilled Beverages: Origin and Development*, Piggott JR (ed), Ellis Horwood Chichester, 192–202.

DUMOT V, LACOUTURE J, MAZEROLLES G, LURTON L, CHAUMET J *et al.* (1993), 'Les principaux types de sols viticoles charentais: observations sur leurs caractéristiques analytiques et agronomiques', In *Elaboration et Connaissance des Spiritueux*, Édition BNIC, Diffusion Lavoisier Tec et Doc, 37–47.

DUMOT V, BOITAUD L, MENARD E, LAMBERT F, CATTÉ A *et al.* (2010), 'Sélection clonale de l'Ugni blanc pour la production du Cognac', Poster. *XXXIIIème Congrès Mondial de la Vigne et du Vin*. 20–27 Juin 2010 Tbilisi (Géorgie).

EUROPEAN UNION (2008), 'Commission Regulation (EC) 110/2008 of 15 January 2008 on the definition, description, presentation, labelling and the protection of geographical indications of spirits drinks and repealing', Council Regulation (EEC) n° 1576/89.

FERRARI G (2002), 'Influence of must nitrogen composition on wine and spirit quality and relation with aromatic composition and defect', Review. *J. Int. Sci. Vigne.Vin.*, **36**, 1, 1–10.

FERRARI G, LABLANQUIE O, CANTAGREL R, LEDAUPHIN J, PAYOT T *et al.* (2004), 'Determination of key odorant compounds in freshly distilled Cognac using GC-O, GC-MS, and sensory evaluation', *J. Agric. Food Chem*, **52**, 18, 5670–5676.

FERRARI G, PINEDA C, LABLANQUIE O (2008), 'Identification de nouveaux composés odorants dans les eaux-de-vie nouvelles de Cognac', In *2ème Symposium International sur les Eaux-de-vie Traditionnelles D'Origine Viticole, Bordeaux 25–27/06/2007*, A. Bertrand, (ed), Lavoisier Tec & Doc,125–129.

FERRARI G, GALY B, ROULLAND C, LURTON L (2010), 'Management of yeast issues in Cognac region', In *Worldwide Distilled Spirits Conference, Edinburgh.* Nottingham University Press, United Kingdom, 69–74.

FRENCH MINISTRY OF AGRICULTURE FOOD AND FISHERIES (2009) 'Appellation D'origine 'Cognac' ou 'Eau-de-vie de Cognac' ou 'Eau-de-vie des Charentes'', Décret n° 2009-1146 du 21 Septembre 2009.

GALY B, LOIZEAU A, GIRAUD N, LURTON L (2008), 'Mécanismes de formation et conséquences qualitatives de l'Isobutanal présent dans les eaux–de–vie de Cognac', In *2ème Symposium International sur les Eaux–de–vie Traditionnelles D'Origine Viticole, Bordeaux 25–27/06/2007*, A. Bertrand, (ed), Lavoisier Tec & Doc, 111–115.

GAY-BELLILE F (1981), 'La distillation, ses techniques', *Revue française d'œnologie*, **86**, 87–96.

LABLANQUIE O, SNAKKERS G, CANTAGREL R, FERRARI G (2002), 'Characterization of young Cognac spirits aromatic quality', *Analytica Chimica Acta*, **458**, 1, 191–196.

LÉAUTÉ R (1990), 'Distillation in alambic', *Am. J. Enol. Vitic.*, **41**, 1, 90–102.

LÉAUTÉ R, MOSEDALE JR, MOURGUES J, PUECH JL (1998), 'Barrique et vieillissement des eaux-de-vie', In *Oenologie, Fondements Scientifiques et Technologiques*, Flanzy C. (coordonateur). Lavoisier, Tec & Doc. Paris, France, 1110–1142.

LECLAIRE E, CANTAGREL R, MAIGNIAL L, FERRARI G (1999), 'Essai de caractérisation aromatique d'eaux-de-vie nouvelles de Cognac', *J. Int. Sci. Vigne Vin*, **3**, 33, 133–141.

LAFON R, LAFON J, COUILLAUD P (1964), *Le Cognac, sa Distillation*, J.B. Baillère (ed), Paris France.

LEDAUPHIN J, SAINT-CLAIR JF, LABLANQUIE O, GUICHARD H, GUICHARD E *et al.* (2004), 'Identification of Trace Volatile Compounds in Freshly Distilled Calvados and Cognac Using Preparative Separations Coupled with Gas Chromatography-Mass Spectrometry', *J. Agric. Food Chem.*, **52**, 16, 5124–5134.

LEDAUPHIN J, LE MILBEAU C, BARILLIER D, HENNEQUIN D (2010), 'Differences in the Volatile Compositions of French Labeled Brandies (Armagnac, Calvados, Cognac, and Mirabelle) Using GC-MS and PLS-DA', *J. Agric. Food Chem.*, **58**, 13, 7782–7793.

LEYRAT M, LABLANQUIE O, ESTREGUIL S, URRUTY L, CANTAGREL R *et al.* (2005), 'The first aromatic map of freshly distilled Cognac spirit', Poster. *In Vino Analytica Scienzia*, 2005, Montpellier, France.

LURTON L, MAZEROLLES G, GALY B, CANTAGREL R, VIDAL JP (1991), 'Influence de la technologie de vinification sur la qualité des eaux-de-vie de Cognac: exemple des norisoprénoïdes et des alcools supérieurs', In *1er Symposium International sur les Eaux-de-vie Traditionnelles D'Origine Viticole, Bordeaux 26–30/06/1990*, A. Bertrand, (ed), Lavoisier Tec & Doc, 127–136.

LURTON L, SNAKKERS G, ROULLAND C, GALY B, VERSAVAUD A (1995), 'Influence of the fermentation yeast strain on the composition of wine spirits', *Journal of the Science of Food and Agriculture*, **67**, 4, 485–491.

MARCHÉ M, JOSEPH E (1975), 'Etude théorique sur le Cognac, sa composition et son vieillissement naturel en fûts de chêne', *Revue Française d'œnologie*, **57**, 3–107.

MAZEROLLES G, CANTAGREL R, VIDAL JP (1991), 'Application des méthodes d'analyse de données à la caractérisation analytique des Cognacs et de quelques autres spiritueux', In *1er Symposium International sur les Eaux-de-vie Traditionnelles D'Origine Viticole, Bordeaux 26–30/06/1990*, A. Bertrand, (ed), Lavoisier Tec & Doc, 174–182.

MAZEROLLES G, VIDAL JP, CANTAGREL R, GASCHET J, FAURE A *et al.* (1993), 'Application des méthodes d'analyse des données à la caractérisation analytique des Cognacs', In *Elaboration et Connaissance des Spiritueux*, Édition BNIC, Diffusion Lavoisier Tec et Doc, 410–416.

NEDJMA M (1997), 'Influence of Complex Media Composition, Cognac's brandy, or Cognac, on the Gas Chromatography Analysis of Volatile Sulfur Compounds – Preliminary Results of the Matrix Effect', *Am. J. Enol. Vitic.*, **48**, 333–338.

NOUET M (2010), 'Le Cognac fait la roue', *Cognac Magazine, supplément à Whisky Magazine*, n°38, 18–19.

PARK YH (1974), 'Contribution à l'étude des levures de Cognac', In *Etude et classification des levures de Cognac. Conn. Vigne Vin*, **8**, 253–278.

PRIDA A, PUECH JL (2006), 'Influence of geographical origin and botanical species on the content of extractives in American, French, and east European oak woods', *Journal of Agricultural and Food Chemistry*, **54**, 8115–8126.

PUECH JL, LÉAUTE R, CLOT G, NOMDEDEU L, MONDIES H (1984), 'Evolution de divers constituants volatils et phénoliques des eaux-de-vie de Cognac au cours de leur vieillissement', *Sciences des aliments*, **4**, 1, 65–80.

RAVAZ L (1900), *Le Pays du Cognac*, L. Coquemard (ed), Angoulême, France.

REAZIN GH (1981), 'Chemical mechanisms of whiskey maturation', *Am. J. Enol. Vitic.*, **32** 283–289.

RIBES P (1986), 'Identification de la flore lévurienne du moût de deux chais de la région de Cognac. Etude des propriétés biochimiques de quelques souches en vue de leur sélection', Thèse, Université P. Sabatier, Toulouse, France.

ROULLAND C, VERSAVAUD A, GALY B, LURTON L, HALLET JN (1995), 'Analyse de la biodiversité des populations levuriennes fermentaires et mise en œuvre d'une stratégie de sélection de souches pour l'élaboration du Cognac', In *Oenologie 95: 5e. Symposium International d'Œnologie*, Bordeaux, Lonvaud-Funel A (coordonateur).

SNAKKERS G, NEPVEU G, GUILLEY E, CANTAGREL R (2000), 'Variabilités géographiques, sylvicole et individuelle de la teneur en extractibles de chênes sessiles français (Quercus petraea Liebl.): polyphénols, octalactones et phénols volatils', *Ann. For. Sci.*, **57**, 3, 251–260.

SNAKKERS G, BOULESTEIX JM, ESTREGUIL S, GASCHET J, LABLANQUIE O *et al.* (2003), 'Incidence du niveau de chauffe du bois sur une eau-de-vie de Cognac logée en fût neuf: Etude pilote', *Journal International des Sciences de la Vigne et du Vin*, **34**, 4, 243–256.

SNAKKERS G, BOULESTEIX JM, ESTREGUIL S, GASCHET J, LABLANQUIE O *et al.* (2008), 'Incidence de la chauffe du bois de chêne sur la composition d'eaux-de-vie de Cognac logées en fut – Modèle pilote', *In 2ème Symposium International sur les Eaux-de-vie Traditionnelles D'Origine Viticole, Bordeaux 25–27/06/2007*, A. Bertrand, (ed), Lavoisier Tec & Doc, 137–141.

TER HEIDE R (1986), 'The Flavour of Distilled Beverages', In *Developments in Food Science 3B (Food Flavours, Part B, The Flavour of Beverages)*; Morton I D, MacLeod A J, (Eds); Elsevier: Amsterdam, 239–336.

TNO – CIVO (1989), 'Volatile Compounds in food, alcoholic beverages', Qualitative and quantitative data. H. Maarse (ed).

URRUTY L, GAUDIN M, SNAKKERS G, PINEDA C, ESTREGUIL S *et al.* (2008), 'Recherche de marqueurs analytiques de *Botrytis cinerea* dans les eaux-de-vie de Cognac', *In 2ème Symposium International sur les Eaux-de-vie Traditionnelles D'Origine Viticole, Bordeaux 25–27/06/2007*, A. Bertrand, (ed), Lavoisier Tec & Doc, 69–71.

VERSAVAUD A, COURCOUX P, ROULLAND C, DULAU L, HALLET JN (1995), 'Genetic diversity and geographical distribution of wild *Saccharomyces cerevisiae* strains from the wine-producing area of Charentes', France, *Appl. Environ. Microbiol.*, **61**, 3521–3529.

VIDAL JP, CANTAGREL R, MAZEROLLES G, LURTON L, GASCHET J (1991), 'Mise en évidence de l'influence du TDN, triméthyl-1-1-6-dihydro-1,2-naphtalène sur la qualité organoleptique des eaux-de-vie de Cognac', In *1er Symposium International sur les Eaux-de-vie Traditionnelles D'Origine Viticole, Bordeaux 26–30/06/1990*, A. Bertrand, (ed), Lavoisier Tec & Doc, 165–173.

VIDAL JP, MAZEROLLES G, ESTREGUIL S, CANTAGREL R (1993), 'Analyse quantitative de la fraction carbonylée volatile des eaux-de-vie de Cognac', In *Elaboration et connaissance des spiritueux*, édition BNIC, diffusion Lavoisier Tec et Doc, 529–537.

WATTS VA, BUTZKE C, BOULTON R (2003), 'Study of Aged Cognac Using Solid Phase Microextraction and Partial Least-Square Regression', *J. Agric. Food. Chem*, **51**, 7738–7742.

12

Gin: production and sensory properties

M. Riu Aumatell, University of Barcelona, Spain

Abstract: Gin is a colourless distilled beverage widely consumed around the world, produced by a number of different methods, but always with juniper as the predominant taste. The volatile content of gin is made up of terpenoid compounds, mainly from juniper berries, but also from other botanicals such as coriander seeds, angelica root or citrus peel. Although some authors have used various methods to develop a vocabulary to describe the sensory characteristics of gin, no uniform vocabulary has been widely adopted. The most usual attributes defined by the tasters were juniper and coriander but spice, liquorice, aniseed, floral and fruity attributes were also reported.

Key words: gin, sensory analysis, volatile composition, botanicals.

12.1 Introduction

Spirits flavoured with juniper, the most famous of which are the beverages known as gin, are now widely consumed across the world and are very important in the market. Gin was produced for the first time by Franciscus de la Boe, a Dutch physician at the University of Leiden (the Netherlands) in 1650, who carried out research into inexpensive medication for the treatment of kidney problems. He worked with neutral grain spirit flavoured with juniper to give it a pleasant taste (Greer *et al.*, 2008).

There are various types and formulations of gin, but in all cases the predominant flavour is juniper and it is typically colourless. According to the most recent European regulation (European Union, 2008), spirits with a predominant flavour of juniper are categorised principally according to their strength, ingredients and method of production as 'juniper-flavoured spirit drinks', 'gin', 'distilled gin', or 'London gin'. Throughout this chapter, gin and distilled gin refer to these products as defined in Regulation (EC), no. 110/2008 (European Union, 2008). Gin refers to the wider family of products. Gin, distilled gin and London gin are required to have an alcoholic strength of at least 40% in the United States, but only 37.5% in the European Union (EU) (Greer *et al.*, 2008, European Union, 2008). Juniper-flavoured spirit drinks have a minimum alcoholic strength of 30% (European Union, 2008).

Gin can be produced simply by flavouring organoleptically suitable ethyl alcohol with natural or artificial flavouring or by redistilling organoleptically suitable spirits of agricultural origin with juniper berries and other botanicals. Gins made by the former method are often referred to as compound gins; those produced by the latter as distilled gins. According to Regulation (EC), no. 110/2008 (European Union, 2008) distilled gin and London gin are exclusively produced by redistilling organoleptically suitable ethyl alcohol of agricultural origin and appropriate quality to a minimum of 96%, in the stills traditionally used for gin. Distillation is carried out in the presence of juniper berries (*Juniperus communis* L.) (Fig. 12.1). Other botanicals may also be added, provided that the juniper taste is predominant. For London gin, recently recognised as an especially high-quality beverage, the ethyl alcohol used should be of higher quality than that used to produce distilled gin, with maximum methanol values being 5 g/Hl of 100% volume alcohol, and no flavourings can be added after distillation.

Gin, on the other hand, may be produced by flavouring ethyl alcohol with the natural and/or nature-identical flavouring and flavouring preparations permitted by the regulation. In the EU, juniper-flavoured spirit drinks, which may also bear the sales denominations of Wacholder or genebra (but which may not be called gin), are produced by flavouring ethyl alcohol of an agricultural origin and/or grain spirit and/or grain distillate with juniper (*Juniperus communis* L. and/or *Juniperus oxicedrus* L.) berries (European Union, 2008).

Finally, some juniper-flavoured spirit drinks and gins bear geographical indications which, according to recent regulation, identify not only the origin of a spirit but also its quality or some characteristic of the beverage. These include:

- Genièvre/Jenever/Genever (Belgium, the Netherlands, France and Germany),
- Jonge jenever, Oude jenever (Belgium, the Netherlands),
- Genièvre Flandres Artois (France),

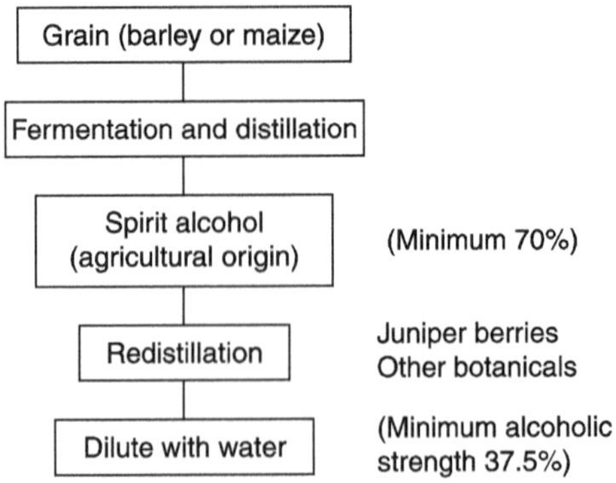

Fig. 12.1 Production of London Dry Gin.

- Ostfriesischer Korngenever and Steinhäger (Germany),
- Plymouth Gin (United Kingdom),
- Gin de Mahón (Spain),
- Vilniaus Džinas/Vilnius Gin (Lithuania),
- Spišká borovička (Slovakia).

12.2 Gin production

The production process for compound gin simply involves the addition of flavourings to ethyl alcohol. However, for distilled gins and London gin, the spirit is instead redistilled in the presence of juniper and other natural flavourings, known as botanicals, which are discussed in detail in Section 12.2.2.

12.2.1 Distillation process for distilled gins and London gin

The distillation process for distilled gins takes place in traditional copper stills that typically hold up to 20 000 l. The details of the distillation process vary between producers, but a typical method used is as follows. The spirit is first diluted with pure water in order to obtain the required alcoholic strength (around 45% abv), before being added to the stills along with the selected botanicals and left to steep. Heating (using a steam coil or jacket) to a temperature of 70–80 °C then causes the release of essential oils from the botanicals, which give the spirit its flavour. The early part of the distillate and the end of the distillate, which are of lower quality, are redistilled. The middle part of the run is used to produce high-quality gin (Lea and Piggott, 2003).

In distilled gin production, the flavourings permissible according to Regulation (EC), no. 110/2008 (European Union, 2008) may be added after this distillation process, while for London gin, only water and further alcohol may be added once distillation is complete. Pure demineralised water is then added to the spirit to achieve the correct level of alcoholic strength, which must be at least 37.5% but can also be higher.

One common method of improving distillation is the use of a vacuum, which allows lower temperatures of around 60 °C to be used (Greer *et al.*, 2008). Some authors have studied the use of a vacuum in the distillation of essential oils such as *Curcuma longa* L. leaf (Babu *et al.*, 2007), *Citrus sinensis* (Lopes *et al.*, 2003), and bergamot oil (Fang *et al.*, 2004). In all cases, using a vacuum increases the level of oxygenated monoterpenes and decreases the levels of heat-stimulated monoterpenes. The low quantity of monoterpenes indicates a high level of stability in the essential oil. In the study conducted by Greer *et al.* (2008) in which the traditional gin distillation method was compared with a novel vacuum distillation method, it could be observed that the level of monoterpenes is lower in gin produced with high vacuum distillation. Monoterpenes can cause the formation of off-flavours in the final product, rendering the use of a vacuum desirable. A trained panel described the gin obtained using this novel method as spicy and more floral, while also presenting less nasal pungency. Furthermore, according to some authors, oxygenated compounds are considered responsible for the aroma while terpenes act as a flavour carrier.

12.2.2 Botanicals

Although the primary flavourings used in gin are juniper (*Juniperus communis* L.) and coriander (*Coriandrum sativum* L.), a large number of other botanicals can be used in order to obtain a beverage with a particular character. Some authors have described more than 100 different botanicals that can be added in order to provide a particular character to the product. Some of these botanicals are discussed below.

Juniper berries (Juniperus communis L.)

Juniperus communis L belongs to the Cupressaceae family and is an evergreen shrub commonly found in Europe and North America. The berries are often used in flavours, perfumes and pharmaceutical compounds (for example, for their diuretic, hypoglycaemic, anti-inflammatory, antibacterial and antifungal properties) and in the aromatisation of alcoholic beverages.

According to EU regulation, juniper must be the predominant taste in all types of gin, and has been identified as the sensory characteristic that distinguishes gin from other alcoholic beverages (Mc Donnell *et al.*, 2001). The composition of juniper essential oil has been widely studied and has been shown to be affected by the age and geographical origin of the plant, as well as the ripeness of the berries (Vichi *et al.* 2007). Juniper berries have been shown to contain between 0.2–3.42% of essential oil (Barjaktarović *et al.*, 2005). The oil consists mainly of monoterpenes (58–85%), sesquiterpenes (2–10.2%), aldehydes, alcohols and other oxygenated compounds.

In the study performed by Vichi *et al.* the use of head space-solid phase microextraction (HS-SPME) for the quality control of juniper berries was tested by comparison of simultaneous distillation-extraction (SDE). The results obtained led to the identification of 77 compounds, representing 80% of the total gin flavour (Vichi *et al.*, 2005). The main group identified was oxygenated monoterpenes (31.2%) followed by monoterpenes (29.9%), sesquiterpenes (22%) and oxygenated sesquiterpenes (5.2%). Monoterpenoid compounds have a high impact on juniper aroma due to their greater volatility.

Barrero *et al.* (2005) also studied the diterpenic composition of juniper berries. Diterpenic compounds are widely studied as a result of their biological activity: they can function as antimicrobial, antiulcer, antioxidant and antitumor agents, and can have an effect on cardiovascular activity (Barrero *et al.*, 2005). Abietane, totarane, sempervirane and labdane have previously been identified in juniper berries. Vichi *et al.* (2008) further evaluated the diterpenoid composition of beverages aromatised with juniper. These compounds can be used to characterise different types of gin and can act as a guarantee of the quality of a product. Ten diterpenoids were identified; three labdane structures (manool, manoyl oxide and epi-manoyl oxide), five abietane derivatives (abieta-8,13(15)-dien-18-ol, dehydroabietal, abieta-8,11,13-trien-7-one, *trans*-ferruginol and 4-epi-dehydroabietol) and two totarane derivatives (*trans*-totarol and *cis*-totarol). The diterpenes identified mainly originate from juniper extract while three of them, manoyl oxide, epi-manoyl oxide and *cis*-totarol, may originate from other botanicals used in the elaboration of gin. Diterpenoid compounds in gin samples were determined using the simple and reliable direct immersion-solid-phase microextraction (DI/SPME) method.

Coriander seeds (Coriander sativum L.)

Coriander sativum L. is the second ingredient added to distilled gins. It belongs to the Umbelliferae (Apiaceae) family, typically found in the Mediterranean region, and the seeds are widely used in perfumery, tobacco products and liqueurs (Kiralan *et al.*, 2009). The weight of coriander essential oil accounts for between 0.3–1.2% of the total weight of the seeds: 60–70% of this weight is linalool. This essential oil is used in the pharmaceutical industry for its antibacterial, bacteriostatic, fungicidal and insecticidal properties (Lo Cantore *et al.*, 2004; Kiralan *et al.*, 2009).

Coriander seeds can be added to distilled gin to provide complexity; gin produced using coriander was described by the panellists as having citrus, floral and fruity attributes (Phelan *et al.*, 2004).

Other botanicals

Other products that can be used to give character and complexity to gins include:

- Orange peel (*Citrus sinensis*),
- Purging cassia (*Cassia fistula*),
- Orris root (*Iris florentina* L.),
- Cardamom seeds (*Elettaria cardamomum* L.),
- Angelica root (*Angelica archangelica* L.),
- Cinnamon bark (*Cinnamomum zeylandicum*),
- Calamus (*Acorus calamus* L.),
- Fennel (*Foeniculum vulgare*),
- Aniseed (*Pimpinella anisum*),
- Lemon peel (*Citrus limon* L.),
- Cumin (*Cuminum cynimum* L.),
- Almond (*Prunus amygdalus* L.),
- Liquorice root (*Glycyrrhiza glaba*).

All of these species contain essential oils that add to the flavour of gins. The carefully balanced mixture of botanicals selected depends on the producers and the characteristics that they are trying to obtain.

12.3 Flavour chemistry

Although gin is widely consumed, few studies have been carried out on its composition and characteristics (Clutton and Evans, 1978; Villalón-Mir *et al.*,

1984a, 1984b; Martin-Alvarez and Herranz, 1991; Vichi *et al.*, 2005; MacNamara *et al.*, 2007; Greer *et al.*, 2008). The volatile composition of gins, like all alcoholic beverages, depends on the botanicals used in its production. All of these botanicals contain essential oils, which are themselves complex mixtures made up of hundreds of individual compounds. The volatile composition of gin samples principally includes both oxygenated and non-oxygenated terpenoid compounds, which play an important role in determining the aroma of the final product. Their high volatility and low perception threshold cause a considerable aromatic impact, but there is limited data available on the sensory notes of these compounds. Usually, the volatile composition of gins is analysed by gas chromatography coupled with mass spectrometry. The extraction of the volatile fraction is performed by liquid–liquid extraction (Martin-Alvarez and Herranz, 1991) or solid phase microextraction (SPME) (Vichi *et al.*, 2005).

According to the literature, the volatile compounds of gin samples can be divided into monoterpenes, oxygenated monoterpenes, sesquiterpenes and oxygenated sesquiterpenes. In most of the studies, the main group that was found were non-oxygenated monoterpenes (Table 12.1). In the study performed by Vichi *et al.*, (2005) six distilled gins (labelled G1–G6), which were amongst the most popular commercial beverages, were compared. Four of them were London dry gins while two were gins with geographical indications (Mahón gin and Plymouth gin).

The main non-oxygenated monoterpenes detected in the gin samples were α-pinene, β-myrcene, limonene, γ-terpinene and *p*-cymene. These compounds have previously been identified as the main compounds found in juniper extracts (Angioni *et al.*, 2003). Additionally, limonene and γ-terpinene are very important compounds in the essential oils of citrus fruits while *p*-cymene is found in cumin essential oil.

The oxygenated monoterpenes were an aromatic group composed of alcohols, esters, ketones and aldehydes. The main compounds identified were linalool and geranyl acetate. Linalool was the most important compound found in coriander essential oil (accounting for more than 60% of its composition). This volatile compound can be used as a marker in other botanicals, especially coriander (Lo Cantore *et al.*, 2004; Greer *et al.*, 2008), but also in cinnamon (Kaul *et al.*, 2003), angelica root (Doneanu and Anitescu, 1998) and citrus fruits (Angioni *et al.*, 2003). Other quantitatively important compounds found in gin and described in juniper berries were 1-4-terpineol and bornyl acetate. Geranyl acetate has previously been detected in gin (Clutton and Evans, 1978) but has never been detected in the more typical botanicals used in gin production.

Sesquiterpenes are less important compounds in terms of sensory perception because they have higher perception thresholds. Sesquiterpenes are usually defined as balsamic, grassy or woody. These compounds may have an important role in the chemical characterisation of a product. The sesquiterpenic profile of a gin must be highly representative of the botanicals that it contains. The main sesquiterpenic compounds detected in the gins were δ and γ-cadinene, caryophyllene, β and γ-elemene, α-humulene and germacrene D. All of these components were typical of juniper berries. The oxygenated sesquiterpenes were

Table 12.1 Minimum and maximum values of the volatile compounds identified in gins

Volatile compounds	Min		Max	Volatile compounds	Min		Max
α-Pinene	1.95	±	6.12	Cuminal	0.07	±	0.96
α-Thujene	0.20	±	0.70	β-Citronellol	0.05	±	1.97
Camphene	0.06	±	0.16	Myrtenol	0.02	±	0.37
β-Pinene	0.37	±	1.53	Nerol	0.01	±	0.10
Sabinene	0.09	±	2.53	t-Carveol	0.05	±	0.54
Verbenene	0.05	±	0.32	t-Geraniol	0.19	±	0.63
δ-3-Carene	0.03	±	0.11	α-Cubebene	0.07	±	0.24
l-Phellandrene	0.06	±	0.19	α-Copaene	0.04	±	0.10
β-Myrcene	2.38	±	11.09	β-Cubebene	0.04	±	0.14
α-Terpinene	0.31	±	0.65	t-β-Caryophyllene and β-Elemene	0.41	±	0.93
dl-Limonene	1.22	±	17.21				
β-Phellandrene	0.20	±	0.64	γ-Elemene	0.22	±	0.92
γ-Terpinene	1.16	±	2.87	α-Humulene	0.25	±	0.90
t-Ocimene	0.04	±	0.11	t-β-Farnesene	0.02	±	0.09
p-Cymene	0.53	±	1.74	γ-Muurolene	0.07	±	0.19
α-Terpinolene	0.25	±	0.59	Germacrene D	0.13	±	0.62
(Z)-Rose oxide	nq	±	nq	α-Selinene	0.06	±	0.15
Verbenyl ethyl ether	2.55	±	24.43	α-Muurolene	0.08	±	0.15
Citronellal	0.06	±	0.41	δ- and γ-cadinene	0.32	±	1.15
Campholenal	0.51	±	4.55	Cadina-1,4-diene	0.02	±	0.11
Camphor	0.83	±	1.54	Germacrene B	0.08	±	0.18
Linalool	1.93	±	36.99	α-Calacorene	0.02	±	0.19
cis-Sabinene hidrate	nq	±	nq	Caryophyllene oxide	0.09	±	5.13
Bornyl acetate	0.18	±	2.01	Torreyol	0.08	±	1.52
1-4-Terpineol	0.22	±	0.79	Elemol	0.01	±	0.32
Myrtenal	nq	±	nq	Spathulenol	0.05	±	1.29
α-Terpineol	1.13	±	9.03	t-Cadinol	0.03	±	0.97
Terpenyl acetate	0.08	±	0.71	t-Muurulol	0.03	±	1.27
Neryl acetate	0.05	±	0.31	Eudesmol	0.01	±	0.18
Geranyl acetate	0.25	±	2.09	α-Cadinol	0.06	±	1.45

Notes: volatile compounds are listed following wax column elution order in wax column. nq not quantified.

Source: Vichi *et al.*, 2005.

mainly alcohol derivatives of the sesquiterpenes. All of the compounds detected have previously been found in juniper berries, but this is the first time that these compounds have been detected in gins (Vichi *et al.*, 2005). The main compounds that were quantified are caryophyllene oxide, torreyol, elemol, spathulenol, t-cadinol, t-muurolol, eudesmol and α-cadinol.

The volatile composition may account for the differences observed between gins in terms of their sensory profile (Fig. 12.2). According to Vichi *et al.* (2005), when an analysis of the principal components was performed, gin G6 (which had

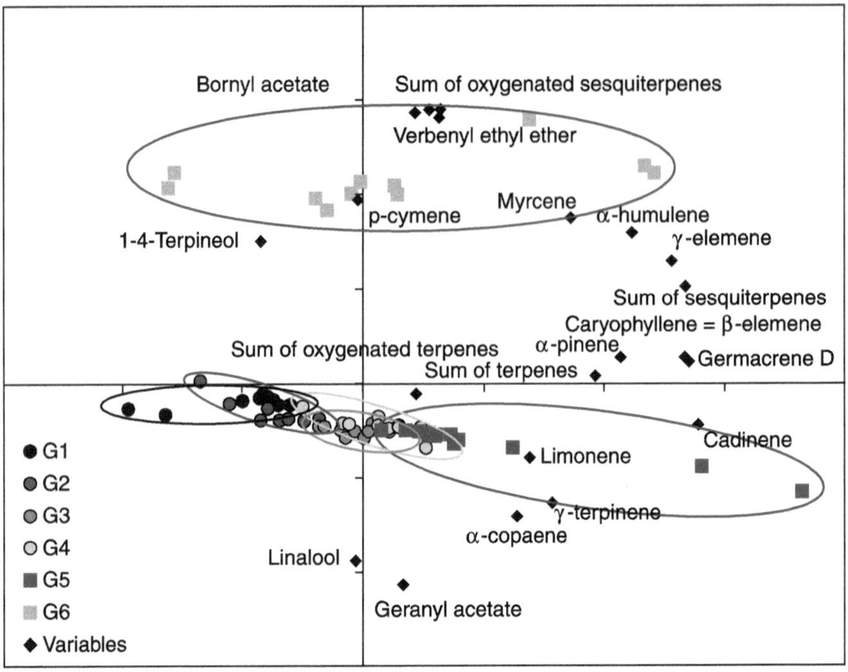

Fig. 12.2 Principal components analysis of volatile compounds of gin.

a geographical indication) was shown to contain higher levels of terpenic compounds typically found in juniper berries and lower levels of volatiles typically found in other botanicals such as coriander or citrus fruits. The other gin with a geographical indication (G5 in Fig. 12.2) showed higher levels of limonene and γ-terpinene and linalool, probably due to the use of citrus species during its production. Moreover, the levels of compounds typically found in juniper berries were significantly lower. The other gins (G1–G4 in Fig. 12.2) had more similar volatile compositions.

12.4 Sensory properties of gin and sensory analysis practice in the industry

The differences between types of gin both in terms of composition and elaboration method make it very difficult to obtain a standardised sensory evaluation. Furthermore, the gin production process requires rigorous sensory controls in order to obtain gin with optimal sensory characteristics. Sensory analysis is a quality control method using human senses as measuring instruments. The methods used in sensory analysis can be divided into difference tests (which include overall difference tests and attribute difference tests), preference tests such as a consumer test, and descriptive tests. Other authors such as Muñoz *et al.* (1992) categorised the most usual sensory methods used in the industry as follows:

- Overall difference test,
- Difference from control,
- Descriptive test,
- In/out of specifications,
- Preference test,
- Typical measurements,
- Qualitative description of typical production,
- Quality grading.

Discrimination methods (difference tests) are used to evaluate differences or similarities between samples, demonstrating whether two samples are different or sufficiently similar (Meilgaard *et al.*, 1990). Discrimination techniques include the triangle test, the duo-trio, the two-out-of-five tests, and the difference for control test, for example, and require between 20–100 semi-trained panellists, depending on the test.

In addition, the preference or acceptance test aims to assess the personal response of current or potential customers to a product or product characteristic (Meilgaard *et al.*, 1990).

The consumer test constitutes the main sensory analysis carried out by producers and has been proven to be highly effective in the development of new products. The consumer test requires a large number of people (a panel consisting of anywhere between 100–500 people is usual). The affective methods can be qualitative or quantitative.

Finally, the most sophisticated tools of sensory analysis are descriptive methods. Descriptive analysis involves the detection and description of both qualitative and quantitative sensory components by a trained panel of judges. Descriptive analyses are widely used in the industry for quality control, in order to understand consumer responses to the sensory attributes of products, to study packaging effects, the effect of ingredients or processing variables or to investigate consumer perception regarding a product (Murray *et al.*, 2001). The training of the panel is described in detail in order to obtain a high quality panel of judges (ISO, 1994). There are several descriptive analysis methods, including the Flavour Profile Method (FPM), Quantitative Descriptive Analysis™, the Spectrum™ Method, the Texture Profile Method, Quantitative Flavour Profiling, Free-Choice Profiling and Generic Descriptive Analysis (GDA). Each method has different requirements and determined specificities. However, GDA can combine multiple different approaches found in each method, and is commonly used in the food industry.

The different processes and different combinations of botanicals used in gin production lead to differences in flavour profile between the various types of gin. There is therefore no standard method for its sensory evaluation. The first step in describing a type of gin is to draw up a list of terms (Mc Donnell, *et al.*, 2001). Three studies on the creation of gin-related vocabulary can be found in the literature: Piggott and Holm (1983), Phelan *et al.*, (2004) and Riu-Aumatell *et al.* (2008). In the study by Piggott and Holm (1983), 21 descriptive terms were developed for gin flavour. The character of gin is a result of both the alcohol content and the juniper essential oils, while each of the botanicals makes a sensory contribution depending on its source, concentration and relationship with other botanicals. Some of the particular notes detected were citrus, spicy and earthy characteristics from angelica (Piggott and Holm, 1983). The vocabulary used in this study requires some further refinement, as well as chemical reference standards to define the meaning of each descriptive term. In the study conducted by Phelan *et al.* (2004), the vocabulary was generated by a panel of experts from the Scotch Whisky Research Institute. The tasters produced a large list of terms which was then reduced and refined for a final list of 16 attributes: pungent, spicy, herbal, citrus, floral, aniseed, juniper, fruity, sweet, sour, buttery, stale, oily, solventy, soapy and sulphury. The panel then underwent training with botanical distillates, in order to recognise all of the attributes that had been analysed.

In the study performed by Riu-Aumatell *et al.* (2008) the development of a sensory vocabulary for gin was performed according to ISO 11035. First, a triangle test was performed with two objectives. The aims were firstly to detect sensory differences between samples and secondly to familiarise the panel with gins and gin vocabulary. The results obtained showed significant differences between gins. ISO 11035 is focused on the selection and generation of vocabulary using a multidimensional approach. This International Standard allows verification of the relevance and independence of each descriptor, and is based on the use of at least six tasters, on the use of products (botanicals for example) to train the panel, a very important role for the panel leader, and on the use of multivariate statistical methods. The panel was extensively trained with hydroalcoholic solutions with standards previously found in gins (myrcene, limonene, linalool, γ-terpinene and α-pinene), natural sensory references such as juniper, coriander, aniseed and lemon peel and gin enriched with ethanolic extracts of juniper, aniseed, coriander and angelica root. The vocabulary was generated by tasting three London Dry gins with the use of a perfumery wheel, natural sensory references and a discussion with the panel leader to eliminate synonyms and hedonistic and irrelevant terms. Finally, a list of ten attributes was drawn up: angelica root, aniseed/fennel, cardamom, citrus peel, coriander, floral, fruity, juniper, liquorice and spice. According to ISO 11035, the first list obtained was too large and had to be reduced. The first reduction was performed using the geometric mean (M). M is the square root of the product of the frequency quotation (F) and the relative intensity (I) of each descriptor.

$$(M = (F \times I)^{1/2}) \qquad [12.1]$$

An analysis of the principal components was then carried out and the discriminative capacity of each descriptor was evaluated using the one-way analysis of variance (ANOVA).

The initial list of ten descriptors was reduced to five descriptors (juniper, citric, liquorice, aniseed and spice) using multivariate analysis, after which the descriptive analysis was performed. In the study conducted by Phelan *et al.* (2004) and Mc Donnell (2001), quantitative descriptive analysis (QDA) was used, while the study performed by Riu-Aumatell (2008) employed generic descriptive analysis (GDA). When a sensory analysis of distilled beverages is performed, the drink must be diluted in order to avoid taster fatigue resulting from alcohol content, and to standardise the alcoholic strength. The samples must be presented in randomised order coded by three digit numbers.

The results obtained showed that the gins could be clearly differentiated according to their sensory flavour profile. Moreover, the flavour profile is in accordance with the previously determined volatile composition (Vichi *et al.*, 2005). The aroma profiles of gin obtained in the study comparing six different distilled gins are shown in Fig. 12.3. G5 and G6, which are gins with geographical indication, are clearly different with respect to London Dry gins. G6 was judged to have the highest intensity of juniper note. At the same time, according to the study performed by Vichi *et al.* (2005) and Riu-Aumatell *et al.* (2008) it showed significantly higher levels of juniper monoterpenes such as sabinene, myrcene and *p*-cymene. The panel judged G5 to have the greatest intensity of citrus attributes and the lowest juniper note. The levels of juniper compounds in G5 were similar to those in G6, but the concentration of limonene and γ-terpinene was clearly higher. It is likely that the higher citrus attribute masks the juniper note in this gin. Similar levels of liquorice were judged to be present in all the

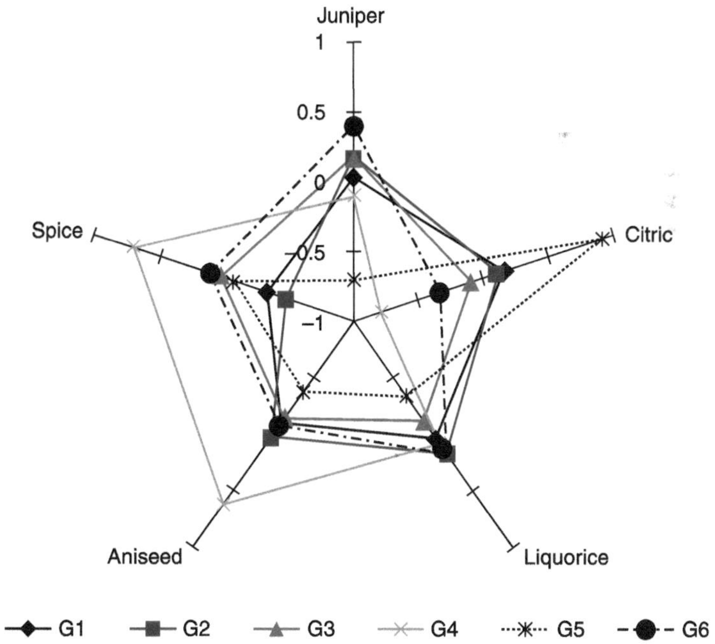

Fig. 12.3 Aroma profiles of gin samples.

samples. Spice and aniseed attributes were reported to be significantly higher in G4 gin. According to Riu-Aumatell *et al.* (2008) the presence of spice attributes could be explained by the levels of sabinene, δ-3-carene and *p*-cymene, compounds which were described as peppery and resinous. G2–G4 gins showed moderate intensities of all the tested notes with no dominant attribute. The results obtained are in accordance with the study conducted by Mc Donnell *et al.* (2001). These authors developed a sensory vocabulary to describe colourless distilled beverages, including gin, in 12 panel hours and to monitor the performance of the panel and the performance of each assessor throughout this procedure. Again, the panel associated the sensory notes of gin with citrus and juniper attributes.

In conclusion, the sensory description of a distilled gin beverage can be assessed using sensory descriptive analysis. Juniper, aniseed and citrus are the attributes defined in all the published studies, while other notes should also be taken into account. Liquorice, spice, pungent, herbal, floral, fruity, buttery, solventy and soapy attributes could all be used with a discriminative function to define the sensory profile of gin. The results obtained from the sensory description are in accordance with the results obtained from the volatile profile that was found in the literature. The volatile composition of gin principally consisted of oxygenated and non-oxygenated terpenoids, found in the composition of juniper essential oil. The other compounds identified originate from other botanicals such as citrus peel, aniseed or angelica root.

12.5 Future trends

The beverage industry, including the gin production sector, is dynamic and has a bright future. However, counterfeiting could prove to be a global problem in the alcoholic drink industry, particularly in the case of gin. The most effective way to obtain a product with the highest quality is to have a detailed understanding of the chemical and sensory composition of a product. One development required to increase the quality of an alcoholic beverage such as gin is the use of new methodologies for the distillation process. The use of vacuum distillation is likely to prove important in obtaining a volatile profile with optimal characteristics.

Even with the advanced sophistication of modern distilleries, product tasting remains necessary, as does the creation of a unified vocabulary list to evaluate the odour of distilled gins. To achieve this, suitable reference materials for the training of a panel are required. The training of a panel in the tasting of spirit drinks is usually a long process that employs authentic samples of spirits with distinctive flavour characteristics or the addition of pure flavour compounds to clean spirits. The use of reference materials that are stable over time, produced during food grade production, and guaranteed safe for tasters is extremely useful in this respect. For example, microencapsulated flavour standards are useful for the training of tasters, the validation of taster performance, and the development of consumer interest.

12.6 Sources of further information and advice

This chapter was written based on the author's own experience and with the aid of some references. One of the most important sources of information was The Gin and Vodka Association, a non-profit and non-trading association dedicated to promoting and protecting UK gin and vodka around the world. The description of sensory analysis followed ISO 11035, (International Organisation for Standardization http://www.iso.org/iso/catalogue_detail.htm?csnumber=19015). Finally, the book *Distilled Spirits: Tradition and Innovation* edited by J.H. Bryce and G.G. Stewart, Nottingham University Press, United Kingdom, 2004 is an interesting and relevant book that provided more information on distilled spirits, including not only gin but also other beverages such as vodka, whisky and rum.

12.7 References

ANGIONI, A., BARRA, A., RUSSO, M.T., CORONEO, V., DESSÍ, S. *et al.* (2003), 'Chemical composition of the essential oils of *Juniperus* from ripe and unripe berries and leaves and their antimicrobial activity'. *J. Agric. Food Chem.*, **51**, 3073–3080.

BARJAKTAROVIĆ, B., SOVILJ, M., KNEZ, Z. (2005), 'Chemical composition of *Juniperus communis* L. fruits supercritical CO_2 extracts: dependence on pressure and extraction time'. *J. Agric. Food Chem.*, **53**, 2630–2636.

BARRERO, A.F., QUÍLEZ DEL MORAL, J.F., HERRADOR, M.M., ARTEAGA, J.F., AKSSIRA, M. *et al.* (2005), 'Abietane diterpenes from the cones of *Cedrus atlántica*'. *Phytochem.*, **66**, 105–111.

CLUTTON, D.W., EVANS, M.B. (1978), 'The flavour constituents of gin'. *J. Chromatogr.*, **167**, 409–419.

DONEANU, C., ANITESCU, G. (1998), 'Supercritical carbon dioxide extraction of *Angelica archangelica* L. root oil'. *J. Supercrit. Fluids*, **12**, 59–67.

EUROPEAN UNION (2008), 'Regulation (EC) No 110/2008 of the European parliament and of the council of 15 January 2008 on the definition, description, presentation, labelling and the protection of geographical indications of spirit drinks'.

FANG, T., GOTO, M., SASAKI, M., HIROSE, T. (2004), 'Combination of supercritical CO_2 and vacuum distillation for the fractionation of bergamot oil'. *J. Agric. Food Chem.*, **52**, 5162–5167.

GREER, D., PFAHL, L., RIECK, J., DANIELS, T., GARZA, O. (2008), 'Comparison of a novel distillation method versus a traditional distillation method in a model gin system using liquid/liquid extraction'. *J. Agric. Food Chem.*, **56**, 9030–9036.

ISO (1994). *11035. Sensory analysis-identification and selection of descriptors for establishing a sensory profile by a multidimensional approach*. International Organization for Standardization, Geneva.

KAUL, P.N., BHATTACHARYA, A.K., RAJESWARA RAO, B.R., SYAMASUNDAR, K.V., RAMESH, S. (2003), 'Volatile constituents of essential oils isolated from different parts of cinnamon (*Cinnamomum zeylanicum* Blume)'. *J. Sci. Food Agric.*, **83**, 53–55.

KIRALAN, M., CALIKOGLU, E., IPEK, A., BAYRAK, A., GURBUZ, B. (2009), 'Fatty acid and volatile oil composition of different coriander (*Coriandrum sativum*) registered varieties cultivated in Turkey'. *Chem. Natural Comp.*, **45**, 100–102.

KIRAN BABU, G.D., , SHANMUGAM, V., RAVINDRANATH, S.D., JOSHIS, V.P. (2007), 'Comparison of chemical and antifungal activity of *Curcuma longa* L. leaf oils produced by different water distillation techniques'. *Flavors Fragrance J.*, **22**, 191–196.

LEA, A.G.H., PIGGOTT, J.R. (2003), *Fermented Beverage Production*. 2nd ed., New York, Kluwer.

LO CANTORE, P., IACOBELLIS, N.S., DE MARCO, A., CAPASSO, F., SENATORE, F. (2004), 'Antibacterial activity of *Coriandrum sativum* L. and *Foeniculum vulgare* Miller var. vulgare (Miller) essential oils'. *J. Agric. Food Chem.*, **52**, 7862–7866.

LOPES, D., RAGA, A.C., STUART, G.R., DE OLIVIER, J.V. (2003), 'Influence of vacuum distillation parameters on the chemical composition of five fold sweet orange oil (*Citrus sinensis* Osbeck)'. *J. Essent. Oil Res.*, **15**, 408–411.

MACNAMARA, K., HOWELL, J., HUANG, Y., ROBBAT Jr. A. (2007), 'Analysis of gin essential oil mixtures by multidimensional and one-dimensional gas chromatography/mass spectrometry with spectral deconvolution'. *J. Chromatogr.*, **1164**, 281–290.

MARTÍN-ALVAREZ, P.J., HERRANZ, A. (1991), 'Application of multivariate statistical methods to the differentiation of gin brands'. *J. Sci. Food Agric.*, **57**, 263–272.

MC DONNELL, E., HULIN-BERTAUD, S., SHEEHAN, E.M., DELAHUNTY, C.M. (2001), 'Development and learning process of a sensory vocabulary for the odor evaluation of selected distilled beverages using descriptive analysis'. *J. Sens. Stud.*, **16**, 425–445.

MEILGAARD, M., CIVILLE, G.V., CARR, B.T. (1990), *Sensory Evaluation Techniques*, 3rd ed. Boca Raton, New York, CRC Press.

MUÑOZ, A.M., CIVILLE, G.V., CARR, B.T. (1992), *Sensory Evaluation in Quality Control*, New York, Van Nostrand Reinhold.

MURRAY, J.M., DELAHUNTY, C.M., BAXTER, I.A. (2001), 'Descriptive sensory analysis: past, present and future'. *Food Res. Int.*, **34**, 461–471.

PHELAN, A.D., JACK, F.R., CONNER, J.M., REID K.J.G., PRIEST, F.G. (2004), 'Sensory assessment of gin flavour'. In *Distilled Spirits: Tradition and Innovation*. J.H. Bryce and G.G. Stewart (eds), Nottingham University Press.

PIGGOTT, J.R. HOLM, A.M. (1983), 'Descriptive sensory analysis of gin flavour'. In Piggott, J.R. (ed), *Flavour of Distilled Beverages*, Chichester, Ellis Horwood, pp 145–153.

RIU-AUMATELL, M., VICHI, S., MORA-PONS, M., LÓPEZ-TAMAMES, E., BUXADERAS, S. (2008), 'Sensory characterization of dry gins with different volatile profiles'. *J. Food Sci.*, **73**, 286–293.

VICHI, S., RIU-AUMATELL, M., MORA-PONS, M., BUXADERAS, S., LÓPEZ-TAMAMES, E. (2005), 'Characterization of volatiles in different dry gins'. *J. Agric. Food Chem.*, **53**, 10154–10160.

VICHI, S., RIU-AUMATELL, M., MORA-PONS, M., GUADAYOL, J.M., BUXADERAS, S. *et al.* (2007), 'HS-SPME coupled to GC/MS for quality control of *Juniperus communis* L. berries used for gin aromatization'. *Food Chem.*, **105**, 1748–1754.

VICHI, S., RIU-AUMATELL, M., BUXADERAS, S., LÓPEZ-TAMAMES, E. (2008), 'Assessment of some diterpenoids in commercial distilled gin'. *Anal. Chim. Acta*, **628**, 222–229.

VILLALÓN-MIR, M., LÓPEZ DE LA SERRANA, H., LÓPEZ MARTÍNEZ, M.C., GARCÍA VILLANOVA, R. (1984a), 'Estudio comparativo por cromatografia en fase gaseosa de los alcoholes, aldehídos, ésteres y componentes esenciales que integran una ginebra tipo con los de otras ginebras comerciales'. *Anal. Bromatol.*, **36**, 61–69.

VILLALÓN-MIR, M., LÓPEZ DE LA SERRANA, H., LÓPEZ MARTÍNEZ, M.C., GARCÍA VILLANOVA, R. (1984b), 'Determinación de terpineol-4 y alfa-pineno en ginebras comerciales por espectrofotometría y cromatografía en fase gaseosa'. *Anal. Bromatol.*, **36**, 133–136.

13

Grape-based brandies: production, sensory properties and sensory evaluation

L. Louw and M. G. Lambrechts, Distell Ltd, South Africa

Abstract: This chapter focuses on the sensory properties and volatile composition of grape-based brandies and the most important considerations for sensory evaluation of such products. The first part of the chapter provides a definition for brandy and a description of the production process and some important brandy styles. The main part of the chapter discusses typical brandy aromas, their origin in the production process and the associated flavour compounds. The final part of the chapter provides some practical considerations for the sensory evaluation and consumer testing of brandy and is concluded with some possible future trends in the sensory research of brandy.

Key words: brandy, aroma, volatile composition, sensory evaluation.

13.1 Introduction

The word 'brandy' is derived from the Dutch word 'brandewijn' which can be literally translated as 'burnt wine'. This refers to brandy being a distilled spirit originating from fermented grapes. For a grape-based spirit to be considered a brandy, it must have been aged for at least six months in oak casks smaller than 1000 litres in capacity or twelve months with oak receptacles. However, the term brandy is sometimes incorrectly used to refer to other fruit-based spirits, e.g. plum brandy. Another variation on brandy is the so-called pomace brandies that are produced from grape skins, a by-product of wine production, e.g. grappa. These are discussed in other chapters of this book. However, throughout this chapter the term brandy will always refer to its proper meaning, i.e. a grape-based spirit (Robinson, 1999).

Brandies are produced in many countries across the globe. Famous brandies include Brandy de Jerez from Spain and South African brandies. South African brandies are noted for their outstanding quality and have won the title of Worldwide

Best Brandy at the prestigious International Wine and Spirits Competition nine times over the past twelve years. South American Pisco is also by definition a brandy, but will be discussed in Chapter 16. Several European nations produce brandies, including France, Germany, Bulgaria, Italy, Portugal and Cyprus. Brandy is also produced in the New World wine-producing countries such as the USA and Australia. Each production area has its own production regulations and some brandies of origin are very strictly controlled (Robinson, 1999).

The rest of this chapter will begin with a description of the brandy production process. Thereafter, the range of typical brandy aromas will be discussed in terms of their origin and the flavour compounds most likely responsible for their presence. This will be followed by some practical considerations for brandy sensory evaluation and consumer testing.

13.2 Brandy production

13.2.1 Raw materials

Grapes are the basic raw material of brandy production. As with any product, the quality of raw materials determines the quality of the final product. Grapes that are rich in fruit acids and fruit-associated aroma compounds are especially desirable for the production of fine brandy. Grape acids, predominantly tartaric acid, act as a natural preservative. This is especially important in South Africa where brandy base wines may not be preserved with the usual table wine preservative, sulphur dioxide, due to the reactions of this compound during the distillation process. Another consideration for the selection of grapes for brandy is yield. Five litres of base wine is required for the production of one litre of brandy. Thus, grape varieties with a high yield potential are preferred for brandy production (Toerien, 2008).

The most popular grapes for the production of South African brandies are Chenin blanc and Colombar d. Ugni blanc is another popular grape cultivar and is used for brandy production in North America, Bulgaria and Italy, among others. North American distillers traditionally used table grape varieties such as Thompson Seedless and Flame Tokay due to grape shortages, but modern distillers now also use wine cultivars such as Ugni blanc, Colombar d, Gamay and Pinot Noir for brandy production. Some countries use local grape varieties for brandy production such as Rkatsiteli and Dimiat in Bulgaria and Baga grapes in Portugal (Robinson, 1999; Toerien, 2008).

13.2.2 Base wine production

Once the grapes are harvested, they are fermented to produce the brandy base wine.

During alcoholic fermentation the brandy base wine is protected against oxidation by the carbon dioxide that is released by the yeast during fermentation. However, after fermentation is completed, oxygen management is critical as many of the desirable volatile compounds in wine are susceptible to oxidation. As mentioned above, sulphur dioxide is traditionally used as a preservative against oxidation in wine production, but is not allowed in brandy base wine production due to the corrosive effect of sulphur dioxide on stills and the risk of sulphuric acid formation during oak maturation (Du Plessis *et al.*, 2002; Jackson, 2008; Toerien, 2008). According to legal requirements, South African brandy base wines must be fermented to a final alcoholic strength of between 10–12% abv. The addition of sugar and/or alcohol is not allowed for the production of brandy base wine in South Africa (Toerien, 2008).

13.2.3 Distillation process

After alcoholic fermentation, the base wine is distilled to obtain the brandy distillates. Distillation can take place in either a continuous or a discontinuous process. Continuous distillation is performed using a column still (Fig. 13.1). Discontinuous distillation are most often performed using a copper potstill (Fig. 13.2) but can also be done using a column still. Discontinuous distillation tends to provide a more aromatic end product than continuous distillation due to the rectification of alcoholic vapours that occur during the latter method. Brandies may be distilled to up to 85% abv with discontinuous distillation and up to 94.8% abv with continuous distillation, provided that the latter does not exceed more

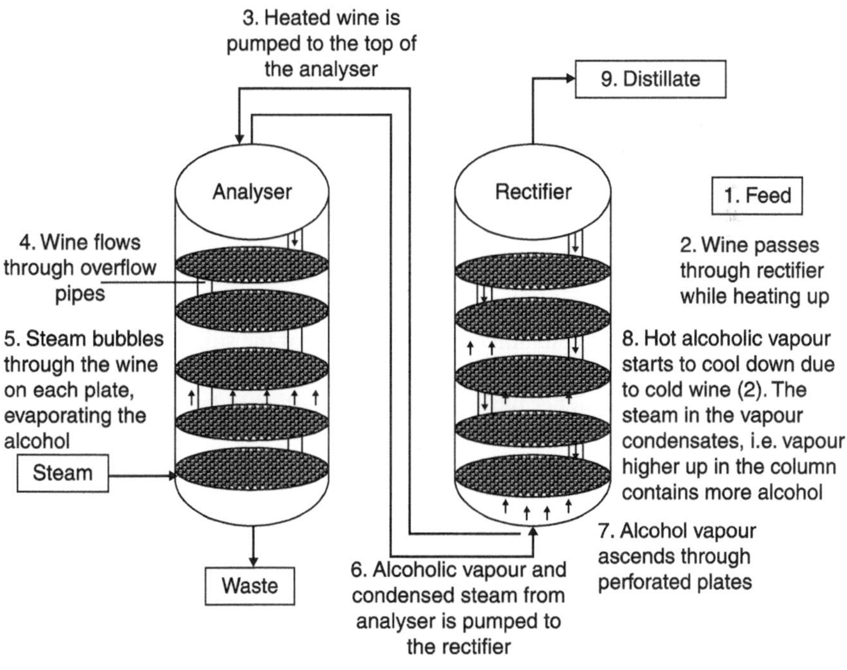

Fig. 13.1 Continuous distillation using a column still.

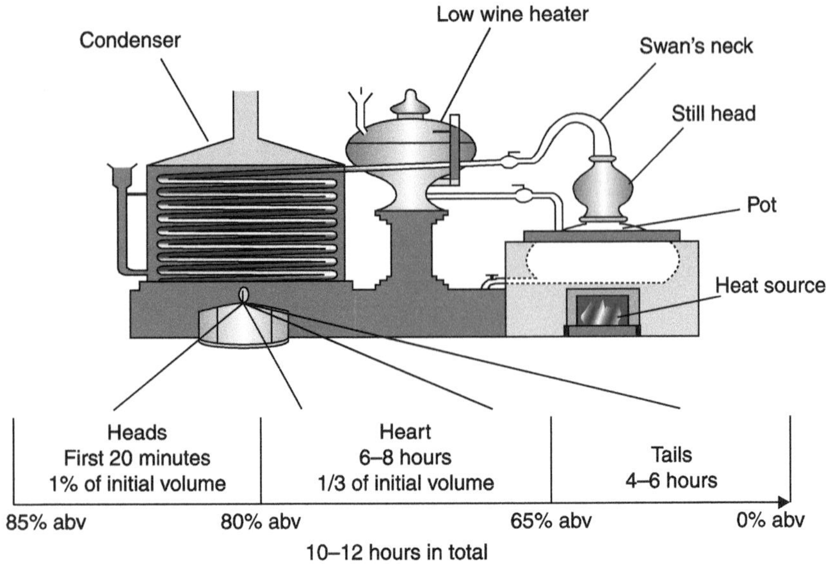

Fig. 13.2 Discontinuous distillation using a potstill.

than 50% of the alcoholic content of the finished product. North American brandies are mostly produced with column stills while distillers from most other countries make use of both potstills and column stills. South African legislation requires that at least 30% of the distilled spirits used in brandy must be distilled in potstills (European Union, 2008; Toerien, 2008; Snyman, 2004).

Potstill distillation

The same distillation process that is used to distill French Cognac is also used to distil brandies, namely *Methode Charentaise*. This method requires that the base wine is distilled twice in copper potstills. Copper is well suited for beverage distillation due to its heat conductivity and that it does not react with the boiling liquid. In addition it removes residual sulphur compounds in the base wine, forming an odourless precipitation on the inner wall of the still. Smaller potstills, where a larger portion of the liquid is exposed to the copper surface area, tend to produce distillates with higher purity than those distilled in larger pots (Snyman 2004; Toerien, 2008).

In the first step the base wine is distilled in to low wines, containing approximately 26–30% ethanol by volume. A low wine can be regarded as a concentrated version of the base wine. The first distillation has a duration of 6–8 hours (Snyman, 2004; Toerien, 2008).

The low wines are redistilled to obtain three fractions; the head, heart and tails. The heads and tails fractions are removed and further distilled to recover ethanol. Since these fractions contain a large amount of unwanted brandy congeners, they are not used for the brandy itself. The heads fraction is cut at approximately 1%

of the distilling volume (Fig. 13.2). The highest concentration of ethanol is distilled in the first part of the distillation process. As the distillation process continues the alcohol strength of the distillate decreases. The heart fraction is cut when the alcohol content of the distillate has decreased to approximately 70% abv. This heart fraction is then subjected to wood maturation, while the remaining tails fraction is discarded (Snyman, 2004; Toerien, 2008).

Column distillation

The distillation of base wines in a column still result in a pure, odourless liquid with an alcohol content of approximately 94% abv. Fractional distillation can also be performed with a column still (Fig. 13.1), resulting in low-strength grape spirits of approximately 85% abv that can be matured as a brandy. The neutral result of column distillation means that poorer quality grapes can be used to produce the base wines. Column stills can consist of between two and six columns; stills with more columns result in purer distillates. The resulting distillate can either be used as-is as a blending component in the production of brandy, or it can be matured in oak barrels. In South Africa, all brandies must contain at least 30% pot distilled brandy. However, in some other countries, a brandy can be made entirely of the distillate obtained from column distillations, provided that it is distilled to less than the EU legal limit of 86% abv (Snyman, 2004; Robinson, 1999).

13.2.4 Maturation

According to EU regulations, a brandy must be matured for at least six months in oak casks (European Union, 2008). Oak maturation procedures differ from country to country. Some countries may have additional, stricter regulations regarding the duration of oak maturation. For example, South African brandies must be aged for at least three years, as do Spanish Gran Reserva brandies and Bulgarian brandies (Robinson, 1999). Although the minimum maturation period is defined as six months, some brandies are matured for up to twenty years. Maturation in the presence of oak alternatives is also allowed by the EU, provided that the brandy is matured for at least one year (European Union, 2008).

Generally French oak barrels are used for brandy maturation. However, American brandies are often aged in charred American oak casks that were previously used for bourbon maturation. These casks are believed to contribute to the richness of the flavours of the brandies. Other types of European oak are also used; extensive studies have been conducted on the effect of Portuguese oak and Portuguese chestnut wood on brandy quality (Caldeira *et al.*, 2002, 2006, 2008, 2010).

In Spain, the solera ageing system is used for the maturation of brandies. The system consists of a range of barrels, previously used for sherry production, containing brandies of varying ages. A fraction of brandy from the first and oldest barrel is removed and replaced by a brandy from the second oldest barrel (one year younger). The second barrel is refilled with brandy from the third oldest barrel and so on until the youngest barrel is refilled with fresh brandy (Robinson, 1999). The solera system has also been successfully used in other countries, for example the South African Nederburg Solera brandy.

After maturation the brandies can be blended with neutral spirits and rectified with approved additives. The regulations regarding rectification vary from country to country. In general, 1–2.5% of natural sweetening agents such as dessert wine, grape must concentrate or honey may be added. Flavourless caramel may be added during the production process to impart a golden colour. The flavouring of brandy is prohibited by EU regulations (Robinson, 1999; European Union, 2008).

The brandies are subsequently diluted with distilled or de-ionised water to the desired alcohol content in a step-wise manner to ensure proper unification with the brandy flavour compounds. The final alcohol strength of brandies may vary according to style. According to EU regulations, a brandy must contain at least 36% abv. South African potstill brandies are bottled at 38% abv while blended brandies are bottled at 43% abv (Toerien, 2008; European Union, 2008).

13.3 Types of brandies

Three types of brandies are found in South Africa. Potstill brandies are made from a minimum of 90% pot distilled spirits that have been aged for a minimum of three years. Vintage brandies must contain at least 30% potstill brandy and must have been aged for a minimum of eight years, and a maximum of 60% wine spirits matured for a minimum of eight years. Vintage and potstill brandies are generally consumed neat. The third type of South African brandy is blended brandies, which must contain at least 30% potstill brandy but can be blended with up to 70% column distilled spirits. This type of brandy is typically consumed with a mixer such as cola or in cocktail (Toerien, 2008). Other well-known and protected styles of brandy include the Slovak 'Karpatske brandy special', Portuguese Lourinhã brandy, Spanish Brandy de Jerez, Wachauer Weinbrand and Weinbrand Durnstein from Austria as well as the Deutcher Weinbrand and Pfälzer Weinbrand from Germany.

13.4 Understanding brandy flavour

Brandy flavour is influenced during the entire production process, from the inherent aromatic composition of the grapes, to the volatile compounds produced during alcoholic fermentation, the concentration of volatiles during distillation, the extraction of flavour molecules from wood during maturation, and finally the evolution of all the above-mentioned compounds during ageing. The formation, degradation and interaction of aroma compounds during the entire brandy production process influence the final sensory characteristics of brandy. Several characteristic brandy aromas have been identified. Jolly and Hattingh have compiled these in an aroma wheel for South African brandies (Jolly and Hattingh, 2001). The two-tiered brandy wheel (Fig. 13.3) consists of two sections, i.e.

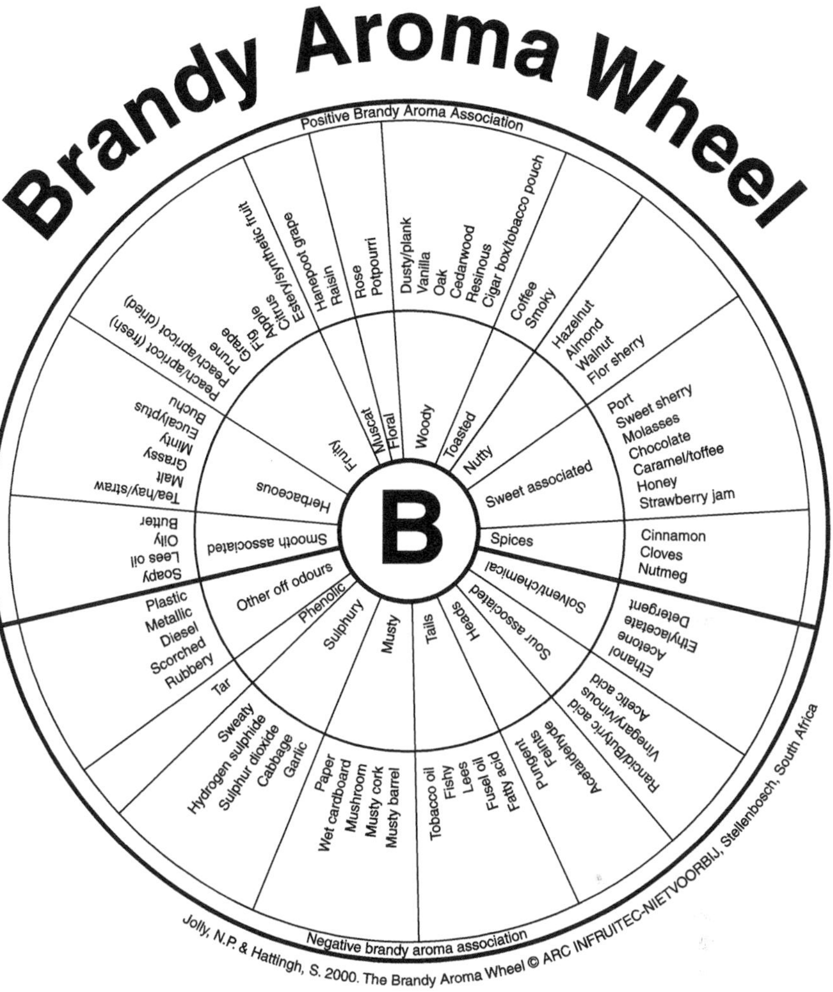

Fig. 13.3 An aroma wheel for South African brandies (Source: Jolly and Hattingh, 2000).

positive and negative attributes. The positive attributes are grouped in the first tier as fruity, muscat, floral, woody, toasted, nutty, sweet associated and spices. The negative attributes are grouped as solvent/chemical, sour associated, heads, tails (referring to the tails section of the distillation process), musty, sulphury, phenolic and other off-odours. Each section is divided further in a second tier. Although the brandy wheel was developed specifically for South African brandies, it also covers the most important characteristics published on Slovak and Portuguese brandies (Table 13.1) and could be considered an international brandy aroma wheel.

Table 13.1 Important aroma attributes for Portuguese and Slovak brandies as reported in the literature

Origin	Attributes	Author
Portugal, Slovak	Alcohol	Caldeira *et al.*, 2006; Caldeira *et al.*, 2010; Janáčová *et al.*, 2008
Portugal	Burned	Caldeira *et al.*, 2002
Portugal	Caoutchouc (rubber)	Caldeira *et al.*, 2002; Caldeira *et al.*, 2006
Portugal, Slovak	Caramel	Caldeira *et al.*, 2002; Caldeira *et al.*, 2010; Janáčová *et al.*, 2008
Portugal	Coffee	Caldeira *et al.*, 2002; Caldeira *et al.*, 2010
Portugal, Slovak	Dried Fruit	Caldeira *et al.*, 2002; Caldeira *et al.*, 2006; Janáčová *et al.*, 2008
Slovak	Flower	Janáčová *et al.*, 2008
Slovak	Forest	Janáčová *et al.*, 2008
Portugal, Slovak	Fruity	Caldeira *et al.*, 2002; Caldeira *et al.*, 2006; Caldeira *et al.*, 2010; Janáčová *et al.*, 2008
Portugal	Glue	Caldeira *et al.*, 2010; Caldeira *et al.*, 2002
Portugal	Green	Caldeira *et al.*, 2006; Caldeira *et al.*, 2010; Caldeira *et al.*, 2002
Slovak	Herbal	Janáčová *et al.*, 2008
Slovak	Honey	Janáčová *et al.*, 2008
Portugal	Rancid	Caldeira *et al.*, 2002; Caldeira *et al.*, 2010
Portugal	Rubber	Caldeira *et al.*, 2010
Portugal	Smoke	Caldeira *et al.*, 2002; Caldeira *et al.*, 2006; Caldeira *et al.*, 2010
Portugal, Slovak	Spicy	Caldeira *et al.*, 2002; Caldeira *et al.*, 2006; Caldeira *et al.*, 2010; Janáčová *et al.*, 2008
Portugal	Sweet	Caldeira *et al.*, 2010
Portugal	Tails	Caldeira *et al.*, 2002; Caldeira *et al.*, 2010
Portugal	Toasted	Caldeira *et al.*, 2006; Caldeira *et al.*, 2010
Portugal, Slovak	Vanilla	Caldeira *et al.*, 2002; Caldeira *et al.*, 2006; Caldeira *et al.*, 2010; Janáčová *et al.*, 2008
Portugal	Woody	Caldeira *et al.*, 2002; Caldeira *et al.*, 2006; Caldeira *et al.*, 2010

With gas chromatography coupled to olfactometry (GC-O), many odour-active compounds have been identified in brandies along with their aroma descriptions (Caldeira *et al.*, 2008; Janáčová *et al.*, 2008). However, it is well known that aromas in a final product can rarely be attributed to a single odour-active molecule but are rather the result of a complex interaction of volatile substances within the mixture. In this section it will be attempted to explain the presence of important aroma compounds that have been identified in brandy as well as the possible impact they may have on brandy aroma and quality.

Aroma compounds in brandy can be roughly divided into two categories: those that are formed during the production of the base wine and subsequently concentrated or removed during distillation, and those that are extracted or formed during wood maturation. The first category is mostly responsible for the aromas in the fruity, floral, muscat tiers in the aroma wheel, but also for aromas in the negative tiers such as tails and solvent/chemical-like aromas. The wood-derived aroma compounds are generally associated with the spicy, woody, sweet-associated and nutty notes found in brandy.

13.4.1 Volatile compounds and aromas obtained from the base wine and distillation process

The distillation process has a major influence in the volatile composition of brandy. Prior to maturation, brandy distillates consist largely of esters, higher alcohols and volatile fatty acids which are derived from alcoholic fermentation and concentrated during the distillation process.

Ethyl acetate and ethyl lactate are the most abundant esters in brandy distillates. Ethyl acetate is mostly found in the heads section of the distillate, and ethyl lactate is mostly found in the tails section. Therefore the distillation process can significantly influence the presence of these compounds in the final brandy. Ethyl lactate has a negative influence on brandy quality and its presence can also be enhanced if the base wine has undergone malolactic fermentation. Ethyl lactate is often accompanied by diethyl succinate, which is regarded as a spoilage compound that can be removed along with the tails section (Cantagrel, 1992; Von Adam *et al.*, 1996).

Longer chain ethyl esters also contribute to the total ester content of brandy. These esters are generally associated with fruity aromas and contribute to overall brandy quality. The ethyl ester content of brandy is influenced by the amount of lees present during distillation. By including a large portion of good quality, fresh yeast lees with the base wine during distillation, the amount of odorous esters can be significantly increased to improve the fruity, floral bouquet of the brandy samples (Cantagrel, 1992; Von Adam *et al.*, 1996).

Almost all esters are low-boiling-point compounds and are distilled over early in the distillation process. Thus, the point at which the heads are cut in order to remove undesirable compounds must be controlled in such a way as not to remove too many desirable esters. Similarly, the tails must not be cut too early in the distillation process to preserve more esters. It has been found that cutting the tails at 60% abv can result in a decrease of between 30% and 60% of ethyl esters (Snyman, 2004).

The occurrence of malolactic fermentation (MLF) during base wine production has a significant impact on the ester composition of brandy distillates. The application of MLF can increase the presence of ethyl lactate and diethyl succinate and decrease isoamyl acetate, ethyl acetate, ethyl hexanoate and 2-phenethyl acetate. The effect of MLF is also observed in the sensory attributes of the distillates. The decrease in esters explains the decrease in fruity aroma observed in distillates for which the base wine has undergone MLF compared to those for which no MLF has taken place. The application of MLF also increased the presence of solvent/chemical aromas (Du Plessis *et al.*, 2002).

An increase in ester levels were reported in brandies aged for six months in various wood ageing systems compared to a control. The authors attributed the higher levels of acetate esters in brandies aged in barrels to increased levels of acetic acid that is formed through the extraction of acetyl groups from the wood xylans. The higher ester content did not result in a significant increase in fruit aroma intensity (Caldeira *et al.*, 2010).

Higher alcohols are by-products from yeast fermentation during the production of the base wine and form an important part of the volatile composition of brandy. The most abundant higher alcohol in brandy is isoamyl alcohol. Like most other higher alcohols, isoamyl alcohol is associated with unpleasant aromas in its pure form. However, no direct correlation has been found between brandy quality and isoamyl alcohol (Steger and Lambrechts, 2000). At lower concentration levels higher alcohols may add to brandy complexity due to synergistic interactions with other odour compounds. In addition, higher alcohols may be esterified during the maturation process, resulting in esters that impart more pleasant aromas. Therefore, some amounts of higher alcohols are desirable in brandy distillates but not at overly concentrated levels.

13.4.2 Aromas and volatile compounds derived from wood

Another source of aroma compounds is the barrels in which the brandy distillates are matured. There are several ways in which wood maturation can influence brandy flavour (Van Jaarsveld *et al.*, 2009a): the direct extraction of aroma compounds from the wood lignin into the spirit, the formation of compounds by reaction of the alcohol with lignin, further conversion of extracted compounds in the spirit and the modification of original spirit congeners in the presence of wood.

The most important aromatic compounds that can affect brandy aroma as a result of wood maturation are oak lactones, phenolic aldehydes and furanic aldehydes. The quantities at which these and other aromatic compounds occur in barrels can be influenced by the species of wood that was used as well as the seasoning and toasting treatments to which the wood has been subjected during the manufacturing of the barrel.

Influence of wood species

It was found that a significant difference in vanilla, woody, caramel, burned/toasted, green, tails and rubbery aromas existed between brandies aged in barrels from different wood origins. Brandies aged in Portuguese chestnut and Portuguese oak barrels had higher intensities of vanilla aroma which the author attributed to possibly higher vanillin content. Portuguese oak wood also resulted in the highest 'woody' aroma while American oak casks and French oak casks imparted lower intensities of this aroma. Vanillin has been found to have a significant correlation to the intensity of vanilla and woody aromas in brandies. Vanillin can occur in brandy due to both wood extraction and the oxidation of coniferaldehylde during ageing (Caldeira *et al.*, 2008).

Woody notes have also been associated with the presence of *cis*- and *trans*-β-methyl-γ-octalactones. These two isomers are only found in oak wood and are therefore commonly referred to as oak lactones.

Portuguese chestnut barrels also imparted higher intensities of caramel, coffee and toasted aromas than French and American oak barrels. 2,4-dihydroxy-2-5-dimethyl-3-furanone, 2-hydroxy-2-cyclopenten-1-one, furaneol, 5-methyltetrahydro-2-furanmethanol, 5-hydroxymethylfurfural and hydroxymaltol have been described as caramel-like using GC-O technology (Janáčová, 2008). Hydroxymaltol has also been described as coffee-like. A significant correlation was observed between toasted aroma and the compounds guaiacol, 4-methylguiacol and syringol.

Portuguese oak barrels seemed to impart less intense green and tails aromas in comparison to American oak and French oak. Portuguese chestnut also resulted in less intense green aromas than French oak. Compounds such as 2-methyl-1-propanol and *trans*-2-hexenol have been associated with green notes through GC-O technology, but no significant correlations have been found between these compounds and the green note in brandy. Similarly, the tails character that is sometimes noticed in brandies has not been linked to a specific compound. Fatty acids constitute a considerable portion of the tails section of the distillation process, but no direct correlations have been reported between tails aroma and any specific fatty acid in brandy (Caldeira *et al.*, 2002, 2010).

Influence of toasting

An increase in toasting level has been found to result in an increase in vanilla, woody, spicy, caramel, burnt, toasted, dried fruits and smoke notes, and a decrease in the intensity of fruity notes as well as negative notes such as green, tails and glue (Caldeira *et al.*, 2002, 2006; Van Jaarsveld *et al.*, 2009b).This follows on the observation that the concentrations of wood-derived congeners are generally higher in brandies matured in medium- and high-toasted barrels than untoasted and lightly-toasted barrels.

Some aromatic wood compounds, notably oak lactones and furanic aldehydes, which form during the toasting process, peak in concentration at medium toast after which they decrease. The decrease of furfural derivatives during heavy toasting is due to the polymerisation of these compounds at high temperatures. The presence of oak lactones has been found to decrease in high-toasted barrels, resulting in a lower intensity of woody aroma in the brandy (Van Jaarsveld *et al.*, 2009b).

Eugenol is frequently associated with spicy notes, especially cloves. However, a weak correlation between eugenol and spicy aroma in brandy has been reported by Caldeira *et al.* (2008). The same authors reported three unknown compounds that could be described as spicy using GC-O, and these may contribute to the overall perception of spiciness in brandy.

Smoky aroma has been reported as an important contributing attribute of brandy quality. Smoky aroma reportedly has a linear correlation with syringol and 4-allylsyringol. Other compounds such as hexanoic acid, guaiacol, 4-propylguaiacol and 4-methylsyringol, acetylpyrazine, γ-crotolactone and several unknown compounds have been associated with smoky notes through GC-O technology (Caldeira *et al.*, 2010; Janáčová *et al.*, 2008).

It appears as if the rate at which sensory attribute intensities increase or decrease with an increase in toasting level is influenced by wood origin. The effect of toasting on the permeability of the wood, and subsequently on the extraction rate of wood congeners during ageing, vary according to oak species (Caldeira *et al.*, 2006; Van Jaarsveld *et al.*, 2009b).

Effect of alternative wood products

Wood alternatives can also affect the presence of aromas and flavour compounds in brandy. Brandies aged with wood tablets contained higher levels of guaiacol, 4-methylguiacol and syringol and subsequently had a more intense toasty aroma than their barrel-matured counterparts. (Caldeira *et al.*, 2010).

13.4.3 Brandy flavour evolution during ageing

Leauté *et al.* (1998) reported that attributes like fruity and green are common for younger brandies, while attributes like vanilla, woody, rancid and spicy increase in intensity during brandy ageing. Other studies have reported an increase in the intensity of aroma notes such as vanilla, woody, toasted and smoke over time (Caldeira *et al.*, 2006), and that some of the odour compounds responsible for caramel aroma may only be present after two years of maturation (Janáčová *et al.*, 2008). There was evidence that the increase in intensity of these attributes may reach equilibrium after four years of maturation. Nevertheless, qualities such as body and flavour complexity seem to consistently increase over time for at least the first five years of maturation in new oak barrels. Astringency, alcohol and glue-like attributes decrease during maturation. Other attributes (dried fruits, smooth, tails and caoutchouc/rubber) may be uninfluenced by maturation and can follow their own evolutionary patterns (Caldeira *et al.*, 2006). More importantly, research has shown that most sensory modifications occur during the first three years of maturation in wood.

13.5 Sensory evaluation of brandy

13.5.1 Practical considerations for dealing with high-alcohol beverages

The high alcohol content of brandy increases the risk of sensory fatigue. To reduce this risk, samples are usually diluted with odourless water to an alcohol content ranging from 20–23% abv prior to sensory analysis (Peña y Lillo *et al.*, 2005; Du

Plessis *et al.*, 2002). Brandy samples are ideally served at room temperature. The dilution of an alcoholic liquid with water causes an exothermic reaction, and therefore a short resting period should be allowed between dilution and serving. It was found that the pungency of whisky samples increases at higher temperatures (Jack, 2003). Although the effect of temperature on brandy aroma has not been tested, it is possible that the volatility of brandy aromas may also be affected by rises in temperatures. Even so, it has been suggested that brandy samples may also be nosed at full strength, as the aldehydes are more prominent at this alcohol strength, while fatty acids and esters are more prominent at approximately 20% abv (Quady and Guymon, 1973).

Another method of reducing sensory fatigue is to evaluate samples by nosing only. A study conducted on Pisco evaluation (which by definition also qualifies as a brandy) showed that there was a strong correlation between orthonasal and retronasal odour perceptions. The authors suggested that the olfactory analysis of Pisco can be conducted based on orthonasal perceptions only. By excluding the ingestion of the product, sensory fatigue can be limited (Peña y Lillo *et al.*, 2005). However, some gustatory attributes such as mouth coating can differ significantly between brandies and may be interesting to measure. Palate cleansers such as spring water and soda biscuits (Peña y Lillo *et al.*, 2005) or just water (Caldeira *et al.*, 2010) can be used when brandy products are tasted for gustatory attributes.

The question of the number of samples to be served in one sitting is often raised during the sensory evaluation of alcoholic beverages. The risks of sensory adaptation, fatigue and inebriation that are coupled to large sample sets can reduce the quality of the responses obtained. It is difficult to determine a maximum number of samples that can be tasted; the ease with which the panel will be able to evaluate the samples will depend on the extent of the differences between the samples, the complexity of the lexicon as well as the level of training of the panellists. However, reports on the descriptive sensory analysis of brandy vary from four to twelve samples, both full strength and diluted (Caldeira *et al.*, 2002, 2006, 2010; Du Plessis *et al.*, 2002; Janáčová *et al.*, 2008; Van Jaarsveld *et al.*, 2009). Du Plessis *et al.* (2002) also reported the evaluation of brandy distillate samples at 23% abv in a series of 12 triangle tests, amounting to 36 distillate samples. The panel were able to detect significant differences in 9 of the 12 triangle tests, which suggests that they were not affected by sensory fatigue.

In small sample sets, comparative sample presentation can be considered instead of monadic sequential presentation. Monadic sequential tests are generally preferred because they provide more independent responses than comparative sample presentations, in addition to providing better control over time delays between samples (Stone and Sidel, 2004). With comparative sample presentation, the evaluation process remains essentially the same as with monadic sequential presentation, i.e. the taster evaluates each sample completely before moving on to the next sample. However, the taster has the advantage of having the entire sample range in front of him and can thereby orientate himself in the sensory space represented by the sample set. This may cause the taster to feel more in control of the task when presented with the entire sample set, which may provide a subtle psychological advantage in a difficult sample set. It has been observed on several occasions that more reliable data can be obtained with comparative sample presentation than with sequential monadic presentation of brandy, provided that the sample set is not too large, i.e. five or six products at most.

13.5.2 Evaluation of brandy colour

Final brandy samples can differ in colour as brandy colour is rectified with flavourless caramel to obtain the desired hue and intensity. Wood maturation can also affect brandy colour. Typical brandy colour hues include straw-yellow, golden, topaz and greenish (Canas *et al.*, 2009). In some research scenarios it may be beneficial to remove the bias caused by colour perception. Blue glasses can be used to deal with this problem (Du Plessis *et al.*, 2002; Peña y Lilo *et al.*, 2005).

13.5.3 Evaluation of brandy quality

Apart from the usual sensory analyses focussing on single attributes as perceived by one of the five senses, brandy is sometimes also evaluated for organoleptic quality. Reported quality measurements include 20-point scales (0 = without quality to 20 = maximum quality) and 6-point category scales (0 = completely unacceptable to 5 = outstanding quality) (Caldeira *et al.*, 2010; Steger and Lambrechts, 2000). Quality scores can be correlated to sensory attributes. For example, significant correlations ($r > 0.7$) have been found between Lourinhã brandy quality and toasted, woody, vanilla, body, flavour persistence and spicy. Furthermore it was found that flavour complexity has a highly significant impact on brandy quality. A negative correlation between brandy quality and the tails, green and caoutchouc (rubber) characters were reported by the same authors. Bitterness and burning did not appear to have an impact on quality (Caldeira *et al.*, 2006).

13.5.4 Sensory evaluation of brandy over extended periods of time

Brandy can be matured for anything from six months to twenty years, which can mean that brandy evaluation projects can last for several years. One example is the study by Caldeira *et al.* in 2006 that reported on flavour evolution that takes place in brandy over five years of maturation (Caldeira *et al.*, 2006). Ideally, one would want the trained panel that participates in the evaluation to be constant to ensure consistent results. However, in reality this is not always possible.

It is therefore very important that the sensory language used to describe brandies remains constant within a working environment. The existence of a common flavour language will ease the incorporation of new brandy panellists. A common flavour language will also make it easier to relate results from independent studies within a working environment. The Brandy Aroma Wheel for South African brandies (Jolly and Hattingh, 2001) provides a standardised list of descriptors that can also be used for brandies from other countries. A very comprehensive list of aroma reference standards was published with the aroma wheel. The reference standards act as a supplement to the aroma wheel and can be used as a tool to align the word–flavour associations between panellists. Unfortunately, many of the references were commercial flavour keys from suppliers that have merged with other companies since the publication of the article. The most important factor is to implement a standardised descriptor list in a working environment with appropriate reference standards. Reference standards can be replaced by alternatives by matching the alternative with the control, as long as the reference standard provokes the same word–aroma association.

Panel performance monitoring is vital during long-term sensory studies. Panellists' reliability can be tested by including blind replicates into a training session or by using statistical tools such as Tucker plots or Pearson's correlations (Caldeira *et al.*, 2006; Tomic *et al.*, 2010). Regardless of which method is used, it is very important to monitor panel performance on a regular basis to ensure that the quality of results remain consistent over time.

13.6 Brandy and consumer research

Published data on quantitative consumer research regarding brandy intrinsics are not common as it is usually funded by the industry. Brandies are sometimes included in conceptual consumer studies testing for drinking habits of alcoholic beverages in general (Scriven *et al.*, 1989; Askegaard and Madsen, 1995). There are several possible reasons why brandies may not commonly be subjected to consumer preference or acceptability testing. The limitations in sample servings due to the high alcohol content may be one reason. Another may be that companies and research institutes are reluctant to invest in consumer research due to the long-term implications of product re-formulation.

Nevertheless, should intrinsic consumer research be conducted on brandy, the basic principles of best sensory practices should still be taken into account. For consumer testing, samples should generally be presented in such a way to mimic as closely as possible the way the consumer would normally consume the product (Meilgaard, 2007). In the case of blended brandies that are rarely consumed neat, it may be necessary to present the product in a soft drink mixer with ice. The question of carriers can become quite complicated as consumers in the same target market may have different preferences for mixers, e.g. cola, ginger ale or even orange juice. Of course, the dilution that occurs with ice may be a further complication. On the one hand, the consumer may not be able to relate to a neat, blended brandy at room temperature, which may mean that acceptability scores may be skewed towards the negative part of the scale. A further consideration is the time of day at which the tasting is conducted. The tasting of spirit products is generally not recommended for early in the morning (Meilgaard, 2007).

13.7 Future trends

There is massive scope for sensory research in brandy, judging by the limited number of publications in this area. One area of interest would be the application of rapid profiling techniques such as flash profiling (Dairou and Sieffermann, 2002) and Napping® (Perrin, *et al.*, 2008). These techniques are valuable screening tools with widespread applications in industrial environments. However, the effect of the high alcohol content of brandy on the effectiveness and reliability of such techniques must be investigated.

Furthermore, there are opportunities to investigate the effect of interactions between odour compounds on the aroma of brandy. Novel fractionating olfactometric methods that enable the capturing of multiple or single peaks obtained during chromatography has shown great potential for investigating the contribution of compounds and combinations of compounds on specific aroma attributes in Pinotage wines (Rohwer *et al.*, 2010). Such techniques may provide further insight into the origin of brandy aromas that the distiller can use to craft his product.

13.8 Sources of further information

FAITH N (2000), *Classic Brandy*, London, Prion Books Ltd.

NYKÄKEN L (1983), *Aroma of Beer, Wine and Distilled Alcoholic Beverages*, Berlin, Akademie-verlag.

RESURRECCION AVA (1998), *Consumer Sensory Testing for Product Development*, Gaithersburg, Aspen Publishers Inc.

TOERIEN W (2008), *Fire Water*, Cape Town, Quivertree publication in association with the South African Brandy Association.

13.9 References

ASAUKEGAARD S, MADSEN TK (1995), 'European food cultures: an exploratory analysis of food related preferences and behaviour in European regions', MAPP Paper no 26, Århus. Available from: http://130.226.203.239/pub/mapp/wp/wp26.pdf [accessed 31 January 2011].

CALDEIRA I, ANJOS O, PORTAL V, BELCHIOR A, CANAS, S (2010), 'Sensory and chemical modifications of wine-brandy aged with chestnut and oak wood fragments in comparison to wooden barrels', *Anal Chimic Acta*, **660**, 43–52.

CALDEIRA I, BRUNO DE SOUSA R, BELCHIOR AP, CLÍMACO MC (2008), 'A sensory and chemical approach to the aroma of wooden aged Lourinhã wine brandy', *Ciência Téc Vitiv*, **23**(2), 97–110.

CALDEIRA I, MATEUS AM, BELCHIOR, AP (2006), 'Flavour and odour profile modifications during the first five years of Lourinha brandy maturation on different wooden barrels', *Anal Chimic Acta*, **563**, 264–273.

CALDEIRA I, BELCHIOR AP, CLÍMACO MC, BRUNO DE SOUSA, R (2002), 'Aroma profile of Portuguese brandies aged in chestnut and oak woods', *Anal Chimic Acta*, **458**, 55–62.

CANAS S, CALDEIRA I, BELCHIOR P (2009), 'Comparison of alternative systems for the ageing of wine brandy, wood shape and wood botanical species effect', *Ciência Téc Vitiv*, **24**(2), 91–99.

CANTAGREL R, MAZEROLLES G, VIDAL JP, GALY B, BOULESTEIX, JM (1992), 'First International Symposium on Cognac', Cognac.

DAIROU V, SIEFFERMANN JM (2002), 'A comparison of 14 jams characterized by conventional profile and a quick original method, the flash profile', *J Food Sci*, **67**(2), 826–834.

DU PLESSIS HW, STEGER CLC, DU TOIT M, LAMBRECHTS MG (2002), 'The occurrence of malolactic fermentation in brandy base wine and its influence on brandy quality', *J Appl Microbiol*, **92**, 1005–1013.

EUROPEAN UNION (2008), Regulation (EC) No 110/2008 of the European Parliament and of the Council of 15 January 2008, *Official Journal of the European Union*, L 39/30–31.

JACK J (2003), 'Development of guidelines for the preparation and handling of sensory samples in the Scotch whisky industry', *J I Brewing*, **109**, 114–119.

JANÁČOVÁ A, SADECKA J, KOHAJDOVA Z, SPANIK I (2008), 'The identification of aroma-active compounds in Slovak brandies using GC-sniffing, GC-MS and sensory evaluation', *Chromatographia*, **67**, S113–S121.

JACKSON RS (2008), *Wine Science: Principles and Practices*, 3rd ed., Oxford, Elsevier.

JOLLY NP, HATTINGH S (2001), 'A brandy aroma wheel for South African brandy', *S Afr J Enol Vitic*, **22**, 16–21.

LEAUTÉ R, MOSEDALE JR, PUECH JL (1998), In C Flanzy (ed.), *Oenologie-Fondements Scientifiques et Technologiques*, Lavoisier, Paris, Tec and Doc, p 1130.

MEILGAARD MC, CIVILLE GV, CARR BT (2007), 'Controls for test room, product and panel'. In *Sensory evaluation techniques*, London, CRC press.

PEÑA Y LILO M, LATRILLE E, CASAUBON G, AGOSIN E, BORDEU E *et al.* (2005), 'Comparison between odour and aroma profiles of Chilean Pisco spirit', *Food Qual Prefer*, **16**, 59–70.

PERRIN L, SYMONEAUX R, MAITRE I, ASSELIN C, JOURJON F, *et al.* (2008), 'Comparison of three sensory methods for use with the Napping procedure: case of ten wines from Loire valley', *Food Qual Prefer*, **19**, 1–11.

QUADY AL, GUYMON JF (1973), 'Relation of maturity, acidity and growing region of Thompson Seedless and French Colombar d grapes to wine aroma and quality of brandy distillate', *Amer J Enol Viticult*, **24**(4), 166–175.

ROBINSON J (1999), *Oxford Companion to Wine*, 2nd ed., Oxford, Oxford University Press.

ROHWER E, NAUDE Y, GORST-ALLMAN P (2010), 'The analysis of Pinotage wine aromas by novel fractionating olfactometric method combined with comprehensive two dimensional gas chromatography-mass spectrometry'. In: *Proceedings of IUFoST 15th World Congress of Food Science and Technology*, Cape Town.

SCRIVEN FM, GAINS N, GREEN SR, THOMSON DMH (1989), 'A contextual evaluation of alcoholic beverages using the repertory grid method', *Int J Food Sci Tech*, **24**(2), 173–192.

SNYMAN CLC (2004), 'The influence of base wine composition and wood maturation on the quality of SA brandy', Doctoral dissertation, Stellenbosch University, South Africa.

STEGER CLC, LAMBRECHTS MG (2000), 'The selection of yeast strains for the production of premium quality South African brandy base products', *J Ind Microbiol Biot*, **24**, 431–440.

STONE H, SIDEL JL (2004), *Sensory Evaluation Practices*, San Diego, Elsievier Academic Press.

TOERIEN W (2008), *Fire Water*, Cape Town, Quivertree publication in association with the South African Brandy Association.

TOMIC O, LUCIANO G, NILSEN A, HYLDIG G, LORENSEN K *et al.* (2010), 'Analysing sensory panel performance in a proficiency test using the PanelCheck software', *Eur Food Res Technol*, **230**, 497–511.

VAN JAARSVELD FP, HATTINGH S, MINNAAR P, BLOM M (2009a), 'Rapid induction of ageing character in brandy products – Part 1. Effects of extraction media and preparation conditions', *S Afr J Enol Vitic*, **30**, 1–15.

VAN JAARSVELD FP, HATTINGH S, MINNAAR P (2009b), 'Rapid induction of ageing character in brandy products – Part III. Influence of toasting', *S Afr J Enol Vitic*, **30**, 24–37.

VON ADAM L, HAUG L, KOLB M, SCHWARTZ CH (1996), 'Contribution to the knowledge of volatile compounds present in wine distillates from the years 1990 to 1993', *Die Branntweinwirtschaft*, **6**, 82–90.

14

Grappa: production, sensory properties and market development

C. Da Porto, University of Udine, Italy

Abstract: Grappa is the traditional Italian spirit derived from marc. The quality of grappa is influenced by the following factors: grape varieties, vinification, system of marc storage, fermentation, distillation, maturation and blending. Among the alcoholic beverages, grappa is the richest in volatile compounds (about 1 per cent by volume). Grappa quality is discussed in terms of its visual, olfactory/retro-olfactory and taste/flavour perceptions. Data on both consumer research and market analysis are reported and discussed in this chapter as well as the future trends of the sector.

Key words: distilled beverage, marc, volatile compounds, sensory properties, consumer research, market analysis.

14.1 Introduction

The name 'grape marc' or 'grape marc spirit' may be replaced by the designation 'grappa' solely for the spirit drink produced in Italy from marc distillation. The Geographical Indications registered for grape marc spirits include the Spanish *Orujo*, the Greek *Tsipouro* from the northern part of the country, named *Tsikoudia* on the isle of Crete, the French *Eau-de-vie de marc*, the Portuguese *Aguardente Bagaceira*, the Cypriote *Zivania* and the Hungarian *Törkölypálinka*. The technical definition and requirements for grappa are established by Regulation No 110/2008 of 15 January 2008 (EC, 2008). The Regulation recognises nine Geographic Indications for grappa: Grappa di Barolo, Grappa piemontese o del Piemonte, Grappa lombarda o di Lombardia, Grappa trentina o del Trentino, Grappa friulana o del Friuli, Grappa veneta o del Veneto, Südtiroler Grappa o dell'Alto Adige, Grappa siciliana o di Sicilia and Grappa di Marsala.

The production of grappa is fragmented among many producers. Grappa demand is mostly domestic. In the last decade exports have been increasing but they still represent only 10–15% of the total production, and their share of total spirits and liqueurs is around 10%. The main export market is Germany, with 28% of the total, followed by Switzerland (8%). Outside Europe, small volumes are sold in the US and Canada. In 2009, the production of grappa was about 11 million litres of pure alcohol (p.a.), 34% of Italian spirit production. Grappa sales have been *ca.* 17 million litres, equally distributed between traditional and large-scale retail (LSR) and hotels, restaurants and cafeteria (Ho.re.ca) channels (AssoDistil, 2010).

14.2 Production process

14.2.1 Marc storage and fermentation

Grape marc, a by-product of the winemaking process, is the main raw material of grappa. Generally, the quality of marc depends on the vinification process used for grapes; mild and soft technologies are the best. Both unfermented marc and fermented marc (at *ca.* 2–4% v/v ethanol) are stored for some months before distillation. This is a traditional practice carried out to promote the most economic management of the distillation plants. However, at present many distilleries distil marc as soon as possible to obtain half-processed spirit at low alcoholic strength (*ca.* 20% v/v) which is then re-distilled. Various storage systems can be used for grape marc: rectangular concrete silos (capacity 400–500 tons), underground concrete silos (capacity 50–100 tons), cylindrical-shaped plastic containers (salami-like containers) (capacity *ca.* 300 tons), stainless steel containers (capacity 200–300 tons) and plastic containers (bins) (capacity 500–700 kg) (Da Porto, 1998). Storage systems affect the concentrations of volatile compounds of the future distillate (Da Porto, 2002; Da Porto *et al.*, 2004). Generally, processing of marc during storage takes advantage of spontaneous anaerobic fermentation. Only a few distillers inoculate the unfermented marc with selected yeast. Storage can cause difficulties such as marc preservation by anomalous fermentation, and bacterial contamination leading to the production of off-flavours. Technological approaches such as pH adjustment, temperature control, and anaerobic environment setup have been adopted by distillers to avoid off-flavours and production of toxic compounds.

14.2.2 Distillation and rectification

The grappa distillation process can be batch or continuous. Generally, both batch and continuous distillation are carried out in two steps. First, marc is introduced into a copper still (batch distillation) or into a de-alcoholising distillation plant (continuous distillation) in which ethanol is steam-stripped to give vapours at about 10–20% (v/v). The condensed vapours then feed a column (generally equipped with a series of bubble-cap plates) for ethanol concentration and separation of undesirable compounds (rectification). When batch distillation is carried out rectification is made by selection of the cut points at the start (heads) and the end (tails) of spirit collection. This is a critical operation for the final quality of grappa. These fractions contain respectively the undesirable highly volatile and less-volatile compounds together with a large amount of ethanol. These can be recycled at the end of distillation. When continuous distillation is used, rectification is carried out using a multi-column system. Since the raw distillate is very rich in methanol, to lower its content down to the legal specification (1 g/100 ml p.a.), a further column, the demethylising column, is also included in distillation plants (Da Porto, 1998; Da Porto and Longo, 2005). Currently, continuous distillation represents about 80% of Italian production. It is fast and cheap, and results in grappa without defects or distinction. The traditional method, practised by 100 or so distillers (20% of Italian production), is slower, involving a batch process and various kinds of specialised stills. Although the styles and abilities of the traditional distillers vary widely, the artisanal grappa is generally highly flavoured and benefits from a distinctive character.

14.2.3 Maturation and ageing

Freshly distilled grappa has unacceptable sensory characteristics and is usually matured in vats of inert material (stainless steel, etc.) for five to six months to produce an acceptable product. Maturation produces a loss of the harsh characteristics of the new distillate. During the maturation period grappa is pumped-over at least once a month to promote oxidation and acetal formation. Acetal formation is important because aldehydes frequently have sour and pungent odours, while acetals are pleasant and fruity. Acetal formation is also important for removing acrolein, which has an acrid smell. Traditionally, grappa is a distillate ready to drink. Ageing it in wood is more due to recent requests from consumers than a real need. The ageing of grappa in barrels is carried out not only in classical oak casks (Profumo *et al.*, 1988), but also in light woods such as acacia, ash, mulberry, almond, apple, pear and cherry. They release little colour and provide an aroma with fine flavours. Juniper, very aromatic, and chestnut, very rich of phenolic acids, can also be used for cooperage. At the end of the maturation period, a well-aged grappa must be clear, slightly yellow in colour, and characterised by great smoothness and persistence. Caramel can be used as a means to adjust colour. Sugar solutions (maximum 2% w/v) are sometimes added to attenuate the 'burn' of the alcohol.

14.2.4 Blending and bottling

Prior to being marketed, different batches of grappa are blended. When the blend is correct (evaluation is made by sensory assessment), deproofing water is added to reduce the strength for bottling. Grappa is subject to the formation of cloudiness when the alcohol concentration is 'reduced' to approximately 40% vol. with demineralised water, and for this reason it is chill filtered in the temperature range −15–0 °C. The choice of the filtration temperature depends on the content of less-volatile compounds (especially esters of long-chain fatty acids) and provides physical stabilisation (elimination of cloudiness) and eliminates fatty and rancid notes, thus improving its fineness and balance.

14.3 Chemical composition and sensory properties

Among alcoholic beverages, grappa is the richest in volatile compounds (about 1 per cent by volume) because it is distilled from a raw material with a high percentage of solids (70–80% w/w). From 100 kg of marc about 3.8 l of pure alcohol can be obtained, corresponding to about ten bottles of grappa at 40% vol. ethanol content. Grappa is a complex mixture of some hundreds of flavour compounds in an ethanol–water matrix. The flavour of grappa originates from a number of sources. The first contribution is derived from the grape variety (e.g. terpenols, nor-isoprenoids, etc.), the second from compounds formed by metabolism of yeast and bacteria during fermentation (e.g. alcohols, organic acids, ethyl esters, acetates, etc.), the third from distillation by possible transformation of the above compounds and precursors, favoured by high temperature and alcohol concentration. The last contribution is due to compounds extracted from wood during ageing, when it is carried out. The naturally occurring relative concentrations of these compounds can vary from high mg/l to ng/l. Each compound has a flavour contribution that is determined by both its concentration and its unique sensory threshold value. Table 14.1 shows a set of average, medium and maximum values found for the volatile compounds detected in 51 commercial samples of white (not aged) grappa, produced from non-aromatic marc and coming from six Italian regions (Versini *et al.*, 1991).

14.3.1 Alcohols

The average content in ethanol is 45% (v/v). Currently, the majority of the distilleries produce grappa at 40% v/v ethanol content, and some at the minimum permitted by the standards of the EC regulation of 37.5% v/v, following present-day consumer requirements of drinks with a lower alcoholic content.

Methanol concentration depends on the storage system and conditions used for marc. Methanol is a toxic chemical and for this reason its limit is fixed by the EC regulation at 1000 g/Hl of 100% vol. alcohol. Methanol is separated by rectification usually carried out using the demethylising column. However, this column has the disadvantage of eliminating not only methanol but also volatile compounds of organoleptic interest (specially terpenes). Methanol has no direct sensory effect on grappa.

Amyl alcohols (3-methyl-1-butanol and 2-methyl-1-butanol), formed during fermentation by deamination and decarboxylation reactions from isoleucine and leucine, respectively, range between 84–472 mg/100 ml p.a (0.3–2.8 g/l) and constitute quantitatively the greater fraction of the higher alcohols. Amyl alcohols are reported to have an 'alcoholic', 'sweet' and 'choking' aroma (Falqué *et al.*, 2001). Versini and Margheri (1979), by sniffing the grappa fraction outgoing from

Table 14.1 Average values of volatile compounds of 51 analysed samples of commercial grappa produced with non-aromatic marc and not aged.

	Average (%vol)	Average (ml/100 ml p.a.)	Standard deviation	Maximum (% vol)	Minimum (% vol)	Maximum (ml/100 ml p.a.)	Minimum (ml/100 ml p.a.)
Ethanol	45.04		3.6985	60.00	39.44		
Methanol		0.51	0.2244			0.92	0.12
1-Propanol		50.02	22.3040			117.50	18.50
2-Butanol		30.22	52.9223			212.00	1.00
2–Methyl–1-propanol		69.98	18.6580			119.00	25.50
1-Butanol		2.55	1.3002			7.50	0.90
2-Methyl-l-butanol		48.04	14.4020			95.00	19.00
3-Methyl–1-butanol		181.52	60.0078			377.00	65.00
Acetaldehyde		114.91	128.3713			540.00	2.00
Ethyl acetate		162.99	133.4181			529.00	4.50
Isoamyl acetate		2.54	1.9418			9.20	0.39
Hexyl acetate		0.20	0.1693			0.72	0.01
2-Phenylethylacetate		0.18	0.1613			0.69	0.01
Ethyl hexanoate		1.85	0.9449			4.14	0.10
Ethyl octanoate		7.08	4.7826			23.00	0.86
Ethyl decanoate		11.97	9.2318			42.90	1.35
Ethyl dodecanoate		3.42	3.2513			15.30	0.09
Ethyl myristate		0.30	0.4972			2.40	0.01
Ethyl palmitate		0.69	1.3010			7.73	0.01
Ethyl oleate		0.08	0.2130			1.33	0.01
Ethyl linoleate		0.35	0.7130			3.70	0.01
Ethyl linolenate		0.08	0.1837			0.96	0.01
Ethyl lactate		32.16	7.7991			98.80	0.01
Isoamyl lactate		0.30	0.3276			1.24	0.01
Diethyl succinate		2.98	2.8238			11.80	0.01

(Continued overleaf)

Table 14.1 Continued.

	Average (%vol)	Average (ml/100 ml p.a.)	Standard deviation	Maximum (% vol)	Minimum (% vol)	Maximum (ml/100 ml p.a.)	Minimum (ml/100 ml p.a.)
1-Hexanol		8.97	3.4959			17.00	3.69
trans–3-Hexen-l-ol		0.18	0.1223			0.44	0.01
cis–3-Hexen-l-ol		0.38	0.2325			1.21	0.06
trans–2-Hexen-l-ol		0.19	0.1591			0.63	0.01
1-Eptanol		0.14	0.0951			0.45	0.01
1-Octanol		0.12	0.0667			0.25	0.01
1-Nonanol		0.12	0.0817			0.37	0.01
Benzyl alcohol		0.07	0.0789			0.40	0.01
2-Phenylethanol		2.30	2.1702			7.20	0.01
3-Ethoxyl-l-propanol		0.12	0.2111			0.88	0.01
Linalool		0.07	0.0847			0.52	0.01
Nerol		0.06	0.2082			1.50	0.01
α-Terpineol		0.08	0.1586			0.70	0.01
Furfural		0.45	0.2944			1.20	0.01
Benzaldehyde		0.81	0.8460			3.28	0.04
1-Decanol		0.10	0.0586			0.25	0.01
Isoamyl octanoate		0.14	0.1872			0.10	0.01
Isoamyl decanoate		0.15	0.1629			0.88	0.01

Source: Versini *et al.*, 1991

the gas chromatograph, reported that only amyl alcohols, especially 2-methyl-1-butanol, have a 'medicinal' note. Overall, the higher alcohols form a sort of desirable 'alcoholic' base note for the other flavour compounds of spirit. Usseglio-Tomasset (1985) showed that higher alcohols do not increase during marc storage, except 2-butanol, which is not present in the fresh marc, being produced by lactic acid bacteria (*Lactobacillus* spp.) during storage. At present, thanks to the introduction of shorter periods of marc storage, the 2-butanol content is generally medium–low. Some aliphatic long-chain alcohols consisting of six or more carbon atoms occur in grappa. The four aliphatic alcohols, 1-hexanol, *trans*-3-hexen–1-ol, *cis*-3-hexen-1-ol and *trans*-2-hexen-1-ol originate during the process of crushing, de-stemming and maceration of grapes. The formation of C_6 alcohols from grape lipids of skin and seeds occurs by the action of grape berry endogenous enzymes. In grappa, 1-hexanol content ranges from 4–17 mg/100 ml p.a. (16–102 mg/l). This alcohol has a positive influence on the aroma of the distillate when its concentration is up to 20 mg/l. At higher levels 1-hexanol (having an aromatic description of 'coconut-like', 'harsh' and 'pungent') can contribute negatively to the aroma of the product. Among the aromatic alcohols, 2-phenylethanol can be contained between traces to 7 mg/100 ml p.a. (0–42 mg/l) in grappa. This compound, which introduces a pleasant aroma to distillates, described as 'rose-like', 'sweet' and 'perfume-like', has a positive influence on the aroma of grappa. The distillation technique plays an important role in the content of this alcohol.

14.3.2 Acetates and esters

High concentration of ethyl acetate is an indication of aerobiosis in the raw material during fermentation or the result of an incorrect separation of the first fraction (heads) during distillation. Ethyl acetate is, in general, the most abundant ester in spirits. In grappa, its average content is 163 mg/100 ml p.a. (734 mg/l). The presence of this ester at levels above 250 mg/100 ml p.a. contributes unpleasant odours of glue and solvents to grappa. The average concentration of isoamyl acetate is 2.5 mg/100 ml p.a. (11 mg/l), well above its perception threshold of 1.6 mg/l. It contributes a fruity note of banana to grappa aroma. 2-phenylethyl acetate and hexyl acetate, highly favoured for their floral and fruity aroma, are also present. Their average concentrations are 0.18 and 0.20 mg/100 ml p.a. (0.8 and 0.9 mg/l), respectively.

The average content of ethyl esters in grappa is 58 mg/100 ml p.a. (260 mg/l). It is higher than in other distilled drinks (Nykänen and Suomalainen, 1983). Ethyl lactate and diethyl succinate derive mainly from bacterial spoilage of the distillate marc. Ethyl lactate is able to stabilise the grappa flavour and softens the harsh flavour characteristics when present at concentrations lower than 150 mg/100 ml p.a. The average content of the ethyl esters of the C_6–C_{12} monocarboxylic acids is 24 mg/100 ml p.a. (110 mg/l). They are generally associated with pleasant fruity flavours (apple, pear, banana, exotic fruits) and their presence is relevant because they harmonically synergise to produce the characteristic beverage aroma. The content of the ethyl esters of long-chain fatty acids (C_{14}–C_{18}) varies between

0.05–16 mg/100 ml p.a. (0.2–96 mg/l). All of these esters are relatively high-boiling and their aroma begins to diverge widely from that of the fruit esters. Some of their odours resemble that of vegetable oils, while others are more or less stearine-like.

14.3.3 Aldehydes

Aldehydes, generally characterised by low sensory thresholds, are important for grappa aroma. The average content of acetaldehyde in grappa is 115 mg/100 ml p.a (518 mg/l). Concentrations of acetaldehyde higher than 100–150 mg/100 ml p.a. give a pungent irritating odour. Acetaldehyde reacts with ethyl alcohol to form acetal, which has an herbaceous aroma. Williams and Strauss (1977) first identified unsaturated carbonyl compounds present in trace amounts – *trans*–2-octenal, *trans,trans*-2,4-heptadienal, *trans*–2-nonenal, *trans,trans*,2,4-nonadienal, *trans*-2-undecenal and *trans,trans*-2,4-decadienal – originating from oxidation of polyunsaturated fatty acids in the grape-seed oil, to be the most pronounced off-odours in spirit derived from heap-fermented grape marc. Versini and Margheri (1979) confirmed the presence of the same aldehydes in grappa. Associated with microbial development during storage and favoured by a high pH in the medium, the content of benzaldehyde can vary between 0.15–20 mg/l. Low concentrations of furfural, ranging from trace to 1.2 mg/100 ml p.a. (7.2 mg/l) can be detected in grappa. This compound, formed during distillation by thermal degradation of sugars, is responsible for a bitter almond and cinnamon aroma.

14.3.4 Terpenes

Terpenes are another very important chemical group which has an important effect on flavour and aroma by supplying floral nuances to grappa. These compounds can be used for varietal differentiation of grappa. Terpene content (linalool, nerol and α-terpineol) in non-aromatic grappa ranges between 0.03–3 mg/100 ml p.a. (trace–18 mg/l).

14.4 Grappa categories

Grappa is classified according to its age, the type of ageing technique used during production, the grape or grapes used to produce it, as well as the plant essences used to aromatise it. Grappa is classified according the following categories:

- Young or white: grappa which is bottled soon after production and is not kept in wooden containers. This kind of grappa has no colour and is transparent, and has a delicate and typical aroma as well as a dry and clean taste.
- Aged: grappa which is bottled after having been aged for a period not longer than 12 months in wooden containers. Its colour, smell and taste depend on the type and capacity of the wooden containers.

- Old: grappa which is bottled after having been aged for a period of time from 12 to 18 months in wooden containers.
- Very old or Reserve: grappa bottled after having been aged for a period of time longer than 18 months.
- Aromatic: grappa produced with aromatic or semi-aromatic grapes, such as Muscat blanc, Gewürztraminer, Malvasia, etc.
- Mono-varietal: grappa produced with the marc of a single variety of grape. The name of the grape used is usually written on the label. This type is a recent addition to the grappa family.
- Poly-varietal: grappa produced with the marc of many varieties of grapes.
- Aromatised: grappa with one or more added aromatic plant essences. The most common aromatic grappas are flavoured with fruits (raspberry, apricot, blackberry, peach, etc.), honey, radicchio, as well as officinal and aromatic herbs (juniper, rue, liquorice, etc.). On the basis of Regulation (EC) No 110/2008, grappa shall not be flavoured but this shall not exclude traditional production methods.

Besides the previous categories, the grappa sector includes also the distilled grape or grape-spirit category. Indeed, grape-spirit does not belong to the grape marc spirit or grape marc category but it is a fruit spirit obtained by distilling fermented whole grapes. In Italy the production and marketing of grape-spirit was authorised in 1988 (Gazzetta Ufficiale Della Repubblica Italiana, 1988) and even now the consumer confuses it with a 'light' grappa.

14.5 Grappa tasting

Young and aromatic grappa should be tested at 8–12 °C, whilst aged, old and reserve grappa should be tested at 15–18 °C. Low temperatures diminish the perception of smells and alcohol; high temperatures accentuate them. In order not to get 'burned' by the alcohol's aggressiveness when smelling the grappa bouquet, a proper glass should be used to obtain a better perception of aromas. In Italy, the Association which represents experienced grappa tasters is ANAG (Associazione Nazionale Assaggiatori Grappa e Acquaviti). Figure 14.1 shows the novel ANAG tasting-card. Tasters fill in a multiple choice question tasting-card and tick off the boxes corresponding to their perception; there are also questions related to 'hedonic' sensations.

First comes the visual sensation. This is the evaluation of clarity and tonality of colour. Examination for clarity verifies the absence of veiling or of powdery suspensions in the distillate, which, if present, would lend evidence to production process problems. Examination for tonality of colour provides evidence of the category to which the grappa belongs. Young grappa is devoid of any colour and is crystal-clear, aged grappa tends towards a yellowish/gold colour, while flavoured grappa is in part influenced by the colour pigment of the herb or fruit used in its flavouring.

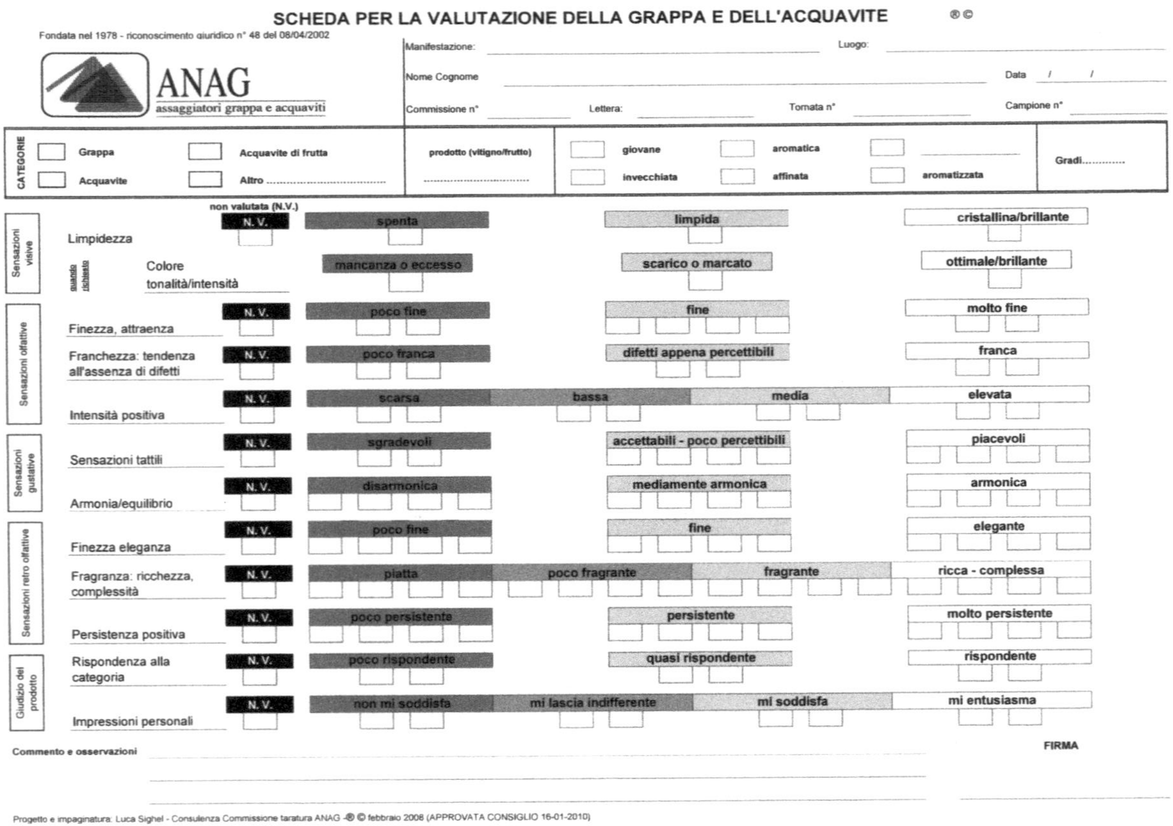

SCHEDA PER LA VALUTAZIONE DELLA GRAPPA E DELL'ACQUAVITE ®©

Fondata nel 1978 - riconoscimento giuridico n° 48 del 08/04/2002

ANAG
assaggiatori grappa e acquaviti

Manifestazione: Luogo:
Nome Cognome Data / /
Commissione n° Lettera: Tornata n° Campione n°

CATEGORIE: Grappa; Acquavite; Acquavite di frutta; Altro ....
prodotto (vitigno/frutto) ....
giovane; invecchiata; aromatica; affinata; aromatizzata
Gradi..........

non valutata (N.V.)

Sensazioni visive	Limpidezza	N. V.	spenta	limpida		cristallina/brillante
	quando richiesto: Colore tonalità/intensità		mancanza o eccesso	scarico o marcato		ottimale/brillante
Sensazioni olfattive	Finezza, attraenza	N. V.	poco fine	fine		molto fine
	Franchezza: tendenza all'assenza di difetti	N. V.	poco franca	difetti appena percettibili		franca
	Intensità positiva	N. V.	scarsa	bassa	media	elevata
Sensazioni gustative	Sensazioni tattili	N. V.	sgradevoli	accettabili - poco percettibili		piacevoli
	Armonia/equilibrio	N. V.	disarmonica	mediamente armonica		armonica
Sensazioni retro olfattive	Finezza eleganza	N. V.	poco fine	fine		elegante
	Fragranza: ricchezza, complessità	N. V.	piatta	poco fragrante	fragrante	ricca - complessa
	Persistenza positiva	N. V.	poco persistente	persistente		molto persistente
Giudizio del prodotto	Rispondenza alla categoria	N. V.	poco rispondente	quasi rispondente		rispondente
	Impressioni personali	N. V.	non mi soddisfa	mi lascia indifferente	mi soddisfa	mi entusiasma

Commento e osservazioni

FIRMA

Progetto e impaginatura: Luca Sighel - Consulenza Commissione taratura ANAG -® © febbraio 2008 (APPROVATA CONSIGLIO 16-01-2010)

Fig. 14.1 Tasting sheet for grappa (Source: ANAG, undated).

The olfactory and retro-olfactory evaluation is the second phase. To avoid irritation of the olfactory mucous membrane, it is advisable to draw the glass slowly to the nose, several times successively. The olfactory examination is based on sensations generated by the volatile compounds contained in grappa. It is of utmost importance to verify the absence of defects or unpleasant odours. Among the off-odours there are those of fodder (livestock feed), of wax or of acridity – all symptoms of a bad marc storage. Particularly pungent sensations lend evidence to problems during distillation – poor cut of the heads.

When the absence of all defects has been verified, the final step can begin with the search for pleasant aromas that recall floral, fruity, herbaceous or spicy notes. These positive sensations are successively amplified and integrated together with the sensations that have reached the olfactory area of the mucous membrane, eventually evaporating in the oral cavity following either the tasting or swallowing of the distillate.

The taste evaluation is last. To avoid sensations of excessive burning, a small sip should be taken. The first perceived taste is the burning sensation of alcohol that is perceived as a 'pseudo-caloric' sensation, more or less strong according to the quantity of alcohol contained in the liquid. After that, the flavours of grappa come out and are easily perceivable and recognisable: sweet and bitter. Acid flavour is affected by alcohol and therefore it is difficult to perceive, whereas the perception of a salty flavour is always considered as a defect. Fruit flavours will also be perceived, usually the same ones perceived by the nose.

The examination of grappa ends with swallowing. The grappa is warmed to body temperature in the oral cavity, and therefore new aromas and odours are perceived as 'retro-olfactory', that is on the inside and back part of the nose. These odours contribute to the finesse and to the quality of grappa. The amount of time these aromas last and the capacity of grappa to leave a 'clean' mouth are both quality factors, sign of a good distillation practice as well as a sign of good storage and care of the marc. Examples of sensory profiles for grappa are shown in Fig. 14.2.

14.6 Grappa and the consumer

According to research on the knowledge and preferences of grappa consumers (Demoskopea, 2003), they are still mainly men (even if the proportion of women is increasing), between 35–50 years old and of good socio-economic status. In particular, three types of grappa consumers can be identified:

- People over 60 years old, with medium–low education appreciating traditional or aged grappa,
- People between 35–40 years old, with medium–high education, appreciative of good food, quality products and also rare/fine grappa,
- The third segment includes young people of around 18–20 years, both male and female, appreciating mostly flavoured and soft grappa.

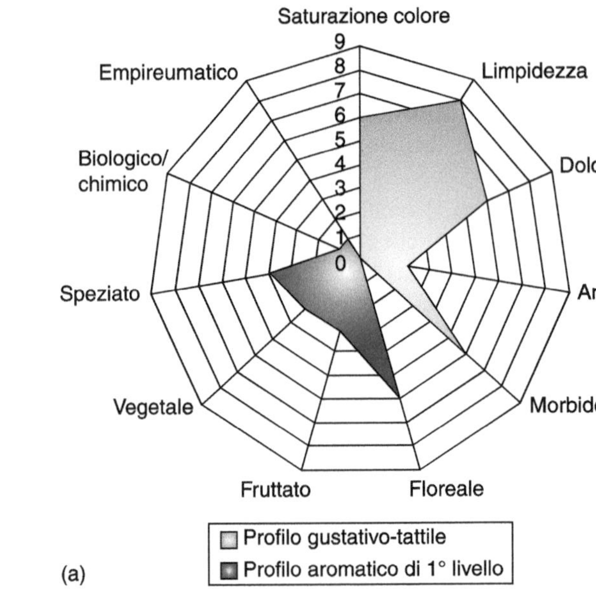

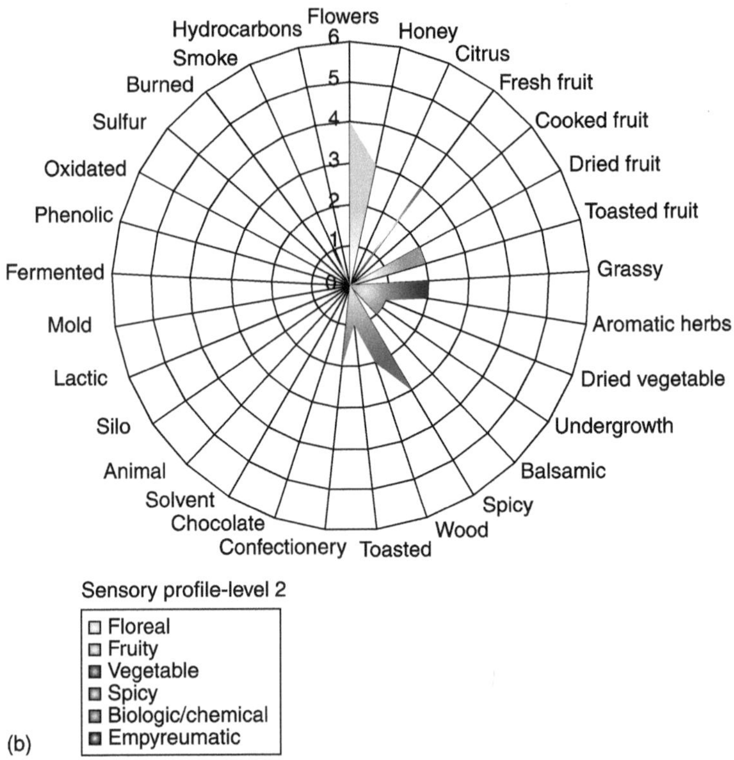

Fig. 14.2 Sensory profiles for an aged grappa. (a) Sensory profile level 1. (b) Sensory profile level 2.

Grappa has thus lost its old image of a traditional but poor and low-class product, and has gained recognition and inclusion as a high-class alcoholic drink. Conversely, the traditional type remains in demand in certain areas: the typical consumer lives in the north-east of Italy. Only a small proportion (8%) think of themselves as 'grappa experts' and the majority have a limited knowledge of brands and types of grappa. About half of the consumers drink young or white grappa because of its taste (for foreigners it *is* an acquired taste) and another fifth because it has good digestive properties. Grappa is often mixed with an espresso and this is called '*Caffè Corretto*', correcting the coffee.

The grappa buyer (either for their own consumption or as a gift) considers first and foremost the taste and smell of the drink (47% of respondents), then the brand (41%) and only afterwards price (24%) and Geographic Indication (GI) (17%). Consumers show an increasing interest not only in quality and taste but also in packaging and bottle shape, especially when grappa is bought as a gift. Actually, packaging includes both bottle (size and shape) and label. Generally, 0.7 and 1 l bottles are definitively prevalent. Indeed, a strong differentiation can be shown across grappa segments. The multiple vine variety bottle size is mostly 0.7 l or greater while the single vine variety bottle is smaller, reaching 0.7 or 0.5 l; flavoured grappa is largely sold in bottles of 0.7 l, as well as aged ones. In the medium–short term, the bottle size has decreased in all segments. The shift from big to small bottles is stronger in single-vine varieties than multiple-vine varieties, while aged and flavoured examples do not show a significant bottle size change. One of the most amusing and innovative packagings for grappa is that developed by a famous and ancient distillery of the north-east of Italy, consisting of 10 bottles of 5 ml single doses in cigarette-like boxes.

The survey carried out by Demoskopea has shown that one-third of interviewees drink grappa out of the home and more than 50% only at home, while the remainder drinks grappa both at home and outside. Consumption seasonality and taste preferences have also been investigated: grappa is drunk especially in winter, white grappa is the most requested while delicate and medium–low grade grappa is preferred to the highly alcoholic one. Lastly, almost 50% of people know some major grappa brand, but they do not know how to taste it. Grappa reaches the final consumer mostly through two marketing channels: hotels, restaurants and café (Ho.re.ca) and traditional and large-scale retail (LSR). Generally, sales have shifted from low price to medium-high price grappa in line with increasing quality and product differentiation. In the single-vine variety, for instance, high-price grappa sales (over 20 euro/l) have strongly increased; flavoured grappa shows increasing price sales as well, especially for new flavours while prices of aged grappa are stable around 12–16 euro/l.

14.7 New product development

Originally, producers' strategies were focused on cost while neglecting a marketing approach. In the last two decades, grappa producers have focused their marketing strategies on consumer demand and brand repositioning by changing marketing channels and product image. This marketing philosophy has lead to a deep differentiation and specialisation in order to target the preferences of rich consumer segments. In fact, grappa has definitively changed its image, though without losing its tradition. The consumption of grappa is growing slightly but steadily (about +3%) thanks to the consumer target expanding to involve both female and young persons, the opening of new markets in Italian geographical areas still 'virgin' (Southern Italy) and greater consumer knowledge of grappa culture. Nowadays, as the grappa sector is characterised by many niche brands, the marketing strategies are mostly oriented to integrated marketing communications including public relations, direct marketing, event marketing and internet marketing rather than to classic and expensive advertising (TV, radio, newspapers, etc.). Currently, the grappa sector is focused on the protection and valorisation of Geographical Indication (GI) in the world as a marketing tool. Indeed, despite the assurances of the EU on strengthening GI for grappa in the new Regulation (EC) No 110/2008 and its notification to the World Trade Organization, there are still numerous cases of counterfeiting, particularly in those countries having a 'wine vocation', for example Brazil, Argentina, United States and South Africa. The GI of grappa can be expected to play an important role in marketing, since the area of origin is one of the most distinctive factors of food products. The Commission of the European Communities should formalise in the reference methods, as it did for the wine sector (Regulation (EC) No 2676/1990 (European Union, 1990) and following amendments), the introduction of stable isotope analysis for spirit drinks. In this way, grappa could be discriminated on the basis of the GI, using already available official isotopic data banks of Italian wines. Authenticity and traceability are taken into consideration by a large enough proportion of grappa buyers and are important factors in establishing price. Besides, the correct implementation of these methods would avoid disputes and ensure unitary control of spirit importation.

Recently, the grappa market has been revitalised by some product innovation like grappa with *kosher* certification produced following the Biblical and Rabbinic regulations, grappa obtained by distilling marc coming from grapes of 'organic agricultural origin', and 'peated' grappa obtained by smoking the grape marc before distillation. In addition, some distillers have worked at blending varieties of grappa to match chocolate, cigars and cheeses. Others have produced liquors based on grappa but having other components such as honey, blueberry, apple or pear juice.

14.8 Future trends

Today, the grappa market shows encouraging positive trends. A marketing philosophy focused on strong differentiation and diversification has successfully targeted even the more demanding consumers. This notwithstanding, the grappa market is already mature and should be revitalised by further developing the product and setting a wider price range to reach potential consumers, especially outside Italy. However, the great challenge for the future is to produce 'grappa' with low ethanol content. Drink-driving legislation and the awareness of health risks from excessive alcohol intake have increased interest in the production of low-alcohol beverages in general. The technologies now available for the production of low- and reduced-alcohol beverages include membrane separation, reverse osmosis, pervaporation and supercritical fluids, but the quality of the beverage is still affected by the incidental removal, with the ethanol, of a proportion of the other flavour compounds.

14.9 Sources of further information and advice

- Istituto Nazionale Grappa, Via Domokos 8–40100 Bologna. Available at: http://www.istitutograppa.org (accessed 30 November 2010).
- Associazione Nazionale Industriali Distillatori di Alcoli e di Acquaviti (ASSODISTIL), Via Torino, 146 – 00184 Roma. Available at: http://www.assodistil.it (accessed 30 November 2010).
- Associazione Nazionale Assaggiatori Grappa e Acquaviti (ANAG). Available at: http://www.anag.it (accessed 30 November 2010).
- FEDERVINI, Via Mentana, 2/B – 00185 Roma. Available at: http://www.federvini.it (accessed 30 November 2010).
- Centro Documentazione Grappa Luigi Bonollo, Località Citille, 43/b, 50022 Greve in Chianti (FI), email: greve@bonollo.com.

14.10 References

ASSODISTIL 2010. Associazione nazionale industriali distillatori di alcoli e di acquaviti, Relazione 2009.

DA PORTO, C., 1998. 'Grappa and grape-spirit production'. *Critical Reviews in Biotechnology* **18**, pp. 13–24.

DA PORTO, C., 2002. 'Volatile composition of 'grappa low-wines' using different storage methods and conditions on an industrial scale'. *International Journal of Food Science and Technology* **37**, pp. 395–402.

DA PORTO, C., CORTELLA, G. and FRESCHET, G., 2004. 'Preliminary study on a cooling practice of grape pomace during storage on an industrial scale.' *Italian Journal of Food Science* **16**, pp. 87–95.

DA PORTO, C. and LONGO, M., 2005. 'Grappa: Part 1, Part 2'. In: *Proceedings of Worldwide Distilled Spirit Conference*, Edinburgh, September 21–25, 2004.

DEMOSKOPEA, 2003. 'Barometro Grappa: Consumi e distribuzione in Italia'. Research presented at *Mondo Grappa Symposium*, Conegliano Veneto (TV), October 18–19.

EC 1990. Council Regulation (EEC) No. 2676/1990 of 17 September 1990. *Official J. Eur. Coummunities* L 272/1–192.

EC 2008. Council Regulation (EEC) No.110/2008 of 15 January 2008. *Official J. Eur. Coummunities* L39/16.

FALQUÉ, E., FERNÁNDEZ, E. and DUBOURDIEU, D., 2001. 'Differentiation of white wines by their aromatic index'. *Talanta* **54**, pp. 271–281.

GAZZETTA UFFICIALE DELLA REPUBBLICA ITALIANA, 1988. n. 270 del 17 novembre 1988 – D.M. 3 novembre 1988.

NYKÄNEN, L. and SUOMALAINEN, H. 1983. 'Occurrence of esters of monocarboxylic acids in distilled beverages'. In: *Aroma of Beer, Wine and Distilled Alcoholic Beverages*. Akademie-Verlag, Berlin in co-publication with D. Reidel Publishing Company, Dordrecht, Holland, pp.148–199.

PROFUMO, A., RIOLO, C., PESAVENTO, M. and FRANCOLI, A. 1988. 'Evolution of the Italian distillate "Grappa" during aging in wood: a gas chromatographic and high performance liquid chromatographic study'. *American Journal of Enology and Viticulture* **4**, pp. 273–278.

USSEGLIO-TOMASSET, L., 1985. 'La grappa nella sua composizione e nei suoi sistemi di elaborazione'. In: *Atti del IX Convegno Nazionale della Grappa*, Treviso, 20–21 settembre.

VERSINI, G., MONETTI, A., DALLA SERRA, A. and INAMA, S., 1991. 'Caratterizzazione analitico-statistica su base regionale della grappa'. In: *Atti del Simposio Internazionale "Le denominazioni di origine dei vini e delle acquaviti viticole: qualità e autenticità"*, Bollettino del CIDEAO, Alessandria (Italy), 30 novembre –1 dicembre, WR Editoriale & Commerciale s.a.s, Alessandria, pp 134–150.

VERSINI, G. and MARGHERI, G., 1979. 'Rapporto fra i costituenti volatili della grappa e le caratteristiche organolettiche'. *Vini d'Italia* **21**, pp. 269–277.

WILLIAMS, P.J. and STRAUSS, C.R., 1978. 'Spirit recovered from heap-fermented grape marc: nature, origin and removal of the off-odour'. *Journal of the Science of Food and Agriculture* **29**, pp. 527–533.

15

Moutai (Maotai): production and sensory properties

Y. Xu, Jiangnan University, China and K. Ji, Moutai Distillery Co. Ltd, China

Abstract: Kweichow Moutai is not only a famous alcoholic beverage popular in China, but is also an excellent example of a typical Chinese liquor, whose production involves unique fermentation and distilling technology, and blending processes. Much of the science and technology behind these processes is relatively unknown in detail. This chapter surveys China's national liquor, Kweichow Moutai, discussing its unique production environment and production process, from the raw materials, sorghum and wheat, to finished product. Information on the sensory evaluation and flavor chemistry of Moutai liquor is also provided. The chapter reviews the historical background of Kweichow Moutai, then discusses the desirable eco-environment for Moutai production and the raw materials required, then focuses on liquor production processes such as *qu* production, solid state fermentation, distillate production, blending, filling and packaging, with a brief discussion of marketing.

Key words: Kweichow Moutai, sorghum, *daqu, qu*, solid-state fermentation, distilling.

15.1 Historical background

Kweichow Moutai (or *Guizhou Maotai* in Mandarin Chinese), was the first of the typical Chinese liquors with a soy sauce aroma, made from grains and *daqu* (also shortened to *qu*, a leavening agent that forms the starting point for liquor production, in a similar fashion to Japanese *koji*). Legend has it that the original inhabitants of Maotai Town on the banks of the *Chishui* or Red River (which marks the boundary between the Sichuan and Guizhou Provinces of Southwest China) were adept at making this liquor during the reign of Emperor Dayu.

According to historical sources, Kweichow Moutai was first developed in 135 BC during the Western Han Dynasty (206 BC~ AD 25). At that time, Maotai Town and its neighboring areas produced a great deal of 'wolfberry sauce aroma liquor', the predecessor of present-day Kweichow Moutai. This liquor had been used as a royal tribute since Emperor Wu (156~87 BC) of the Han Dynasty (206 BC~AD 220) sampled it and acclaimed it as 'sweet and delicious', as recorded in *Shi Ji* or the Records of the Grand Historian.

All through the Song (960~1279), Yuan (1279~1368), Ming (1368~1644) and Qing (1644~1911) dynasties, it was offered to the emperors and kings as a major tribute. It is now regarded as the country's national liquor, and was officially given this status at the Founding Ceremony of the People's Republic of China (PRC). At the Panama Pacific International Exposition in 1915, it won a gold medal, achieving the same international status as Scottish Whisky and French Cognac.

The state-owned Kweichow Moutai Distillery was founded in 1951 and has since made rapid progress. Kweichow Moutai has won both national and international awards more than ten times in succession and was judged to be among the Top 4, Top 8, Top 13, and Top 17 long-established liquors in China in a series of liquor tasting and appraisal meetings.

From its origins in a primitive still to its development into a contemporary super-sized Chinese enterprise group, Kweichow Moutai's success can be attributed to its unique production processes, consistent high quality, distinctive attributive properties and brand culture, and to the unique ecological environment in which it is produced.

15.2 Ecological environment

Kweichow Moutai is produced in Maotai Town, Renhuai County, in the northwest of the Guizhou Plateau. The Kweichow Moutai Distillery Co. Limited is situated on the eastern bank of the Red River at an altitude of 450 m. With long summers and short winters, the region enjoys an agreeable climate and frosts are rare: temperatures range from −2–40 °C with an annual average of 18 °C. The relative humidity of the region is high, at about 78%, while its wind speed is about 1.2 m/s on average through the year. All of these natural elements constitute the unique ecological environment for making Kweichow Moutai.[1]

The unique flavor of Kweichow Moutai relies largely on the equally unique climate, water and environment found in Maotai Town (Figs. 15.1 and 15.2); as a result, it is impossible to distill Kweichow Moutai anywhere else in the world. The successful production of Kweichow Moutai results from a combination of three factors: firstly, the nature of the environment in the Maotai area, which contains a vast community of active microorganisms essential for the production of the liquor; secondly, the climate and water supply from the Red River, which, as mentioned above, provide perfect conditions for distillation; and thirdly, the long-standing distillation processes practised by the original inhabitants of Maotai Town, which led to the development of the ideal production process for a highly distinctive aromatic liquor such as Kweichow Moutai.

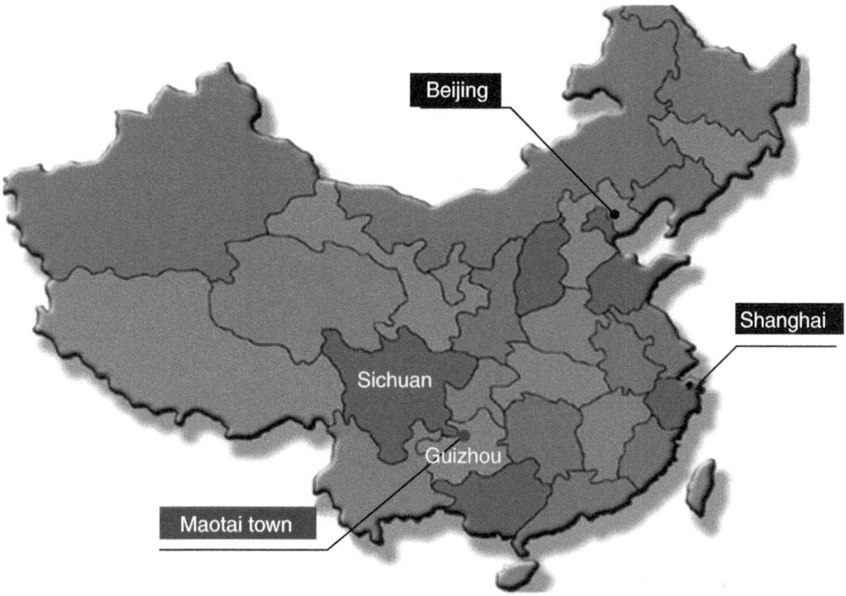

Fig. 15.1 Location of Maotai Town.

Fig. 15.2 Bird's eye view of Maotai Town and the Red River.

15.3 Production of Kweichow Moutai

15.3.1 Raw materials

The basic raw materials for making Kweichow Moutai are *hongyingzi* (meaning 'with red-tasselled ears') glutinous sorghum and wheat. These are produced locally, with the use of only natural organic fertilizers and no chemical fertilizers or pesticides. The sorghum grown locally in Maotai Town and its neighboring areas is perfectly suited to the production of Kweichow Moutai: firstly, it has thick husks and small kernels, with a high proportion of germ and amylopectin starch; secondly, it is resistant to steaming and boiling and, thanks to its low rate of water absorption, is slow to gelatinize. This low rate of water absorption is another result of its high starch content of more than 60% and specifically, its amylopectin starch content of more than 88%.

The raw materials for *qu* production are all selected from the locally-grown wheat from Maotai Town or the surrounding area. The wheat is compact and evenly-ripened, with a thin hull and plump kernel. It has no worm infestations or mould growth, and is high in purity and light yellow in color, with no browning at the end of the kernel. The wheat also has a starch content of over 60%.

The water used for distilling Kweichow Moutai comes from the Red River and it is clean, colorless and transparent. Both its quality and purity level are high, and it is free from deposited solids and flavor taints. Its temperature is on the low side and its pH range fluctuates between 6.5–7.5.

15.3.2 Processes used in Kweichow Moutai production

The production of Kweichow Moutai, including fermentation and distillation, is carried out in approximately year-long cycles, and involves several typical Chinese liquor-making processes. These include stack solid-state fermentation (pre-fermentation), and *qu* production, and all require higher temperatures than those used for other Chinese liquors. For example, the maximum recorded temperature for *qu* production is 67 °C.

The following provides an overview of the production processes, which are discussed in detail in subsequent sections. Each year around the time of the Dragon Boat Festival (falling upon 5 May of the Chinese Lunar Calendar), *qu* or starters are formed into brick shapes and then subjected to fermentation and gradually to high temperatures to encourage natural fermentation. This is followed by 40 days of cultivation. After at least six months in storage, the cultured *daqu* can then be used as saccharification and fermentation agents. Solid-state fermentation in a stone-pit is typical of Moutai production, and is carried out in nine batches over the course of a year. Around the Double Nine Day (on 9 September of the Chinese Lunar Calendar), two lots of sorghum are added to the first two batches. The batches are then steamed and cooked, *qu* is added, and they are double-fermented. The *qu* and fermented grains are then piled together in a heap for an initial two days of fermentation to promote yeast growth and propagation, followed by a further 30 days of fermentation in pits. The same processes, from the piled fermentation to pit fermentation, are carried out for the remaining seven batches. Except in the case of the first two batches, distillates from the initial two batches are added during fermentation in order to achieve a higher quality spirit; distillates from the following seven batches are collected for storage. These seven batches then undergo distillation to produce the spirit, while the initial two batches do not. The method of batch distillation, carried out with solid-state fermented grains in a cylindrical pot, is also typical of Chinese spirit production processes.

When compared to other Chinese liquor-making processes, the production processes used for Moutai can be described as 'three highers, three lowers, three greaters and one smaller'. This means: higher-temperature *qu* production, higher-temperature fermentation and higher-temperature distillate-cooling; lower saccharification of *daqu*, lower water content of *daqu* and lower liquor yield; greater usage of *qu*, a greater number of repeated batches and greater usage of grains; and finally smaller quantities of auxiliary materials. The distilled basic liquor, sorted by batch, type, alcoholic strength and age, is stored for a long period before it is ready to be blended.

There are a total of four processes involved in making Kweichow Moutai: *qu* production, distillate production, blending, filling and packaging.

Qu production

The process used to make *qu* for Kweichow Moutai is illustrated in Fig. 15.3.

The raw materials for *qu* production are ground roughly into scoured grist, which should contain pieces of husks but not too much fine powder, and should not have either an obviously coarse or obviously smooth appearance. The amount of *qu* inoculum and water accounts for about 6–8% and 40% respectively of all the *qu* materials in the mixing process. This ensures the perfect combination of water, *qu* materials and wheat powder, so that no dry powder remains and the mixture can easily be moulded into a ball but separates when flung. The *qu* is then shaped into brick chunks, with no cracks or dust remaining. The bricks, which should be firm at the sides but loose in the centre, are then stacked on their ends in the fermentation barn with two horizontal and three vertical chunks interlacing with each other. During the fermentation period of at least 40 days, the *qu* chunks must be turned twice and the temperature in the barn must be monitored so that any remaining moisture is removed. The fermented *qu* chunks have a strong aroma, thin skin, evenly-distributed color and a chrysanthemum-shaped center. They should not be greasy, have a sour smell, or show signs of mould or rot caused by bacteria.

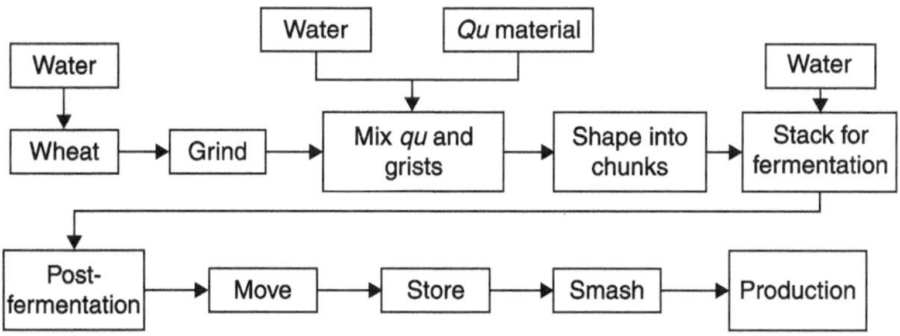

Fig. 15.3 *Qu* production process for Kweichow Moutai.

Distillate production

The distillate production process is illustrated in Fig. 15.4.

Grains are added twice during the process. These additions are known as *xiasha* (first) and *zhaosha* (second) deposits. The raw materials should be ground until the fine powder content reaches 27–30%. The moistening temperature must be above 90 °C and the water used accounts for about 50–54% of the total amount of raw materials. The pressure for steaming grains ranges from 0.08–0.12 mPa, and 80–100 minutes are needed for steaming the second deposits. Fermented grains should make up 8–12% of the total amount of raw materials. The mixing process requires a temperature of between 28–32 °C, a dose of *qu* appropriate for the batch in question, and 3–5% feints. The mixture is then stacked for approximately two days and is then moved into a stone pit for a further 30 days of fermentation. The optimum temperature for the stack fermentation stage, designed to promote the propagation of yeast and other strains is between 50–53 °C, while the temperature for pit fermentation should be around 30–43 °C. For the distilling process, the pressure should remain at 0.08 mPa for a duration of longer than 40 minutes. A spirit distillation is divided into three fractions: foreshots, middle cut and feints. The main running of spirit distillation make up the middle cut, are known as the basic distillate and are stored for blending. Such main distillates are high in the characteristic flavor compounds of Moutai, meaning that the blending process is better able to supply consistently high-quality liquor.

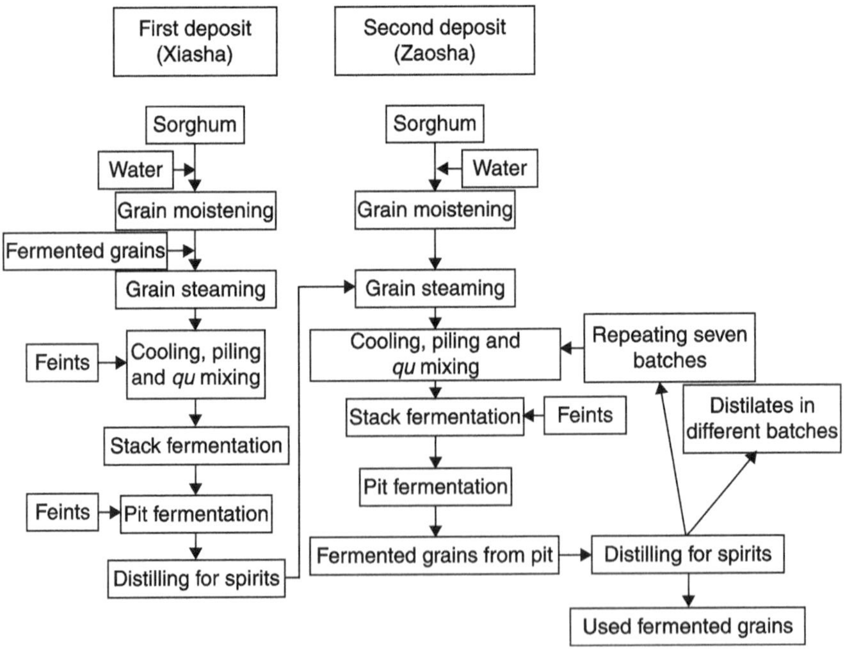

Fig. 15.4 Distillate production process for Kweichow Moutai.

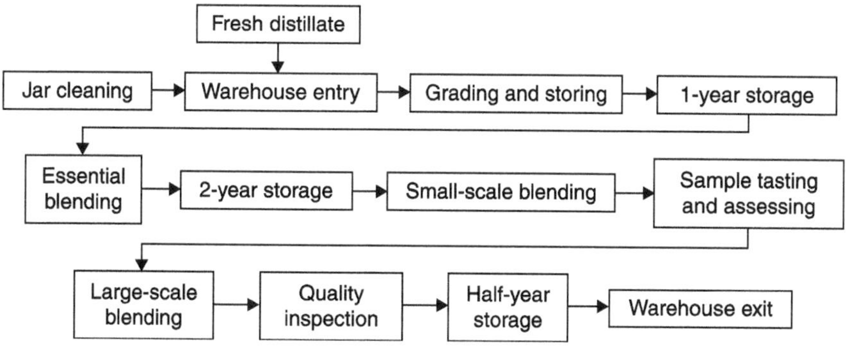

Fig. 15.5 Blending process of Kweichow Moutai.

Blending

The blending process used in Kweichow Moutai production is illustrated in Fig. 15.5.

Before elaborate blending and packaging can take place, the freshly-made Kweichow Moutai basic distillates are stored in jars for at least three years according to their round of distillation, type and alcoholic strength. Jars must be cleaned before the fresh distillate is stored. As a rule, jars are first filled with clean water and then left for two to three days to check for any leakages, cracks, impurities or dust. Only when they are confirmed to be clean and sanitary can they be filled with the fresh distillates. The jars are then put into barns for preservation, and labeled with information such as the year of production, date, barn number, jar number, workshop, team staff, round and quantity.

After they have been stored for a whole year, distillates of the same round, aroma and class can be used for essential blending. Two years after that, small-scale blending can be carried out in accordance with production standards. Large-scale blending is then carried out with the same proportional parameters as small-scale blending. Generally speaking, blending involves basic distillates of different types, alcoholic strength and duration of storage, taken from various batches of production, on the condition that the standards of the blended samples meet the requirements of Kweichow Moutai. Each batch of Kweichow Moutai blends is finished by using at least 80–100 basic distillates. When this has been completed, the blended liquor will be stored for another half year before being packaged ready for sale. The whole process, from the start of production to preparation for sale, takes at least five years.

Filling and packaging

The filling and packaging process used for Kweichow Moutai is illustrated in Fig. 15.6.

The filling and packaging process includes such procedures as blend intake, filling, marking and so on. The bottles used are expected to be flawless and to meet the requirements for uniform thickness and size. Prior to the filling process,

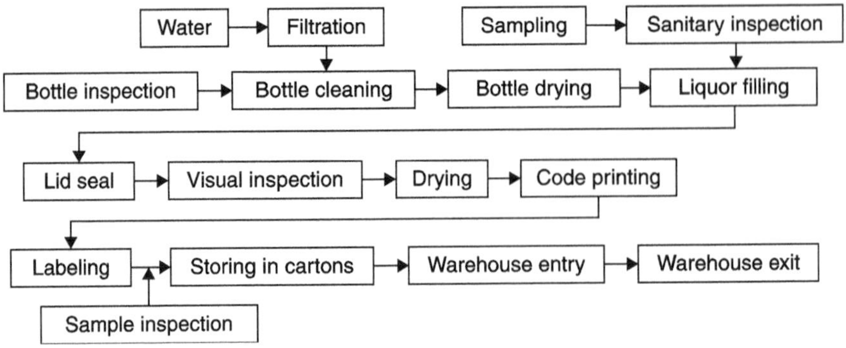

Fig. 15.6 Filling and packaging process of Kweichow Moutai liquor.

bottles should be washed with clean, clear water with no suspended solids, and inspected to ensure that both the interior and exterior of the bottles are clean, without impurities and with no remaining excess water. The liquor blend to be put into the bottles should first undergo quality inspection to ensure that it is pure, free from turbidity and suspended solids, and then checked against the specifications required for each specific variety and batch, including with regard to the net weight of liquor. The lid used for sealing the bottles should meet with the following standards: it should be smooth, clean and tight, with no dust, deformation or leaks. It should fit the bottle well enough that no water can escape even if the bottle is turned upside down. Once the lid is sealed, a ribbon is tied around the neck of the bottle, with information including the brand name 'Kweichow Moutai', date of production, batch number, and serial number of the bottle and production venue codes marked on the lid. A specially designed tag coupled with an anti-counterfeit label is then attached to the bottle symmetrically and smoothly. The bottles of liquor are then put into cartons or boxes, ensuring that they are packaged the right way up, with the appropriate accessories for the stipulated variety and quantity. The ends of the cartons are then neatly sealed with tapes and, finally, straps are looped around the cartons for transportation.

15.4 Sensory properties and flavor chemistry

15.4.1 Properties of flavor compounds

Kweichow Moutai is produced with *daqu* as the saccharifying and fermenting agent, simultaneous fermentation together with saccharification as the technique and sorghum starch as the raw material. In these respects, it differs from both Scottish Whisky, which uses malt as the saccharifying agent, and from French Cognac, which uses grape must containing mainly sucrose and other saccharides as the raw material and added yeasts for fermentation.

The *qu* production and liquor production processes used for Kweichow Moutai make full use of the ecological advantages and the unique natural conditions and resources of the region. Such processes as *qu* production, stack fermentation and anaerobic fermentation all take place at high temperatures, which constantly promote the growth of liquor-making micro-organisms. Due to the variety of micro-organisms present in the region and through heredity, mutation and evolution, extreme microbial communities have developed around the Kweichow Moutai workshops that are resistant to high temperatures, acidity and alcoholicity. During the Kweichow Moutai production process, the metabolism of the functional micro-organisms gives rise to highly stable heat- and acid-resistant enzymes such as amylases, proteases, glucoamylases, cellulases, glucosidases, xylanases as well as various dehydrogenases and phosphoenol-pyruvate carboxykinases involved in oxidation–reduction reactions, etc.[3] All kinds of extreme micro-organisms and enzymes, encouraged by the long production process, thus help to produce the rich flavor compounds of the liquor. These flavor compounds enter into the basic distillates by means of different processes and further assist in the formation of the unique taste and perfect quality of Kweichow Moutai. To date, functional micro-organisms such as *Bacillus licheniformis*[4] and *Bacillus subtilis*,[5–7] which produce the soy sauce aroma, have been isolated from Moutai *daqu* that is resistant to high temperatures, and their fermentation and metabolism can further produce flavor substances such as acetoin and 2, 3, 5, 6-tetramethylpyrazine and furaneol.[8–11] The unique fermentation environment involved in the production of Kweichow Moutai provides the liquor with its distinctive and unique flavor attributes.

Analysis and comparison of the micro-components in Chinese soy sauce aroma spirits (Kweichow Moutai-flavor liquors), strong aromatic spirits (Luzhou-flavor liquor), faint-scented spirits (Fen-flavor liquor), brandy and whiskey has been carried out via the highly-sensitive full two-dimensional GC-GC/TOFMS. The peaks detected were as follows: 963 for soy sauce aroma liquors; 674 for strong aromatic spirits; 484 for faint-scented spirits; 440 for brandy; and 264 for whiskey, confirming that Kweichow Moutai is the distillate with the richest flavor micro-components worldwide (Fig. 15.7). Qualitative analysis in Chinese Spirit 169 Project also shows that there are 873 chemical components in Moutai including

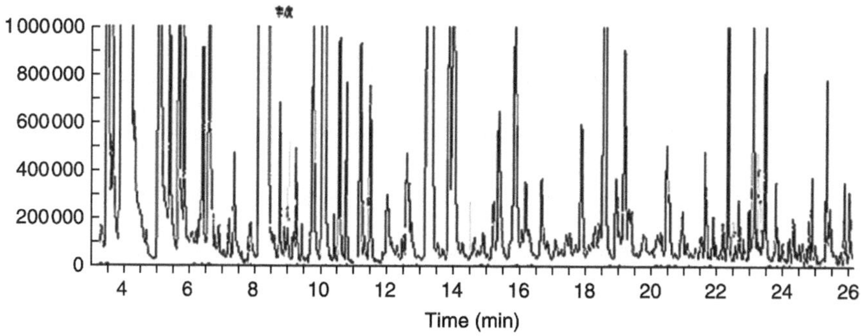

Fig. 15.7 Capillary column gas chromatogram of a Moutai-flavor liquor extract.

380 esters, 85 acids, 155 alcohols, 96 ketones, 73 aldehydes, 36 nitrogenous compounds and 48 other chemicals. In particular, there is greater variety in the nitrogenous compounds found in Kweichow Moutai than in other types of liquors, with up to three and seven times as many as those in Luzhou-flavor liquor and Fen-flavor liquor, respectively.[12–14]

The findings from research on the flavor impact compounds of Kweichow Moutai using gas chromatography-olfactometry (GC-O), gas chromatography–mass spectrometry (GC-MS) together with OSME, a specific gas chromatography-olfactometry technique, and aroma extract dilution analysis (AEDA) show that there are over 300 flavor impact compounds that contribute to the unique taste of Kweichow Moutai, 126 of which have a major influence on the formation of Kweichow Moutai's aroma while a further 65 also play important roles in the process. Some of the impact volatile aroma compounds found in Moutai-flavor liquor and their descriptors are shown in Table 15.1.

Table 15.1 Volatile constituents and descriptors of Chinese Maotai-flavor liquor

Compound		Descriptor
Esters		
	Ethyl acetate	Pineapple
	Propyl acetate	Fruity
	Ethyl propanoate	Banana, fruity
	Ethyl 2-methylpropanoate	Fruity, sweet
	2-Methylpropyl acetate	Floral, fruity
	Ethyl butanoate	Pineapple
	Ethyl 2-methylbutanoate	Berry, fruity
	Ethyl 3-methylbutanoate	Apple
	3-Methylbutyl acetate	Fruity
	Ethyl pentanoate	Fruity
	Ethyl hexanoate	Sweet, fruity
	Hexyl acetate	Floral, fruity
	3-Methylbutyl butanoate	Fruity
	Propyl hexanoate	Fruity
	Ethyl heptanoate	Fruity
	Ethyl 2-hydroxypropanoate	Fruity
	3-Methylbutyl pentanoate	Fruity, floral
	2-Methylpropyl hexanoate	Apple, sweet
	Butyl hexanoate	Pineapple, fruity
	Hexyl butanoate	Fruity
	Ethyl 2-hydroxy-3-methylbutanoate	Floral
	Ethyl 2-hydroxybutanoate	Fruity, floral
	Ethyl octanoate	Fruity
	3-Methylbutyl hexanoate	Fruity, apple, green
	Pentyl hexanoate	Fruity
	Ethyl nonanoate	Floral, fruity
	Ethyl 2-hydroxyhexanoate	Floral, jasmine
	3-Methylbutyl heptanoate	Sweet, fruity
	3-Methylbutyl lactate	Floral, fruity

	Hexyl hexanoate	Apple, peach
	3-Methylbutyl octanoate	Fruity, pineapple
	Ethyl decanoate	Fruity, grape
	Diethyl butanedioate	Fruity, sweet
	Heptyl hexanoate	Sweet, fruity
	Ethyl undecanoate	Floral
	Ethyl dodecanoate	Leaf
Alcohols		
	2-Butanol	Fruity
	1-Propanol	Alcoholic, fruity
	2-Methylpropanol	Wine, solvent
	2-Pentanol	Fruity, alcoholic
	1-Butanol	Pungent, alcoholic
	3-Methylbutanol	Rancid, nail polish
	2-Hexanol	Fruity
	Cyclopentanol	Fruity, plant
	2-Heptanol	Fruity
	3-Methyl-1-pentanol	Pungent, alcoholic
	1-Hexanol	Floral, green
	3-Hexen-1-ol	Floral
	3-Octanol	Woody
	2-Octanol	Fruity
	1-Heptanol	Fruity, alcoholic
	2-Ethylhexanol	Green
	2-Nonanol	Fruity
	1-Octanol	Fruity
	1-Nonanol	Green
Acids		
	Propanoic acid	Vinegar
	2-Methylpropanoic acid	Acid, rancid
	Butanoic acid	Rancid, cheesy
	3-Methylbutanoic acid	Rancid, acidic
	Pentanoic acid	Sweaty, rancid
	2-Methylpentanoic acid	Sweaty, rancid
	3-Methyl-2-butenoic acid	Floral
	4-Methylpentanoic acid	Sweat, sour
	Hexanoic acid	Sweaty, cheesy
	Heptanoic acid	Sweaty
	Octanoic acid	Sweaty, cheese
	Nonanoic acid	Fatty
	Decanoic acid	Fatty
	Dodecanoic acid	Cheese
Aromatic compounds		
	Benzaldehyde	Fruity, berry
	Phenylacetaldehyde	Fruity
	Ethyl benzoate	Fruity
	Acetophenone	Floral, sweet
	(2, 2-Diethoxyethyl)benzene	Vegetable
	Naphthalene	Mothball
	Ethyl phenylacetate	Rosy, honey

(Continued overleaf)

Table 15.1 Continued

Compound		Descriptor
	2-Phenylethyl acetate	Rosy, honey
	1-Methylnaphthalene	Fatty
	Ethyl 3-phenylpropanoate	Fruity, floral
	Benzenemethanol	Floral, sweet
	2-Phenylethanol	Rosy, honey
	1-Phenylethanol	Rosy, honey
	2-Phenylethyl butanoate	Fruity
	Benzenepropanol	Sweet, floral
	2-Hydroxy-3-phenylpropanoic acid	Goaty, smoky
	Benzoic acid	Fruity, cherry
	Vanillin	Sweet, floral
	2-Phenylacetic acid	Rosy
	3-Phenylpropanoic acid	Fruity, floral
Aldehydes and ketones		
	2-Methylbutanal	Malt, green
	3-Methylbutanal	Malt, green
	2-Pentanone	Fruity
	2-Heptanone	Fruity, spicy, cinnamon
	3-Octanone	Vegetable, green
	3-Hydroxy-2-butanone	Buttery
	2-Hydroxy-3-pentanone	Floral, fruity
	2-Nonanone	Fruity
Nitrogen-containing compounds		
	2, 5-Dimethylpyrazine	Baked
	2, 3-Dimethylpyrazine	Nutty
	2-Ethyl-6-methylpyrazine	Nutty, roasted
	2, 3, 5-Trimethylpyrazine	Roasted, nutty
	2, 6-Diethylpyrazine	Nutty, baked
	2, 3-Dimethyl-5-ethylpyrazine	Baked
	3, 5-Dimethyl-2-ethylpyrazine	Roasted, nutty
	2, 3, 5, 6-Tetramethylpyrazine	Baked
	2-Acetyl-6-methylpyridine	Roasted, baked
	2, 5-Dimethyl-3-pentylpyrazine	Nutty, baked
	2-Acetylpyrrole	Herbal, medicine
Furans		
	Furfural	Sweet
	2-Acetylfuran	Sweet, caramel
	5-Methyl-2-furfural	Roasted
	2-Acetyl-5-methylfuran	Roasted
	Ethyl 2-furoate	Balsamic
	2-Furanmethanol	Burnt sugar
	Furfuryl hexanoate	Caramel, fruity
Phenols		
	Guaiacol	Clove, spicy
	Phenol	Phenol, medicinal
	4-Methylphenol	Animal, phenol
	4-Ethylphenol	Smoky

Acetals		
	1, 1-Diethoxyethane	Fruity
	1, 1-Diethoxy-3-methylbutane	Fruity
	1, 1, 3-Triethoxypropane	Vegetable, fruity
	1, 1-Diethoxynonane	Fruity
Sulphur compounds		
	Dimethyl disulfide	Onion, cabbage
	Dimethyl trisulfide	Sulfur, rotten cabbage
Lactones		
	Γ-Hexalactone	Nutty

(Reproduced by permission of Jiangnan University)

Research carried out on the aroma diffusion compounds of Kweichow Moutai and Luzhou-flavor liquors found that there were up to 163 kinds of these compounds present. Furthermore, 28 aroma compounds were found to linger in empty glasses.

Studies have also shown that Kweichow Moutai liquor contains a large number of nutritious and restorative micro-components. For example, more than 17 of Kweichow Moutai's flavor micro-components have been found to be beneficial to human health. The international joint program Influences of Kweichow Moutai Liquor on Gene Expression of Mice Livers, (Cherian and Liu Jie)[15] studied the potential health benefits of Kweichow Moutai. The results confirmed that an appropriate intake of Kweichow Moutai can contribute to good health, as when consumed in moderation, the liquor encourages the liver to produce elevated amounts of beneficial metallothionein.

15.4.2 Sensory properties

The basic distillates of Kweichow Moutai are produced by making full use of the unique natural and ecological resources in Maotai Town, while high-quality Moutai liquor is produced using various scientific and complex blending processes, depending on the different features of the base distillate and final products. Kweichow Moutai varieties differ in their quality and sensory attributes due to, and the duration of, storage and alcohol content. For example:

- 53% vol aged Kweichow Moutai. The age of the base distillate is no less than 15 years. This is then blended into Kweichow Moutai aged liquor, whose sensory characteristics are as follows:
  - Appearance: light yellowish and transparent, without suspended or deposited solids,
  - Smell: featuring soy sauce aroma and obvious maturation aroma, fine and strong tasting, with an enduring aroma lingering in empty glasses,
  - Taste: with notable maturation flavor and lingering aftertaste, tasting fine and smooth, pleasantly strong and full,

  - Style: featuring soy sauce aroma, refined and delicate, pleasantly strong and full, with obvious agreeable maturation aroma and enduring aftertaste lingering in empty glasses.
- 53% vol Kweichow Moutai. Its sensory characteristics are as follows:
  - Appearance: colorless (or yellowish) and transparent, without suspended or deposited solids,
  - Smell: featuring soy sauce aroma, fine and strong tasting, with enduring aroma lingering in empty glasses,
  - Taste: tasting pleasantly strong and full, with lingering aftertaste,
  - Style: featuring soy sauce aroma, refined and delicate, pleasantly strong and full, with enduring aftertaste and lingering aroma in empty glasses.
- 43% vol Kweichow Moutai. Its sensory characteristics are as follows:
  - Appearance: clear and transparent, without suspended or deposited solids,
  - Smell: fine and strong tasting, with obvious soy sauce aroma and enduring aroma lingering in empty glasses,
  - Taste: tasting full and pleasantly strong, with lingering aftertaste,
  - Style: with the unique style of Kweichow Moutai.
- 38% vol Kweichow Moutai. Its sensory characteristics are as follows:
  - Appearance: clear and transparent, without suspended or deposited solids,
  - Smell: fine and tasteful, with obvious soy sauce aroma and enduring aroma lingering in empty glasses,
  - Taste: tasting mild, mellow, with lingering sweet aftertaste,
  - Style: with the unique style of Kweichow Moutai.
- 33% vol Kweichow Moutai. Its sensory characteristics are as follows:
  - Appearance: clear and transparent, without suspended or deposited solids,
  - Smell: quite fine and strong tasting, with obvious soy sauce aroma and enduring aroma lingering in empty glasses,
  - Taste: tasting mild, mellow, with lingering sweet aftertaste,
  - Style: with the unique style of Kweichow Moutai.

15.5 Marketing

As the national liquor of China, Kweichow Moutai enjoys the joint protection of China's alcohol industry and the local authorities of its native region and is recognized as being both environmentally-friendly and organic. It occupies a special position in China, and has been used internationally to establish friendship and facilitate cultural exchanges. Within China, Kweichow Moutai has come to represent the image of the Chinese people and the Chinese nation, while internationally it is known as a typical example of a Chinese liquor, and has become the preferred liquor for consumption in international political, economic and cultural events.

Kweichow Moutai has been consumed, collected and well-received in more than 100 countries and regions and has become one of the top choices for liquor consumers at home and abroad, thanks to its organic, cultural and scientific traits.

Its organic properties offer health benefits that attract consumers; its cultural heritage makes the product appealing; and its scientific background leads to a superior quality which makes it enormously popular.

The supply of Kweichow Moutai currently falls considerably short of domestic demand at present, even during periods of low demand. In some cities of China, Moutai stocks are at risk of running out. Despite this, consumers still maintain a high level of enthusiasm for Kweichow Moutai, whose sales tend to show a notable increase every year, and imbalance between supply and demand continues, especially in terms of aged Moutai liquor. This is largely due to the brand value of Kweichow Moutai as China's national liquor and the improvement in the living standards of Chinese people. With respect to the overseas market, publicity campaigns that emphasize the cultural heritage of Moutai, along with increasing global recognition and understanding of traditional Chinese culture, will lead to greater worldwide demand for Kweichow Moutai and an increase in global consumer acceptance. It is therefore likely that a further shortage in supply will arise, given the ever-increasing demands of the international market.

15.6 Sources of further information

ZHU B R, ZHANG K C (2000), *Scripture of Chinese Alcoholic Beverages* (*ZhongGuoJiuJing*), Shanghai Culture Publishing House.

ZHANG KC (1995), *Technology of Alcoholic and Distilled Beverages*, China Light Industry Press.

15.7 Acknowledgements

Many thanks to Yihua Zhu and Yongguang Huang from Jiangnan University who gave valuable work on language translation and editorial work. Their efforts are very much appreciated. Acknowledgments are due to the Ministry of Science and Technology, P. R. China, under the National Key Technology R & D Program (2007BAK36B02 and 2008BAI63B06), and to the Ministry of Education, P. R. China under the Programme of Introducing Talents of Discipline to Universities (111-2-06), and to the Natural Science Foundation of China (NSFC) (No. 31000806, 30470046, 20872050), and 169 Project for their financial support.

15.8 References

1. XU Y, WANG D, FAN W, MU X, CHEN J. 'Traditional Chinese biotechnology'. *Advances in Biochemical Engineering/Biotechnology*, 2010, **122**: 189–233, Springer.
2. CHEN X, JI K. 'General introduction to the individualities of Moutai. *Liquor-Making Science & Technology*, 2006, **2**: 79–84.
3. YUAN R. *More Difficult to Produce Moutai than Atomic Bombs* N/OL. [2007-04-20 2007-04-20]. http://finance.qq.com/a/20070421/000086.htm.

4. ZHANG R and XU Y. 'Research on *Bacillus licheniformis* from Daqu of soy sauce type and its metabolites with flavor characteristics'. *Industrial Microbiology*, 2010, **40**(3): 1–7.
5. ZHU B, XU Y, FAN W. 'Tetramethylpyrazine production by fermentative conversion of endogenous precursor from glucose by *Bacillus* sp.' *Journal of Bioscience and Bioengineering*, 2009, **108**: S122 (doi: 10.1016/j.jbiosc.2009.08.358).
6. ZHU B, XU Y, FAN W. 'Study of tetramethylpyrazine formation in fermentation system from glucose by *Bacillus subtilis* XZ1124'. *New Biotechnology*, 2009, 25S: S237.
7. ZHU B and XU Y. 'High-yield fermentative preparation of tetramethylpyrazine by *Bacillus* sp. using an endogenous precursor approach'. *Journal of Industrial Microbiology and Biotechnology*, 2010, **37**(2): 179–186.
8. HUANG Y, XU H, HUANG P. 'Extreme liquor-making environment and extreme liquor-making microbes of Moutai liquor'. *Liquor-Making Science & Technology*, 2006, **12**: 47–50.
9. ZHU B and XU Y. 'A feeding strategy for tetramethylpyrazine production by *Bacillus subtilis* based on the stimulating effect of ammonium phosphate'. *Bioprocess and Biosystems Engineering*, 2010, **33**(8): 953–959.
10. ZHU B and XU Y. 'Production of tetramethylpyrazine by batch culture of *Bacillus subtilis* with optimal pH control strategy'. *Journal of Industrial Microbiology and Biotechnology*, 2010, **37**(8): 815–821.
11. FAN W, XU Y., ZHANG Y. 'Characterization of pyrazines in some Chinese liquors and their approximate concentrations'. *Journal of Agricultural and Food Chemistry*, 2007, **55**(24): 9956–9962.
12. XU Y, FAN W, WANG H, WU Q. 'Advance in flavor-directed analytic technology for Chinese liquor (*baijiu*)'. *Liquor-Making Science & Technology*, 2010, **11**: 73–76.
13. JI K, GUO K. 'Investigation on microconstituents in Moutai liquor'. *Liquor-Making Science & Technology*, 2006, **10**: 98–100.
14. JI K, GUO K, ZHU S, LU X, XU W. 'Analysis of microconstituents in liquor by full two-dimensional gas chromatography/time of flight mass spectrum'. *Liquor-Making Science & Technology*, 2007, **3**: 100–102.
15. CHERIAN, M.G. and LIU JIE 'Kweichow Moutai Liquor on Gene Expression of Mice Livers'.

16

Pisco: production, flavor chemistry, sensory analysis and product development

E. Bordeu, E. Agosín and G. Casaubon, Pontificia Universidad Católica de Chile, Chile

Abstract: Pisco is a young brandy produced predominantly from Muscat wines in the north of Chile and also in Peru. The chapter describes the areas of production in Chile, wine production and distillation techniques and gives an overview of production figures and the main markets nowadays. The sensory properties and flavor chemistry of Pisco are described emphasizing the properties that make it different from other young or non-aged wine brandies. Sensory practice and consumer research in particular for product development in the Chilean Pisco industry are finally discussed.

Key words: Pisco, brandy, muscat, wine distillation, odour activity value, aroma, flavor, sensory properties, gas chromatography, mass spectrometry, gas chromatography-olfactometry, aroma extract dilution analysis, free choice profiling.

16.1 Introduction

According to Chilean legislation Pisco is defined as a brandy produced and bottled in units of consumption in the III (Atacama) and IV (Coquimbo) Regions, through the distillation of genuine wine, from *Vitis vinifera* varieties specified by law and planted in those regions.

Pisco is by far the most important distilled beverage produced and consumed in Chile. It corresponds to a non-aged or slightly aged kind of brandy, produced in the north of Chile and protected by one of the oldest denominations of origin of the country. The other particularity of Pisco as a young brandy is the inclusion of Muscat grapes to obtain the wines that are distilled. This coupled with a traditional distillation gives most Piscos a distinctive Muscat fruity aroma that clearly differentiates it from other spirits. Along with wine and beer, Pisco in Chile is clearly one of the most popular drinks, consumed mostly in different cocktails or mixed with sodas and fruit juices. Both in Chile and in Peru, Pisco is the most important distilled beverage. However, it is almost unknown elsewhere and exports are still minimal, mostly to neighbor countries or to Chilean colonies residing outside Chile.

According to Chilean tradition the production of this kind of brandy is even older than independent Chile having originated during the colony, when Peru and Chile belonged to the same Spanish Viceroyalty, which explains why Pisco exists in both countries. Information about bottled brandy labeled Pisco in Chile was already available in the second half of the nineteenth century, and as early as 1931 the official Denomination of Origin was created, specifying the area of production and grape varieties that can be used. Pisco also exists in Peru, and a similar brandy called Singani exists in Bolivia. According to the Peruvian position, the use of the word Pisco to designate a brandy would have started earlier in Peru from the name of the Peruvian port, Pisco, from where it was exported.

16.2 Pisco production

The area of production is restricted to five valleys: Copiapó, Huasco, Elqui, Limarí and Choapa. They run from the Andes to the Pacific Ocean in an extremely dry and luminous environment in the southern border of the Atacama Desert, the driest place on earth. Snow melting in the Andes provides the limited water supply that allows viticulture in this very arid condition. The region is located between the 27th and 32nd parallels of latitude south. The climate that characterizes the central portion of the valleys can be described as sub-desertic with a strong influence of the Pacific anticyclone, determining minimal cloud cover and no rainfall nine to ten months of the year, giving a high level of solar radiation during the growing season. This area, with higher temperatures and easy grape maturation, is where Pisco production started and where most of the acreage of vineyards was concentrated. Currently, viticulture has also extended closer to the Pacific Ocean, which strongly modulates the climate, lowering heat accumulation and provides a great diversity of 'terroirs' depending on the distance to the sea. This makes it possible to find optimal conditions for different grape varieties and styles. Currently, high-quality wine production for direct consumption in this area is under strong development. Extremely dry conditions, with no rain at all during the growing season, assure an excellent maturation and the best sanitary conditions, making it very easy to obtain organic products.

Grape varieties for Pisco production are mostly characterized by their strong fruity aromas, including several Muscat-type grapes like the world renowned Muscat of Alexandria, known locally also as Italia, White, Yellow, Orange and Black Muscats, Chasselas Musque Vrai, Frontignan Muscat, Hamburg Muscat, Moscato de Canelli, Torontel, and Moscatel Rosada (Pink Muscat), a local selection also with a strong terpenic aroma. Besides these Muscat-style varieties, Pedro Jimenez and a variety known as Austria Muscat, both with low Muscat aroma, are also considered for Pisco production providing the more neutral spirits necessary for blending. Most vineyards are trained as overhead 'pergolas' in the valleys and as traditional vertical trellises in the hillsides. Irrigation is essential considering the absence of rain. Enough water supply is provided by the Andes and a system of reservoirs in each valley.

Winemaking has evolved from very traditional techniques, still seen in some small enterprises (Bordeu and Pszczolkowski, 1982), to very modern facilities in the larger companies with stainless steel tanks, up-to-date equipment and highly technological enology providing the high-quality, aromatic white wines to be distilled. Winemaking is typical of modern white wine styles with low-temperature fermentation of low-solid clarified musts protected from oxygen. The use of sulfur dioxide is minimized as in other wines for distillation to avoid the distillation of sulphur dioxide and high levels of acetaldehyde. Wine conservation before it is distilled is basically assured by high alcohol and hygiene because pH tends to be high. The storage period is shortened by early distillation.

As soon as wines are finished with fermentation, the distillation process is started using traditional copper alambics, similar to those used for Cognac production in France. Alcohol is usually obtained in a single-batch distillation process with the classic separation of heads, heart or brandy and tails. Short rectification columns are generally used at the end of the heart fraction and the raw alcohol obtained is around 60°. Some new distillation techniques have evolved recently such as double distillation, looking for more neutral products to compete with spirits such as vodka.

An important difference with other types of brandy is that Pisco is not usually aged significantly, being bottled after a short period of maturation in wood, basically to refine the aroma. However, some types are aged longer in oak barrels looking for styles where wooden and age-related aromas are predominant. This style, found in some of the more expensive products, represents only a small segment of the market. Not all high-quality and expensive Pisco undergoes this ageing process; they rely mostly on the quality and intensity of the Muscat aromas. After maturation or aging Pisco is adjusted to the final consumption alcohol content (35–50°), clarified and bottled. The resulting distillate ranges from completely colorless to light amber, and is characterized by fruity Muscat aromas combined in different degrees with aromas coming from aging and oak.

There is no legal classification of Pisco types; however, four types are usually found on the market with different alcohol contents: traditional, 30–35°; special, 35°; reserve, 40° and 'Gran Pisco' 43–46° alcohol.

Four commercial categories of Pisco can also be described as relatively typical:

- Artisanal Pisco: small producers using traditional technologies like skin contact during winemaking, originating complex products requiring some maturation or aging.
- White Pisco: young distillates with less than six months aging, in general no contact with wood, frequently coming from single varietal wines, only 'heart distillates' low in non-alcohol fractions. Can be triple distilled.
- Mature Pisco: aging between six and twelve months in used cooperage with a combination of fruity aromas and some wooden notes and light color. Usually coming from a mixture of grape varieties.

- Aged Pisco: over twelve months aging in young American oak barrels. Clearly amber in color and with an important participation of wood-derived aromas combined with fruity aromas coming from grapes. Multi-varietal also.

The total area dedicated to Pisco production was 10 508 hectares in 2007; most was concentrated in the Coquimbo Region (9816 hectares) while only 693 hectares exist in the dryer northern region of Atacama. Around 215 000 tons of Pisco grapes are produced every year. The resulting 170 million liters of wine are used for distillation (116 000 000 l) and direct consumption as wine (54 000 000 l) (SAG, 2007).

The industry has a strong social impact in the regions of production. There are 2734 grape growers, most of them very small, 83% have less than 5 hectares and 36.6% less than 1 hectare, while only 3.3% have more than 20 hectares. Most growers also have other agricultural sources of income and only 18.8% of them are dedicated exclusively to Pisco grapes. It is estimated that over 30 000 people live from the production of Pisco, including industrial workers (INE, 2007).

The Pisco industry in Chile is clearly organized around two very large companies with a cooperative structure in terms of grape provision. These companies process over 90% of the 120 millions of tons of grapes used for Pisco production. Both companies sell several Pisco brands and are also involved in other alcoholic beverages like bottled 'Pisco Sour', a cocktail with lemon juice. Other smaller companies also exist, usually selling one brand of Pisco in the domestic market.

Pisco production in Chile is now around 45 000 000 l, above 5 million 9 l cases. The production is basically consumed locally where Pisco represents around 65% of the distillates market. Annual consumption in Chile is around 3 l per capita. Pisco exports are still limited, with a volume of only 296 000 l representing US$ 1 115 250 in 2007. Exports are concentrated in the two large producers, Capel and Compañía Pisquera de Chile, plus two smaller enterprises, Aba and Bauzá. The main export markets are the United States, Argentina and Germany. The industry is interested in increasing exports as local demand is stagnant and competition from imported spirits, in particular rum, is increasing.

16.3 Pisco flavor chemistry

16.3.1 Impact odorant compounds in Pisco

The concentration range of odor-active compounds from several commercial Piscos is illustrated in Table 16.1. Among the more than 120 volatile compounds present in these distillates, identified by gas chromatography-mass spectrometry (GC-MS) analyses, only a subset of less than 30 are sensory detectable. Several of these compounds show an olfactory impact despite being present in very low concentrations (mg/l or ng/l) because of their low sensory detection thresholds (Belitz *et al.*, 2004; Guth, 1997). Odor activity values (OAV), defined as the ratio between the concentration of the compound and its detection threshold, were used to evaluate the potential sensory contribution of the volatiles found in the distillates; only those showing OAVs higher than 0.5 were reported in Table 16.1.

Table 16.1 Concentration range of odor active compounds (mg/l) in commercial Piscos

Compounds[b]	Concentration range (μg/l)	Olfaction threshold[†] (μg/l)	Odor activity value (OAV) range	Description	Reference
Acids					
Octanoic	2 420–6 833	8 800 (a)	0.3–0.8	Fatty acid	Cuillere *et al*. (2004)
Dodecanoic	409–2 155	1 000 (a)	0.4–2.2	Dry, metallic	Cuillere *et al*. (2004)
Isovaleric	78–411	33.4 (c)	2.3–12.3	Sweat, acid, rancid	Delfini *et al*. (2001)
Alcohols					
1-Hexanol	1 647–4 785	8 000 (a)	0.2–0.6	Toasty, green, dry	Cuillere *et al*. (2004)
Isoamyl alcohol	58 699–84 203	30 000 (a)	2.0–2.8	Whiskey, malt, burnt	Delfini *et al*. (2001)
Isobutyl alcohol	5 174–14 320	7 000 (b)	0.7–2.1	Sweet, chemical	Benn and Peppard (1996)
1-Heptanol	75–240	3 (b)	25.0–80.0	Green	Curionia and Bosset (2002)
C_6 compounds					
cis-3-hexenol	68–112	400 (a)	< 0.5	Grass	Cuillere *et al*. (2004)
Terpenols					
Linalool	672–2 910	15 (a)	45.0–194.0	Flowery, muscat	Cuillere *et al*. (2004)
Citronellol	9–264	100 (a)	0.0–2.6	Citrus	Peter *et al*. (1999)
α-terpineol	446–1 091	250 (c)	1.8–4.4	Anis	Cuillere *et al*. (2004)
Geraniol	18–158	30 (a)	0.6–5.3	Fruity	Peter *et al*. (1999)
Nerol	34–177	300 (b)	0.1–0.6	Rose	Semikolenov *et al*. (2003)
Ho-trienol	513–1 920	100 (a)	5.1–19.2	Sweet, honey	Rocha *et al*. (2005)
Farnesol	8.0–27	20 (b)	0.4–1.4	Lemon, anis	Cuillere *et al*. (2004)
C_{13}-Norisoprenoids					
β-Damascenone	0.0–17.0	0.05 (a)	0–340	Baked apple	Cuillere *et al*. (2004)
β-Ionone	4.0–38.0	0.09 (a)	44–422	Violets	Cuillere *et al*. (2004)
Furans					
5-Hidroxymethyl furfural	1242–2 914	15 000 (a)	0.3–6.5	Aldehyde, caramel	Câmara *et al*. (2004)

(Continued overleaf)

Table 16.1 Continued

Compounds[b]	Concentration range (μg/l)	Olfaction threshold† (μg/l)	Odor activity value (OAV) range	Description	Reference
Vainillins					
Vainillin	7–354	200 (a)	0–1.8	Vanilla	Peter *et al.* (1999)
Lactones					
γ-Butyrolactone	19–122	100 (a)	0.2–1.2	Sweet aromatic	Tateo and Bononi (2003)
cis Oak-lactone	13–171	35 (a)	0.4–4.9	Coconut, flower	Delfini *et al.* (2001)
Phenolic derivatives					
4-Vinyl guaiacol	12–29.0	40 (a)	0.3–0.7	Clove, curry	Delfini *et al.* (2001)
Esters					
Isoamyl acetate	538–3 192	30 (a)	18.0–106.0	Banana	Delfini *et al.* (2001)
Ethyl hexanoate	669–16 448	14 (a)	48.0–1175.0	Fruity, anis	Cuillere *et al.* (2004)
Ethyl octanoate	1759–5 241	2 (a)	880.0–2620.0	Fruity, anis	Cuillere *et al.* (2004)
Ethyl decanoate	1 416–6 412	200 (c)	7.0–32.0	Grape	Delfini *et al.* (2001)
Isobutyl acetate	39–140	30 (a)	1.3–4.7	Flowery	Cuillere *et al.* (2004)
Ethyl hexadecanoate	5.0–2 341	2 000 (b)	0.0–1.2	Fatty, rancid, fruity, sweet	Yong-Sheng Tao and Hua Li (2009)
Ethyl heptanoate	17–326	2.2 (b)	7.7–148.0	Banana, strawberry	Giri (2010)
cis-3-hexenyl acetate	43–49	8 (b)	5.4–6.1	Banana	Greger and Schieberle (2007)

Notes: Commercial Piscos analyzed: Horcon quemado, Capel 35, Alto del Carmen 35, Alto del Carmen envejecido, Artesanos de Cochiguaz, Campanario; Mistral, Ruta Norte.

†Olfaction thresholds determined in (a) water/ethanol (10–15% v/v), (b) water, (c) water/ethanol (13%) with glycerol and pH 3–3.4

Although the latter only provides a rough evaluation of the real contribution of each compound to the global aroma, this 'odor activity scale' is much closer to reality than the concentration scale.

The most important odorants in Pisco are the terpenes linalool and Ho-trienol; the C_{13} norisoprenoids β-ionone and β-damascenone; ethyl esters, in particular ethyl hexanoate, heptanoate and octanoate, and acetate esters, such as isoamyl acetate. Other compounds that increase the aroma complexity of the distillate are lactones and phenolic derivatives, furans, heptanol and isovaleric acid.

16.3.2 Origin of the odor-active compounds found in Pisco

Pisco is obtained after distillation of white wines produced from Muscat grape varieties. Therefore, most of the aroma in the distillate results from a unique mixture of volatile compounds originated from grapes (varietal aromas), secondary products formed during the wine fermentation (fermentative aromas) and, to a lower extent, from a short aging process (post-fermentative aromas) (Lambrechts and Pretorius, 2000; Swiegers and Pretorius, 2005). Moreover, winemaking from Muscat grapes has progressively evolved from red to white winemaking techniques to retain as much aroma as possible (Agosin *et al.*, 2000). Finally, during wine distillation, several aroma compounds are also released or degraded as a consequence of the harsh environmental conditions prevailing in the still.

Varietal aromas

Monoterpenols, such as linalool, geraniol, nerol or ho-trienol, among others, represent the basis of Muscat typicity which distinguishes Pisco from other young distillates (Baumes *et al.*, 1994); surprisingly, few C_{13}-norisoprenoids also contribute significantly to the fruity aroma of the distillate. It is worth noticing that these flavor compounds are also present in significant amounts in Muscat wines as sugar-bound non-volatile – i.e. odorless – precursors, including glycosides. This is important in the case of wine distillates because, during distillation, the latter undergo fast acid-catalyzed hydrolysis, yielding free volatiles.

Both free and bound odorants are particularly abundant in the grape varieties Muscat de Alexandria and Moscatel Rosada (Belancic *et al.*, 1997). Other grape varieties also grown for Pisco, such as Pedro Jimenez in Chile or Quebranta in Peru, contribute little to aroma (Bordeu *et al.*, 2004).

Fermentative aromas

The aroma complexity dramatically increases during alcoholic fermentation as a result of the synthesis of important volatile compounds by the wine yeast *Saccharomyces cerevisiae* (Swiegers *et al.*, 2006). The nature and amount of the synthesized volatile compounds depends on multiple factors, such as the nitrogen content of the must, the temperature of fermentation, and the yeast strain (Molina *et al.*, 2009; Lambrechts and Pretorius, 2000; Swiegers *et al.*, 2006).

The volatile compounds synthesized by wine yeasts include higher alcohols, like the isoamylic alcohol (fusel, marzipan and floral aromas), medium- and long-chain volatile acids (fatty, cheesy and sweaty aromas), acetate esters and ethyl esters (fruity and floral aromas), and aldehydes (buttery, fruity and nutty aromas), among others (Delfini *et al.*, 2001; Lambrechts and Pretorius, 2000; Stashenko *et al.*, 1992). The capacity of yeast to synthesize these compounds varies between different strains (Lambrechts and Pretorius, 2000; Wondra and Boveric, 2001).

It is worth noticing that the varietal bound precursor fraction, i.e. non-volatile aroma compounds linked to sugar moieties, is not hydrolyzed during wine fermentation and remains odorless.

Fate of free and bound aroma compounds during wine distillation

Rigorous control and operation of the distillation process is critical to retain in the final distillate the complex aroma reached after the fermentation. Unfortunately, information about the reactions occurring during this operation is scarce. Ohta (1995) studied the transformation of geraniol and nerol in linalool and alfa-terpineol during distillation of *shochu*, a traditional Japanese distillate. He also reported a significant hydrolysis of the glycosidic conjugates into the free aglycones.

We also determined a significant increase in free terpenols during distillation of Muscat wines, resulting in more than twice the concentration initially present in the original wine. The latter results from the hydrolysis of a significant portion of the glycosidic fraction as a consequence of the harsh conditions – acidic pH and high temperatures – prevailing in the still; however, linalool oxides increased and some key terpenols were partially degraded as well (Agosin *et al.*, unpublished), suggesting that complex reactions are taking place during this operation, which deserves further study.

Aging

An increasingly common practice in the Pisco industry is to expose the distillates to a short aging process or to add caramel. The presence of lactones, furans and, eventually, of phenolic derivatives in commercial Piscos (Table 16.1) is a direct indication of these treatments.

16.3.3 Off-flavors

Off-flavors in Pisco could be generated either by concentration of wine volatiles over their detection threshold after distillation – alcohol content increases around three-fold in distillate compared to wine – or by faulty distillation. Isovaleric acid and isoamylic alcohol are examples of the former; hydroxymethylfurfural of the latter (Table 16.1). Other off-flavor compounds commonly determined in Pisco are acetaldehyde and a few fatty acids, as well as unknown compound(s) responsible for a sulfury, artichoke-like taste.

16.4 Main sensory properties of Pisco

An important group of sensory properties of Pisco are similar to those of young wine distillates. However what differentiates it from other wine spirits is the use of Muscat varieties, providing their typical terpenic aromas.

A set of descriptors to be used by the industry was proposed in 2004 (Bordeu *et al.*, 2004) and part of the terminology was later validated in a study about odor and aroma profiles of Chilean Pisco (Peña y Lillo *et al.*, 2005). The terminology proposed (see Fig. 16.1) was organized in the typical wheel format similar to those for wine and whiskey. Like that for whiskey, the Pisco wheel incorporates aroma descriptors and also mouth-feel and primary taste terms. A set of standards for most descriptors was also proposed. Similar to the wine wheel, standards correspond to easily available products and materials or chemicals easy to obtain. The wheel is organized in three concentric levels going from general to more specific descriptors in the outer ring. A total of 48 terms is proposed in this outer ring; far less than those proposed in other wheels. Several of the terms proposed are similar to those proposed for other spirits; however, some terms are specific to a young spirit with Muscat aroma, like geraniol, linalool, raisin and honey. Even

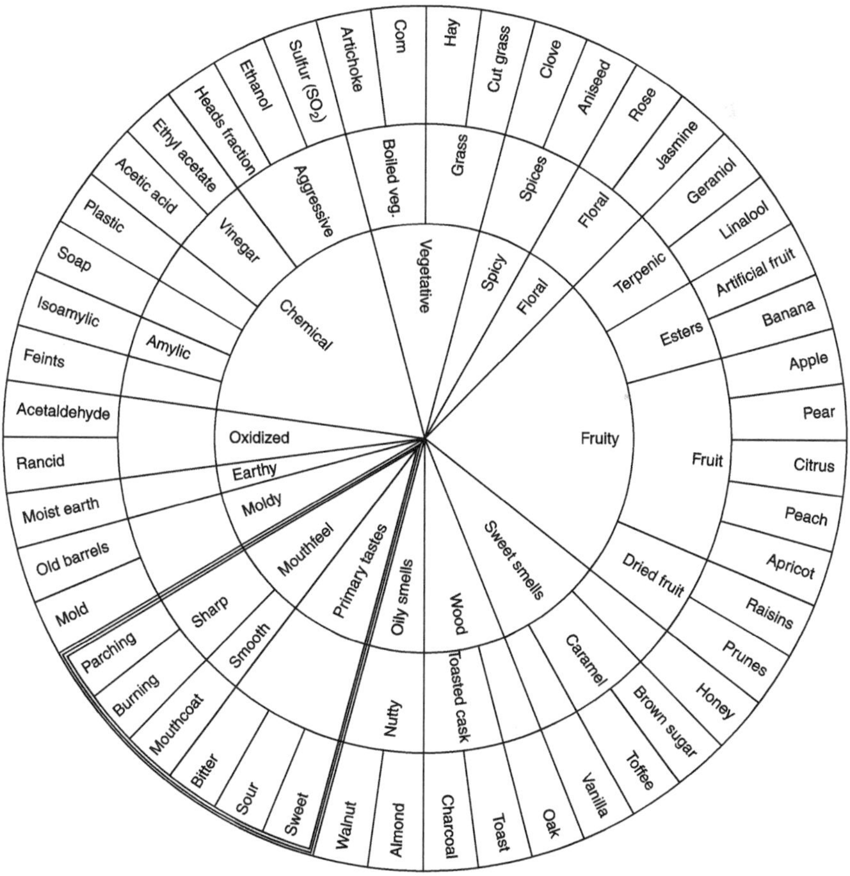

Fig. 16.1 Pisco and mouth-feel wheel, displaying its 48 outer-ring descriptors. (Source: Bordeu *et al.* 2004. © 2004 ASEV. Reprinted with permission.)

if Pisco is not aged for long periods in wood, descriptors like vanilla, oak and toasted are important in describing products with some aging. The aroma descriptors with higher frequency of use in the development of the Pisco wheel were vanilla, raisins, linalool charcoal and prunes (Bordeu *et al.*, 2004).

In another study relating odor perception and physicochemical properties of volatiles in Pisco, terms as 'distillation tails' and 'oak' correlated with acids (hexanoic, octanoic, decanoic) and esters (ethyl lactate, diethyl succinate) while 'artificial fruit' correlated with total and individual ester content, 'linalool odor' correlated with linalool content and 'vanilla', 'caramel' and 'citric' correlated with higher alcohol and terpene (linalool, geraniol and nerol) content. Linalool and esters like ethyl octanoate, ethyl hexanoate and isoamyl acetate, respectively correlated with linalool and artificial fruit descriptors and were pointed out as possible odor impact molecules in Pisco (Peña y Lillo *et al.*, 2003a, 2003b).

A quantitative descriptive analysis (QDA) was done over odor and aroma profiles of 15 Pisco spirits and 14 distillation fractions by a panel of 12 trained subjects by the use of a list of 13 attributes: Acetaldehyde, Tails, Acetic acid, Ethyl acetate, Oak, Caramel, Vanilla, Raisins, Citrus, Pear, Artificial fruit, Linalol, Hay (Peña y Lillo *et al.*, 2005). Chilean Piscos were provided by two major distilleries and by smaller, independent ones. Fourteen distillation fractions from five different distillation runs (non-commercial products) were also included. Tasters were trained in the use of the terminology and the scale during 24 sessions before starting the formal tastings. Odor (orthonasal perception obtained by sniffing the samples) and aroma (retronasal perception obtained by sipping the samples) sensory profiling were performed in separate sessions.

The results of two-factor ANOVA showed a significant product effect for all odor attributes except pear ($P = 0.40$) and all aroma attributes except caramel ($P = 0.22$) and ethyl acetate ($P = 0.09$). Principal component analysis (PCA) was applied to odor and aroma data separately, considering only the discriminant attributes. Correlation between sensory attributes in each mode proved to be extremely high. The first two principal components covered 86% of the total variance for odor and 80% for aroma, giving practically the same amount of information in both cases.

For the commercial samples, PCA of odor-significant attribute results showed that samples were differentiated according to the type of distillation fraction and maturation process. In general, the commercial samples were separated by the PC1 (70% of the total variance) and PC2 (15%) in three main groups, one related to fruity/floral Muscat derived aromas, a second associated to oak and the maturation process and finally samples characterized mostly by off-flavors (Fig. 16.2).

Distilled fractions were grouped also in three major groups, the head fractions were related to artificial fruits, associated with esters; heart fractions were close to linalool and citrus attributes, and finally the fractions close to the tail were associated with the off-flavors.

Partial least square (PLS) models gave a good prediction of aroma from odor scores, except for the oak attribute which exhibited a non-linear behavior. Based

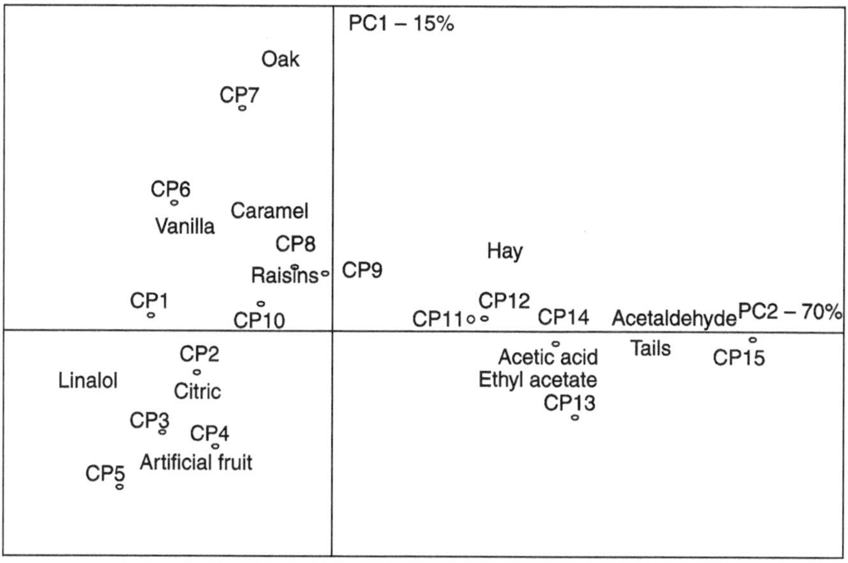

Fig. 16.2 Principal component analysis of 15 Pisco commercial samples, 11 attributes, 12 judges, 4 neplicates.

on the results of this study, Pisco olfactory perception can be limited to odor assessment.

16.5 Sensory analysis practice in the Pisco industry

Some enologists, with five-year university studies as agricultural engineers with a specialization in viticulture and enology, work in the industry and are responsible for the base wines and distillates quality. Sensory training of enologists is concentrated in wine tasting so experience to taste alcohols is gained at the industry level. Sommeliers, who work almost exclusively in the wine industry in Chile, are also beginning to work with Pisco, in particular in relation to cocktails and the relations between Pisco and Pisco-derived products and food that is consumed with this product. A third group of professionals involved in sensory evaluation of products are food engineers, also with formal training in sensory analysis.

Sensory practice in the Pisco industry differs mostly depending on company size. Small Pisco producers usually rely on tastings done by one or two technicians to follow distillation and assess quality of base alcohols and to prepare commercial lots. In large companies, sensory issues are handled by larger groups of technicians and professionals. Teams of enologists are responsible for wine and alcohol quality but panels are also used for quality control and product development. Typically, the technician in charge of each distillation plant first evaluates each lot of distillate and then a sample is sent to a trained panel that ranks it to accept or to reprocess it. Panelists are professionals, technicians and quality control personnel trained by the enologists in charge. Expert panels, with participation of enologists, are in general responsible for assigning each base alcohol to product lines and mixing lots to obtain the final commercial products. Consumer panels with no training are also used for acceptance studies of new products. Collaboration with regional universities offering food engineering is also used to evaluate Piscos and Pisco-based cocktails.

New product development is usually started by enologists or expert panels and then marketing personnel and independent market research companies are used to organize focus groups and evaluate acceptance at consumer level.

For more specific situations or problems, the industry also relies on specialized sensory and analytical studies commanded to universities or laboratories offering this service in Chile or sometimes abroad.

Even if Pisco is a denomination of origin, there is no agreement about the tasting of Piscos, as in some European wine denominations of origin. Potability of the final product is checked first by the laboratories of the Pisco producers and then it is controlled by analysis practiced by official laboratories under supervision of the Servicio Agricola y Ganadero: Agricultural and Livestock Service (SAG), belonging to the Ministry of Agriculture. All exported Piscos must comply with an equivalent potability export analysis that includes the parameters proposed internationally by Organisation Internationale de la Vigne et du Vin (OIV) for wine distillates. The contents of total acids, aldehydes, furfural, and methanol are limited to international standards. However, a minimum content of 2.5 g/l (based on 100° alcohol) of non-alcohol volatiles is required to avoid excessive rectification and to guarantee the use of traditional distillation practices.

16.6 Consumer research and product development in the Pisco industry

Among different studies commanded by the Pisco industry to specialized sensory laboratories, two of them have a particular interest because of the methodologies used. One of them deals with the production of a Pisco with a profile competitive with rum and the other was aimed to solve problems with 'Pisco Sour' a ready-to-drink (RTD) product derived from Pisco spirits.

16.6.1 Design of a new category of spirit: Pisco with rum flavor profile

In the last decade, rum has increasingly became a major competitor of Pisco, growing at rates over two digits, while Pisco sales have been stagnant over recent years. The largest Chilean Pisco producer decided to innovate through new product development, to better compete with the aggressive penetration of rum (Casaubon *et al.*, 2009).

Therefore, it was decided to design a new product based on consumer preferences and the application of senso-chemometrics. A multidisciplinary team was established, including the production and marketing managers of the company, researchers from Centro de Aromas y Sabores DICTUC and Compusense Inc.

In 2006, a consumer sensory test was carried out with 110 male drinkers of Pisco and rum, encompassing four socio-demographic segments (ABC1 and C2 crossed with two age ranges, 18–25 and 25+). The sensory space was covered by eleven products: eight Piscos and three rums, including large and small Pisco producers and the major rum brands. Consumers evaluated samples after individually mixing the spirit with cola and ice, as typically consumed by Chileans. A complete balanced block design was used to evaluate preferences using a 10 cm hedonic scale. Simultaneously, an ethnographic questionnaire was applied.

Descriptive analysis was carried out with a trained panel composed of 15 assessors. ANOVA showed that fourteen flavor attributes were discriminant. Principal component analysis (PCA) produced a sensory map in which Pisco samples were characterized by pear, citrus, apple and raisins, while rums were described by woody, vanilla, caramel, toffee and spicy attributes.

Internal preference mapping and cluster analysis were performed on the data. Four consumer sensory preference's segments were clearly clustered. The relationship between consumer preferences and descriptive analysis revealed a business opportunity to compete in the rum flavor sensory segment.

Chemical analysis of volatile compounds of the same samples using GC-MS showed that Piscos had the highest concentration of terpenes (i.e. linalool), esters and norisoprenoids (grape varietal compounds), responsible for the floral-fruity flavors that characterized Pisco; while rums were richer in oaky aromas, like vanilla, contributing to the woody sensory profiles.

The outcome of this study resulted in a new product category, a Pisco with rum flavors, the product is produced with neutral Muscat grapes and quickly aged with new, oak barrels. Finally, the marketing department developed a specific strategy to communicate the sensory attributes of this new product. The practical success of the sensometric design of this novel product has been proven as it is currently sharing a proportion of the shelf with the regular products in the Chilean market.

16.6.2 Sensory optimization of ready-to-drink products (RTD) from Pisco

Ready-to-drink products represent today more or less 25% of the total Pisco spirit commercialized in Chile. These are elaborated with Pisco spirit mixed with different flavors and include a wide range of types as citrus, mango, berries or coconut.

The addition of lemon juice gives a bitter and acidic flavor to the homemade Pisco Sour that is widely accepted by the Chilean population. The problem is the low stability of this product. The bottled product is affected mainly by oxidation problems. The sensorial properties of the Pisco Sour do not have a long lifetime because the raw materials of the lemon essence have low stability in the Pisco matrix.

The main objective was to improve the Pisco Sour stability to keep the fresh Pisco Sour aroma and flavor. To improve the stability of Pisco Sour on time it was necessary to study each compound in the Pisco Sour matrix. The main techniques used for this work were GC-MS, gas chromatography-olfactometry (GC-O), aroma extract dilution analysis (AEDA) and sensorial panels.

Free choice profile studies by an expert panel composed of researchers from Centro de Aromas y Sabores DICTUC and the Pisco company, using GC-O-AEDA, were used to correlate the sensory properties and volatile chemical composition of a 'good Pisco Sour' (bottling time 0), a 'bad Pisco Sour' (bottled after six months), competitors' commercial products and the fresh cocktail produced with natural lemon and the traditional recipe.

Generalized Procrustes analysis applied to the free choice profile data acquired allowed definition of the sensory properties expected from a natural Pisco Sour and compare it with the commercial products and new prototypes of RTD. The first two dimensions accounted for a total of 60.2% of the variance. Dimension 1 (45.4%) was described as a continuum from artificial flavor, oxidized and chemical to fresh lemon, citric, foam and 'sutil' lemon. Dimension 2 (14.8%) was described by the terms sweet and lemon at one end, to sour and floral at the other end (see Fig. 16.3). Natural Pisco Sour samples were related to fresh lemon, citric and foam and few commercial products were close to them, while most of the commercial products were associated with artificial flavor, chemical and oxidize attributes.

The olfactometric technique, using a sensory panel composed of eight panelists trained in this method, allowed description of the odorant zones of the olfactogram and then correlation with the chromatogram obtained from GC-MS analysis of the sample (Cuillere *et al.*, 2004; Curionia and Bosset, 2002; Francis and Newton, 2005; Greger and Schieberle, 2007).

GC-MS analysis of a natural Pisco shows a series of 27 chemical compounds affecting the aroma and flavor impact, according to the odorant activity value (OAV). Correlation analysis of GC-MS and GC-O/AEDA of a natural Pisco Sour found that 35 chemical compounds affect the impact of natural aromatic Pisco

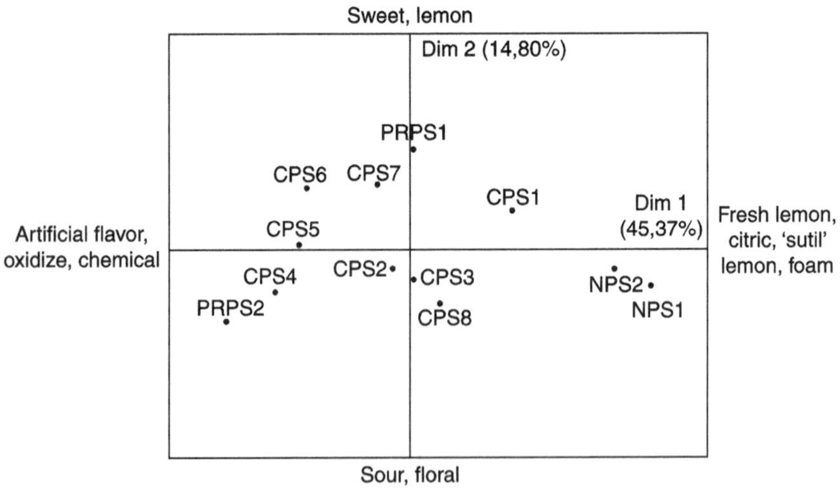

Fig. 16.3 Generalized Procrustes analysis group average, 12 products, 6 expert panelists.

Sour. These results have been useful in the design of new premium RTD, with long shelf life and properties similar to natural Pisco Sour.

16.7 Future trends

Distillation appears as a critical operation in the production of distillates that should preserve the fruity aromas coming from Muscat grapes. All the upstream efforts like viticultural practices, harvest date, crushing, maceration, white style winemaking and others can be annihilated by an improper distillation.

It is worth mentioning that the efforts of the Pisco industry have focused on obtaining a wine retaining most of the potential aromas. This has allowed a significant improvement in Pisco quality. Clearly this strategy is a necessary condition but is not sufficient. To be able to obtain in a reproducible way a range of alcohols differentiated in aromatic content, it is fundamental to understand and optimize the distillation process with the particularities of a batch multi-component system with chemical reactions similar to that of Pisco. This system is not only complex but also most of the components to be separated are in a very low concentration; something very different from studies reported in literature, dealing with binary mixtures or multi-component systems with similar concentrations.

Another aspect that needs to be faced by the Pisco industry is the need for new premium products, allowing it to better compete in international markets. From this point of view, sensory evaluation will be used to select the best raw materials and to define the 'premium profile' from the sensory point of view and to introduce a sensory definition in quality control of the process and blending.

On the other hand, the need to export Pisco to international markets, considering that local markets are stagnant, will require consumer research studies on the target markets. Defining the sensory characteristics that consumers expect in the category in which Pisco will compete is essential: white or aged brandies from *Vitis vinifera*, vodkas or RTD. Sensory consumer research should play a key role on the success of internationalization of Pisco.

16.8 Acknowledgments

Several studies were supported by Fontec CORFO, Fondef D97I1013 CONICYT, ECOS-CONICYT C00B08, DIPUC, MECESUP, Colegio de Ingenieros de Chile-BBVA Banco Bhif, French government among others. The authors also would like to thank Capel for their research involvement and funding.

16.9 Sources of further information

- J FARIA, E LOYOLA, M LOPEZ and JP DUFOUR, 'Cachaça, Pisco and Tequila', In *Fermented Beverage Production*, Andrew Lea and John Piggott (eds), Second edition, Kluwer Academic/Plenum Publishers, New York, 2003, pp 335–363.

- Centro des aromas y sabores http://www.centroaromas.cl/en/pisco_en
- Capel Pisco http://www.piscocapel.cl/
- Nuestro Pisco http://www.nuestropisco.cl/esp/

16.10 References

AGOSIN E, BELANCIC A, IBACACHE A, BAUMES R, BORDEU E *et al.* (2000), 'Aromatic potential of certain Muscat grape varieties important for Pisco production in Chile', *Am J Enol Vitic*, **51**, 404–408.

BAUMES R, BAYONOVE C and GUNATA Z (1994), 'Connaissances actuelles sur le potentiel aromatique des Muscats', *Progres Agric et Vitic*, **111**, 251–256.

BELANCIC A, AGOSIN E, IBACACHE A, BORDEU E, BAUMES R *et al.* (1997), 'Influence of sun exposure on the aromatic composition of Chilean muscat grape cultivars Moscatel de Alejandría and Moscatel Rosada', *Am J Enol Vitic*, **48**, 181–186.

BELITZ H, GROSCH W and SCHIEBERLE P (2004), *Food Chemistry*, Berlin, Springer.

BORDEU E and PSCZCZOLKOWSKI PH (1982), 'Elaboración del Pisco', *El Campesino*, 38–47 Abril.

BORDEU E, FORMAS G and AGOSIN E (2004), 'Proposal for a standardized set of sensory terms for Pisco, a young Muscat wine distillate', *Am J Enol Vitic*, **55**, 104–107.

CASAUBON G, AGOSIN E, FINDLAY C, AZOCAR P and BASTIAS P (2009), 'Design of a new category of spirit: Pisco with rum flavor profile'. *Sensometrics 2008, Discover a New World of Data*, 20–23 July, Brock University, St. Catharines, Ontario, Canada.

CUILLERE L, ESCUDERO A, CACHO J and FERREIRA V (2004), 'Gas chromatography-olfactometry and chemical quantitative study of the aroma of six premium quality Spanish aged red wines', *J Agric Food Chem*, **52**, 1653–1660.

CURIONIA PMG and BOSSET JO (2002), 'Key odorants in various cheese types as determined by gas chromatography-olfactometry', *International Dairy Journal*, **12**, 959–984.

DELFINI C, COCITO C, BONINO M, SCHELLINO R, GAIA P *et al.* (2001), 'Definitive evidence for the actual contribution of yeast in the transformation of neutral precursors of grape aromas', *J Agric Food Chem*, **49**, 5397–5408.

FRANCIS D L and NEWTON J L (2005), 'Determining wine aroma from compositional data', *Australian Journal of Grape and Wine Research*, **11**, 114–126.

GREGER V and SCHIEBERLE P (2007), 'Characterization of the key aroma compounds in apricots (*Prunus armeniaca*) by application of the molecular sensory science concept', *J Agric Food Chem*, **55**, 5221–5228.

GUTH H (1997), 'Quantitation and sensory studies of character impact odorants of different white wine varieties', *J Agric Food Chem*, **45**, 3027–3032.

INE (2007), *VII Censo Agropecuario*, Ministerio de Economía, Chile.

LAMBRECHTS M and PRETORIUS I (2000), 'Yeast and its importance to wine aroma – A review', *S Afr J Enol Vitic*, **21**, (Special Issue), 97–129.

MOLINA A, GUADALUPE V, VARELA C, SWIEGERS J, PRETORIUS I *et al.* (2009), 'Differential synthesis of fermentative aroma compounds of two related commercial wine yeast strains', *Food Chemistry*, **117**, 189–195.

OHTA, T (1995), 'Transformations of geraniol, nerol and their glucosides during Schochu distillation'. In R. Cantagrel (ed.), *Elaboration et Connaissance des Spirituex.* Vient de Paraitre sous l'egide du Bureau National Interprofessionel du Cognac, pp. 313–315.

PEÑA Y LILLO M, AGOSIN E, ATHES V, BORDEU E, CASAUBON G *et al.* (2003a), 'Relation between odour perception and physicochemical properties of volatile compounds in Chilean Pisco spirit'. In J. L. Le Quéré and P. Etiévant (eds), *Flavour Research at the Dawn of the Twenty-First Century*. Cachan, France, Lavoisier.

PEÑA Y LILLO M, LATRILLE E, CASAUBON G, AGOSÍN E, BORDEU E *et al.* (2003b), 'Panel and individual performance in orthonasal and retronasal profiles of Chilean Pisco Spirit', *Abstracts of papers of the American Chemical Society*, **226**, 162–AGFD Part 1.

PEÑA Y LILLO M, LATRILLE E, CASAUBON G, AGOSIN E, BORDEU E *et al.* (2005), 'Comparison between odour and aroma profiles of Chilean Pisco spirit', *Food Quality and Preference*, **16**, 59–70.

SAG (2007), *Catastro Vitícola 2007*. Ministerio de Agricultura, Chile.

STASHENKO H, MACKU C and SHIBAMATO T (1992), 'Monitoring volatile chemicals formed from must during yeast fermentation', *J Agric Food Chem*, **40**, 2257–2259.

SWIEGERS J and PRETORIUS I (2005), 'Yeast modulation of wine flavour', *Adv Appl Microbiol*, **57**, 131–175.

SWIEGERS J, FRANCIS I, HERDERICH M and PRETORIUS (2006), 'Meeting consumer expectations through management in vineyard and winery: The choice of yeast for fermentation offers great potential to adjust the aroma of Sauvignon Blanc wine', *Australian and New Zealand Wine Industry Journal*, **21**, 34–42.

WONDRA M and BOVERIC M (2001), 'Analyses of aroma components of Chardonnay wine fermented by different yeast strains', *Food Technol Biotechnol*, **39**, 141–148.

17

Sugar cane spirits: cachaça and rum production and sensory properties

J. B. Faria, Univ Estadual Paulista – Unesp, Brazil

Abstract: The production of cachaça, a distilled spirit made directly from sugar cane juice and rum obtained by fermenting molasses, a by-product of the sugar industry, are comparatively presented. The main differences, from raw material to fermentation, distillation and aging process are pointed out, in order to explain the resulting composition and distinct sensory characteristics of these two beverages and to show the importance of the sensory quality control, during all the production process.

Key words: cachaça, rum, sugar cane, molasses, fermentation, distillation, sensory quality control.

17.1 Introduction

Cachaça, the most traditional and popular alcoholic beverage produced in Brazil since the middle of the sixteenth century, is produced by distilling fermented fresh sugar cane juice. Rum, produced in various countries, is already recognized as one of the most traditional spirits derived from sugar cane.

According to Brazilian legislation, the term cachaça is reserved for sugar cane spirit having 38–48% alcohol (v/v) at 20 °C, made by distilling fermented fresh cane juice, only in the Brazilian territory (Brasil, 2002). Rum, according to the same legislation, is a distilled alcoholic beverage, having 38–54% alcohol, produced by distilling fermented molasses, or molasses mixed with sugar cane juice, and aged in oak or other wooden casks (Brasil, 1997).

17.2 Raw materials

Cachaça and rum are both derived from the same sugar cane plant (*Saccharum officinarum*).

The raw material used to produce cachaça is the juice squeezed from the sugar cane stalks, while rum is mainly obtained from the molasses, a by-product of the industrial sugar process. The juice extracted to make cachaça can also be heat-evaporated to obtain a concentrate, similar to industrial molasses, to make so-called agricultural rum. Thus rum and cachaça, though coming from the same plant, are in fact obtained from distinct raw materials: cachaça is made directly from the sugar cane juice, while rum production is mainly based on sugar cane molasses.

17.2.1 Must preparation

The sugar cane juice as extracted from the stalks, the raw material for cachaça, is itself a natural must. However, to achieve better fermentation and more economic results, some procedures are carried out before fermentation. The juice is first left standing in a settling tank, to separate the dense impurities by gravity. It is then diluted with water, in order to adjust the sugar concentration and the fermentation yeast capacity, preventing inhibition of the yeast by high alcohol levels, as well as reducing waste and preventing bacterial fermentation of the remaining sugar.

Usually, the juice temperature is adjusted to 30 °C, sulphuric acid is added to lower the pH (5.5–4.5), and nutrients are used to promote fermentation. In some cases, disinfectants and antibiotics are also used in the must preparation, to eliminate undesirable microorganisms.

The molasses provided by the sugar industry, as already mentioned, is the main raw material for rum production. Its most important constituent is sugar (55–56% w/w), but other components, such as sulphated ash, nitrogen compounds, gums and unfermentable sugars, may also affect the quality of the molasses (Nicol, 2003). Given that molasses, coming directly from the sugar refinery, is delivered hot at the distilleries, it is recommended that its stored temperature never exceeds 45 °C, to prevent Maillard reactions and occasional spontaneous combustion (Chen, 1985).

In order to partially remove the suspended solids and avoid contamination, the molasses is first diluted to 45 °Bx and the temperature raised to 70 °C. At this stage sulphuric acid is added to lower the pH and encourage sedimentation. The diluted and acidified molasses is then transferred to a settling tank and the supernatant (clarified molasses) is decanted from the sludge, diluted to the desired working specific gravity (16–20 °Bx), pumped to the fermenter, enriched with nutrients and finally pitched with yeast (Nicol, 2003, Delavante, 2004).

17.3 Fermentation

In the old traditional technique for cachaça and rum production, the fermentation process occurred naturally by the action of yeasts and bacteria present in the air, contaminating the dilution water and the sugar cane juice or lying dormant in the molasses. Nowadays, besides this traditional technique, pure or mixed cultures are being more and more employed (Rosa *et al.*, 2009).

The sugar cane juice fermentation for cachaça production is a very robust process that can occur even under technically adverse conditions, owing to the high adaptative capacity of the yeast used (mainly *Saccharomyces cerevisae*). The use of selected yeasts is common in large distilleries, to ensure predictable flavor and process time, but many small-scale production units still use baker's yeasts or the so called 'fermento caipira', natural or wild yeast mixtures, developed in the distilleries themselves (Faria *et al.*, 2003a).

Rum fermentation is mainly conducted by *Saccharomyces cerevisae, Saccharomyces bayanus, Schizosaccharomyces pombe* and various strains of competing bacteria in the 'dunder', the natural ferment foam residue of a previous rum batch, which is used as the inoculum.

17.3.1 Fermentation process

The fermentation process is usually carried out in cylindro-conical fermenters with domed tops or in cylindrical open-topped vessels with sloping bases. At the end of the process, there are a lot of suspended solids from the raw materials and the yeast cells that need to be separated before distillation.

Cachaça fermentation, depending on several interfering factors, can take from 20–30 hours to be completed, while the rum (molasses) fermentation takes around 24 hours and yields 5–7% (v/v) alcohol, depending on the original settling gravity (Rosa *et al.*, 2009).

In both cases, stainless steel is the material of choice, but fermenters made of wood, mild steel and other materials, are still often used.

17.4 Distillation

When the compositions of sugar cane juice and molasses are compared, it is easy to understand why the products resulting from the two fermentations are quite distinct.

The fermented molasses, due to its more complex composition, certainly requires a more carefully controlled distillation to separate the undesirable compounds from the distillate, which can explain why the distilling apparatus and procedures used for rum and cachaça are not the same.

17.4.1 Pot still distillation

Pot still cachaça or 'cachaça de alambique'

In view of the composition of the sugar cane 'wine', the use of one simple pot still, also called alembic and usually made of copper or stainless steel and copper, is enough to separate and concentrate the desired distilled compounds from the residual wine components.

Fig. 17.1 Cachaça pot stills.

The alembics used for cachaça distillation (Fig. 17.1) may have distinct shapes and sizes, but in all cases the main objective is to separate the head fraction, which has the more volatile compounds, from the heart and tail fractions. The heart, which contains about 80–85% of the total distilled volume and has from 40–48% (v/v) alcohol, can already be bottled and consumed (Faria *et al.*, 2003a), while the head (first) and tail (last) fractions should be discarded.

A double distillation process as shown in Fig. 17.2, or even using the same alembic, has also been proposed, to obtain a spirit with a lighter aroma and taste, which would be further aged (Novaes, 1999; Bizelli, 2000; Rota and Faria, 2009).

Pot distilled rum

In order to obtain a distillate of fine sensory characteristics from fermented molasses, the traditional approach is double distillation, an effective way of removing undesirable components and bringing the strength of the spirit to about 65% (v/v) alcohol.

However new systems of distillation, such as the retort still developed in Jamaica can, in a single distillation, bring the strength up to 85% (v/v) alcohol, with the desired character (Delavante, 2004).

The cachaça, distilled in alembics, like the pot still rum, usually has a more complex composition than distillates obtained by column distillation, mainly due to the reactions among the volatile compounds that occur during the distillation process, catalyzed by the copper of the still (Cole and Noble, 1995).

Fig. 17.2 Double cachaça pot stills.

17.4.2 Column distillation

After the invention of the original Coffey still, the use of continuous column distillation became a practical alternative for spirit production. The main advantages of this kind of spirit distillation are: fine distilling selectivity, increased production, energy savings, decrease of sensory defects and better standardization of the distillate.

Cachaça distilled in columns or 'cachaça de coluna'

As mentioned above, column distillation allows greater distilling control than the single alembic units, making it easier to remove the undesirable compounds (produced by defective fermentations) and resulting in lighter, standardized and cheaper products (Fig.17.3).

The sensory characteristics of cachaça distilled in a column are different from those of pot-still cachaça. This fact, together with the diversity of regional and technical practices, may explain the great range of characteristics of cachaça, readily perceived all over Brazil.

Rum column distillation

The two-column still, patented by Coffey in the nineteenth century, is still used to produce lighter-flavored rums with up to 95% (v/v) alcohol, as well as a new three-column still, used to produce an even lighter spirit (Delavante, 2004).

A new column patent French Savalle still can also be used to produce very distinctive products from light- to medium-flavored rums, as well as rectified spirits for vodka and gin production (Nicol, 2003).

Fig. 17.3 Column distillery with capacity to produce 300 000 l of cachaça per day. (Source: reproduced with kind permission of the Ypioca Group.)

17.5 Maturation

Aging distilled spirits in oak casks (180–700 l) favors the acquisition of good color and taste from the oak and the loss of the harsh sensory characteristics of the new distillates. In some processes the aging or maturation step forms an integral part of the production of distilled beverages.

17.5.1 Cachaça aging

Very few spirits (such as cachaça) have a satisfactory sensory quality straight after being distilled, which could explain why aging cachaça was not a normal practice in the past. However, this process may produce a significant improvement in its flavor (Cardello and Faria, 1998, 1999).

Currently, besides the use of remade oak whiskey casks from North America and Scotland, a great number of Brazilian wood species have been tested and used for aging cachaça (Faria *et al.*, 2003b).

17.5.2 Rum aging

The aging process is normally associated with the processing and definition of rum, and its effects are already part of the sensory characteristics expected in rum.

17.6 Cachaça and rum: similarities and differences

Although obtained from the same raw material and in a very similar way, some features of cachaça and rum production must be pointed out, to explain the observed differences between these two spirits.

The sugar cane juice and the molasses, used to prepare the must to be fermented, already have quite different chemical and microbiological compositions, mainly due to the heat treatment to which the sugar cane juice is subjected, during processing in the sugar industry.

The chemical reactions and the concentration of juice components that occur during heating and even afterwards, will certainly change the original sugar cane juice into a quite distinct by-product: molasses. Also, the fermentation is done with quite distinct yeasts and processing conditions, as already mentioned.

Lastly, in view of the different wine compositions that result, the distillation of cachaça and rum must also be conducted in distinct ways, resulting in distillates that, aged or not, will have very distinct chemical and sensory profiles.

The overall effect of all the differences observed during the production of cachaça and rum may explain their distinct and variable compositions, shown in Fig. 17.4 and 17.5 (Cardoso *et al.*, 2004), where significant differences can be observed among the compositions of commercial samples of cachaça and rum.

Thus, based on the production differences and on the final compositions, it is possible to state that cachaça and rum, while having a common origin, are distinct beverages with well-defined sensory characteristics (Magnani, 2009).

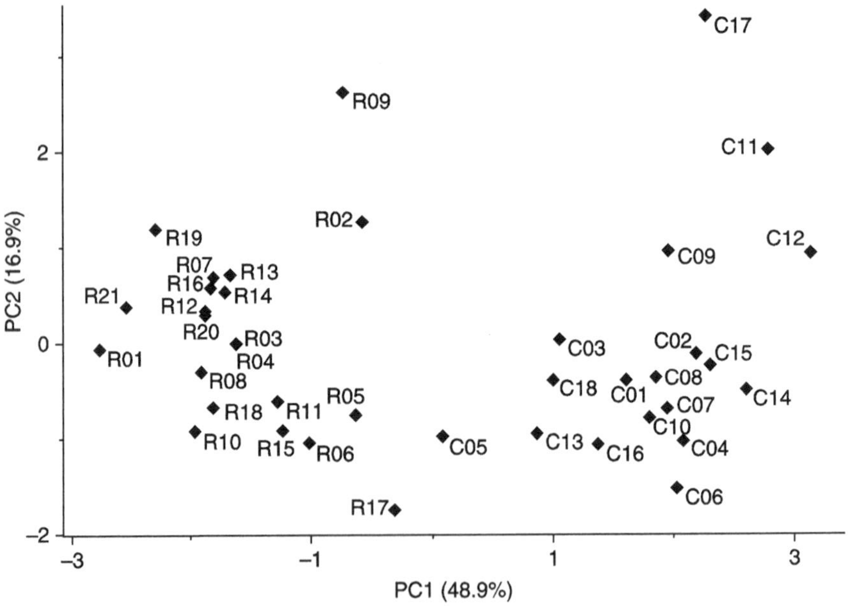

Fig. 17.4 PCA scores plot of cachaça and rum samples using the variables: protecatechuic acid, propanol, isobutanol, copper, magnesium and manganese (Source: Cardoso *et al.*, 2004).

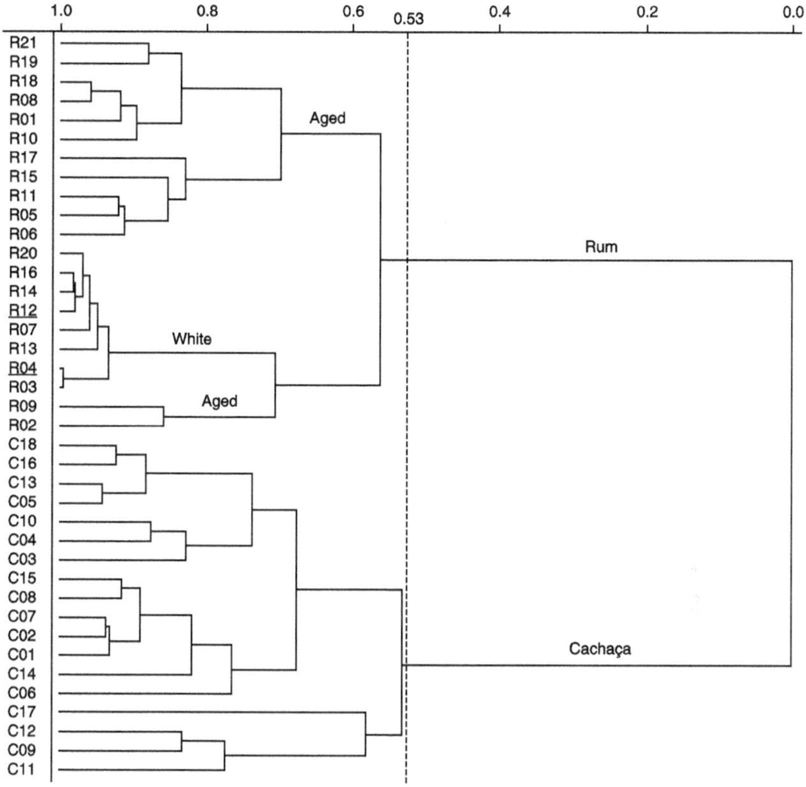

Fig. 17.5 HCA dendrogram of cachaça and rum samples (Source: Cardoso *et al.*, 2004).

17.7 Sensory quality of cachaça and rum

Any distilled beverage is a mixture of all the compounds derived from the raw material and production process that will help define its sensory identity.

Some of these compounds are already known, as well as their positive or negative sensory effects on the quality of the drink (Janzantti, 2004).

The quality control of the process must ensure that the expected good sensory characteristics are present in the final product, and the best and most effective way to achieve this objective must certainly be based on methods of sensory analysis (Maçatelli, 2007). Thus, even during the production process, some sensory tests may be used to indicate whether the fermentation is going well or, for example, when it is time to cut the tail fraction.

The matured spirits, like whiskey, cognac, rum and aged cachaça, have some common sensory characteristics, mainly related to the fermentation and aging processes, but some special and distinct compounds generally define the unique aroma of each spirit.

The sensory profile of cachaça and rum are usually described in terms of the harsh characteristics of new distillates, such as sour, grassy, oily, sulphury, and also to the mature flavors developed during the aging process, including vanilla, spicy, floral, wood and smooth flavors, as well as some variations in vocabulary.

Some compounds with strong sensory characteristics are often related to the sensory quality or defect of rum or cachaça: aldehydes, higher alcohols, ethyl and higher esters, acids, phenols and acetals, as well as some particular compounds found in specific spirits, such as 2-ethyl 3-methyl butyric acid in rum and dimethyl sulfide, directly related to a sensory defect of cachaça distilled in the absence of copper (Faria *et al.*, 2003c).

Figure 17.6 shows the sensory attributes related to cachaça and rum commercial samples and how it is possible to recognize and compare their sensory profiles.

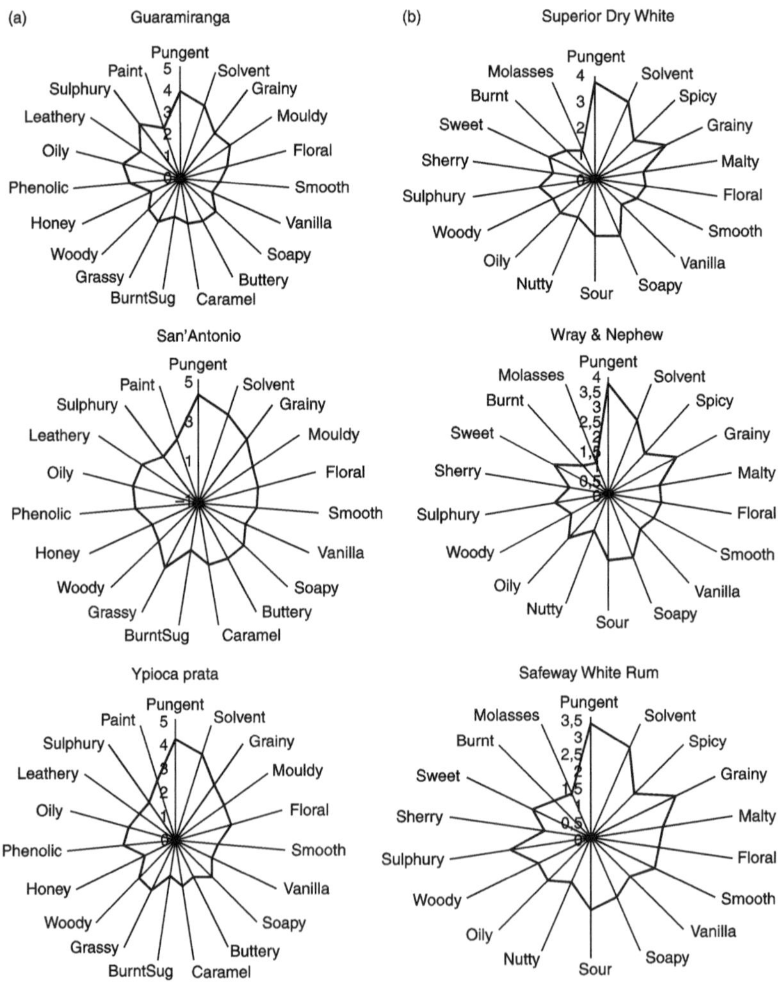

Fig. 17.6 Sensory attributes related to cachaça and rum commercial samples. (a) Commercial cachaça samples. (b) Commercial rum samples.

17.8 The challenge of new markets

Besides the very well established international market for traditional and well-known distilled beverages like rum, spirits like cachaça, new to this market, are showing a great potential for export (Faria *et al.*, 2004).

Unfortunately most of the very large volume of Brazilian cachaça is still produced by traditional techniques and directed to the internal market, where the competition is mainly based on the price and the profit primarily linked to quantity, not quality.

To change this situation, efforts must be made to develop the great sensory quality potential of cachaça. Changes in production technology, research into new sugar cane varieties and selected yeasts, as well the development of new distilling and aging techniques, always supported by sensory analysis (Janzantti, 2004; Maçatelli, 2007), will certainly provide good quality products, the best way to recognize the social and economic importance of this historical Brazilian beverage and to profit with its exportation.

17.9 Sources of further information

CARVALHO, M.P.M. and SILVA, S.P. (1988). *Cachaça, uma alegre História Brasileira.* São Paulo, Brasil: Caninha 51 – Indústria e Comércio de Bebidas Ltda.

NICOL, D.A. (2003). 'Rum'. In: *Fermented Beverage Production.* 2nd ed. Edited by Lea, A.G.H. and Piggott, J.R., Kluwer AcademidPlenum Publishers, New York, pp. 263–287.

FARIA, J.B., LOYOLA, E., LOPEZ, M.G. and DUFOUR, L.P. (2003) 'Cachaça, Pisco and Tequila'. In: *Fermented Beverage Production.* Second Edition, Edited by Lea, A.G.H. and Piggott, J.R, Kluwer AcademidPlenum Publishers, New York, pp. 335–363.

PEREIRA, JOSÉ ANTONIO MARQUES, ROSA, CARLOS AUGUSTO and FARIA, J. B. 'Cachaça de alambique'. Brasília – DF: L K Editora e Comunicação, 2006, v.1. 180p.

FARIA, J. B. 'Rum and Cachaça' In: *Distilled Spirits – Production Technology And Innovation.* Nottinghan: Nottinghan University Press, 2008, pp. 327–334.

ROSA, C. A., MORENO, A. and FARIA, J. B. 'Cachaça Production' In: *The Alcohol Textbook.* 5th ed. Nottinghan: Nottinghan University Press, 2009, pp. 481–490.

Universidade Estadual Paulista 'Júlio de Mesquita Filho' website: http://www.fcfar.unesp.br/portaldacachaca

17.10 Acknowledgement

Dr John Piggott, University of Strathclyde, kindly provided the data for Fig 17.6.

17.11 References

BIZELLI L.C (2000). 'Influência da condução da dupla destilação nas características físico-químicas e sensoriais da aguardente de cana'. Piracicaba – SP, (Dissertação de Mestrado) Escola Superior de Agricultura 'Luiz de Queiroz' – USP.

BRASIL (1997). *Decreto n. 2314 de 4 de setembro de 1997.* Diário Oficial da União, 5 de setembro de 1997.

BRASIL (2002). *Decreto n. 4072 de 3 de janeiro de 2002*. Diário Oficial da União, 4 de janeiro de 2002.

CARDELLO, H.M.A.B. and FARIA, J.B. (1998). 'Análise descritiva quantitativa de aguardentes de cana durante o envelhecimento em tonel de carvalho (*Quercus album*, L.)'. *Revista de Ciência e Tecnologia de Alimentos,* **18**: 169–175.

CARDELLO, H.M.A.B. and FARIA, J.B. (1999). 'Análise tempo-intensidade de características sensoriais de aguardente de cana durante o envelhecimento em tonel de carvalho (*Quercus* sp.)'. *Boletim da Sociedade Brasileira de Ciência e Tecnologia de Alimentos,* **33**: 27–34.

CARDOSO, D.R., ANDRADE-SOBRINHO, L.G., LEITE-NETO, A.F., RECHE, R.V., ISIQUE, W.D. *et al.* (2004). 'Comparison between Cachaça and Rum using pattern recognition methods'. *Journal of Agricultural and Food Chemistry*, **52**, pp. 3429–3433.

CHEN, J.C.P. (1985). *Cane Sugar Handbook*. Wiley Interscience, New York.

COLE, V.C. and NOBLE, A.C. (1995). 'Flavor chemistry and assessment'. In: *Fermented Beverage Production*. Edited by Lea, A.G.H. and Piggott, J.R., Blackie Academic and Professional, Glasgow, UK, pp. 361–381.

DELAVANTE, M.P. (2004). 'Rum – the commercial and technical aspects'. In: *Distilled Spirits Tradition and Innovation*. Edited by Bryce J.H. and Stewart G.G., Nottingham University Press, Nottingham, UK, pp. 209–221.

FARIA, J.B., LOYOLA E., LOPEZ, M.G. and DUFOUR J.P. (2003a). 'Cachaça, Pisco and Tequila'. In: *Fermented Beverage Production*. Second Edition, Edited by Lea, A.G.H. and Piggott, J.R, Kluwer AcademidPlenum Publishers, New York, pp. 335–363.

FARIA, J.B., CARDELLO, H.M.A.B., BOSCOLO, M., ISIQUE, W.D., ODELLO, L. *et al.* (2003b). 'Evaluation of Brazilian woods as an alternative to oak for cachaça aging'. *European Food Research and Technology,* **218**: 83–87.

FARIA, J.B., FERREIRA, V., LOPES, R. and CACHO, J. (2003c). 'The sensory characteristic defect of "cachaça" distilled in absence of copper'. *Alim. Nutr.*, **14**(1): 1–7.

FARIA, J.B., FRANCO, D.W. and PIGGOTT, J.R. (2004). 'The quality challenge: cachaça for export in the 21st century'. In: *Distilled Spirits – Tradition and Innovation.* Edited by Bryce J.H. and Stewart G.G., Nottingham University Press, Nottingham, UK, pp. 215–221.

JANZANTTI, N.S. (2004). 'Compostos voláteis e qualidade de sabor da cachaça'. SP, 179p. Tese de Doutorado. Faculdade de Engenharia de Alimentos – UNICAMP.

MAÇATELLI, M. (2007). 'Determinação do perfil sensorial de marcas comerciais de cachaça'. SP, 126p. Dissertação de Mestrado. Faculdade de Ciências Farmacêuticas – UNESP.

MAGNANI, B.D. (2009). 'Estudo comparativo das características sensoriais do rum e da cachaça'. SP, 88p. Dissertação de Mestrado. Faculdade de Ciências Farmacêuticas – UNESP.

NICOL, D.A. (2003). 'Rum'. In: *Fermented Beverage Production*. 2nd ed. Edited by Lea, A.G.H. and Piggott, J.R., Kluwer AcademidPlenum Publishers, New York, pp. 263–287.

NOVAES, F.V. (1999). 'Processo de dupla destilação'. In *Programa de fortalecimento do setor de aguardente de cana-de-açúcar e seus derivados do Estado do Rio de Janeiro*. Sebrae, Rio de Janeiro-RJ, p. 118.

ROSA, C.A., MORENO, A., FARIA, J.B. (2009). 'Cachaça production' In: *The Alcohol Textbook*. 5th ed. Edited by Ingledew W.M. *et al.*, Nottinghan: Nottinghan University Press, p. 481–490.

ROTA, M.B. and FARIA, J.B. (2009). 'Efeito do processo de bidestilação na qualidade sensorial da cachaça'. *Alimentos e Nutrição* (UNESP), V.**20**, N.1., P. 121–127.

18

Tequila and mezcal: sensory attributes and sensory evaluation

S. Villanueva-Rodriguez and H. Escalona-Buendia, Centro de Investigación y Asistencia en Tecnología y Diseño del Estado de Jalisco, Mexico

Abstract: Tequila and mezcal are both Mexican spirits obtained from the distillation of fermented agave juice. Currently, they are produced in different regions of Mexico and each product possesses a distinctive sensory character. This chapter summarizes the authors' experience in applied research and technological development regarding sensory properties of tequila and mezcal.

Key words: tequila, mezcal, agave spirit, flavour, taste, sensory, preference.

18.1 Introduction

Tequila and mezcal are both Mexican spirits obtained from the distillation of fermented agave juice. Currently, tequila is mainly produced in the state of Jalisco and mezcal in other regions. In the past, though, mezcal was the name given to the agave spirits produced in a vast region of the country including the state of Jalisco. This review summarizes the results of more than a decade of research on the sensory properties of tequila and mezcal from the perspective of the Centro de Investigación y Asistencia en Tecnología y Diseño del Estado de Jalisco A.C (CIATEJ A.C.), which belongs to a network of public research centres run by the National Council of Science and Technology of Mexico (CONACYT). Because of its location and mission, CIATEJ A.C. has supported the tequila industry in particular, as well as producers of other agave spirits such as mezcal. Important contributions from CIATEJ A.C. are considered together with those from other research institutions in the present review.

Two types of tequila familiar to consumers are the 'white' or 'silver' varieties (i.e. unaged tequilas), and the varieties matured in oak barrels. Recently, flavoured tequilas, often containing citrus essences, have been introduced to the market and accepted by consumers. Mezcal is usually sold either as 'silver' (i.e. unaged), cask matured, or flavoured, or as containing other ingredients, which are added after or during distillation. The spirit is commonly flavoured through the addition of a maceration of herbs or some other regional ingredients such as worms. It is important to highlight that the practice of adding a worm to a bottle only occurs in the production of some mezcal products and is never practised in tequila production.

While the tequila industry has grown during the last two decades and there are now large international companies that produce the product on a large scale and market it all over the world, mezcal production remains on a small scale. It is produced either by small enterprises or by local producers who have joined forces to form societies. It is common practice for these societies to have bottling facilities where the spirit from the member producers is collected, blended, bottled and labelled ready for sale.

18.1.1 Definition of tequila and mezcal

As previously described, tequila and mezcal are both Mexican spirits produced by distillation of fermented agave juice. Tequila was first made in the town of Tequila and its surrounding area in the state of Jalisco. Nowadays, tequila is produced throughout the state of Jalisco, and in some territories of four other states (Table 18.1). Tequila was known originally as mezcal from Tequila, but through the years it gained popularity due to its distinctive flavour attributes and the spirit became known solely as tequila after its region of production. Only one agave species, *Agave tequilana* Weber, blue variety, is permitted for use in tequila production whereas several agave species can be used for the production of mezcal.

Mezcal retains its reputation as a good quality traditional spirit, and current regulations allow the use of this title for agave spirit produced throughout the territories of several Mexican states such as Oaxaca and Guerrero (Table 18.1).

According to the current standards for tequila (NOM-006-SCFI-2005) and for mezcal (NOM-070-SCFI-1994), both spirits are defined as regional beverages obtained by the distillation and refinement of musts prepared directly by the extraction of agave heads previously cooked or hydrolyzed and fermented with yeasts (cultured or not). The definition for each spirit establishes the agave species that are permitted for use in their production, and their respective territories for the denomination of origin. It also stipulates the minimum amount of agave sugars required to be present in the must for the use of the name tequila or mezcal to be permitted, as well as the prohibition of the use of cold mixtures (blends of agave spirit with other sources of distillates).

18.1.2 Regulation of safety, authenticity and denomination of origin

The Mexican Ministry of Economy has established specific regulations for each spirit. The regulation of the Mexican Ministry of Health (NOM-142-SSA1-1995)

Table 18.1 Summary of parameters regulated for tequila and mezcal

	Tequila*	Mezcal§
Denomination of origin region†	Jalisco (whole state) Some municipalities of Michoacan Nayarit Guanajuato Tamaulipas	Oaxaca (whole state) Guerrero (whole state) San Luis Potosi (whole state) Durango (whole state) Zacatecas (whole state) Some municipalities of Guanajuato Tamaulipas Michoacan (under evaluation)
Agave species allowed‡	*Agave tequilana* Weber, blue variety	Agave angustifolia Agave potatorum Agave salmiana Agave esperrimia Jacobi Agave weberi cela Agave duranguensis Other species so long as they are not restricted by other protected beverages in the same state
Wood maturation‡	Silver (not wood matured) Gold (not wood matured, blended or flavoured) Reposado (rested at least two months) Añejo (aged at least one year) Extra Añejo (aged at least three years)	Joven (silver, not wood-matured, blended or flavoured) Reposado (rested, at least two months) Añejo (aged at least one year)
Types‡	Tequila 100% agave (all fermentable sugars are from agave) Tequila (at least 51% fermentable sugars from agave)	Mezcal 100% agave (Type 1, all fermentable sugars from agave) Mezcal (Type 2, at least 80% sugars from agave)
Ethanol strength‡	35–55% (v/v)	36–55% (v/v)

Notes: †According to the declaration of Denomination of Origin (Instituto Mexicano de la Propiedad Industrial). ‡NOM-006-SCFI-2005 for tequila and NOM-070-SCFI-1994 for mezcal.

Sources: *Diario Oficial de la Federacion (1977), §Diario Oficial de la Federacion (1994)

states in an obligatory standard the maximum levels allowed of some congeners such as higher alcohols and methanol. The Regulatory Council of Tequila and the Regulatory Council of Mezcal are the organisations responsible for ensuring compliance with the respective standards.

The standards for tequila as stated in the regulation (NOM-006-SCFI-2005) and for mezcal (NOM-070-SCFI-1994) are summarized in Table 18.1 according to the denomination of origin, the classes of spirit defined by the maturation process, and the types of spirit produced relating to the proportion of agave sugars used in the must prior to fermentation. To be able to use the name tequila, the amount of sugar extracted from *Agave tequilana* Weber blue variety in the must should to be at least 51%. If this limit is not achieved, the spirit is normally labelled as agave distillate.

The standard for tequila also regulates the permitted amount of higher alcohols, methanol, esters, acetic acid and furfural in the final product. All of these compounds contribute in some degree to the flavour of the drink, but cannot exceed an established maximum limit for safety reasons. The place of bottling is also regulated; tequila 100% agave has to be produced, bottled and labelled in the region stated by the granted denomination of origin, while tequila containing 51% fermentable sugars from agave, although it also has to be produced in the specific region stated in the regulations, can be bottled and labelled outside the restricted area.

The minimum proportion of sugars from agave in the must used for mezcal production has to be at least 80%. The denomination of origin for mezcal is related to the region of production, which is generally a political state, in the same way that wine is named according to the region in which it is produced. Therefore, the protected name of the drink is linked to the area of its manufacture: Mezcal from Oaxaca, Mezcal from Guerrero, Mezcal from San Luis Potosi, etc. The manufacturers of Mezcal from Oaxaca have been proactive in developing the image and status of mezcal with this particular denomination of origin, and Mezcal from Oaxaca is the mezcal that has accomplished the largest growth in sales both inside and outside of Mexico in relation to other mezcal brands. According to the Mexican Institute of Industrial Property, some territories of Tamaulipas and Guanajuato regions are currently considered to be inside the mezcal denomination of origin, while the state of Michoacan has been conducting research to support the future incorporation of some of its municipalities in the mezcal-producing areas (Gallardo 2008).

18.1.3 The production process for tequila and mezcal

The production processes for tequila and mezcal are broadly very similar, which supports their common origin. Harvesting and cutting agave, cooking, mashing or milling to obtain agave juice rich in sugar, fermentation, first distillation, second distillation, cask maturation (if the product is rested or aged), and fining and bottling are carried out in a similar way for both spirits.

The production process for tequila has been previously discussed by Cedeño (1995) and Faria *et al.* (2003). For tequila production, agave harvesting takes place when the blue agave required for tequila manufacture is mature enough to provide sufficient sugars in the must prior to fermentation. The main reserve carbohydrate in agave is a branched-chain fructan polysaccharide which, if needed, can be hydrolyzed in the plant yielding fructoligosaccharides and fructose. The agave heart (also known as the head or pine) is extracted manually in an operation called jima in which leaves, roots and the flower structure (called quiote) are cut from the plant and its heart is removed from the soil. The person who carries out this operation is known as a Jimador and this is a symbol rooted in the culture developed around this spirit.

Cooking the agave pines is generally the second step in the production process and is commonly performed either in retorts under vacuum pressure or in brick ovens where steam is the heat source. Cooking is carried out with the objective of hydrolyzing the fructans in the agave into fructose, but it also transforms the texture of the agave, softening the fibres and facilitating crushing and juice extraction. Moreover, cooking promotes browning reactions.

Crushing the agave and extraction of juice is commonly performed in a mill similar to that used for sugar cane extraction. Agave chunks are crushed, and washed with hot water in several steps to extract the sugar-rich juice from the fibrous material. A new technological development is to use a diffuser to extract the carbohydrate-rich juice either from the raw or the cooked agave heart. Because cooking is a necessary step for fructan hydrolysis and flavour development, raw juice obtained by diffusers is cooked afterwards in retorts or heated tanks before fermentation, which is the next step in the process.

Fermentation of agave juice used to be a spontaneous process using the natural flora from the agave plant and from the local factory environment, and was mainly performed by yeasts. In addition, it was a common practice to use either baking or wine yeast as a fermentation starter. In modern production plants, because of the drive to improve processing operations and product consistency, a more scientific approach is now taken. The tequila industry has collaborated with research institutions to carry out studies to isolate and partially identify the natural flora used in agave juice fermentation. CIATEJ A.C. has developed the technology for isolating, preparing and preserving an inoculum for the industrial fermentation of tequila (Gschaedler *et al.* 2004). A collection of native tequila yeasts has been identified, including strains of *Saccharomyces cereviseae*, which are adapted to grow in agave juice. Fermentation normally takes place in open containers and stops when alcohol levels inhibit yeast growth, which relates directly to the consumption of fructose and other fermentable sugars.

Once fermentation is complete, the must is transferred to a distillation device, which consists either of a pot still (made from copper or stainless steel) or a fractional distillation column. The spirit obtained from the first distillation is known as 'ordinary', and is subjected to a second distillation, again either in a pot still or in the same fractional distillation column. This second distillation is known as 'rectification', and the spirit obtained should be approximately 50% ethanol (v/v).

If the tequila is to be matured, it is then transferred into oak casks, either made from new wood, or wood reused from the production of American bourbon whisky, and the spirit is left to mature in cellars. Both the white (not matured) and the matured tequila are fined using cellulose or charcoal filters to remove particles and to optimize the clarity and appearance of the spirit. Afterwards, it is bottled, labelled and packed for sale.

Mezcal, as mentioned above, is produced using the same basic process as tequila, but on a smaller scale using artisanal methods. Mezcal production starts with the manual harvesting of agave to obtain their hearts or pines; the leaves can also be recovered. The hearts are used whole, or cut in half (according to size) and are cooked in a hole, or oven in the ground that has been heated for the purpose by burning wood (normally oak or mesquite wood) to heat stones that retain the heat. Once this oven is heated; agave hearts and sometimes leaves are placed inside and covered with grass, mud and stones to prevent the heat from escaping.

Once cooking has finished (the cooking time is determined by an experienced mezcal producer), the agave hearts are removed and mashed, either manually with hammers and hatchets, or using a millstone in a device known as a Tahona.

In contrast to tequila, mezcal fermentation is carried out using the whole mash from the agave heart, including the fibres. The fermentation vessel can vary depending on the different regions, and is made of either wood or leather, or some other material adapted for this process. Fermentation is carried out by a very diverse range of yeasts, where some yeasts are involved in the production of ethanol and higher alcohol fractions, and other yeasts strains and also bacteria are involved in the production of different volatile components.

Once fermentation is finished, the must is transferred to a small scale (sometimes handmade) still made of copper for the first distillation. The second distillation is carried out in the copper pot still at which point some other ingredients can be added to the macerate or extract to obtain flavoured mezcal, such as 'Mezcal with damiana' or 'Mezcal with worm'. This artisanal distillation has to be controlled very carefully in order not to exceed the legal limit for copper and methanol in the product. This limit is set due to their toxic effects.

After distillation, the spirit containing 50% ethanol (v/v) is normally taken to the blending and bottling facility where cask maturation may take place, or if there is no ageing step, the blending is carried out, and ethanol strength adjusted, followed by bottling and labelling of the product for sale. Mezcal is also subjected to cellulose and active charcoal fining that, besides removing turbid particles, also blends and softens the smoky flavours generated during cooking.

The diversity of regional practices for mezcal production results in a product that is characteristic of each region; therefore, the processing operations described are just a general overview of the steps taken in mezcal production. Some mezcal producers from Oaxaca have introduced more industrial processing methods such as using retorts for cooking; in Zacatecas, mezcal is produced in a way more similar to that used for tequila, while the other states may follow a more traditional procedure with regional variations.

18.2 Origin of the sensory attributes of tequila and mezcal

Just as for any other spirit, the distinctive sensory character of tequila and mezcal is provided by the flavour compounds that come from the raw material and its processing prior to fermentation. Tequila and mezcal's most abundant components are ethanol and water; therefore, their sensory identities are provided by small amounts of volatile compounds, and extracted components from wood or from other flavourings. Age, region, species and variety of each agave batch, as well as the way the must is prepared for fermentation, will impact sensory attributes.

Local flora present in the factory will also impact the final character of the product and contribute to the generation of distinctive products for different brands, regions, factories and spirits. The distillation process, either in copper or stainless steel vessels, will determine the recovery of the flavour volatiles and the reduction of compounds that are undesirable in high quantities. The effect of region and raw material on the volatile profile have been reported by Aguilera-Rojo *et al.* (2003), showing that the place where tequila is produced and the presence of other sugar sources (comparing tequila 100% agave and tequila) will impact the final character of the spirit.

18.2.1 Tequila flavour origin

Tequila flavour volatiles have been widely studied since the 1990s, when the spirit became very popular worldwide (Larios Medrano 1995, Estarron 1997, Martin del Campo 1998, Benn and Peppard 1996, Lopez 1999, Vallejo-Cordoba *et al.* 2004, Martin del Campo *et al.* 2010). Hundreds of flavour volatiles have been identified in tequila, and olfactometry has detected some important odour-active components (Benn and Peppard 1999; Lopez and Dufour 2001; Cantor 1999). Groups of volatile compounds found in tequila are alcohols, esters, aldehydes, furans, lactones, ketones, terpenes, etc. Much of this work has been summarized in a previous review on tequila quality (Escalona-Buendia *et al.* 2004). Tequila flavour depends, like any other food or drink, on the interaction of the whole mixture of volatile compounds in the product, and not on the individual low-threshold components. Tequila flavour is built throughout the production process, so that every step of its production contributes to create a specific chemical mixture for the product of each region or even each factory (Prado-Jaramillo 2002).

18.2.2 Tequila flavour contribution of raw materials and production process

The cooking of mature agave hearts provides not just the release of fermentable sugars, but also the formation of compounds derived from Maillard and caramelizing reactions that provide furanic volatiles (Mancilla-Margalli and Lopez 2002; Waleckx *et al.* 2008). The agave plant is rich in terpenic compounds and the presence of both vanillin and syringaldehyde have been reported in agave juice, which are normally provided by wood used in the production of other spirits. Other spice compounds such as eugenol are also present in agave.

Some of the volatile compounds from agave are transformed through processing, but some are retained in the final product. The effect of different juice extraction conditions on volatile composition deserves further investigation, and this is particularly relevant because new diffuser technology has changed the way that cooking is performed. Juice extraction and cooking are very relevant steps relating to the sensory character of the product, since they impact the way that volatiles from agave are produced, modified and recovered before fermentation and, as previously mentioned, these groups of volatiles provide the distinctive character in the same way as for other spirits such as whisky, rum, or brandy.

Methanol is an important volatile compound that is produced in mezcal and tequila processing. It comes from the demethylation of pectin that is abundant in agave. Even though methanol is toxic in large quantities, it contributes to the whole volatile flavour impact and could be an indicator of the presence of agave in the process.

Ethanol and higher alcohols are mainly produced by yeasts during fermentation. Yeast fermentation also produces esters, aldehydes and some organic acids. Moreover, some yeast strains may act to increase ester production, and the release of some terpene compounds which provide some herbal and floral notes (Fiore *et al.* 2005). Yeast can also affect vanillin and, depending on the conditions, is sometimes chemically transformed before distillation.

Distillation is also relevant to the flavour of the product as it is in this step that the recovery of volatiles takes place by means of selective separation. Furthermore, some chemical transformation of volatiles may take place when copper is used. A paired comparison test showed significant differences in the sensory attributes between tequila made using a stainless steel pot, and tequila made using a copper pot, this last having more fruity and flowery notes (Prado-Ramirez *et al.* 2005).

Including maturation in wood in the production process has a special impact on the flavour profile of the spirit, but in contrast to some other spirits such as whisky that has to be cask aged, tequila that has not been matured has a very pleasant character. This is especially true of tequila 100% agave, which is richer in the raw material volatiles. These volatiles blend with those created during fermentation after a properly-conducted distillation process (Fig. 18.1).

Rested tequila remains in wooden casks for at least two months and up to one year. Depending on the type of wood (new or reused, or the type of oak) a different character will be generated. Lignin-derived compounds such as vanillin and syringaldehyde, as well as oak lactones, are leached from the wood into the spirit. Also, phenolic non-volatile compounds from the wood are extracted by the spirit, providing colour and structural changes in the alcoholic matrix which can modify flavour release of some compounds, as has been showed in model solutions (Conner *et al.* 1994, 1999a, 1999b). Tequila that is classified as aged has to be held in a cask for maturation for more than a year, which makes the character in the spirit derived from the wood stronger as there is more extraction from the barrels, more phenolic polymerization and a longer period for oxidation. Aged tequila may be more similar in composition to other spirits aged for long periods in casks, such as whiskies or cognac, because some of the characteristic agave compounds could be lost by either oxidation or evaporation.

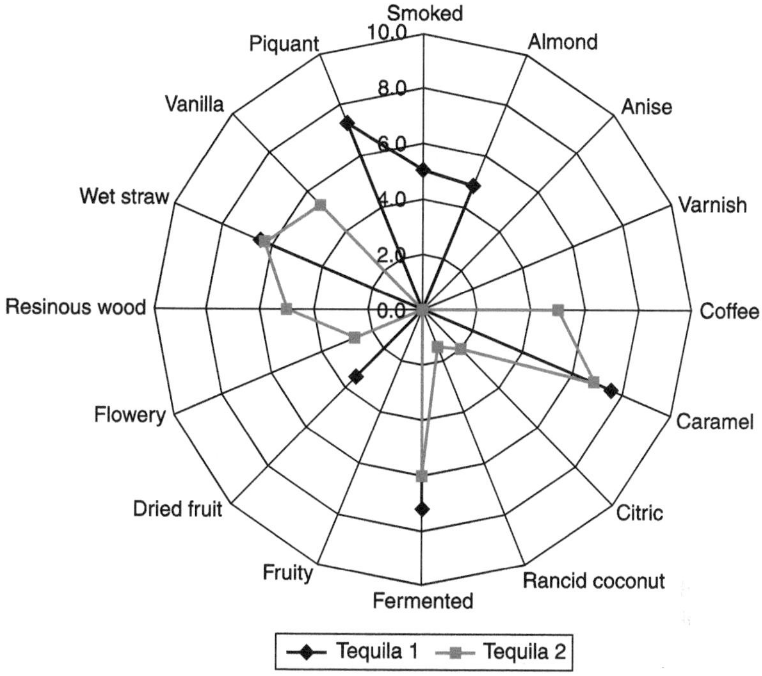

Fig. 18.1 Odour profile for selected silver tequilas from producers near Tequila town.

18.2.3 Mezcal flavour origin

Several studies on mezcal have been recently published (Molina-Guerrero *et al.* 2007; De Leon-Rodriguez *et al.* 2006; Vallejo-Cordoba *et al.* 2005; Vera-Guzman *et al.* 2009, 2010); however, the high diversity of regions in which mezcal is made, the number of agave species used in mezcal production and wide range of production methods used mean that these studies are very specific to a particular mezcal product.

18.2.4 Mezcal flavour contribution of raw materials and diverse processes

Agave species used in mezcal production provide mostly terpene compounds, which can be either spiced (terpinene and terpineol) or herbal terpenes. The fact that leaves from the agave plant can be used in the production process, as well as the diversity of species used and variations in the ripening stage of the agave results in a wide range of flavour attributes between mezcals. Traditional cooking of agave in ground holes with burning wood and heated stones produces furans and smoky volatiles that are retained by the agave.

Spontaneous fermentation does not only produce alcohols and esters, but it is also involved in the production of organic acids, terpene transformation, and formation of lactones and ketone compounds. The way in which the distillation is performed, just as for tequila, will not only determine the level of recovery of volatiles but also produce some transformation of volatiles, especially when copper stills are used.

Some herbs (eg. damiana) and some other natural materials (eg. worms) are macerated or extracted in the spirit, enriching the volatile profile of mezcal. Maturation of mezcal in wood is optional, either in new or used oak barrels, contributing flavour in a similar way to all the other wood-matured spirits.

18.3 Sensory evaluation of tequila and mezcal from a research perspective

Both tequila and mezcal have distinctive sensory attributes. The differences in raw materials and processes used, and regions from which they originate, mean that regular consumers can easily distinguish between different products. Table 18.2 shows the results of a preliminary test comparing the flavour attributes from two mezcals from Oaxaca and commercial silver tequilas using trained assessors. It is clear that even though there are some similarities, there are important differences. Sensory attributes for each spirit will be discussed.

18.3.1 Flavour attributes of tequila

Tequila producers have recently shown an increased interest in flavour diversification of the product. However, there are few published studies related to sensory descriptive evaluation of this spirit. Cantor (1999) and Garibaldi-Ramos (2004) have reported the development of a sensory vocabulary to evaluate tequilas from different types and regions.

Table 18.2 Results of a comparative test of the flavour of two Mezcals from Oaxaca against the flavour of commercial silver tequila

Mezcal 1		Mezcal 2	
Similar attributes to silver tequila	Attributes different to silver tequila	Similar attributes to silver tequila	Attributes different to silver tequila
Lemon	Solvent	Fruity	Straw
Fruity	Straw	Caramel	Smoky
Agave	Burnt	Guayacol	Solvent
Woody	Smoky	Vanilla	Piquant
	Mud		Herbs
	Rancid		Burnt
	Caramel		
	Stronger (alcohol)		
	Piquant		
	Acetic		

Flavour descriptors are related to the raw material, fermentation aromas and the bouquet obtained by maturation. Cooked agave has aroma notes of browning reactions reminiscent of caramel, molasses or raw brown sugar; it also has a smell redolent of cooked pumpkin and overripe or fermented fruits. Agave can also contain some spice flavours together with eugenol and some compounds also present in wood, as well as terpenes with floral and herbal notes. Fermentation provides the vinous character from higher alcohols; as well as fruity flavours from esters. Terpene transformation and their release given during fermentation also contribute herbal–floral characters.

Vanilla notes and some other woody-like characters may appear in silver tequila, in addition to the cooked agave flavour. The fruitiness and vinous character is derived from fermentation, and volatile acids are responsible for sour flavours. Oaky, nutty, spiced flavours, and an increase in vanilla and caramel notes are found both in rested and aged tequilas. In fact, the longer the ageing process, the more likely that sensory characteristics derived from wood override the agave-derived flavour notes.

An example of sensory flavour profiles, using quantitative descriptive analysis (QDA), for silver tequilas was reported by Cantor (1999), who evaluated selected brands from the two main producing regions: tequila from the surroundings of Tequila town and 'Los Altos' in the highlands of Jalisco State (Fig. 18.2).

Another study on wood-rested tequilas was carried out to explore flavour differences between expensive and low-priced products (Garibaldi-Ramos 2004) also using QDA (Fig. 18.3).

Among the impact compounds identified by gas chromatography-olfactometry (GC-O), phenyl ethanol (floral) and β-damascenone (fruity–winey) were reported as odour-active volatiles (Benn and Peppard 1996). Lopez and Dufour (2001) also reported the presence of these compounds, as well as 3 methyl-butanol, phenylethyl

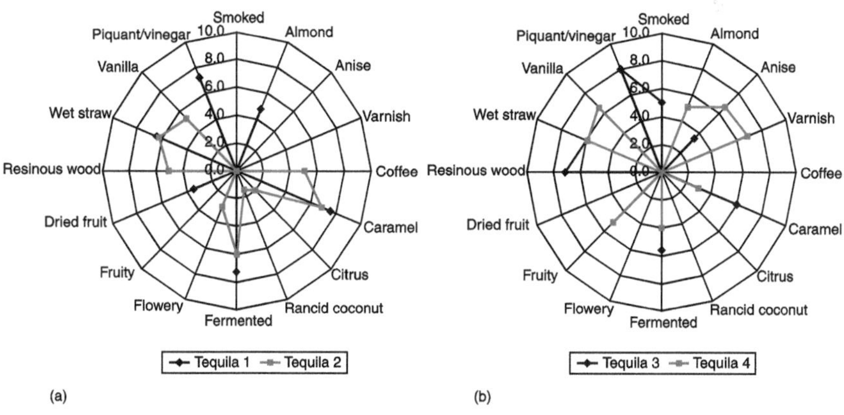

Fig. 18.2 Flavour profile for tequilas from different regions in Jalisco State. (a) Silver tequila from Tequila. (b) Silver tequila from Los Altos.

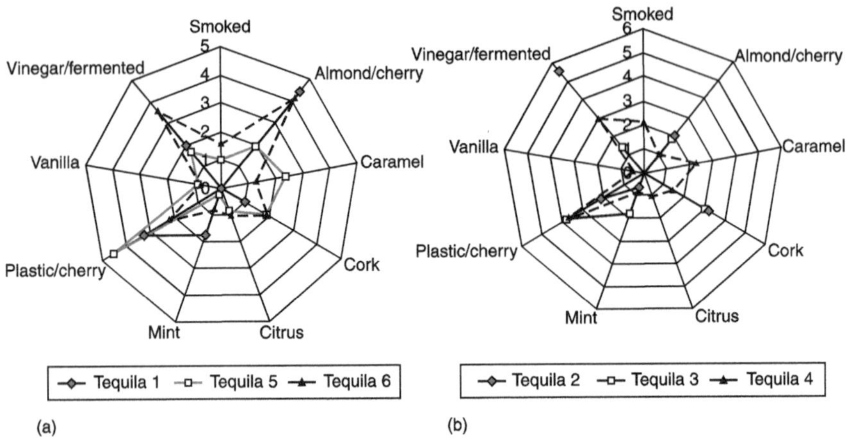

Fig. 18.3 Odour profile for tequilas with different costs. (a) Expensive tequila, aged three months. (b) Cheap tequila, aged three months.

acetate (with a note of fermented fruit), and vanillin. Likewise, Cantor (1999) reported similar flavour-active compounds using qualitative GC-O.

18.3.2 Taste and trigeminal attributes of tequila

Most research on tequila sensory attributes has been directed at flavour, although taste and mouth-feel are also important attributes for consumers. Figure 18.4 shows an example of taste and trigeminal profiles where bitterness and burning are the more intense attributes. Tequila can be sweet, bitter and acid, the last two characteristics being related to volatile organic acids. Furthermore, both the smoothness and harshness of the product are important attributes that can be either enhanced or reduced by adjusting some processing parameters. Trigeminal sensations such as pungency and burning are provided by ethanol and other alcohols. Pungency is also produced by organic acids in the throat, and astringency may be attributed to some wood phenolics. However, resting the spirit in barrels and treatment of the wood used promotes phenolic polymerization and provides the spirit with a smooth mouth-feel. Indeed the mouth-feel of some tequilas is modified to be smoother by placing the spirit in barrels for a few weeks, but not for long enough to achieve the classification of aged or rested. Sharpness or smoothness of the tequila profile can be modified by different ethanol strengths, ranging from 35–55 % (v/v); these are the limits allowed in the Official Standard for tequila, and by different resting times in the barrel.

18.3.3 Flavour attributes of mezcal

The flavour of mezcal is greatly influenced by the diverse range of agave species used in its manufacture, and the fermenting flora related to each region of

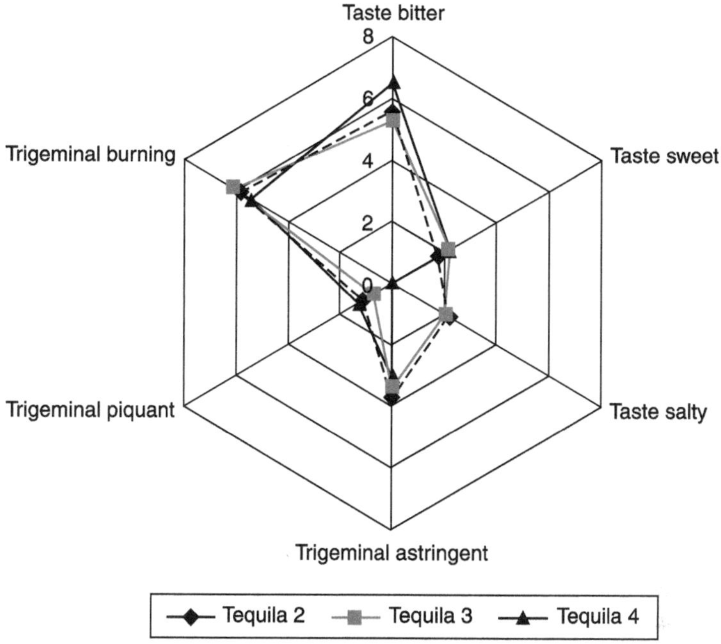

Fig. 18.4 Taste and trigeminal sensory profile for selected tequilas.

production. Sensory notes could vary from sour flavours from mixed fermentation to fresh, fruity and floral notes from specific agave plants with the appropriate fermentative flora. Higher alcohols' contribution to the sensory profile is important as for any alcoholic drink, providing the vinous solvent-like character.

There are two distinctive operations in mezcal production that provide a characteristic flavour to this spirit in comparison to tequila: cooking agave in a ground hole with burning wood, and adding a worm to the spirit. Not every mezcal is produced in this way, but the smoky flavour from cooking and the presence of worms are some attributes that consumers associate with this spirit. Nevertheless, some mezcal producers have moved towards steam cooking (in retorts or ovens) to avoid the smoky character in order to enhance the specific flavours derived from the agave.

De Leon Rodriguez *et al.* (2006) has related the presence of 3-hexen-1-ol, as well as 6,9-pentadecadien-1-ol and some unsaturated hydrocarbons to the presence of the worm, which is actually a larvae that grows in the agave plant (Berumen-Barbosa (2009) reported the worm as the larva of *Hipopta agavis*). Similarly, Molina-Guerrero *et al.* (2007) found (Z-E)-9,12 tetradecadienol, which is also related to insects, in mezcals containing worms. However, further research on these compounds' sensory impact has still to be performed.

The sensory characteristics associated with wood that are found in mezcal, as for other spirits, are related to the oaky, vanilla, nutty and spicy notes, which can

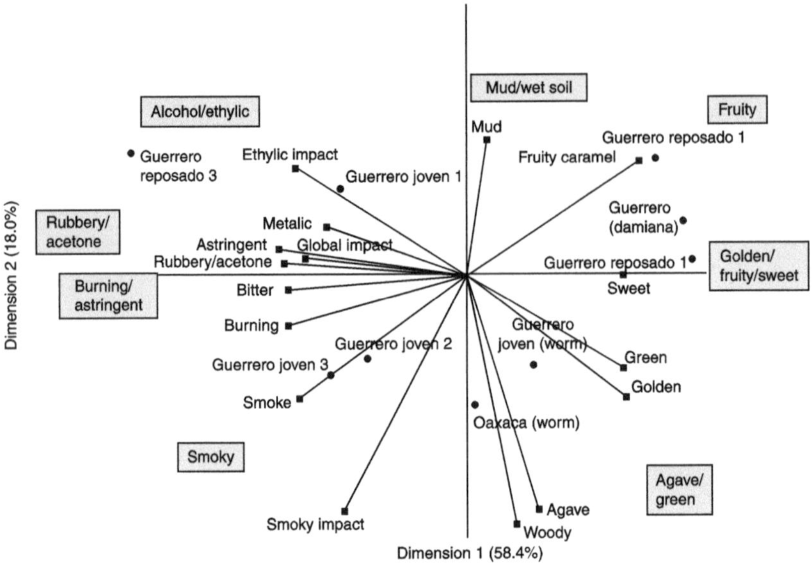

Fig. 18.5 Sensory descriptive map after principal component analysis, for selected Mezcals from Guerrero in comparison to one Mezcal of Oaxaca (Research performed for 'Consejo Estatal del Maguey y Mezcal de Guerrero'. Source: adapted from Pedrero, 2007).

also reinforce the smoky character depending on the wood treatments for the barrels.

Research directed at developing a sensory vocabulary and to profiling descriptive attributes has been documented by Pedrero (2007) who compared selected brands of Mezcal of Guerrero (Fig. 18.5), and by Gschaedler *et al.* (2008) who studied Mezcal of Michoacan (Fig. 18.6) in support of the incorporation of this region to the list of regions in which the spirit can be produced.

18.3.4 Taste and trigeminal attributes of mezcal

Just as for tequila, harshness or smoothness of mezcal are important sensory quality attributes. Sweet, bitter and acid tastes are also developed in the spirit. Pungency, burning and astringency are trigeminal sensations related to mezcal, which is commonly considered by consumers to be a strong, sharp, or harsh spirit.

18.4 Application of sensory evaluation in industry

As long as the tequila industry has grown and the popularity of the product has increased, producers have focussed their interest on diversifying its sensory character and in improving methods for controlling the sensory attributes. Some of the large tequila producers have been working with expert tequila tasters,

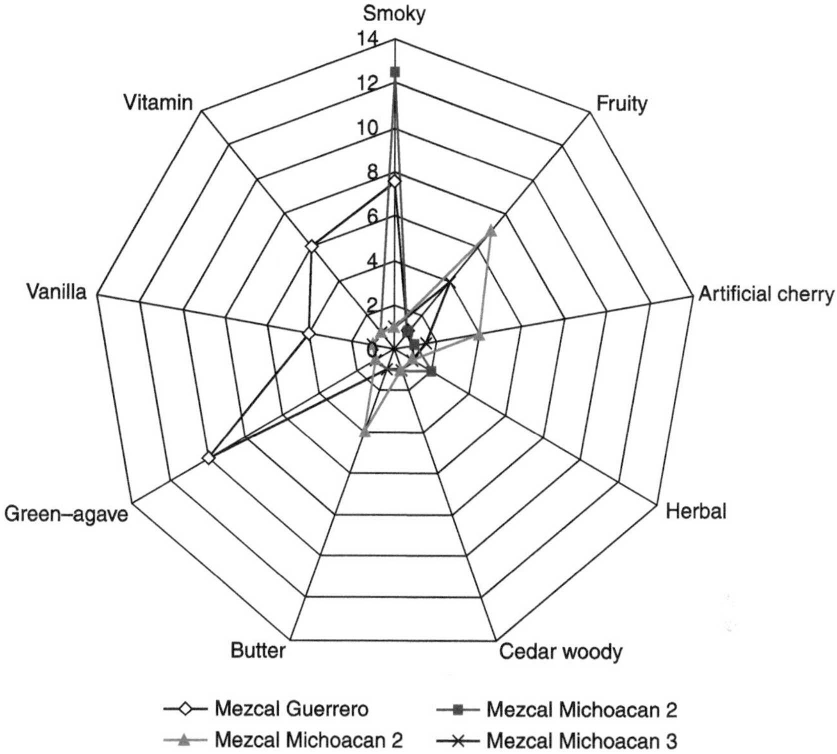

Fig. 18.6 Sensory profile of three Mezcals from Michoacan in comparison to one Mezcal from Guerrero (Source: adapted from Gschaedler *et al.* 2008).

and they have gradually implemented more structured sensory evaluation methodologies in their factories, with trained assessors for quality control and research and development.

Other medium-sized tequila producers, as well as mezcal producers that are growing towards a small industry, have also realized the need for more structured control using sensory methodologies that follow the procedures stated in the International Standards (ISO) as part of their quality management system. However, not all the medium or small producers have a proper research and development department; therefore, they have to search for external consultancy.

18.4.1 Quality control

Discriminative tests such as triangle or duo-trio tests are the most frequently used methodologies for quality control of sensory attributes of tequila. These tests are fast and practical, and use a typical product as reference.

Some of the tequila producers have their own research and development departments where they have the facilities for training descriptive assessors who can participate in more specific tests for sensory evaluation. For creating a sensory analysis unit, support provided by research institutions and other experts in this field has played an important role.

18.4.2 Research and development

Research and development departments can perform sensory studies using descriptive methodologies, and coordinate consumer studies. Research institutions and universities have been carrying out investigations for developing technologies and analytical methods for the tequila industry, as well as training expert sensory panellists to be hired by the industry. CIATEJ and CINVESTAV are among the two Mexican institutions with more established research developed on tequila; however, the number of institutions interested in this subject is increasing more and more. Several government programmes have been developed to facilitate and encourage collaboration between industry and research institutions, for which states that produce either tequila or mezcal are the most interested in obtaining funding.

Sensory research in industry is mainly directed at improving the sensory character contribution of every production step. Furthermore, improvements in handling of raw materials, yeasts and the implementation of new technologies have to be validated by volatile profiling and sensory perception to ensure consumer acceptance.

Research and development on mezcal is mainly developed through consultancy and the support of local universities and research institutions such as CIDIR-IPN Oaxaca, IPICYT from San Luis Potosí, the Technological Institutes of Durango and Guanajuato, as well as the national universities and the institutions working on tequila mentioned above. Programmes from local governments have been the main source of funding for supporting these artisan producers, according to the development strategy established by each regional society, either for exploring in detail their own process or for scaling up their production.

18.4.3 Consumer studies

Most consumer and hedonic tests for tequila are performed by industrial producers in response to the increasing interest in research into preference, hedonic level assessment and understanding consumer expectations. Tequila has diversified towards many market segments; there are products for mixing, citrus flavoured products, others that emphasize a smooth mouth-feel and some others that keep the more traditional character of a strong spirit. Rested tequila has become perhaps the most popular tequila as it blends the agave and fermentation flavours with those generated by wood.

Consumer studies on mezcal are directed at fulfilling the diverse range of strategies followed by producers' societies and some of the larger companies. Pedrero (2007) has documented a very exhaustive investigation about sensory attributes and consumers' perception of Mezcal of Guerrero, under an initiative of the CEMMEZCAL (Guerrero State Council for Maguey-Mezcal). The preference and conceptual image of mezcal, and its relation to tequila was explored using consumers from Mexico City, with ages ranging between 30–40 years, who were frequent spirit drinkers and travellers. This group was expected to be more representative and to provide more information than younger consumers who normally drink spirits in mixers and have less experience in differentiating styles and origins of regional beverages. Results showed that mezcal and tequila have some common attributes with respect to their image as Mexican products that consumers use for socializing, and are related to agave flavour; however, mezcal was distinguished to be a stronger spirit that is reminiscent of the countryside, artisan production, farms or hills. Other important image attributes for mezcal are related to the worm's presence, mud, clay and wood. This study also showed that both tequila and mezcal are accepted for their flavour, but consumers tend to relate tequila for mixing and mezcal to drink on its own.

This investigation showed the different images that both spirits possess in one region of Mexico, but of course these images have to be explored in other regions, countries and market segments, particularly because there is sometimes confusion between tequila and mezcal in other countries.

While the tequila industry is moving forward and developing new technology to increase yields and production capacities, as well as diversifying flavours and tastes for different segments of the market; mezcal producers have kept its image as an artisanal spirit, but are focussed on increasing the product quality, considering technological innovations for improving their production control and for addressing safety issues without destroying the authenticity and specific character provided by each producing region. For both spirits, sensory evaluation is becoming one of the main tools to direct innovation and shape strategies for future growth.

18.5 Sources of further information

18.5.1 Regulation

- Secretary of Economy-Norms Direction (Mexican Government):
  - http://www.economia.gob.mx/swb/es/economia/p_Direccion_General_de_Normas
  - http://www.economia-noms.gob.mx
- Health Secretary (Mexican Government):
  - http://www.salud.gob.mx/unidades/cdi/nomssa.html
- Mexican Institute for Industrial Property (denomination of origin):
  - http://www.impi.gob.mx/wb/IMPI/denominacion_de_origen

18.5.2 Relevant research institutions and universities

- CIATEJ A.C. http://www.ciatej.net.mx

- CINVESTAV- Unidad Irapuato: http://www.ira.cinvestav.mx
- CIDIR-IPN- Unidad Oaxaca. http://www.cidiroax.ipn.mx
- IPICYT A.C.-Molecular Biology: http://www.ipicyt.edu.mx/espanol/biologia-molecular/
- Technological Institute of Durango: http://www.itdurango.edu.mx
- Technological Institute of Celaya, Guanajuato: http://www.itc.mx/

18.5.3 Regulatory councils

- Consejo Regulador del Tequila (CRT): http://www.crt.org.mx
- Consejo Mexicano Regulador de la Calidad del Mezcal (COMERCAM): http://www.comercam.org/
- Consejo Estatal del Maguey Mezcal de Guerrerro (CEMMEZCAL), 21 de marzo no. 18, 2° piso, Col. Benito Juárez., Chilpancingo, Guerrero. Tel (747) 478 4609.

18.5.4 Industrial organizations and chambers

- National Chamber of Tequila Industry: http://www.tequileros.org/home_en.php
- National Chamber of Mezcal Industry. Km. 42.5 (crucero de Mitla), carretera Oaxaca-Tehuantepec, Mitla, Oaxaca.

18.6 Acknowledgements

The authors wish to thank the collaboration of Mr Daniel Pedrero from Penta Sensorial S.A. de C.V. and the Guerrero State Council of Maguey-Mezcal (CEMMEZCAL) for sharing relevant information for this review.

18.7 References

AGUILERA ROJO, D.K., MARTÍN DEL CAMPO, S.T., COSÍO-RAMÍREZ, R., ESCALONA-BUENDÍA, H. and ESTARRON-ESPINOSA M. (2003). 'Identification of distinctive parameters between Tequila mixto and Tequila 100% agave by gas chromatography'. In: Jean-Luc Le Quere and Patrick Etievant (editors). *Flavour Research at the Dawn of the Twenty-first Century*. Lavoissier- Intercept: 761–764.

BENN, S. and PEPPARD, T. (1996). 'Characterization of tequila flavor by instrumental and sensory analysis'. *J. Agric. Food Chem.* **44**: 557–566.

BERUMEN-BARBOSA, M.E. (2009). *Oaxaca: La Actividad Productiva Maguey–Mezcal*, Electronic Edition. Full text at www.eumed.net/libros/2009a/492/2011.

CANTOR, E. (1999). 'Evaluación de los compuestos volátiles representativos del olor del tequila. Correlación de mediciones instrumentales y mediciones sensoriales'. Tesis Ingeniero Químico. CUCEI. Universidad de Guadalajara. Guadalajara. Jalisco México.

CEDEÑO, C.M. (1995). 'Tequila production'. *Critical reviews in Biotechnology* **15**: 1–11.

CONNER, J.M., PATERSON, A. and PIGGOTT, J.R. (1994). 'Agglomeration of ethyl esters in model solutions and malt whiskies'. *J. Sci. Food Agric.* **66**: 45–53.

CONNER, J.M., PATERSON, A. and PIGGOTT, J.R. (1999a). 'Release of distillate compounds in Scotch malt whisky'. *J. Sci. Food Agric.* **79**: 1015–1020.

CONNER, J.M., PATERSON, A., BIRKMYRE, L. and PIGGOTT, J.R. (1999b). 'Role of organic acids in maturation of distilled spirits in oak casks'. *J. Inst. Brewing* **105**: 287–291.

DE LEÓN-RODRÍGUEZ, A., GONZÁLEZ-HERNÁNDEZ, L., BARBA DE LA ROSA, P., ESCALANTE-MINAKATA, P. and LÓPEZ, M. (2006). 'Characterization of volatile compounds of mezcal, an ethnic alcoholic beverage obtained from *Agave salmiana*'. *Journal of Agricultural and Food Chemistry* **54**: 1337–1341.

DIARIO OFICIAL DE LA FEDERACION (1977). *Declaración General de Protección a la Denominación de Origen "TEQUILA"*. http://www.impi.gob.mx/wb/IMPI/tequila (2011)

DIARIO OFICIAL DE LA FEDERACION (1994). *Declaración General de Protección a la Denominación de Origen "MEZCAL"*. http://www.impi.gob.mx/wb/IMPI/mezcal (2011)

ESCALONA-BUENDIA, H., VILLANUEVA-RODRÍGUEZ, S., LOPEZ-RAMIREZ, J.E., GONZALEZ-HERRERA, R.M.C., MARTÍN-DEL-CAMPO-BARBA, S.T. *et al.* (2004). 'Calidad del tequila como producto terminado'. *Ciencia y Tecnología del Tequila: Avances y Perspectivas.* CIATEJ-CONACYT: 172–256.

ESTARRON, M. (1997). 'Identificación de los principales compuestos volátiles que caracterizan al Tequila 100% de Agave'. Tesis de Maestría en Ciencias en Procesos Biotecnológicos. CUCEI, Universidad de Guadalajara, Jalisco, Mexico.

FARIA, J.B., LOYOLA, E., LÓPEZ, M.G. and DUFUOR, J.P. (2003). 'Cachaça, Pisco and Tequila'. In: Lea, A.G.H and Piggott, J.R. (editors). *Fermented Beverage Production.* Kluwer Academic/Plenum Publishers. New York, 335–364.

FIORE, C., ARRIZON, J., GSCHAEDLER, A., FLORES, J. and ROMANO, P. (2005). 'Comparison between yeasts from grape and agave musts for traits of technological interest'. *World Journal of Microbiology and Biotechnology* **21**: 1141–1147.

GALLARDO, J. (2008). *La Producción del Mezcal en el Estado de Michoacán.* CIATEJ-CONACYT y Michoacan State Government. Jalisco-Michoacan, Mexico.

GARIBALDI-RAMOS, N.O. (2004). 'Calidad sensorial y parámetros fisicoquímicos de tequilas reposados relación con el fabricante'. Tesis para obtener el Título de Químico Farmacobiólogo. CUCEI, Universidad de Guadalajara, Jalisco, Mexico.

GSCHAEDLER-MATHIS, A.C., GALLARDO, J. and VILLANUEVA, S. (2008). 'El proceso de elaboración del mezcal en el Estado de Michoacán'. In: Gallardo J. (editor). *La Producción de Mezcal en el Estado de Michoacán.* CIATEJ-CONACYT and Michoacan State Government, Jalisco-Michoacan, Mexico. 75–98.

GSCHAEDLER, A., RAMIREZ, J., DIAZ, D., HERRERA, J., ARELLANO, M. *et al.* (2004). 'Fermentación'. In: *Ciencia y Tecnología del Tequila, Avances y Perspectivas.* CIATEJ-CONACYT. Guadalajara, Mexico: 63–120. www.impi.gob.mx/wb/IMPI/denominacion_de_origen (2011).

LARIOS MEDRANO, I. (1995). 'Caracterización del tequila y su proceso de elaboración'. Tesis de Maestría en Ciencias en Procesos Biotecnológicos. CUCEI, Universidad de Guadalajara, Jalisco, México.

LOPEZ, M.G. 1999. 'Tequila aroma'. In F. Shahidi and C-T Ho (editors). *Flavor Chemistry of Ethnic Foods.* Kluwer Academic/Plenum Publishers. New York, 211–217.

LOPEZ, M.G. and DUFOUR, J.P. (2001). 'Tequila: Charm analysis of blanco, reposado, and añejo tequilas'. In Leland, J.V., Schieberle, P., Buettner, A. and Acree, T.E. (editors). *Gas Chromatography-Olfactometry: The State of Art.* American Chemical Society. Washington D.C.

MANCILLA-MARGALLI and LÓPEZ, M.G. (2002). 'Generation of maillard compounds from inulin during the thermal processing of *Agave tequilana* Weber Var. Azul'. *J. Agric. Food Chem.* **50**: 806–812.

MARTÍN DEL CAMPO, S.T. (1998). 'Validación de un método analítico para la determinación de los compuestos volátiles en el tequila'. Tesis de Maestría en Ciencias en Procesos Biotecnológicos. CUCEI. Universidad de Guadalajara, Jalisco, Mexico.

MARTÍN-DEL-CAMPO, S.T., GOMEZ-HERNANDEZ, H.E., GUTIERREZ, H., ESCALONA, H., ESTARRON, M. *et al.*. (2010). 'Volatile composition of tequila. Evaluation of three extractive methods'. *CyTA-Journal of Food*, in press, DOI: 10.1080/19476337.2010.499569.

MOLINA-GUERRERO, J.A., BOTELLO-ÁLVAREZ, J.E., ESTRADA-BALTAZAR, A., NAVARRETE-BOLAÑOS, J.L., JIMÉNEZ-ISLAS, H. *et al.* (2007). 'Compuestos volátiles del mezcal'. *Revista Mexicana de Ingeniería Química* **6**: 41–50.

NOM-006-SCFI-2005. 'Norma Oficial Mexicana. Bebidas alcohólicas-Tequila-Especificaciones'. Secretaría de Economía. Diario Oficial de la Federación. México 2006.

NOM-070-SCFI-1994, 'Norma Oficial Mexicana. Bebidas alcohólicas-Mezcal-Especificaciones'. Secretaría de Economía. Diario Oficial de la Federación. México 1997.

NOM-142-SSA1-1995, 'NORMA Oficial Mexicana, Bienes y servicios. Bebidas alcohólicas. Especificaciones sanitarias'. Etiquetado sanitario y comercial. *Diario Oficial de la Federación,* México 1995.

PEDRERO, D. (2007). 'Descripción Sensorial del Mezcal de Guerrero'. Research Report, Penta Sensorial S.A. de C.V. www.pentasensorial.com.mx.

PRADO JARAMILLO, M.N. (2002). 'Estudio de los compuestos volátiles en las diferentes etapas del proceso de producción del tequila'. Tesis de Maestría en Procesos Biotecnológicos. CUCEI, Universidad de Guadalajara, Jalisco, México.

PRADO-RAMIREZ, R., GONZÁLES-ALVAREZ, V., PELAYO-ORTIZ, C., CASILLAS, N., ESTARRON, M. and GOMEZ-HERNANDEZ H.E. (2005). 'The role of distillation on the quality of tequila'. *International Journal of Food Science and Technology* **40**: 701–708.

VALLEJO-CORDOBA, B., GONZALEZ-CORDOBA, A.F. and ESTRADA-MONTOYA, M.C. (2004). 'Tequila volatile characterization and ethyl ester determination by solid-phase microextraction gas chromatography/mass spectrometry analysis'. *J. Agric. Food Chem.* **52**: 5567–5571.

VALLEJO-CORDOBA, B., GONZALEZ-CORDOBA, A.F. and ESTRADA-MONTOYA, M.C. (2005). 'Latest advances in the characterization of Mexican distilled *agave* beverages: tequila, mezcal and bacanora'. 229th ACS meeting, San Diego, CA, 2005; AGFD-113.

VERA-GUZMAN, A.M., GUZMAN-GERONIMO, R.I. and LOPEZ, M.G. (2010). 'Major and minor compounds in a Mexican spirit, young mezcal coming from two agave species'. *Czech Journal of Food Sciences* **28**: 127–132.

VERA-GUZMAN, A., SANTIAGO-GARCIA, P. and LOPEZ, M.G. (2009). 'Aromatic volatile compounds generated during mezcal production from *Agave angustifolia* and *Agave potatorum*'. *Revista Fitotecnia Mexicana* **32**: 273–279.

WALECKX, E., GSCHAEDLER, A., COLONNA-CECCALDI, B. and MONSAN, P. (2008). 'Hydrolysis of fructans from *Agave tequilana* Weber var. Azul during the cooking step in a traditional tequila elaboration process'. *Food Chemistry* **108**: 40–48.

19

Whiskies: composition, sensory properties and sensory analysis

F. R. Jack, The Scotch Whisky Research Institute, UK

Abstract: The complex sensory characteristics of whisky make it one of the world's leading categories of spirit drinks, which is widely produced and appreciated across the globe. Sensory evaluation is the primary means of monitoring and controlling flavour that the industry uses to ensure that whiskies have the attributes desired by the consumer. This chapter describes the raw materials and production processes used in the manufacture of whisky and how these influence the composition of the product, ultimately shaping its aroma, taste, mouth-feel and appearance. An in-depth account of the methods used for the sensory evaluation of whisky is given, from the typical evaluations carried out as part of the production process through to describing its role in consumer research. Finally, likely future trends and their impact on whisky character are considered.

Key words: whisky, whiskey, sensory evaluation, flavour, aroma.

19.1 Introduction

Whiskies are spirit drinks produced from the distillation of fermented cereal mash. The resulting distillate then undergoes a period of maturation in oak casks. In relation to many other spirit drinks, whiskies have complex flavour characteristics. These are not added, rather they are created naturally through the various stages of whisky manufacture. Whiskies are produced across the globe, some of the best known whisky-producing countries being Scotland, United States of America, Canada, Ireland and Japan. The spelling, either 'whisky' or 'whiskey', depends where in the world you are. For the purposes of this chapter the nomenclature 'whisky' has been selected for use. As a result of regional differences in production methods, whisky, although clearly distinctive from other spirit categories, encompasses a wide range of flavours and styles. It is vital that manufacturers are able to monitor and control the sensory attributes of their products, in order to ensure that they consistently meet consumer expectations. Although analytical techniques have been developed that can provide data on levels of certain flavour compounds, there are currently no instrumental measures available that can give a full picture of whisky flavour. Consequently, sensory evaluation is used as the primary means of flavour control in the industry. This chapter describes the origins of flavour in whisky and the typical sensory methods used both during production and in consumer research.

19.2 The influence of raw materials and production processes on sensory character

The sensory attributes of whiskies result from compounds that originate from the raw materials, which are further modified by the methods used in production. In the whisky industry the compounds responsible for aroma, taste, mouth-feel and appearance are referred to as congeners. Each step in production influences congener levels and hence sensory character. The key stages involved in the production of whisky are outlined in Fig. 19.1.

The starting point in the manufacture of any whisky is the production of a cereal mash. Malt whiskies must be produced using malted barley, though a wide range of different cereals can potentially be used for making other types of whiskies. Despite the choice available, certain cereals are favoured by the industry. In addition to barley, these include corn (maize), rye and wheat (Miles and Richards, 2009). Cereals are largely unfermentable in their natural form so they must be converted into a fermentable substrate. This involves breaking down the structural components of the grain to release starch, then enzymically converting the starch to sugar. There are two ways in which this can be achieved. The first is to malt the cereal, a process of controlled germination and enzyme development.

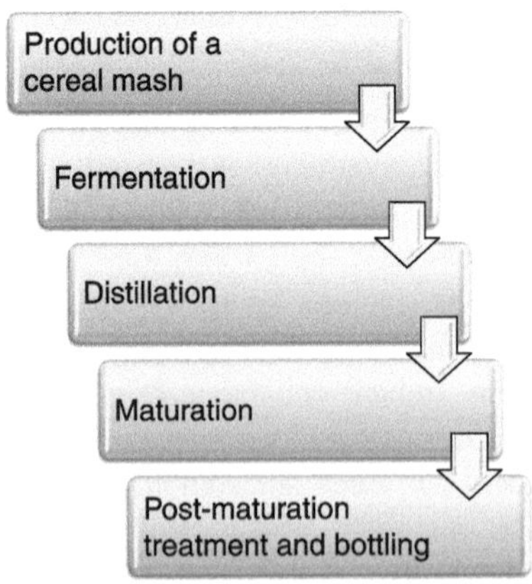

Fig. 19.1 The key stages in whisky manufacture.

Malting is followed by mashing, where hot water is added to the ground malt, gelatinising the starch and making it accessible to enzymic degradation. An alternative to malting and mashing is to cook the cereal before adding a small amount of malt or commercial enzymes to provide the required breakdown.

There are a number of variables in the production of the cereal mash that can potentially influence the sensory character of the whisky. The first is the type of cereal. Using maize, for example, will give a spirit with heavy, more oily notes than wheat. Even different varieties of the same type of cereal may result in subtle flavour differences. Regardless of the method used to produce the mash, a heat treatment step will be required. If the cereals are malted then the main heat input will be during the kilning (drying) stage, while the cooking process requires enough heat input to break down the grain and gelatinise the starch. In both cases the temperatures used are important as a number of flavour-forming reactions are heat-mediated. One of the most important types of heat-induced reactions are the Maillard reactions, which occur between amino acids and sugars. These reactions can create a wide range of flavours, depending on the types of amino acids and sugars that react, though the main flavours produced tend to be cereal, burnt or toasted notes. High temperatures can denature enzymes and break down sugars resulting in reduced alcohol yields. Hence, a balance must be struck between yield and flavour development, with the former generally being of most interest to the whisky producer at this early stage in production. The kilning step during malt production offers an opportunity to modify the flavour profile of the whisky. The malt can be smoked by introducing peat smoke into the kiln during the early stage of drying. Flavour-active compounds from the smoke adhere to the surface of the grain. These carry through the production process, giving distinctive smoky, medicinal aromas to the whisky. Peated malts have typically been used in the production of some Scotch and Japanese whiskies, though other countries are now experimenting with the smoking of malt using peat and other materials. Finally the production of a cereal mash requires the addition of water. Water composition can vary in terms of both organic and inorganic constituents, depending on where the water has been sourced from and the type of supply. Differences in water composition can influence yeast growth and metabolism during fermentation, which has been demonstrated to have a detectable, though subtle, impact on flavour (Wilson, 2010).

The cereal mash is then fermented. Yeast converts the sugars into ethanol giving a 'wash' or 'beer' with an alcohol content of around 8–10% alcohol by volume (abv). During fermentation the yeast also produces a range of secondary metabolites, such as other alcohols, acids, esters, aldehydes, ketones and sulphur-containing compounds. These make an important contribution to sensory character as many of them are highly flavour active. Different yeast strains have slightly different metabolisms. Hence yeast choice can have an impact on flavour, though strain selection is generally based on the ability to give good alcohol yields. Today most whisky producers use commercial cultures of *Saccharomyces cerevisiae*, though in some cases inocula from previous fermentations or distillery strains mixed with spent brewer's yeast are used. Yeast is supplied in a variety of formats, namely dried, pressed and cream yeast. These different forms can have a subtle influence on the progress of fermentation and hence flavour. In addition, fermentation parameters, such as the capacity of the fermentation vessel, its construction material, the ambient temperature and the fermentation time, can also affect congener formation. Non-yeast microorganisms may also be present during fermentation. At low levels certain organisms, such as lactic acid bacteria, can contribute congeners that add to complexity, but high levels of contaminant organisms can result in reduced alcohol yields and potential formation of undesirable and atypical flavours.

The fermented wash is then distilled, concentrating up the level of alcohol and giving a high-strength spirit. During this process solid material is removed and there is a selective extraction of flavour compounds from the wash. In general terms, congeners with similar, or in certain cases higher, volatilities to ethanol are most likely to end up in the spirit, while less volatile compounds are separated out. Distillation is more than just a simple separation process: formation of new flavour compounds will occur, while others are eliminated. Most of the flavour formation is due to heat-mediated reactions, such as the Maillard reactions described previously. In terms of flavour removal, one of the most important reactions during distillation is a reduction in the level of sulphur-containing compounds. These compounds typically have vegetable, rotten egg, rubbery types of aromas. They are present in the wash and are extremely volatile, so are concentrated up by the distillation process. Sulphur compounds in the spirit vapours react with copper in the still system, limiting the levels in the final spirit. This is one of the main reasons why whisky stills are made of, or at least contain, some copper.

The flavour of the distillate depends on the type of still used. There are two main types used in the whisky industry: batch and continuous stills. As the names suggest, batch processes involve the distillation of a fixed amount of wash, while in continuous stills there is a steady feed of wash and simultaneous draw-off of spirit. The main difference between these two types of process is the degree of rectification (concentration of alcohol) that can be achieved. A single-batch distillation will concentrate a wash of 9% abv up to around 23% abv, hence two or three batch distillation steps are needed to give a high-strength spirit. Continuous stills are much more efficient in terms of alcohol concentration, with the spirit typically being drawn from the still at around 94% abv. The high degree of rectification in a continuous still results in a good separation of alcohol from other congeners, giving a light-flavoured spirit. In a batch distillation process the degree of rectification is lower so the spirit is more complex. This greater complexity allows an opportunity to introduce greater diversification of flavour, with parameters such as still shape and design and operating conditions all having an impact on spirit character. Consequently, batch-distilled whiskies can vary from light, fruity and floral through to heavier, complex and even slightly sulphury spirits.

Newly distilled whisky spirit is relatively pungent and harsh and can exhibit undesirable levels of sulphury, feinty, cereal and sour aromas. Maturation, the storage of the spirit in oak casks, mellows these immature characters and results in the development of desirable flavours, such as sweet, dried fruit, woody, spicy and nutty notes. The appearance also changes, with the clear newly-distilled spirit developing a golden, amber colour over time. There are two key types of reaction that occur during maturation: subtractive and additive reactions. Subtractive reactions involve the loss or removal of flavour compounds. Often the internal surface of the cask will be charred during its construction. This layer of char has a subtractive effect, selectively removing sulphur compounds and reducing some of the immature aromas that they impart. Also, due to the porous nature of the cask, a proportion of the most volatile congeners are lost by evaporation. Oxygen can permeate in, allowing oxidation reactions to continue throughout the maturation period. These oxidative reactions are less well understood in flavour terms though they are thought to have a combined subtractive and additive effect, reducing the levels of certain flavour congeners while simultaneously generating others. The main additive reactions, however, involve the extraction of flavour compounds from the oak into the spirit. Some of these compounds are naturally present in the wood, while others are generated by the heat treatment processes during cask construction. Heat treatment breaks down some of the structural components of the oak, such as lignin, forming extractable flavour congeners. In addition to flavour, certain additive components impart colour.

Again there are a number of variables relating to maturation that can have a significant effect on sensory character. The length of maturation is crucial, with a longer maturation time giving more cask-derived flavour and colour and greater reduction of immature character. The type of cask is also important. Some sectors of the whisky industry only use new casks, while others use casks that have held other products. These casks can be reused until they no longer give the desired maturation within an economic timescale. These factors, along with the species of oak, heat treatment practices, cask size and storage conditions all influence the progress of the additive and subtractive reactions that determine the final sensory attributes of the whisky.

After maturation, when the whisky has developed the desired characteristics, it will be bottled. Rather than being bottled individually, a number of casks will generally be combined to give a more representative and consistent product, overcoming natural cask-to-cask variation. This can involve either mixing similar types of whisky, or combining a range of different whiskies to create a unique product. Prior to bottling, the whisky must be reduced to the desired alcohol strength using water. The quality of this water is important from a sensory point of view. It must be free from taints and should be demineralised, as any calcium ions present can react with wood-derived congeners giving a visible, white precipitate. Whiskies are filtered to take out particles such as cask char and, depending on the type of product, chill filtered to remove long-chain fatty acid esters which can cause cloudiness at low storage temperatures. Finally, caramel can be added. This is the only additive generally permitted in the production of many whiskies, such as Scotch. It is used at the minimum levels required to even out natural batch-to-batch colour differences, not to modify flavour.

19.3 Whisky composition and sensory properties

The two main constituents of any whisky are ethanol and water. These account for around 99.9% of the total volume, with the relative proportions varying depending on the selected bottling strength. Although ethanol imparts mouth-feel, these two components have relatively little sensory impact. It is the remaining 0.1% of other congeners that dictate overall character and differentiate one whisky from another (Aylott, 2003).

The sensory attributes of whiskies can be divided into aroma, taste, mouth-feel, nasal effects and appearance. These, with the exception of appearance, are summarised in The Scotch Whisky Flavour Wheel shown in Fig. 19.2.

Although developed to describe the sensory attributes of Scotch whisky, this Flavour Wheel is also applicable to other types of whisky. The wheel is divided into fifteen key flavour categories, and there are three tiers of attributes for each

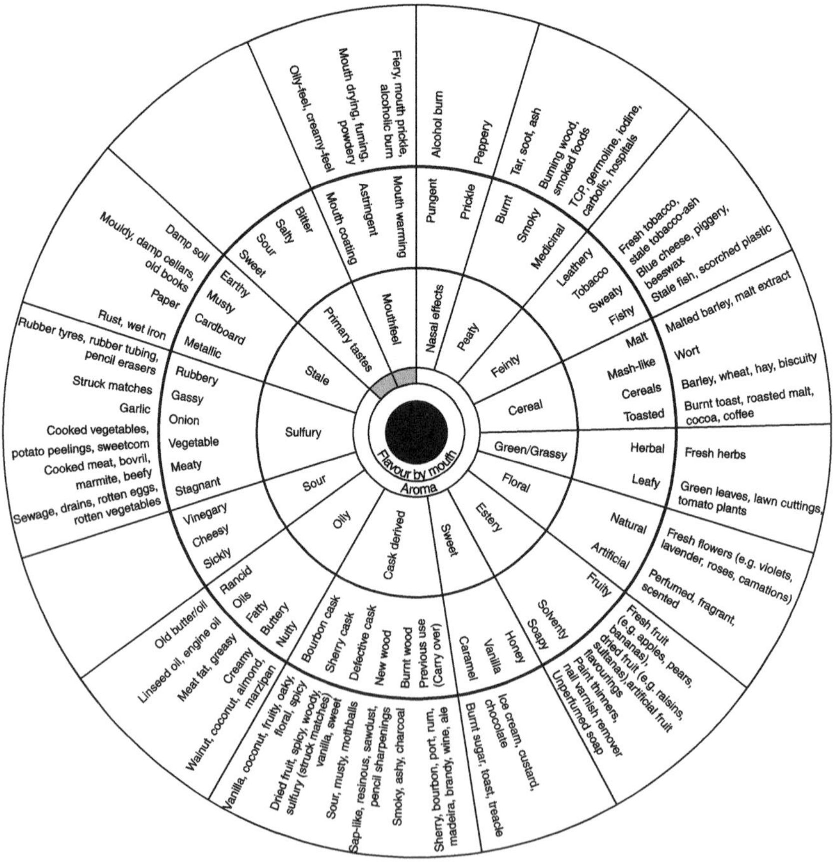

Fig. 19.2 The Scotch Whisky Research Institute's Flavour Wheel.

category. These attributes become more specific as you move from the inside to the outside of the wheel. The Flavour Wheel highlights the importance of aroma in the overall sensory character of a whisky, with twelve of the key flavour categories being aroma attributes.

As described previously, these aroma congeners develop naturally during whisky manufacture. The compounds responsible for certain types of aromas and their origins in the process are well understood. For example, 'peaty' characteristics are due primarily to the presence of phenolic compounds from the smoking of the malt, while sweet associated aromas are due to the presence of vanillin and other cask-derived compounds. In contrast, research is still ongoing to identify the compounds that give rise to cereal and feinty notes. The main technique used in the identification of flavour-active congeners is gas chromatography-olfactometry (GC-O), described in detail in Chapter 5. GC-O has revealed that a typical whisky contains in excess of 80 aroma compounds (Steele *et al.*, 2004). Around two-thirds of these compounds have already been identified.

Taste characteristics are due to the presence of congeners that stimulate the taste buds on the surface of the tongue. In whiskies, bitterness and sourness are the most important taste attributes. Bitterness is due to the presence of compounds such as tannins that are extracted from the wood during maturation. Whiskies can also contain a wide range of acids. Usually the levels of these are too low to have a significant sensory impact, though if levels are higher than normal then sour tastes will be detected.

Mouth-feel and nasal effects, including the burning sensations that arise when smelling and tasting whiskies, are mainly due to the high levels of ethanol in the product, though other alcohols and solventy-type compounds can also contribute to these sensations.

Finally, appearance plays a role in consumer acceptability, with colour and clarity both being important factors. Colour is extracted from the wood during maturation, though the compounds responsible are yet to be fully identified. The clarity of the product is ensured by chill-filtering the whisky to remove haze esters and other components that might cause visible precipitates in the bottled product.

19.4 Sensory analysis in the whisky industry

Sensory evaluation is the key means used by the whisky industry to monitor and control quality, since even the most sophisticated instrumental measures can only provide information on aspects of flavour and other characteristics. The following sections describe the sensory practices employed in the whisky industry.

19.4.1 Methods for evaluating aroma, taste, mouth-feel and appearance

As described in Section 19.3, aroma is by far the most important sensory category, in terms of number of attributes and contribution to complexity. As a result, sensory evaluation of whisky often focuses on aroma. In fact, many of the sensory tests carried out are based solely on the smell of the product, an approach known as 'nosing'. Nosing is often preferred as it has the advantage of overcoming problems associated with tasting samples containing high levels of alcohol, both in terms of having to limit intake and to prevent sensory fatigue, hence allowing a greater number of assessments to be carried out in a day or session. Also, since the product is not ingested, it also allows the same set of samples to be evaluated by several assessors. Aroma assessments are generally used in the early stages of whisky production, prior to maturation, as the spirit at this point tends to be quite harsh.

A whisky must be tasted in order to appreciate its full sensory character. Taste tests will usually be carried out once maturation is underway and continue through the process to the tasting of the bottled product. The advantage of these tests is that they provide information on aspects of taste and mouth-feel that are overlooked by nosing tests.

Sensory assessors can be asked to provide descriptions of appearance though, since accurate instrumental measures of both colour and clarity are available, such evaluations do not typically feature in the routine sensory tests carried out by the whisky industry.

19.4.2 Sample preparation and presentation

The way in which a whisky sample is prepared and presented depends largely on the type of sensory test being carried out. For nosing tests the whisky is typically diluted to 20–23% abv with water. When alcohol comes in contact with the mucous membranes lining the nose it causes mild pain sensations, perceived as burning effects. This causes sensory fatigue, limiting the number of samples that can be nosed at a given time. Dilution of the whisky reduces these effects and also makes it easier to perceive the other aroma characteristics of the product. Samples are presented in tulip-shaped nosing glasses, such as the one shown in Fig. 19.3. These are usually covered using a watch glass, which traps the volatile aroma compounds in the headspace above the liquid. Depending on the aims of the test, coloured glasses, often made from cobalt-blue glass, can be used to mask differences in colour or appearance that could potentially influence the judgement of other sensory attributes.

In nosing tests the samples are not ingested. This means that a single set of samples can be evaluated by a number of assessors. Advanced preparation of samples and repeated nosing can be carried out with minimal impact on the characteristics of the samples; in other words the levels of volatile aroma compounds in the headspace are not significantly depleted (Jack, 2003).

Samples for tasting can again be diluted to overcome the burning effects caused by alcohol. However, depending on the purpose of the test, it may be preferable to prepare the sample in a way that represents the manner in which it would typically be consumed. This may be with or without water, or other additions such as ice or mixers.

In line with standard sensory procedures, described elsewhere in this book, samples are coded before being presented to the assessor. Formal sensory tests are

Fig. 19.3 Nosing glass.

carried out in dedicated sensory rooms, with conditions such as temperature, lighting and extraneous odours all being carefully controlled. In some cases it may be desirable to also carry out tests at the production site where less control over environmental factors is possible.

19.4.3 Sensory assessors

In the past the whisky industry has relied heavily on expert sensory assessors to monitor and control the flavour profiles of their company's products. Today, many companies have supplemented these expert assessments with data produced from panels of trained sensory assessors. Panels often take over some of the more routine tasks, allowing the company's flavour expert to have more input into marketing the product and educating the consumer. Due to the complexity of whisky, substantial training and experience is required in order to become a fully functioning panellist, with additional knowledge on the production process and its impact on sensory character also being useful. In some cases sensory data will also be collected from untrained consumers, which is described in more detail in Section 19.4.6.

19.4.4 Types of sensory test

Different types of sensory test are required for different applications. When monitoring consistency of production, basic screening tests will be carried out. Here the assessor will evaluate a number of samples, looking for off-notes or atypical characteristics. This may involve comparing each sample with a target reference sample, though more often the assessor relies on their knowledge of the product in order to determine whether or not it has the expected flavour profile.

Difference tests, such as triangle, paired comparison and duo-trio tests, are also widely applied. As the name implies, these tests are used to determine whether or not there is a difference between samples. Examples of their use include the comparison of samples taken before and after making a change to the production process or in the evaluation of customer complaints against standard reference samples.

Other evaluations involve the detailed description of a product's sensory attributes. These descriptions are often based on a pre-determined vocabulary of attributes, such as the Flavour Wheel (Fig. 19.2), though in certain circumstances assessors will be asked to describe the product using their own words. The next step up from simply describing flavours is to obtain an intensity measure for each attribute using methods such as quantitative descriptive analysis (Stone *et al.*, 1974). Here, a group of eight or more trained and experienced panellists score samples according to the levels of a number of selected flavour attributes. Average intensity scores are then calculated across the panel. These scores can be graphically represented in the form of a flavour profile, such as the example shown in Fig. 19.4.

This gives a good visual representation of the overall flavour character of a sample, allowing easy identification of the key attributes and their relative intensities. This type of approach has been applied in the evaluation of production samples and in whisky flavour research. More recently flavour profiles have also been used in the marketing of whiskies as they present complex sensory information in a visual way that is easy for the consumer to interpret.

Finally, subjective sensory tests, where information is collected on personal preferences, likes and dislikes, are also important in certain applications. These tend to be consumer tests carried out on the final bottled product, rather than on samples from earlier in the production process.

19.4.5 Sensory evaluation in the whisky production process

Sensory tests are carried out at various stages in the production process to track the progress of flavour development. As a rule, the first of these tests are carried out after distillation, with an evaluation of the newly-distilled spirit. Each distillery will aim to produce a spirit with a particular flavour profile, depending on the raw materials, fermentation and distillation parameters that have been used. These initial sensory tests, which are normally based on aroma, ensure that the spirit has the desired flavour with no off-notes or atypical aromas. If issues are identified, steps can be taken to correct the process and bring flavour back in line. The whisky

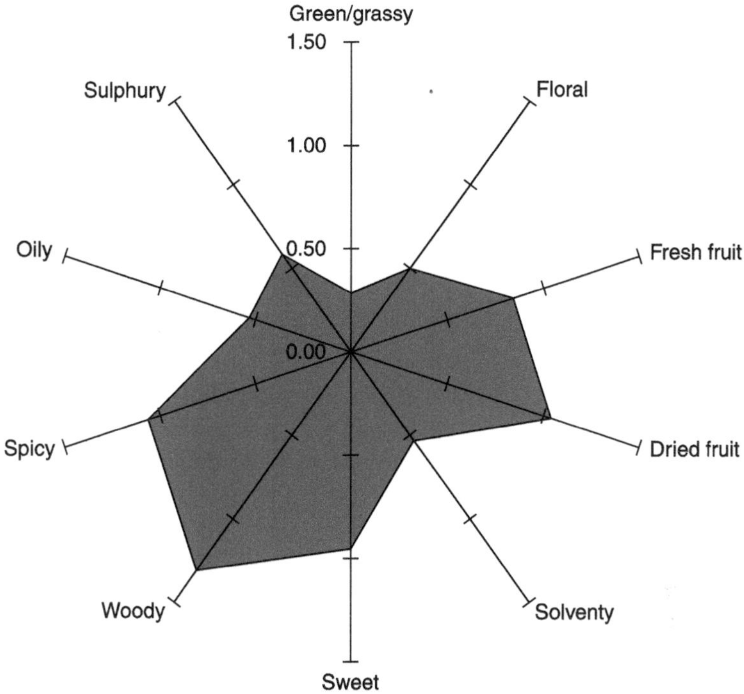

Fig. 19.4 Flavour profile for a Scotch whisky matured in an ex-sherry cask.

manufacturer may also make decisions on subsequent maturation parameters based on the character of the spirit. If the spirit is quite sulphury they may decide to mature it in a heavily charred cask in order to reduce this note. Conversely, a spirit that is quite light in character may be best suited to maturation in a less active cask in order to retain some of the flavour of the original spirit.

During maturation, sensory samples may be taken at regular intervals or at the end of a pre-determined storage period. Here the focus is on the development of mature character, both in terms of flavours extracted from the cask and the reduction of immature aromas. Again assessors will be looking to identify any off-notes, which can occasionally arise during maturation. The decision whether or not to bottle the whisky at this stage will depend on the results of these tests.

Sometimes individual casks will be bottled, but it is more usual for a number of casks to be mixed or blended together first. In some cases groups of similar casks from a single distillery are combined, for example in the production of a single malt whisky. This evens out cask-to-cask variations in sensory character. Each cask will be individually assessed before being emptied into a vat. A final sensory check of the vatted product is required to confirm that the resulting whisky has the desired attributes. Alternatively, a range of different whiskies may be blended together, with the aim of creating a product that is unique in sensory terms from any of the components. Selecting the casks required to achieve the desired final product generally relies on an expert sensory assessor, often referred to in the industry as the Blender. Again, checks will be carried out on both the individual cask samples and the vatted product. If the resulting blend does not have the required flavour profile, the Blender will use his or her expertise to select casks to add to the vat that will give the required adjustment in flavour.

Lastly, tests on the bottled product will be undertaken to make sure that the whisky has not picked up any flavour taints or been cross-contaminated during the final stages of production.

19.4.6 Consumer research

Sensory characteristics are central to any whisky's acceptability and success in the marketplace. They are used to differentiate one product from another and widely applied in promotion and marketing, with descriptions of flavour and other characteristics appearing on packaging and other advertising material. Hence, gauging perceptions of sensory characteristics is an important aspect of the consumer research carried out by the whisky industry. Consumer tests will generally focus on individual preferences and the likelihood of purchasing one product over another. Relationships between these factors and the whiskies' sensory character can then be explored, and if necessary the product adjusted to best match specific target markets.

In consumer sensory tests, samples are usually prepared in a manner that reflects the way in which the whisky is actually consumed. Rather than diluting to a fixed alcohol strength, consumer panellists may be given the option to add water or ice according to their own personal taste. In certain markets consumers mix their whiskies with soft drinks. For example in China, which is an important emerging market, whisky is mixed with green tea. In addition to sensory character, consumer tests often also explore perceptions of other aspects of the product, such as packaging, product or brand image, price and consumption situation, all of which influence purchase intent.

The selection of individuals to take part in consumer tests is generally not based on sensory ability but on their current or intended future consumption of the products being evaluated. Age and other demographic factors will also be taken into account. Often the whisky industry will hold tests in a number of countries in order to better understand regional differences in consumer needs. Tests can either be carried out on an individual basis or alternatively can take the form of focus groups, where a number of consumers are brought together to discuss product characteristics. There is also the option of using either a central location or conducting the tests in the consumer's own home.

19.5 Future trends

The legislation that defines whisky and its production is important to the industry. It put boundaries on what can and cannot be defined as whisky, differentiating it from other spirit drinks. Sensory attributes are an important aspect of this, with manufacturers wanting to ensure that their whiskies have the characteristics associated with this spirit category. In many countries, this legislation has become more stringent over recent years. Since the legislation controls production, which in turn underpins composition and sensory character, it is neither likely nor desirable that there will be any dramatic shifts in whisky flavour in the near future. Nevertheless, within the legislative constraints there is scope for producing a wide range of whiskies which, although clearly different from one another, still exhibit whisky character. The industry will continue to manipulate the attributes of their products through selection of raw materials or modification to the production described earlier in this chapter. The global market for whisky continues to grow, with a number of new emerging markets. In order to capitalise on these, whisky producers must evaluate the needs of the consumers in each market and tailor the sensory characteristics of their whiskies accordingly.

It is also essential that manufacturers have control over the consistency of their whiskies. Continued development in the area of flavour research provides a means of achieving this. A full understanding of the compounds responsible for all sensory characteristics, encompassing aroma, taste, mouth-feel, colour and overall appearance, and where they originate in the production process, will help to reach this goal.

Finally, in recent years the whisky industry, like all manufacturing industries, has to take steps to minimise the environmental impact of its production processes. This work continues and new approaches are being developed. However, in any future modification to the process the impact on the sensory attributes of the whisky must always be a primary consideration.

19.6 Sources of further information and advice

Many of the facts presented in this chapter are based on unpublished research carried out by the Scotch Whisky Research Institute, though some key references have been included.

More in-depth information on the science and technology of whiskies can be found in the following two books: *Whisky – Technology, Production and Marketing* edited by I. Russell, published by Academic Press (2003) and *The Science and Technology of Whiskies* edited by J. R. Piggott, R. Sharp and R. E. B. Duncan, published by Longman Scientific and Technical (1989). Both books are available from http://whiskyjourneys.com.

There are a large number of other books that describe sensory characteristics, both relating to specific brands and describing whisky as a whole, and increasingly information can be found on company websites.

Finally, the Scotch Whisky Association website (www.scotch-whisky.org.uk) is a useful resource for those wanting to find out more about whisky production and whisky industry statistics.

19.7 References

AYLOTT R (2003). 'Whisky analysis'. In: *Whisky – Technology, Production and Marketing.* Edited by I Russell, Academic Press Sussex.

JACK FR (2003). 'Development of guidelines for the preparation and handling of sensory samples in the Scotch Whisky industry'. *Journal of the Institute of Brewing*, **109**, 2, 114–119.

MILES J and RICHARDS C (2009). 'Whiskies around the world'. In: *The Alcohol Textbook,* 5th ed. Edited by WM Ingledew, DR Kelsall, GD Austin and C Kluhspies, Nottingham University Press, Nottingham, UK.

STEELE GM, FOTHERINGHAM RN and JACK FR (2004). 'Understanding flavour development in Scotch Whisky'. In: *Distilled Spirits – Tradition and Innovation.* Edited by JH Bryce and GG Stewart, Nottingham University Press, Nottingham, UK.

STONE H, SIDEL JL, OLIVER S, WOOLSEY A and SINGLETON RC (1974). 'Sensory evaluation by Quantitative Descriptive Analysis'. *Food Technology*, **28**, 24–34.

WILSON CA, JACK FR and PRIEST FG (2010). 'The role of water composition on malt spirit quality'. In: *Distilled Spirits –New Horizons: Energy, Environment and Enlightenment.* Edited by GM Walker and PS Hughes, Nottingham University Press, Nottingham, UK.

Part IV

Consumer research methods: principles and application to alcoholic beverages

20

Consumer research in the wine industry: new applications of conjoint measurement

L. Ettinger Lieberman, D. I. Moskowitz and H. R. Moskowitz, Moskowitz Jacobs Inc., USA

Abstract: We present a marketer-useful system to identify those specific phrases that drive consumer interest. The approach begins with phrases about wine (i.e. features, benefits, emotions, quality statements, brands). These ideas are mixed and matched by a computer program to create small, easy-to-understand concepts about wine. These concepts are evaluated by consumers. Each respondent rates different combinations of ideas. The analysis, done at the individual respondent level, reveals what particular ideas 'drive' interest for that respondent. The full dataset is analyzed by key subgroups defined ahead of time, as well as parsed to identify different mind sets of individuals with radically different responses to these messages. We show the application of the approach for simple attributes such as interest, and then for more cognitive complex attributes such as appropriate price and feeling/emotion.

Key words: consumer, conjoint analysis, segmentation, marketing, wine concepts.

20.1 Background: understanding the mindset of the consumer

Since the end of World War II and the emergence of a consumer economy, management in the food and beverage industries have been forced to focus on the needs and wants of the consumer. In the earlier years before World War II, almost anything edible and reasonably safe would be lapped up by consumers hungering for new products. Peace and the consumer economy afterwards encouraged competition, which brought with it the recognition that although consumer demand was great, it was not infinite. At some point, the consumer would be surrounded by a surfeit of choice. The previous inexhaustible demand of the consumer would give rise to the discerning consumer, one who 'knew' the product, and one who would morph into a new creature. Pine and Gilmore called it the change from the product economy to the experience economy (Pine and Gilmore, 1999).

Given the global marketplace, how can companies communicate with consumers about their products? Can communication become a science? Are there ways by which they can create databases of ideas about products and use those ideas to guide development on the one hand, and communication on the other?

At the more concrete level of today, confronted with increasing imports and exports, line extensions as well as new wine introductions, how does the marketer make a brand make itself stand out from the crowded shelf? Market researchers and product developers can employ a number of methods, such as brain storming, reviewing and updating old, traditional campaigns, and running focus groups, respectively. These methods are not new. They have been popular for almost four decades (Axelrod, 1975). Each method yields mixed results. In addition, the strategies are not much help when faced with increasing, global competition, a larger marketplace with differing local and regional tastes.

The commercial importance of wines and other alcoholic beverages means that marketers, and thus scientists and consumer researchers, cannot stand still, and be comfortable with methods that have served well in the past, and thus should protect one in the future. Researchers are always looking for new ways to better understand the consumer. This chapter presents the application of conjoint measurement to the uncovering of the 'algebra of the consumer mind'. This is a phrase the current authors use to describe knowledge of how messages drive consumer responses. Although the chapter presents new-to-the-world data, researchers have been working with conjoint analysis, in the world of wines, for at least 20 years.

20.2 Experiments and statistical tools

In consumer research, the notion of experimentation has evolved into two distinct forms: (1) testing and (2) experimentation. Testing uses the responses of the consumer/expert panelist to judge the performance of test stimuli on a number of characteristics. From the ratings, the researcher discovers the attributes that perform well versus those that perform poorly. When testing several, unconnected stimuli, the researcher performs additional analyses to discover the characteristics which 'drive' or account for strong versus poor performances, respectively. The bottom line is that in testing single stimuli or an array of unconnected test stimuli, the researcher generates a report card about performance, and may even discover some possible causes of acceptance.

In experimentation, the researcher systematically varies the factors that can be controlled. Experimental design lays out the specific combinations of the variables. Each respondent evaluates one, several, or even all of these combinations, rating each combination on one or more scales. From the ratings, the researcher uses statistical tools, especially least-squares regression, in order to estimate the specific contribution attributable to each element. The scientific discipline of experimental design has made it popular among development scientists in industry, as well as consumer researchers who want to apply it to realms such as the combination of ideas (Box *et al.*, 1978). Furthermore, the widespread use of shrink-wrapped software, e.g., SYSTAT (Systat, 1997) has made experimental design accessible to everyone who works with a personal computer.

20.3 Conjoint measurement (CM): experiments by mixing together ideas

At its very core, conjoint measurement (CM) is an implementation of experimental design. Experimental design in general, and conjoint measurement in particular, operate under the assumption that we learn the most about independent variables when they are presented in combination, the way they would be experienced in real life. Indeed, people often cannot evaluate single elements (ideas) in a valid way. For example, brands of products are important, and would be rated important if evaluated singly. Yet, when evaluated in combination with product features and health benefits, quite often brands do not perform as well.

Conjoint measurement pulls out the effect of the individual elements when they are evaluated in the more 'ecologically-valid' mixtures that constitute the experience of daily life. Conjoint measurement quantifies the responses to those mixtures. From the responses, conjoint measurement attempts to identify the underlying impact or utility of each element in the mixture.

CM has come to mean systematic exploration of combinations of independent variables which are discrete, i.e. either appearing or not appearing in a particular combination. The test stimuli are typically discrete options, present or absent in the test stimulus. These stimuli or elements are often cognitively 'rich', having meaning in and of themselves, such as brand names, benefits, and prices.

The history of conjoint measurement as a guide to understanding the mindset of the consumer is a rich one. As a branch of statistical design, conjoint analysis was prefigured by the methods of experimental layouts, going back to the 1920s and even earlier. In the mid 1960s, mathematical psychologists and statisticians became increasingly interested in principles of measurement. They moved beyond simple methods to creating measurement systems allowing people to act as valid measuring instruments. The notion that a person could generate valid measurements through the use of numbers was, and continues to remain, a source of contention among mathematical psychologists. Conjoint measurement was conceived and born in this period of axiomatization and mathematization of psychological processes. The term 'conjoint analysis' appears to have been used first by mathematical psychologist Luce and statistician Tukey in their seminal paper (Luce and Tukey, 1964).

Somewhat in parallel, psychologist Norman Anderson would recognize the value of this deconstruction in his so-called functional measurement (Anderson, 1977). It remained for Wharton Marketing Professor Paul Green and his colleagues to push conjoint analysis to its limits (Krieger *et al.*, 2004), and then to inspire others to push conjoint analysis even further (Wittink and Cattin, 1989).

In the decades that would follow Green's pioneering effort new applications would emerge. This chapter on the algebra of the consumer mind presents a number of those new applications, specific to studies of wine, but promising far greater applicability. Some of the basic ideas in this chapter can be found in books and articles (e.g., Moskowitz *et al.*, 2005b), but others represent new ideas, based in economics (Galanter *et al.*, 2011), and even newer-to-the-world efforts such as the research on wine and emotion.

20.3.1 Specific advances in CM that allow it to measure the 'algebra of the consumer mind'

By the 1980s, conjoint analysis had evolved into one of the most popular approaches for consumer research (Wittink and Cattin, 1989). Different methods would be tried. Arcane methods of transforming data matrices to achieve linearity, necessary in conjoint analysis and, in the early days, a tedious method, were abandoned in favor of dummy variable regression analysis. Once the computational approach was simplified using well-known procedures, a number of improvements occurred rapidly over the next 30 years, each of which will be relevant to our study of wines:

1. Tests with many elements, not just a few: Early studies with conjoint measurement worked with a limited number of elements, often fewer than 20, due to the nature of the task – to identify promising physical features. In such studies, engineers typically constrained the conjoint study to options that would make sense from a technical point of view. Later studies dealt with positioning, i.e. advertising statements, often alone, but sometimes tested in the same study as the actual technical features. There were often many dozens of these elements, and hundreds of combinations. The combinations had to be created ahead of time, at the start of the experiment, reviewed, and then edited in order to assure that incompatible combinations would never appear. At the same time, it was vital to maintain the integrity of the experimental design (Moskowitz, 1994).
2. Focus on individual-level models with many elements: individual-level models are the goal in statistics. One does not have to worry about sample balancing, as each person generates their own model in conjoint analysis. The approach to creating individual models, and ensuring that each individual evaluated a unique set of combinations, was solved in the early 1990s. The approach, called IdeaMap®, began with a basic experimental design structure. Each concept was created within that basic structure. The key difference between this basic structure and so-called 'fixed design' was that the structure itself sampled elements from the full set of available elements, checking that there were no previously-defined pairs of incompatible elements, and then presenting the test concepts to the respondent. The entire process was done in real time. The key change was that every respondent now evaluated unique combinations. The IdeaMap® strategy allowed the researcher to implement conjoint analysis more easily, without the onerous task of creating all of the test stimuli ahead of time. Furthermore, IdeaMap® changed thinking about sample size of respondents. Previously, sample

size was used to ensure a stable mean; with IdeaMap® the sample size was used to ensure that a lot of the concept 'space' would be covered (Moskowitz *et al.*, 2001).

3. Easier and more intuitive analysis by ordinary least square regression. Earlier forms of conjoint measurement used logistical regression, requiring a great deal of computer time. Although technically a slightly better form of regression because of the nature of the independent variables (binary, rather than truly continuous), logistical regression does not produce results that are user-friendly. The user, and in turn management commissioning the study, cannot easily understand the output of logistical regression. They cannot determine what the elements mean nor can they combine the elements mentally to come up with new combinations. Changing to the more user-friendly, intuitively clearer ordinary least squares regression produces the means by which we can understand the mindset of the consumer.
4. Beyond the general groups to mindsets: We know that people differ from each other. Most researchers work with well-defined groups of individuals. These groups may be created based on geo-demographics (who the people are), on attitudes (what people say they want or value), or on behavior (what people do). Conjoint measurement as an experiment allows us to measure the patterns of responses to test stimuli, and then divide the respondents by the patterns of their responses. When the test stimuli deal with communications about products/services, then the segmentation generates clusters based on mind sets, or revealed decisions about products. We will see how this segmentation helps us understand responses to white and red wines.
5. Beyond simple evaluation to pricing: Pricing has always been a key topic in consumer research. Pricing is one of the five P's of marketing (product, positioning, packaging and promotion being the other four). Experimental design helps us to understand the role of price, whether price be one of the stimulus elements, or price be a response. When we deal with sweet dessert wine, we will see how price can be used as a response variable, and how regression analysis can assign a dollar value to every element.
6. Beyond the rational stimulus element to the emotional stimulus element: Most early conjoint studies worked in the realm of product specifications. Most conjoint measurement studies instructed and continue to instruct the respondent either to choose between/among a set of alternatives, or to rate the alternatives on a scale. Emotions do not come into play here. However, the elements for conjoint analysis can move beyond the rational to the emotional. We will see these different elements, rational and emotional, used in all three studies (white wine, red wine and dessert wine).
7. Beyond the rational response to the emotional response. Recently, one of the authors (HRM) has presented an alternative approach, where the respondent reads a test concept, and selects a feeling/emotion that is experienced when reading the particular concept. The approach was previously used for studying package design (Moskowitz *et al.*, 2009), and is now presented in the study of dessert wines.

20.4 Applying CM to wine: traditional approaches

To give an idea of how conjoint measurement has been used by academic researchers, here are just five of the many different efforts that have appeared in the academic literature, along with a short précis of the paper provided by the journal itself, in the form of an abstract or summary available on Google Scholar®. The purpose of the précis is to show the nature of the different types of studies. Most use simple 'factoids' about the product as bare-bones elements in the test stimulus. The focus of these studies is to find general drivers of acceptance and segmentation, but not to probe more deeply into the mindset of the consumer by finding out the specific messages and tonalities that drive acceptance. That is, the studies deal with the stimuli as almost emotion-free facts, rather than focusing on the richness of the stimuli as something to be studied.

1. Drivers of acceptance: (Perrouty *et al.*, 2006). Previous research suggests that the region of origin equity is significantly moderated by the other wine attributes with which it is combined on the wine label: commercial brand, level of price, type of bottler, grape variety. The authors test whether the moderating effects depend on self-perceived consumer expertise . . . Results show that the region of origin equity is indeed significantly moderated by the other wine attributes . . . These moderating effects are more important for consumers who perceive themselves as 'more expert' than for consumers who feel they are 'novices'.
2. Drivers of acceptance: (Lockshin *et al.*, 2006). Shows how relative purchase rate changes as brand, region, price, and award are changed. The results show that low involvement consumers use price and award to a greater degree than high involvement consumers. A gold medal increases the choice probability the most, but mainly at the lower and middle price points, and a well-known region amplifies the desirability of small brands more than large brands.
3. Drivers of acceptance: Gil and Sanchez (1997) use three attributes in the conjoint design: price, origin and grape vintage year. Among these attributes, Navarra consumers assigned more importance to the origin of the wine, followed by price and grape vintage year.
4. Drivers of acceptance: (Sanchez and Gil, 1997). Two producer regions are considered, Navarra and Aragon, and three wine attributes: price, origin and grape vintage year (wine age). In both regions, differences between urban and rural consumers' preferences are tested. Results indicate that the origin of the grapes is the most important attribute for the wine consumer. The price and the age of the wines are secondary attributes in consumer preferences.
5. Segmentation: A methodology for segmenting the Australian wine market on a behavioural basis is proposed and illustrated (Johnson *et al.*, 1993). The initial segmentation is accomplished by using estimated conjoint part-worths for the attributes of price, variety, area and vintage as criteria in a cluster analysis.

20.5 Understanding white wine using experimental design of ideas

20.5.1 'Mind Genomics': understanding the algebra of the mind regarding wine experiences

Let us now use conjoint analysis to understand white and then other wines, but do it a bit differently than has been done before. We move away from the traditional approaches used in conjoint measurement, namely 'bare-bones elements' and simple concepts comprising just a few words. The traditional approaches use conjoint measurement to understand decision processes. The elements themselves are factoids, often deliberately stripped of emotion, so that they are simplistic ideas. Instead of using conjoint analysis to test these bare-boned ideas, we will use conjoint measurement to test richer, more ordinary phrases, of the type people would encounter in everyday life. Our approach will be that of 'Mind Genomics' (MG). MG is a phrase the current authors use to describe their version of conjoint analysis. The term was first mentioned by Moskowitz *et al.* in their 2005 paper (Moskowitz *et al.*, 2005a) and then used extensively by Moskowitz and Gofman (2007). In its simplest description it can be considered to be the algebra of the mind with respect to everyday life. In other words MG is used to study mindsets with regard to everyday experience (Moskowitz *et al.*, 2006) and is the quantitative psychology of the ordinary. MG tries to quantify the response to everyday stimuli in terms of how those stimuli interest people, drive feeling/emotions, or even drive willingness to pay for specific products and experiences, respectively. The stimuli are important, not just because they represent basic ideas in the mind, but important in and of themselves as the building blocks of normal experience.

Conjoint measurement is a tool of choice for MG. Conjoint measurement allows the researcher to combine ideas into small, easily understood vignettes, to which the respondent can react and which can be rated on attributes. The patterns of reactions of the ratings tell us how important each element is as a driver of interest. To the degree that the elements can be cognitively rich, with evocative phrases that mirror what a person sees in everyday life, the results of the conjoint experiment will provide some of the algebra of the mind. We apply that approach here to understanding communications about wine.

20.5.2 Mechanics of the experimental design and its analysis

The particular studies that we present as examples come from a variation of conjoint analysis known as rule developing experimentation (RDE). The goal of RDE is to create a database, as much for the sake of increasing knowledge about consumers, as for solving a particular business problem.

The specifics for RDE, i.e. the differences from conventional conjoint analysis, are as follows, along with the rationale for those exceptions:

1. We begin with the test stimuli for conjoint analysis into silos and elements. Silos (also known as variables, buckets, categories) are general groups of

related elements. In the parlance of conjoint measurement these silos are known as 'variables.' Elements (also known as levels) are the specific phrases that will be combined into stimuli, responses to which will be measured. The same notion of an underlying structure of silos and elements holds, whether the topic is a description of a food, a professional service, or even a topic in public policy such as health care and nutritional compliance.

2. Each silo comprises an equal number of elements. The rationale for the equal number is that later on the basic experimental design will be permuted, to create a set of mathematically identical designs, but whose combinations will differ from respondent to respondent (Gofman and Moskowitz, 2010).
3. The test concepts comprise 2–4 elements; at most one element from each silo. The incompleteness, dictated by the experimental design, ensures that the ratings from each respondent can be analyzed by ordinary least-squares regression (contribution of the design), and that the coefficients have absolute value (contribution of the incomplete design, with true zero conditions). The majority of other conjoint methods require that each test concept be complete, so that there is one element from each silo. The argument, not necessarily valid, is that the respondent will have a difficult time understanding incomplete concepts. The reality is quite different. Respondents work with any information available to them, and seem to have no trouble making consistent, repeatable decisions in these tasks.
4. The elements are statistically independent of each other, so that regression analysis can be used to pull out the effect of each element.
5. Each element appears against many different backgrounds, preventing pairs of elements from always appearing together, so that their separate contributions cannot be estimated. This is called confounding. Confounding often happens when the researcher does not use a design, but rather 'guesses', or puts together so-called 'rifle shots'.
6. We begin our analysis with the results from the white wine study. For this study, the regression modeling generates two different equations for each of the 212 respondents who participated. These equations are both expressed in the same way:

$$\text{Rating} = k_0 + k_1(\text{Element A1}) + k_2(\text{Element A2}) \ldots k_{36}(\text{Element D4}) \quad [20.1]$$

The parameters of the regression equation are easily estimated by standard regression modeling. The two equations will differ because the dependent variable changes in a subtle but very important way.

1. The first model, the so-called Persuasion Model, uses the 9-point rating as the dependent variable. Each element, A1–D9, appears as either 0 (absent from the particular concept) or as 1 (present in the concept). The additive constant shows the number of rating points on the 9-point scale that can be attributed to the rater, without the presence of elements. Each element, in turn, generates its own contribution, in terms of the fractional number of points.

2. The second model, the so-called Interest Model, uses a transform of the 9-point rating, so that ratings of 1–6 are transformed to 0, and ratings of 7–9 are transformed to 100. The focus of the Interest Model is to relate the presence/absence of the 36 elements to whether the concept is rejected (rating 1–6) or accepted (rating 7–9). The ordinary least squares regression estimates the parameters of a model which is written the same. The additive constant shows the conditional probability or percent of respondents who would accept the concept (i.e. rate it 7–9) in the absence of elements. The additive constant is a baseline. The coefficients or utilities of the Interest Model are the conditional probabilities that the concept would move from not interesting (rating of 1–6) to interesting (rating of 7–9) when the element is introduced into the concept.
3. For both the Persuasion and the Interest Models the coefficients and the additive constant have substantive meaning, and have absolute meaning. That is, they should not depend upon the other elements in the study, at least from a statistical point of view. Thus we can database these parameters across studies, and across time. We can compare the parameters for different beverages, and for different people in the same study.

20.5.3 Early applications of Mind Genomics to wine: the mega studies

In early 2001, author H. Moskowitz and his colleague Jacqueline Beckley at the Understanding and Insight Group were approached by Maryanne Gillette of McCormick & Company to create a system by which to understand craveability. At that time, in late 2000 and early 2001, the notion of craveability for food and drink was just being promoted as a prospective new way to understand foods (Beckley and Moskowitz, 2002).

As part of the solution to the issue of craveability, Beckley and Moskowitz used conjoint analysis to create a set of 30 studies. Each study had the same basic structure of silos and elements, 'raw materials'. The elements fell into four silos or categories, with each silo in turn comprising nine elements. The structure of the silos and elements was such that the elements across different studies paralleled each other. Thus the basic design, called a mega-study, provided a format by which to more profoundly understand responses to messaging.

The Craveability studies used conjoint analysis to identify the contribution of the different elements. The results generated a database that retains its usefulness almost a decade later. One can return to the data to discover how a specific term fared across all of the different foods or how different elements fared in a single food.

A few years later the same approach was used for beverages. This time the effort was sponsored by Wild Flavors. The mega-data from that set of studies was called Drink It!

Drink It! comprised parallel studies for white wine, red wine, low-alcohol drinks, flavored beer, flavored tequila, and carbonated spritzers. We will look at the data for white wine and red wine in this chapter.

20.5.4 Learning about alcoholic beverages – the basic study design for 'Drink It!'

Participants in the Drink It! studies evaluated a unique set of 60 test concepts, created from the 36 elements for the specific beverage being studied. The 60 combinations or test concepts for any individual respondent were constructed to allow statistically valid estimation of the individual-level model for that individual, and the part-worth contribution or utility value of each element.

1. The elements comprise different types of message, ranging from product descriptions to consumption situations, to brands (Table 20.1).
2. A test concept comprised 2–4 elements, with a maximum of one element per silo, but often no element for one or two silos.
3. The fractionalized, incomplete design ensured that each element would appear five times, statistically independently of every other element. Every respondent evaluated a totally unique set of 60 combinations, created by permuting the original experimental design to create new combinations (Moskowitz and Gofman, 2004).
4. The concepts were presented on the computer screen as short sets of phrases, one phrase per line. There were no connectives between the phrases.
5. The rating question was interest in the product.

Table 20.1 The 36 elements for white wine, the four silos, and the utility value for each element from the total panel. The table shows the results from the Drink It! study

Element		Total
Base size		212
Additive constant		33
Silo A: Product description		
A1	Wine with a pale yellow color and a fruity flavor	4
A2	White wine with added antioxidants	−3
A3	A lightly carbonated pale white wine with a hint of raspberries . . . so soft and sweet	1
A4	A light bodied white wine with a fresh, fruity flavor	8
A5	Pale yellow colored wine with a smooth creamy texture and the flavor of green apples, melon and hazelnut	−5
A6	Golden colored wine with a sweet complex fruity flavor and delicate flowery fruit aroma	6
A7	Golden yellow colored wine with a dry flavor of grape and berry with a hint of oak	−4
A8	Bubbly fresh and very sweet white wine, with a hint of citrus lemon, lime, and melon flavors	−1
A9	Bubbly fresh and sweet with a hint of lemony tart flavor	−7
Silo B: Consumption and related emotions		
B1	Hits the spot on a hot summer day	2
B2	Goes down smooth and easy	3

B3	Made in the tradition of the greatest wine producers all over the world	8
B4	Premium quality	4
B5	So refreshing you want to savor how it makes you feel	2
B6	100% natural	4
B7	Dry, Sweet, Semi Sweet . . . whatever you're looking for	7
B8	You can imagine the taste even before you drink it	4
B9	So refreshing . . . you have to drink some more	1
Silo C: General emotions and rewards		
C1	Quick and fun . . . ready to drink, no bartender required	−1
C2	When you think about it, you have to have it . . . and after you have it, you can't stop drinking it	−1
C3	Simply the best	3
C4	Relaxes you after a busy day	3
C5	A great way to celebrate special occasions	5
C6	Looks great, smells great, tastes delicious	3
C7	A wonderful experience . . . shared with family and friends	5
C8	Pure satisfaction	2
C9	It quenches THE THIRST	0
Silo D: Brands, packaging		
D1	From Northern California	5
D2	From Gallo	−3
D3	From Kendal Jackson	2
D4	Imported from France	6
D5	Tastes great . . . without the taste of alcohol	−6
D6	Multi serve containers . . . so you always have enough!	−1
D7	Icy cold	2
D8	Resealable single serve container . . . to take with you on the go	−2
D9	With the safety, care and quality that makes you trust it all the more	0

Source: It! Ventures, LLC

20.5.5 Learning more about the respondent: the self-profiling classification

It is important in any study to learn about the respondent. People vary in what is important. The person-to-person variation in responses to test concepts need not be entirely ascribed to random variability. There may be some variation in responses that can be traced to meaningful differences in people. Unless we find out more about the person beyond the reactions to the test concepts we will never know whether the person-to-person variation we encounter in responses is random, or whether the variation comes from meaningful differences such as age, gender, or even self-stated attitudes about the product.

A lot of information about a person emerges from a self-profiling classification. In the RDE studies, and at the end of the conjoint portion of the study, the participants answered a separate but linked questionnaire about the specific product that was the topic of the study. Across all of the different studies in Drink It! the self-profiling questionnaire for classification was designed to be as similar as possible. This similarity created a valuable database, in and of itself.

We see part of the classification questionnaire in Fig. 20.1, specifically that pertaining to one's attitudes about wine, and consumption of wine. By itself, the self-profiling questionnaire comprises an entirely independent study about how the wine consumers feel about products. When combined with the conjoint

Q1: For the following set of questions, there may be answer options that are relevant to some beverage categories and not others. Please answer each question based on your idea of what options are relevant for WHITE WINE.

Q2 How thirsty are you right now?

Q3 What time is it?

Q4 Which 3 attributes interest you to drink WHITE WINE MOST? [check three]

M1. Appearance	M2. Aroma	M3. Texture
M4. Taste	M5. Lots of new flavor varieties	M6. Beverage temperature
M7. Memories	M8. Associations	M9. Brand
M10. Advertising	M11. Packaging	M12. Portion size
M13. Social situation	M14. Goes with my meal	M15. Mood
M16. Health	M17. Healthy ingredients	

M18. Convenience easy to use

M19. Convenience easy to buy

M20. Convenience easy to carry

M21. Convenience No mixing or preparation required

Q9 What kind of container do you typically drink WHITE WINE from? [check two]

M1. Glass bottle	M2. Plastic bottle	M3. Sport bottle
M4. Can	M5. Wine glass	M6. Beer glass
M7. Drinking glass	M8. Cup	M9. Mug
M10. Styrofoam disposable cup		M11. Waxed paper cup
M12. Juice box with a straw		M13. Pouch with a straw

Q10 How often do you drink WHITE WINE?

S1. Once a day or more	S2. 4–6 times a week
S3. 2–3 times a week	S4. Once a week
S5. 4–6 times a month	S6. 2–3 times a month
S7. Once a month or less often	S8. Only on holidays, special occasions
S9. Only when I have guests over	S10. Never

Fig. 20.1 Some of the questions from the self-profiling classification questionnaire used in the Drink It! project used for wine and other studies involving alcoholic beverages.

analysis, the classification questionnaire identifies subgroups based upon attitudes, with the analysis then discovering how differences in these attitudes translate into differences of response-pattern to concepts.

20.5.6 Learning about white wine: total panel and frequency of drinking wine

Table 20.1 showed us the elements, and presented the different utility values for white wine from the total panel. Let us now look at the strongest versus the weak performing utility values. The data are from the Interest Model, and the respondents are from the total panel. Table 20.2 shows us these results sorted, and eliminates most of the elements between +8 and −8. Table 20.2 also shows us the results for frequent wine drinkers (more than one glass a week) versus infrequent wine drinkers (defined as one glass per week or less). The elements in Table 20.2 are those for which any group generates a very high utility (>8) or a very low utility (<−8).

The rationale for showing only part of the data rather than the full set of data is as follows:

Table 20.2 Strongest and weakest performing elements for white wine; total and frequent versus infrequent wine drinkers

		Several times/week	Once/week or less
Base size		104	108
Additive constant		33	33
Strong performing elements			
B3	Made in the tradition of the greatest wine producers all over the world	**9**	7
A4	A light bodied white wine with a fresh, fruity flavor	6	**9**
A6	Golden colored wine with a sweet complex fruity flavor and delicate flowery fruit aroma	4	**8**
D4	Imported from France	4	**8**
Weak performing elements			
D2	*From Gallo*	***−10***	5
A9	*Bubbly fresh and sweet with a hint of lemony tart flavor*	***−8***	*−6*
A5	*Pale yellow colored wine with a smooth creamy texture and the flavor of green apples, melon and hazelnut*	*−3*	***−8***

Source: Interest Model for white wine courtesy of It! Ventures, LLC

- An analysis of variance shows that an 8-point range separates two elements with a probability less than 0.05. So, to be conservative, we use that 8-point range.
- Observation of performance of elements in real world situations suggests, not necessarily proves however, that beyond +8 or −8 we see correlated behaviors, such as elements being part of a successful advertising campaign for elements whose utilities are 8+.
- The results from the analysis of frequent versus infrequent drinkers teaches us the following about white wine:
  - We saw from Table 20.1 that the additive constant for the total panel is 33, meaning that without any additional information, about one in three respondents is interested in white wine. The elements have to do the work.
  - The frequency of consumption does not affect the additive constant.
  - Most of the elements do not reach significance. They are not 0, but they do not contribute very much. Table 20.2 shows that only seven elements ever perform very strongly. This is important; most elements may contribute, but the truth is that they have very little effect.
  - There are differences in the way frequent and infrequent drinkers respond to elements. One key element is Gallo: strong performer among the infrequent drinkers of white wine, weak performer among the frequent drinkers.

20.5.7 Individual differences in preferences and the notion of Mind Genomics

Our approach to understanding responses to white wine suggest individual differences. There may be more profound differences than the differences based on frequency of consumption. MG will be relevant for the rest of our discussion on individual differences in preference. The notion of MG is inspired by the science of genomics, which itself studies individual differences and the presumption that the differences express more profound differences among people.

MG deals with the everyday aspects of experience. Very simply, we live in a world of concrete experiences which can be broken up into meaningful wholes of everyday life, such as shopping for a specific item, or in our case drinking white wine. In turn, the experience can be dimensionalized into various components by extracting its different aspects, putting them into silos, and then creating alternative elements to represent different parts of the experience. Thus we have the structural connection with conjoint analysis. For example, shopping for a car can be broken up into the different aspects, from the nature of the store, to that of the interaction with salespeople, to the nature of the purchase.

In MG, people are segmented based upon the patterns of their responses to these granular aspects of everyday life. Segmentation of the respondents into groups with similar patterns reveals how different people 'think about' the specific experience. Often there may be as few as two or as many as five to ten segments. The fewer segments the better. The segments represent individuals with different points of view about the same everyday experience, as well as represent theoretical states that the mind could take (e.g. price orientation versus convenience orientation).

20.5.8 What we learn about white wine by segmenting mind sets

RDE studies of the type just presented for white wine suggest that elements may perform differently, that groups may respond differently to the same elements, but that there is no simple, coherent, compelling 'story' behind these differences. They are factoids, observed from the response of people to cognitively rich phrases, such as a brand name.

In various studies, especially those dealing with food and drink, the manifestation of the chemical senses in behavior, individual variability appears again and again, reliably, and often becoming the basis of new products (Gladwell, 2009). It is not that individual differences are the intractable essence and problem of consumer behavior, but rather the variability begs for an organizing approach. Is there some coherent story, some set of rules suggested by the variability? Or is the variability truly intractable, simply something with which the researcher must cope when trying to figure out how 'nature works?'.

For more than 50 years consumer researchers have recognized that beyond the similarity of people as people there are truly different mind sets. The observation should not be surprising; it was well known almost a century ago to experimental psychologists who counted the number of people saying that they liked a particular taste solution (Engel, 1928), or a particular smell (Kenneth, 1928). The notion of individual differences was clear in the world of the chemical senses, and became increasingly clear in the world of consumer research. We saw evidence of individual differences in the short précis about conjoint analysis and wine presented in the first section of this chapter.

It is not the individual differences that should interest us as much as the nature of an organizing principle under those differences. What is nature trying to tell us? Yes, there are differences, but are there rules to accompany those differences and tell us the story?

20.5.9 Making sense of individual differences – the particulars of mind set segmentation applied to the data on white wines

Typically, researchers segment people on the basis of some external criterion. We just saw segmentation when we analyzed the differences between respondents who drink white wine more frequently versus those who drink wine less frequently. These subgroups of consumers are presumed to differ, which they clearly do, when we consider the way we divide them. Researchers have a host of such different criterion, ranging from who the respondents are, to what the respondents do, to what the respondents think. All of these methods generate criteria along which one can divide the consumer population into different, presumably more homogeneous subgroups.

Following in the tradition of trying to make sense of person to person differences, we will also divide the respondents. However, our segmentation will move in a different direction. We are not interested in differences among people in ways that can be measured using external criteria, whether by questionnaire (attitudes), or observations (behavior, or who they are). It may be that we will find differences in the pattern of responses to elements about white wine . . . or, perhaps not.

Let us proceed in a different direction, following the spirit of MG, which focuses on dimensionalizing the everyday experience. . We are going to limit our segmentation to the world of white wine, and even more particularly to the pattern of responses that the respondents show to our different elements dealing specifically with white wine. We are turning the problem around 180 degrees. Traditional segmentation divides people and hopes that the division will manifest itself in different patterns of responses, in our case responses to elements about white wine. We will do the opposite. We will divide people by the pattern of responses to white wine, independent of who the people are. We will then look for the story that this segmentation tells us. We will not worry about finding covariates of this segmentation in other factors that we can measure about the people. It is the segmentation itself, pure and simple, that will interest us.

20.5.10 The mechanics of mind set segmentation and what emerges for white wine

At the individual respondent level we develop the Persuasion Model showing the number of rating points that can be traced to each element. That profile or vector of numbers will be used as the input for clustering. We will divide the respondents by the pattern of their responses to elements dealing with white wine. The tool for such segmentation is known as cluster analysis, and is found in most off-the-shelf statistical packages, such as SYSTAT (1997). Our segmentation is based on *k*-means clustering from SYSTAT.

The foregoing approach to mind set segmentation differs profoundly from the typical segmentation. The differences are the following:

- Mind set segmentation is limited and granular. The segmentation deals only with responses to ideas about white wine, and nothing else. We are attempting to segment on the basis of the simplest of behaviors: response to test stimuli. In this case the test stimuli are a number of statements about white wine.
- The segmentation comes from behavior in an experiment. We look at the response patterns to these elements when the respondents are confronted with the test concepts and instructed to rate them. We are not looking at self-stated attitudes *per se*, but rather at behavior patterns evoked by statements.
- The segmentation generates groups of people who clearly respond differently to the same elements. People in a segment are homogeneous with respect to how they respond to the specific messages. We begin with the pattern of responses and work upwards. There is no need to look for a link between the segmentation and the pattern of responses to messages.
- Segment 1 comprises individuals who are fascinated by statements about the sensory characteristics of wine, and indeed about the sensory experience of white wine (Table 20.3). They begin with a low additive constant (25) so it is the elements that have to do all the work. And there are many elements which do so. Indeed most elements are either positive or only slightly negative.

Table 20.3 Strongest and weakest performing elements for white wine; total and two segments

		Total	Segment 1	Segment 2
Base size		212	92	120
Constant		33	25	39
Total panel				
B3	Made in the tradition of the greatest wine producers all over the world	8	8	8
A4	A light bodied white wine with a fresh, fruity flavor	8	**17**	1
A9	Bubbly fresh and sweet with a hint of lemony tart flavor	−7	8	−18
Segment 1 – The sensory experience				
A3	A lightly carbonated pale white wine with a hint of raspberries . . . so soft and sweet	1	**27**	***−19***
A6	Golden colored wine with a sweet complex fruity flavor and delicate flowery fruit aroma	6	**18**	−3
A8	Bubbly fresh and very sweet white wine, with a hint of citrus lemon, lime, and melon flavors	−1	**17**	***−15***
A4	A light bodied white wine with a fresh, fruity flavor	8	**17**	1
A5	Pale yellow colored wine with a smooth creamy texture and the flavor of green apples, melon and hazelnut	−5	**17**	***−22***
A1	Wine with a pale yellow color and a fruity flavor	4	**13**	−3
D4	Imported from France	6	**10**	3
B3	Made in the tradition of the greatest wine producers all over the world	8	**8**	8
A9	Bubbly fresh and sweet with a hint of lemony tart flavor	−7	**8**	*−18*
D5	*Tastes great . . . without the taste of alcohol*	−6	***−8***	*−4*
Segment 2 – Just the wine				
B7	Dry, Sweet, Semi Sweet . . . whatever you're looking for	7	3	**11**
B3	Made in the tradition of the greatest wine producers all over the world	8	8	8
A7	*Golden yellow colored wine with a dry flavor of grape and berry with a hint of oak*	*−4*	*6*	***−12***
A8	*Bubbly fresh and very sweet white wine, with a hint of citrus lemon, lime, and melon flavors*	*−1*	*17*	***−15***
A9	*Bubbly fresh and sweet with a hint of lemony tart flavor*	*−7*	*8*	***−18***
A3	*A lightly carbonated pale white wine with a hint of raspberries . . . so soft and sweet*	*1*	*27*	***−19***
A5	*Pale yellow colored wine with a smooth creamy texture and the flavor of green apples, melon and hazelnut*	*−5*	*17*	***−22***

Importantly, this segment is not the majority, but rather comprises a substantial minority (38%).

- Segment 2, with 62%, begin at a higher base level of interest (39). However, almost nothing excites them. These are individuals who are simply interested in the 'white wine' and nothing more. Nothing appeals to them. No romancing will have an effect, at least no romancing that we have dealt with in the conjoint study.

20.5.11 Beyond knowledge to application: identifying respondents in the mind set segments

Many segmentations generate interesting stories about the way people divide, whether the topic area is foods, health concerns, cars, and so forth. Based upon the segmentation, the researcher creates a persona, a small picture and story about the individual. The personas and stories, called heuristics, are interesting and make intuitive sense. However, these heuristics soon take over. Respondents begin to believe that the persona, originally a way to make the segmentation come alive, must be real. And, in turn, when an individual who looks like the persona shows up, the natural response is to assume that the individual is a member of the segment. *What needs to be remembered is that the segmentation refers to the way people think, not to who they happen to be. In most cases, it will be virtually impossible to assign a good proportion of people into these mind set segments when the only information available is previous purchase behavior and/or geo-demographics.*

The recurrent conundrum, however, is that the typical segmentation does not correspond to different mind sets. Rather, the typical segmentation classifies people as belonging to similar-appearing groups, such as upper middle class, living in a certain type of neighborhood. There is the unsupported belief that 'birds of a feather both flock together', and think similarly.

However, the segmentation often does not capture groups based upon mind sets, but rather based upon easier to measure variables such as age, gender, income, market, and even stated attitudes to a product category rather than an unedited, non-politically correct pattern of responses to messages from the category. MG using conjoint analysis works at the granular level to segment the respondents, and so emerges with reasonably homogeneous mind sets as we saw in the previous section.

Yet the question still remains: If these segments exist and have practical value as well as scientific interest, then just how does one find these segments, or stated differently, how does one discover the segment to which a particular individual belongs? Data mining exercises may show some relation between segment membership and certain measurable characteristics of an individual, such as gender, age, amount spent on wines, and so forth. Yet the truth of the matter is that the ability to classify a person as belonging to a segment is marginal, when one only knows certain externally measured facts about the person. There has to be a different, better way, which 'taps into' the respondent's mind, and in so doing is more likely to assign the person to the correct segment.

Our metaphor for this better and ultimately more practical way is the practice of today's medicine. Decades ago, before the explosion of capabilities in medicine, the doctor would sit with the patient and take a family history. From that history, and from a not-particularly-technology-enabled personal examination, the doctor would be able to identify certain problems, should they exist. Today, however, things are quite different. The doctor takes a urine sample, a blood sample, perhaps a variety of ultrasound measures, sends these to the laboratory, and receives a printout of automated analyses, and relevant 'warnings' or 'flagged measures'. The key here is an intervention test. One interacts with the patient, obtains a set of measures, and then classifies the patient in terms of health issues.

The notion of 'interaction' or 'intervention' is the operative phrase. It applies to the problem of identifying an individual as being a member of a segment. One has to intervene, to have the respondent participate in a short test or exercise. One common approach, used in pop tests on the web, employs a so-called typing tool, which is nothing more than a series of questions answered by the respondent. The answers are fed into a classification algorithm which generates a value. From that value the researcher can then assign the respondent to one of the different segments. Typing tools work fairly well. They put the respondents into the appropriate segments, as they should. The identification is not perfect, being usually 60–70% correct. However, that percentage is far higher than guessing.

20.5.12 Applying the typing tool to the white wine segmentation

The next issue is to bring the segmentation to life for our white wine data. Let us see how such a typing tool, using an intervention, might be implemented. We know the different utility values for each respondent, from the basic study on white wine. Recall that each respondent evaluated 60 different concepts. We can relate the presence/absence of the 36 elements to the 9-point rating, and generate the coefficients or utilities for the Persuasion Model. Those coefficients summarize what is important to the particular respondent. The coefficients for the Persuasion Model are estimated separately for each respondent using ordinary least-squares regression. The analysis now moves one step further. It uses the coefficients on a respondent-by-respondent basis. The analysis now moves one step further, to use these coefficients as the tool by which to estimate the likely 9-point rating of each of the 36 elements for each respondent, and then to use discriminant function analysis to identify key elements which drive segment membership. We now apply the result to the individual models for our white wine data.

20.5.13 Identifying people who are in the two white wine segments

Our two segments react to different elements. We can estimate how each respondent would react to each of the 36 elements by estimating how each of these 36 elements would have scored on the 9-point scale as a 1-element concept. We do this by using the respondent's own Persuasion Model, which shows the basic response to the concepts (additive constant) and the incremental or decremental part-worth utility of each element. The Persuasion Model suffices for us to estimate how each of the 36 elements would have scored on the 9-point scale, for each of our 212 respondents. When it comes time for a person to 'type himself' as to segment membership, we are going to present the respondent with several, individual elements that appeared in the conjoint study, and which were determined ahead of time to be good separators of respondents into segments. We will use the respondents' ratings to these single elements to help identify the segment to which the respondent belongs.

At a practical level, we can move one step beyond the 9-point scale, and work with a more tractable 3-point scale. We have a simple, practical rationale for shifting from a 9-point scale to a shorter, perhaps less discriminating, 3-point scale. When we want to implement the typing tool, anything that we do to make the respondent's job easier and faster will be welcome. It is one thing to have a respondent use a 9-point scale in a test situation. It is quite another to have the respondent use a scale on the Internet, or in a personal interview. In the latter two situations, simpler scales and shorter scales are always better.

The steps used to create the typing tool come from the well-accepted method of discriminant function analysis (DFA) (Systat, 1997). The inputs to DFA are the segment membership of each respondent, and the rating that the respondent would assign to all 36 elements. In our first DFA analysis we work with the estimated ratings on the 9-point scale and see what emerges. In our second DFA we will work with the estimated ratings on the 9-point scale, but then convert that 9-point scale to a more easily used 3-point scale. The transformation will be

$$(1\text{–}3 \rightarrow 1;\ 4\text{–}6 \rightarrow 2;\ 7\text{–}9 \rightarrow 3). \qquad [20.2]$$

We follow the steps listed below to create the typing tool. The typing tool, using DFA's classification model, classifies new respondents as belonging to Segment 1 or Segment 2. The practical application will be during the stage of product design to work with the sensory-responsive group, as well as during the stage of sales to discover what one should say to the prospect, knowing that the segments respond to different messages.

1. For each respondent, recreate the rating that would have been assigned to each of the 36 elements on the 9-point scale.
2. We now know for each of our respondents both the segment membership based on the Persuasion Model, as well as the 9-point rating that would be assigned to each of the 36 elements.
3. DFA tells us which of our 36 elements predicts segment membership. The prediction is not perfect. We are simply interested in the best that we can do, with the fewest number of predictors.
4. For our white wine we were able to reduce the number of such statements to two (Table 20.4). DFA returns with the F ratio, which in our case is really a measure of signal to noise. The F ratios are quite high (78 and 66), so these two elements truly separate the segments.

Table 20.4 Results of the discriminant function analysis for white wine

	Segment 1	Segment 2	% correct	P1	P2	P3	P4	P5	P6	P7	P8	P9
Part 1 — DFA based on a 9-point scale for typing												
Classification matrix (cases in row categories classified into columns)												
	Segment 1	Segment 2	% correct									
1	70	22	76									
2	22	98	82									
Total	92	120	79									
Classification functions	Segment 1	Segment 2		P1	P2	P3	P4	P5	P6	P7	P8	P9
Constant	−4.52	−4.82										
A lightly carbonated pale white wine with a hint of raspberries . . . so soft and sweet (F=78)	0.70	0.04		1	1	1	5	5	5	9	9	9
So refreshing you want to savor how it makes you feel (F=66)	0.68	1.40		1	5	9	1	5	9	1	5	9
Segment 1 – Wine Sensory				−3.1	−0.3	2.4	**0.1**	**2.8**	5.6	3.3	**6.0**	**8.7**
Segment 2 – Just the wine				−3.4	**2.2**	**7.8**	−3.2	2.4	**8.0**	−3.1	2.5	8.1
Part 2 — DFA based on a 3-point scale for typing												
Classification matrix (cases in row categories classified into columns)												
	Segment 1	Segment 2	% correct									
1	47	45	51									
2	10	110	92									
Total	57	155	74									
Classification functions	Segment 1	Segment 2		P1	P2	P3	P4	P5	P6	P7	P8	P9
Constant	−6.87	−5.74										
White wine with added antioxidants (F=49)	1.06	2.58		1	1	1	2	2	2	3	3	3
Pale yellow colored wine with a smooth creamy texture and the flavor of green apples, melon and hazelnut (F=57)	4.59	2.72		1	2	3	1	2	3	1	2	3
Segment 1 – Wine Sensory				−1.2	**3.4**	**8.0**	−0.2	4.4	**9.0**	0.9	5.5	**10.1**
Segment 2 – Just the wine				−0.4	2.3	5.0	**2.1**	**4.9**	7.6	**4.7**	**7.4**	**10.2**

Note: The table shows the success/error rate in classification, the classification functions, and the segment to which a respondent would be assigned. Part 1 (top half of the table) shows the results when the classification exercise involves a 9-point scale. Part 2 (bottom half of the table) shows the results when the classification exercise involves a 3-point scale

- A lightly carbonated pale white wine with a hint of raspberries . . . so soft and sweet (F = 78)
- So refreshing you want to savor how it makes you feel (F = 66)

5. DFA then creates a classification function for each segment. Since we are dealing with two segments, we end up with two classification functions. Each function provides an additive constant and weight to be applied to the appropriate element.
6. A respondent who goes through the typing exercise (Fig. 20.2) sees the two elements in randomized order, and assigns a number to each.

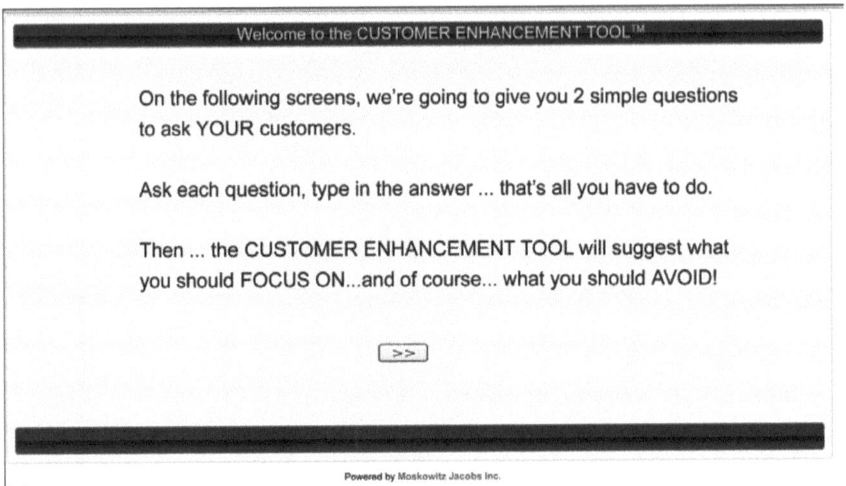

(a)

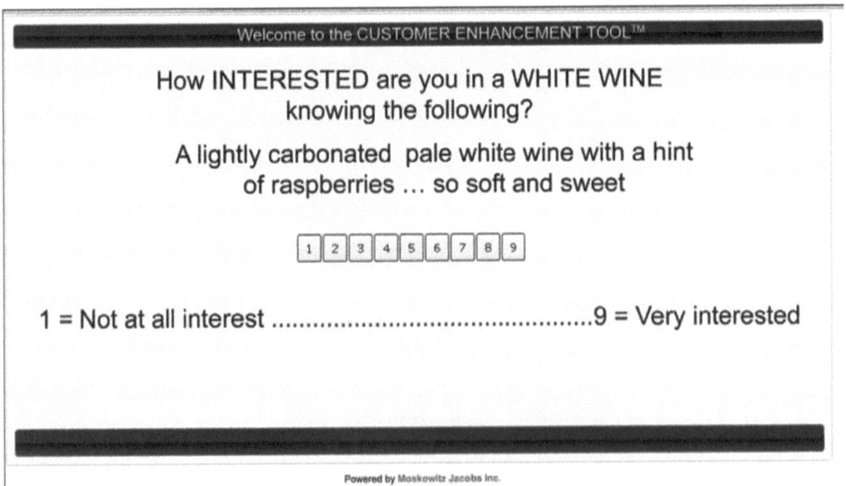

(b)

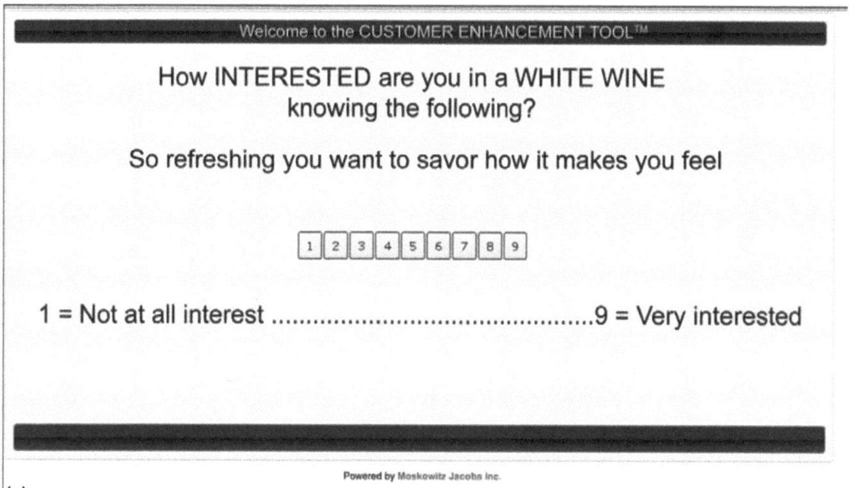

(c)

Fig. 20.2 Typing tool for wines. (a) Introductory screen. (b) White wine typing question 1. (c) White wine typing question 2.

7. Behind the scenes, a computer program 'solves' each of the classification functions, using the two ratings that the respondent assigned.
8. The outcome of Step 7 comprises two numbers, an expected weighted result for the classification function for Segment 1, and a corresponding expected weighted result for the classification for Segment 2. One of the two must be higher, and must be positive. The higher, positive result then shows us the likely segment membership.
9. The higher the positive number the more likely it will be that the respondent belongs to the segment. The greater the difference in the weighted values for the two classification functions, the more confidence we can place on assigning the respondent to one segment, rather than to the other.
10. Table 20.4 shows our success ratio (about 79%). We develop this ratio and our correct and incorrect classifications by applying the classification functions to the data from our 212 respondents.
11. Table 20.4 further shows that when we simplify the respondent's task, using a 3-point rating scale, we drop from 79% correct assignments (of our current respondents) to 74%. This is not a dramatic drop, and makes the field execution easier.
12. Table 20.4 shows the likely segment membership for respondents who assigned different pairs of ratings.

Figure 20.2 shows a screen shot of the typing tool for white wine. Figures 20.3–20.6 are screen shots showing the outcome of typing for each segment.

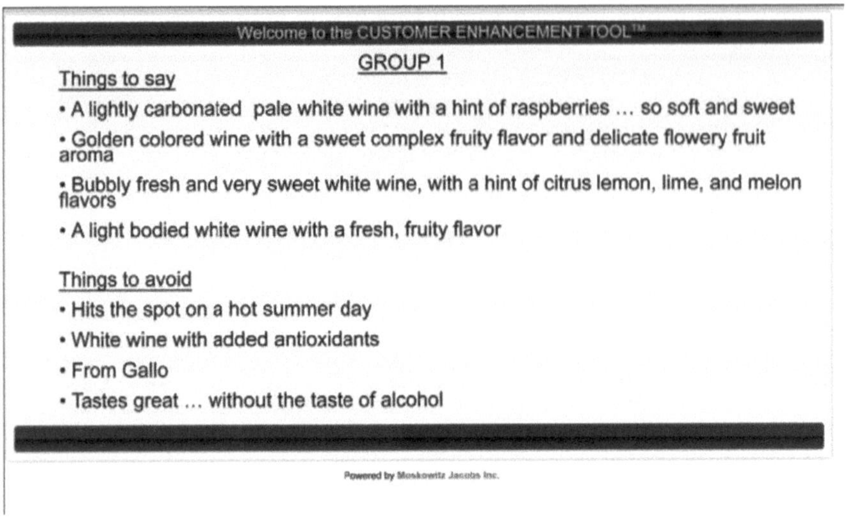

Fig. 20.3 Outcome of white wine typing question 1 for Group 1.

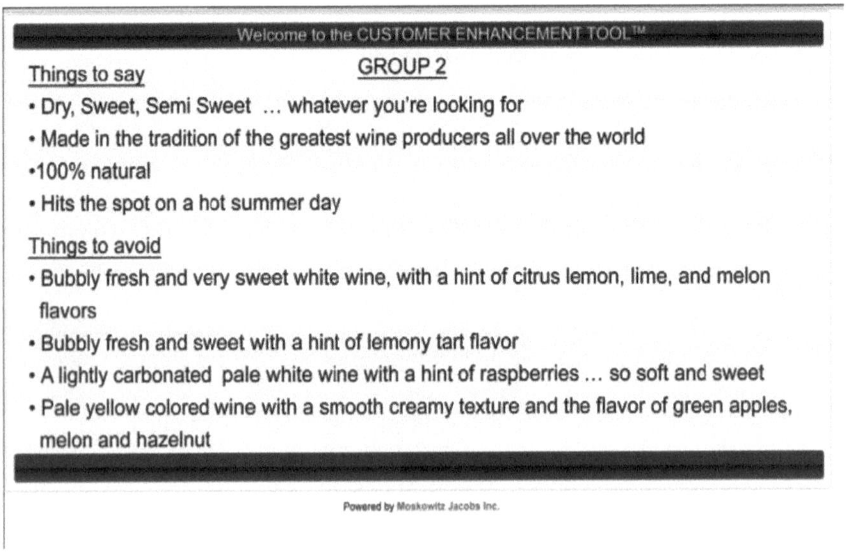

Fig. 20.4 Outcome of white wine typing question 1 for Group 2.

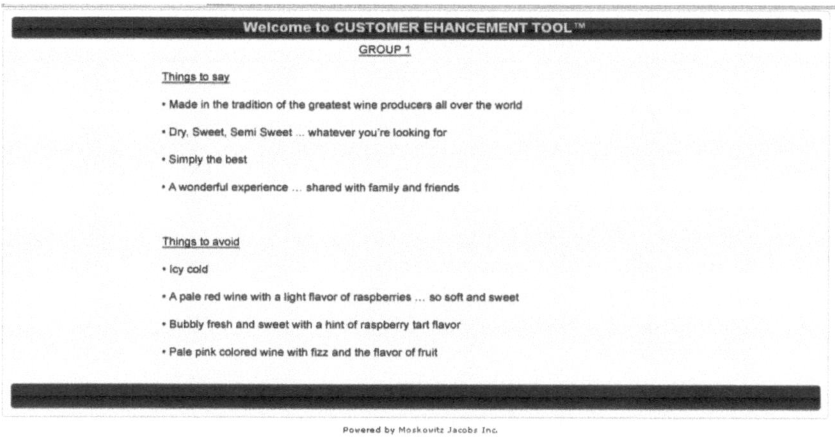

Fig. 20.5 Outcome of white wine typing question 2 for Group 1.

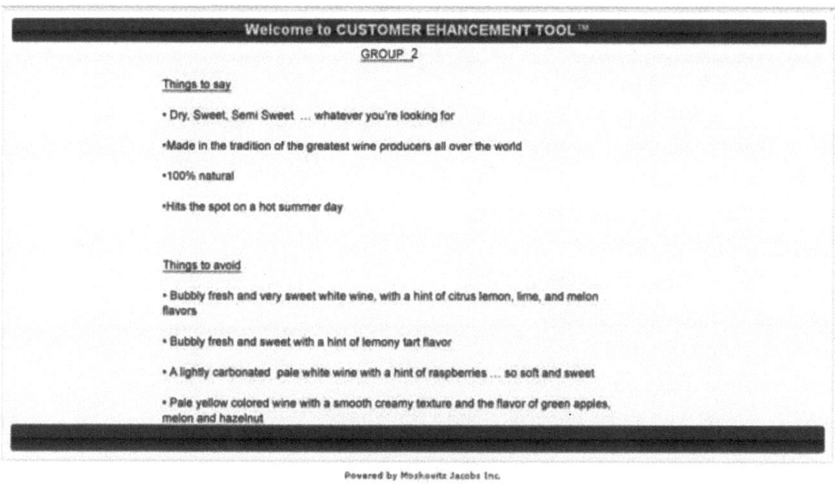

Fig. 20.6 Outcome of white wine typing question 2 for Group 2.

20.6 The case of red wine

Let us turn now to red wine, and analyze the results in the same way, albeit with a shorter report of the results. Table 20.5 presents the results from the total panel and from the two mind set segments which emerged. The actual experiment was virtually identical. Respondents were invited to participate in the Drink It! mega study by an email invitation. Respondents were led to the 'wall' which presented the available studies. Those who selected red wine were led to a similar experiment,

Table 20.5 Performance of the red wine elements, by total panel and by the two mind set segments

Red wine		Total	Segment 1	Segment 2
Base size		220	97	123
Constant		32	18	42
A5	A heavy red wine with a warm plumy fruit flavor and a little bit of dryness	**12**	**26**	1
A3	Deep red colored wine with a sweet complex fruity flavor and delicate flowery fruit aroma	6	**16**	−1
A4	Ruby red color with a blackcurrant flavor and a hint of oak	6	**21**	−6
C5	A great way to celebrate special occasions	6	**10**	3
B3	Made in the tradition of the greatest wine producers all over the world	5	4	5
B7	Dry, Sweet, Semi Sweet . . . whatever you're looking for	4	4	4
C3	Simply the best	4	3	4
B2	Goes down smooth and easy	3	2	3
C7	A wonderful experience . . . shared with family and friends	3	1	4
A7	A chilled light bodied red wine with a fresh fruity flavor	3	**10**	−3
B6	100% natural	3	4	1
C4	Relaxes you after a busy day	3	5	1
B4	Premium quality	2	2	3
D3	From Kendal Jackson	2	2	3
D4	Imported from France	2	4	1
C8	Pure satisfaction	2	2	2
D1	From Northern California	2	2	2
A1	Wine with a pale ruby color and a fruity taste	2	7	−2
C6	Looks great, smells great, tastes delicious	2	1	2
B5	So refreshing you want to savor how it makes you feel	1	−2	3
B8	You can imagine the taste even before you drink it	0	2	−1
B9	So refreshing . . . you have to drink some more	0	1	−1
D9	With the safety, care and quality that makes you trust it all the more	0	−4	3
B1	Hits the spot on a hot summer day	−1	1	−2
D2	From Gallo	−2	−5	1
A8	Red wine with added antioxidants	−3	1	−6
C9	It quenches THE THIRST	−3	−4	−2
C1	Quick and fun . . . ready to drink, on the shelf no bartender required	−3	−4	−3

C2	When you think about it, you have to have it . . . and after you have it, you can't stop drinking it	−3	−2	−4
A6	A pale red wine with a light flavor of raspberries . . . so soft and sweet	−3	**15**	**−18**
D6	Multi serve containers . . . so you always have enough!	−5	−7	−3
D8	Resealable single serve container . . . to take with you on the go	−5	−5	−5
A9	Bubbly fresh and sweet with a hint of raspberry tart flavor	−6	**10**	−19
D5	Tastes great . . . without the taste of alcohol	−8	−13	−3
D7	Icy cold	−8	−7	−9
A2	Pale pink colored wine with fizz and the flavor of fruit	−11	1	−19

albeit with elements relevant to red wine. Each respondent evaluated a unique set of 60 test concepts, created by conjoint analysis. The results from each individual's Interest Model were aggregated to generate an average profile of utility values. Finally the respondents were segmented according to the pattern of their utilities from the Persuasion Model.

For red wine, Table 20.5 shows us the average utility values for all 36 elements, sorted by the utility value for the total panel. The average additive constant is 32, meaning that about one in three respondents is prepared to rate the concept as 7–9 (interested) in the absence of elements. Thus far we have a pattern similar to that shown by white wine. Finally, when it comes to the total panel, only one element performs strongly: *A heavy red wine with a warm plumy fruit flavor and a little bit of dryness*. There is one element which performs as poorly as this element performs well: *Pale pink colored wine with fizz and the flavor of fruit.*

As we saw with white wine, the key to interesting respondents lies in the mind set segments. Although all the respondents chose to participate, and thus demonstrated some minimal level of interest, we see that there are substantial differences between the segments:

1. Segment 1 (wine sensory) begins with a low additive constant (18). For segment 1, it is the elements that must do the work.
2. The strongest performing elements are those that deal with rich flavor.
3. Segment 2 begins with a higher additive constant (42). There are more respondents in Segment 2 than in Segment 1 (123 vs 97).
4. Segment 2 (just the wine) is not interested in much of anything. They want to hear about the wine, but not specifics. For Segment 2, just knowing that the wine is red and good is what matters. Beyond that the messaging is counterproductive.
5. Finally, Table 20.6 shows us the classification functions, Fig. 20.7 and 20.8 show the typing exercise for red wine, and Fig. 20.9 and 20.10 the outcomes.

Table 20.6 Results of the discriminant function analysis for red wine, showing the success/error rate in classification, the classification functions, and to which of the two segments the respondent would be assigned

Classification matrix (cases in row categories classified into columns) – 9-point typing	Segment 1	Segment 2	% correct								
1	73	24	75								
2	19	104	85								
Total	92	128	80								
Classification functions	Segment 1	Segment 2									
Constant	−2.90	−3.75									
Bubbly fresh and sweet with a hint of raspberry tart flavor (F=73)	0.72	−0.02	1	1	1	5	5	5	9	9	9
Tastes great . . . without the taste of alcohol (F = 121)	0.24	1.24	1	5	9	1	5	9	1	5	9
Segment 1 – Wine sensory			−1.9	−1.0	0.0	**0.9**	1.9	2.9	**3.8**	**4.8**	5.7
Segment 2 – Just the wine			−2.5	**2.5**	**7.4**	−2.4	**2.6**	**7.5**	−2.3	2.6	**7.6**

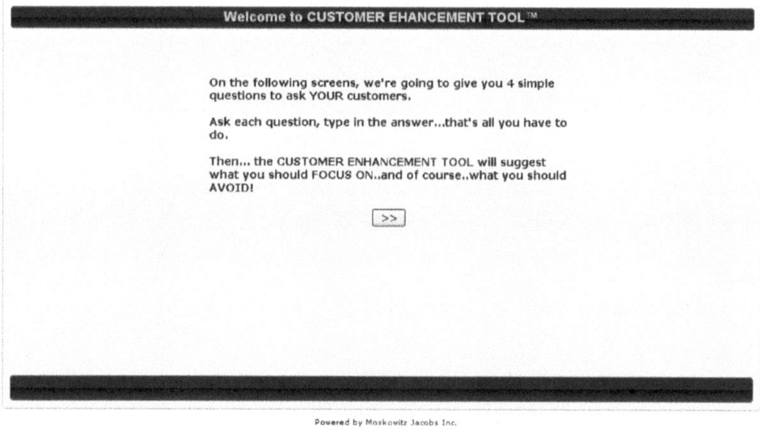

Fig. 20.7 Red wine typing tool.

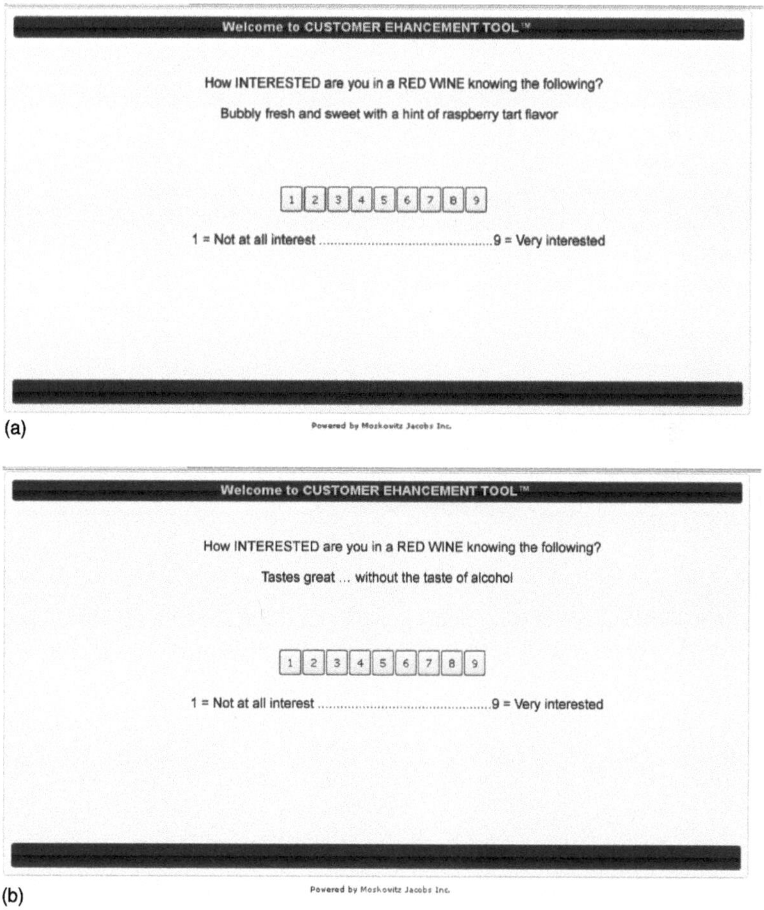

Fig. 20.8 (a) Red wine typing question 1 and (b) Red wine typing question 2.

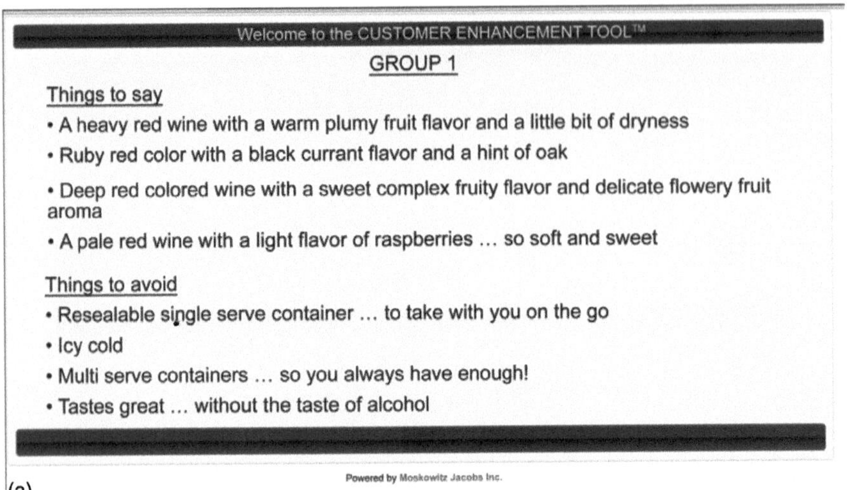

(a)

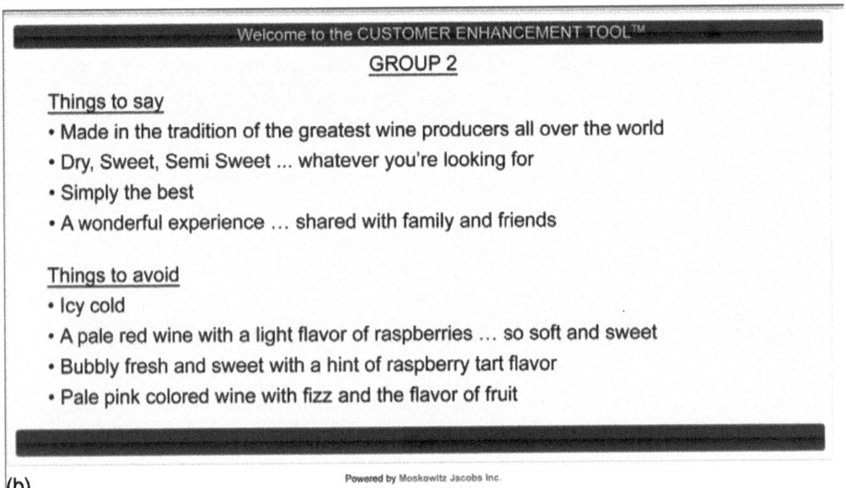

(b)

Fig. 20.9 Outcome of red wine typing question 1. (a) For Group 1. (b) For Group 2.

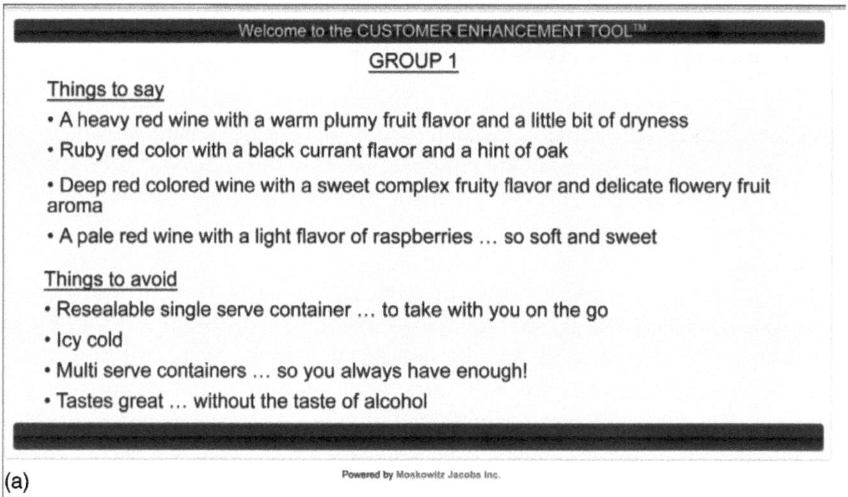

(a)

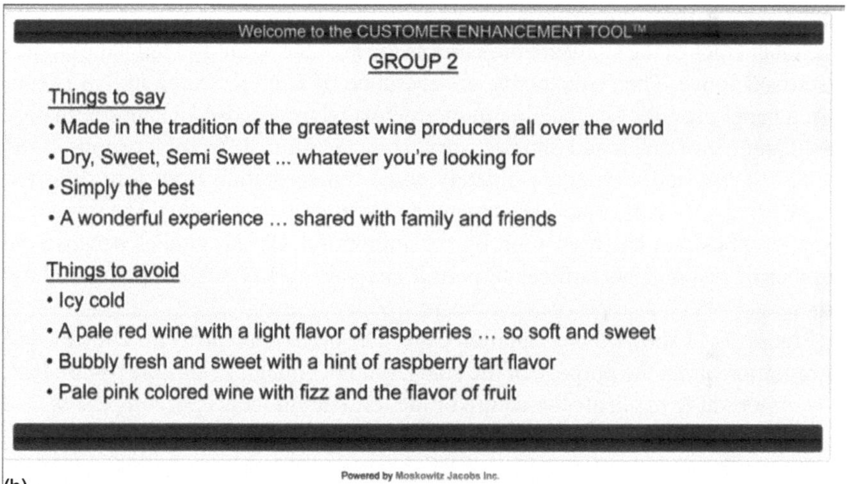

(b)

Fig. 20.10 Outcome of red wine typing question 2. (a) For Group 1. (b) For Group 2.

20.7 Next generation thinking: price and emotion/feeling and its application to dessert wine

We now move beyond simply evaluating the test concepts on an interest rating to more profound understanding. This final section on wine deals with two emerging areas, price and emotion. Both are emerging because they have not been subject to a lot of investigation. Yet both price and emotion are quite relevant to wine; wine varies in price and wine calls up a variety of emotions, especially among aficionados.

Investigating price and emotion do not require much new in the way of experimental design. We are still going to work with elements and with silos. We will change the design a bit, so that instead of working with four silos each comprising nine elements, we will work with six silos, each comprising six elements. This more efficient design forces us to test 48 vignettes rather than 60, less of a burden on the respondent especially when we realize that each vignette will be evaluated on both price and emotion.

20.7.1 Executing the sweet wine study

The mechanics of the study are the same as those of the white and red wine studies discussed above. The respondents, pre-recruited by Luth Research in San Diego, CA, a panel provider specializing in internet panels, comprised an approximately equal number of males and females, varying in age, from different locations of the United States, and with approximately equal representation from four different ethnic groups (Asians, Blacks, Hispanics and Whites, respectively). The foregoing panel composition has been used by the authors for various studies because the number of respondents suffices to permit analysis by key subgroup, e.g., ethnic group.

Figure 20.11 shows the orientation page. The orientation provides a little bit of information about the purpose of the study, but not enough to bias the respondent. It is important to explicate the nature of the test stimuli (concepts comprising 2–4 elements), and the nature of the two scales.

Respondents do not begin most RDE interviews knowing a lot about what is required of them. Pointing out the fact that they must respond to the test stimuli (combinations of elements) as a totality, is critical. Otherwise many respondents get confused and look for the rating scale for each element. The orientation page disabuses them of that response.

The scales for this third study differ from the scale of interest. The study involves two aspects. The first is the selection of the price that the respondent feels to be 'correct', and would be willing to pay (Fig. 20.12). The second is the selection of the single emotion that the respondent feels when reading the test concept (Fig. 20.13). Each of these rating scales is easy to use. Although respondents might hesitate a bit on the first or second stimulus because of the relative novelty of the scale (most are accustomed to scales of 'liking'), rather quickly the respondent accommodates to the task.

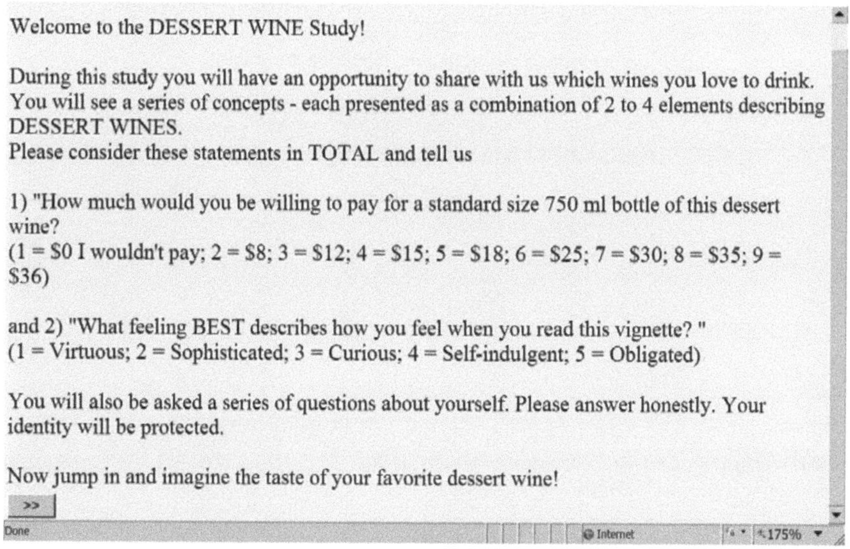

Fig. 20.11 The orientation page for the sweet wine study, showing the instructions to the respondent and the explication of the scale.

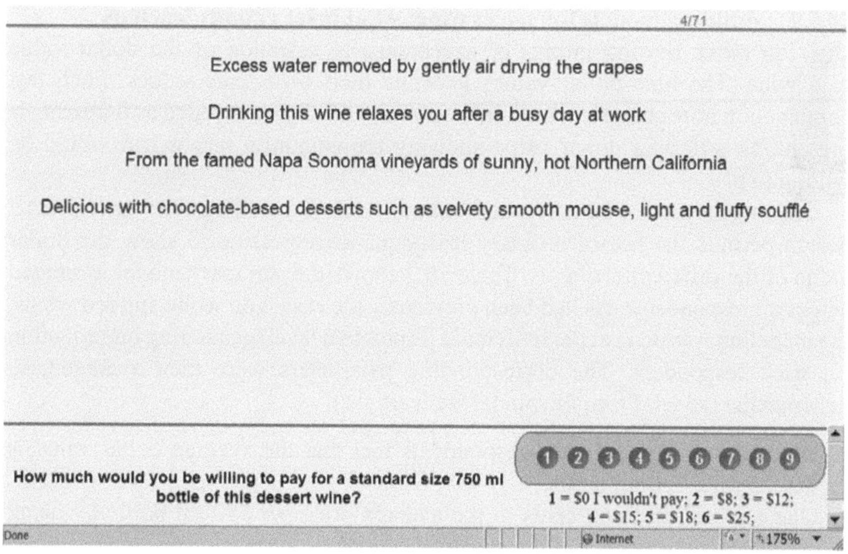

Fig. 20.12 Example of a test concept and the rating scale (price). The respondent selects the appropriate price from the set. It will be the prices, not the scale number, that constitute the actual rating to be analyzed.

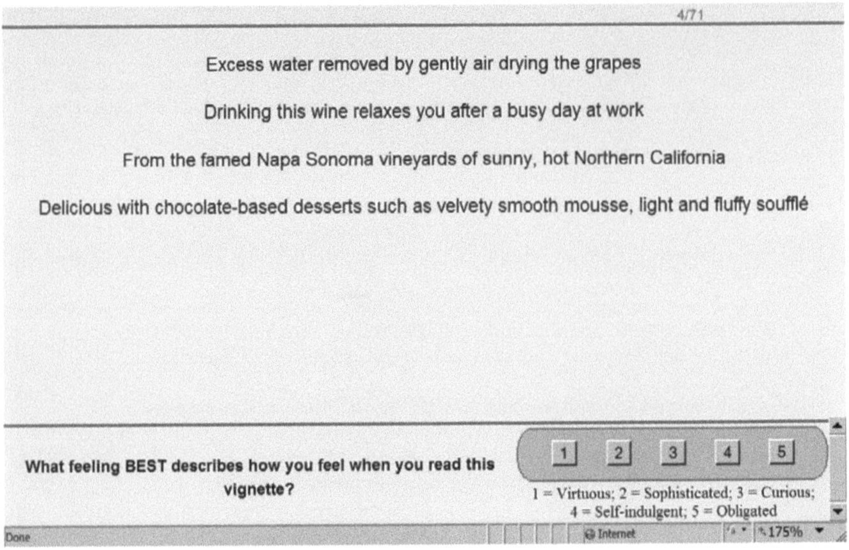

Fig. 20.13 The same test concept as in Fig. 20.12, this time with the five emotions listed. The respondent is to select the emotion that he feels. The subsequent analysis will assess the frequency with which each emotion is associated with each test element.

20.7.2 Building models for sweet wine: what price ratings teach us

One can move beyond ratings of acceptance to selection of the dollar value of a wine. The nine dollar values generate their own scale values. Each test combination now comprises both its own recipe of elements coded as 0/absent, 1/present, as well as a dollar value showing the amount a respondent would be willing to pay.

Creating the test combinations for each respondent according to an experimental design permits the use of ordinary least-squares regression to show the dollar value of the different elements. Table 20.7 shows the summary model, averaged across all respondents. As had been previously the case with white and red wines, the modeling was done at the individual respondent level, generating one equation for each respondent. The corresponding parameters were then averaged, to generate the model. From the model we learn that:

- Without any elements, the respondents feel that the average dollar value is \$19.75 for a 750 ml bottle.
- One can add up to 60 cents to the average price per 750 ml bottle by using certain evocative descriptions, such as:
  - From the gentle sloped ancient vineyards of southwest France,
  - Drink it alone or with luscious dewy fruits and fresh baked sweets,
  - Sweet wine served at a summer picnic or a formal dinner.

- It is not generally a case of elements driving up the price as much as it is elements driving down the price. The wrong elements can drop the selected price by $1.40 or more:
  - Both high levels of sugar and of alcohol,
  - From the wide open areas of down under Australia,
  - Manipulated by the vintner . . . unfermented MUST added after fermentation.

20.7.3 Building models for sweet wine: what selecting feelings/emotions tell us

Moving beyond money as a dependent variable we finish our analysis of sweet wine with an analysis of how the different elements drive 'emotion'. When it comes to understanding how people feel about food, today's research efforts to understand the consumer are moving towards the notion that engagement, feeling, emotional involvement and so forth are as important as the product characteristics. Books about the importance of emotion (Roberts, 2005) underscore the critical nature of understanding the emotional context of, and reaction to, products.

The conjoint approach used throughout this paper can be extended to emotions. One can either have the respondent scale the depth of an emotional response to the different concepts (measure), or select a key emotion that is felt when one reads the concept (classify). We choose the latter approach. The reason is simple. There are many emotions. In a conjoint study the respondent's task is hard enough. Evaluating 48 test stimuli is a long process, lasting about 15 minutes. Forcing the respondent to scale different emotions means several rating attributes for each test concept. A more productive way instructs the respondent to select which emotion from a limited set best applies to how one feels after reading the particular test concept. The approach is not complete by any means, but provides a way to understand the emotional dimension of the concepts. And, in the happiest of situations, the approach generates an 'emotional signature' of each element.

The steps to produce the emotional signature are enumerated below. The results will appear in Table 20.7, in the five final columns to the right (from 1-Virtuous → 5-Obligated).

1. Identify a limited set of responses that can be considered to be feelings or emotions. There are no standard lists. Make the emotions relevant to the product. For the case of sweet wines we chose virtuous, sophisticated, curious, self-indulgent, and obligated. These feelings/emotions need not fall along any underlying dimensions such as virtuous → self-indulgent. The exercise is purely exploratory. At the early stages of science it is not a bad idea to explore different emotions because one does not know what to expect.
2. For each test concept instruct the respondent to select the one feeling/emotion that the respondent believes to describe his reaction when reading the test concept.

Table 20.7 The models for sweet dessert wine, showing the contribution of each of 36 elements to selected price, and to selected emotion

	Dollar	1 – Virtuous	2 – Sophisticated	3 – Curious	4 – Self-indulgent	5 – Obligated
	19.75					
A velvety sweet wine served with dessert	0.70	2	5	8	9	3
Drink it alone or with luscious dewy fruits and fresh baked sweets	0.65	2	9	**10**	3	5
Sweet wine served at a summer picnic or a formal dinner	0.60	1	7	**10**	4	5
Wine with over 14% alcohol by volume	0.44	0	8	6	9	4
Both high levels of sugar and of alcohol	0.43	4	5	9	4	6
Water removed from the grapes to concentrate the sugar	0.35	2	5	**12**	5	3
Grapes grown naturally in gently rolling vineyards	0.20	0	**11**	**10**	6	0
Manipulated by the vintner . . . sugar or honey added before fermentation process begins	0.10	3	7	**11**	4	2
Manipulated by the vintner . . . unfermented MUST added after fermentation	0.07	2	5	**11**	5	3
Contains additional alcohol (typically brandy) . . . which is all natural sugar and is not fermented	0.03	1	7	**12**	5	1
Excess water removed by gently air drying the grapes	−0.12	4	6	**12**	3	1
Some water is carefully removed by freezing the grapes	−0.29	4	5	**17**	1	−1
Delicious wine hits the spot on a summer's day	−0.31	3	3	**16**	4	1
Refreshing wine goes down smooth and easy	−0.32	2	8	**12**	3	2
Wine made in the tradition of the greatest wine producers all over the world	−0.35	3	5	**13**	3	3
Wine so refreshing you savor how it makes you feel	−0.47	2	2	**19**	0	2
Wine is 100% natural	−0.54	4	7	**11**	3	0
Wine can be dry, sweet, semi-sweet . . . whatever you're looking for	−0.55	4	3	**14**	4	1
Both quick and fun . . . it's ready to drink, on the shelf, no bartender required	−0.60	1	2	**15**	7	2

When you think about this wine, you have to have it . . . and after you've tasted it, you simply can't stop drinking it	−0.64	2	2	**10**	8	5
Drinking this wine relaxes you after a busy day at work	−0.67	4	4	7	**10**	1
Drinking this wine is a great way to celebrate special occasions	−0.72	3	3	**13**	5	3
This sparkling sweet wine looks great, smells great, tastes delicious	−0.73	3	7	5	8	3
Drinking this wine is a wonderful experience. . . best shared with family and friends	−0.76	3	8	**11**	3	2
From the famed Napa Sonoma vineyards of sunny, hot Northern California	−0.81	2	5	**14**	3	2
From the mountainous areas of Germany	−0.83	5	2	9	8	2
From the gentle sloped ancient vineyards of southwest France	−0.85	1	**12**	6	6	2
From the foothills of the Alps in northern Italy	−0.87	3	8	**11**	4	2
From the sunny coastal areas of Spain	−0.93	1	6	**11**	6	2
From the wide open areas of down under Australia	−0.94	1	2	**20**	2	1
Alongside brown apple betty, apple crisp or a mixed berry fruit torte	−0.96	3	6	**11**	7	0
Accompanied by aged cheddar, stilton, brie and other cheeses as well as smooth Italian cheesecake	−1.05	1	**10**	7	9	0
Delicious with chocolate-based desserts such as velvety smooth mousse, light and fluffy soufflé	−1.14	3	8	8	8	0
A smooth accompaniment to sweet tiramisu layered with lady fingers or crème brulée	−1.40	1	7	8	**11**	0
Enhances the taste of mouth-watering white, milk or dark chocolate	−1.44	5	2	**12**	8	0
A dessert in itself, no need for pairing it with anything else	−1.66	2	9	7	**10**	0

3. By presenting the respondents with combinations of elements, rather than with single elements, and by forcing a choice, we reduce the likelihood of a politically correct answer such as 'sophisticated'.
4. Afterwards, each respondent generates 48 such selections, one feeling/emotion selected for each test concept.
5. Create five new dependent variables, one dependent variable per feeling/emotion. Code the feeling/emotion variable '0' for a particular concept when the feeling/emotion was not selected; code the emotion variable '100' for a particular concept when the feeling/emotion was selected.
6. Use ordinary least-squares regression (OLS) to relate the presence/absence of each of the 36 elements to the selection of the feeling/emotion. Do the OLS analysis for each respondent, and for each feeling/emotion. Do not use the additive constant when estimating the coefficient because we are interested simply in the percent of respondents (or conditional probability) of selecting a particular feeling/emotion for a specific element. Here we create 252 equations for each of the five feelings/emotions, or a total of 1260 regression models.
7. Average the corresponding coefficients across respondents for each feeling/emotion. The result will generate a set of five averages (one per feeling/emotion) for each element.
8. The numbers in the body of Table 20.7, corresponding to the five feeling/emotions, show the percentage of respondents who we would estimate to assign the particular feeling/emotion to the particular element.

The foregoing exercise generates the feeling/emotion signature for each of our elements. The data suggests a number of patterns:

- In Table 20.7 the numbers show the estimated percent of respondents who would assign each specific feeling/emotion to one of the elements in a concept. Therefore the 5% for an element (one per feeling/emotion) add up to approximately 25% because a typical concept comprises four elements, so each element only constitutes one quarter of the message.
- The most frequent response is 'curious,' but there are other feelings/emotions as well.
- The five responses have to compete with each other to drive selection. The second tier of selection after 'curious' comprises 'sophisticated' and 'self indulgent.'
- On average, 'virtuous' and 'obligated' are rarely selected.
- Yet, for the less-frequent feelings/emotions, 'sophisticated' and 'self-indulgent', some elements emerge quite linked with them.
- 'Sophisticated' is driven by the word picture of 'rolling' and 'slope': *Grapes grown naturally in gently rolling vineyards* and by *From the gentle sloped ancient vineyards of southwest France,* as well as by imported desserts such as *Accompanied by aged cheddar, stilton, brie and other cheeses as well as smooth Italian cheesecake.*
- 'Self-indulgent' is driven by the experience *Drinking this wine relaxes you after a busy day at work*, by the accompanying dessert product: *A smooth*

accompaniment to sweet tiramisu layered with lady fingers or crème brulée, and by some puffery associated with the product itself, *A dessert in itself, no need for pairing it with anything else.*
- The bottom line for feelings/emotions is that they present a new dimension for understanding the algebra of the consumer's mind. In terms of execution one need only have the respondent select the appropriate emotion/feeling from a list. The relevant, perhaps 'correct' feelings/emotions remain a matter for experimentation.

20.8 Concluding remarks

This chapter presents an approach to understanding the mind of the wine consumer. The approach uses experimental design of ideas (conjoint analysis) to identify the specific messages that most appeal to consumers, the response patterns of different groups of consumers, the emergence of mind set segments, and finally new directions dealing with pricing on the one hand, and emotion on the other.

One may walk away from these experiments with at least two different points of view. One is 'why?' What is the value of an experiment with compound elements when perhaps one could just as easily ask the respondents what is important, or how much the respondent would pay. Why do the experiment when direct, simple questionnaires might be just as revealing, and certainly a lot simpler. The second is 'what'. What of lasting value does the experiment provide – beyond simply answering the question, 'what kind of science is being created?'.

20.8.1 Motivations for carrying out an experiment

We began this chapter with an analysis of two experiments on wine, one on white wine, and one on red wine. The objective of the experiment was to understand the way that the consumer respondents trade-off different factors when judging a compound stimulus, a test concept. One might have asked the respondent to rate the different elements, and come up with a rating. In such a case we might see the same results, or we might not. Certainly brand is believed to be very important, yet brand name does not appear as a key driver of interest, either in white wine or in red wine. Thus the experiment, forcing the elements to compete with each other, is more stringent. Faced with a compound stimulus and forced to assign a rating, the respondent must weigh the different alternatives to generate a rating. The respondent does not rate all of the combinations the same, either. So, at the end of the experiment we see both that the respondent discriminates among the different concepts, and that in a constrained situation with only 9 points the respondent makes trade-offs. Regression shows us the underlying algebra. The odds are that we would not have generated these results had we stayed with the tried-and-true questionnaire. We would have seen a lot more elements playing an important role. We would also have seen brand name come up as a lot more important than was actually the case when we put brand names as elements to 'fight against' the other elements.

20.8.2 Lasting value provided by the experiment

As a closing note, let us look back at the elements, and then compare what we learn from the conjoint studies to the studies about wine previously reported in the literature. We cited a few in the opening section of this chapter. The elements from MG, our version of conjoint, move beyond the strategic to the tactical, to the minutiae of every day experience. We do not live in the rarified world of strategy. We live in the world of specifics. We do not live as much in the world of profound deep ideas, bullet points of life, as much as we live in the evocative, image-rich, experience-filled world of the everyday. And so our experiment provides us with data about how the ordinary person responds to the ordinary stimuli of the everyday. It is not that we are testing hypotheses about how the world works, by using our experiments. That would be the hypothetico-deductive system, a good system when we know a lot about the world and then want to understand more. It is not what we are doing here. Here we are mapping the everyday, using tools to understand the basic building blocks of the ordinary. Our experiments provide us with the elements and numerical values for the dimensions of ordinary experience. In other words, the lasting value of these experiments is the inductive understanding of a world through new parallax views, and new lenses.

20.9 References

ANDERSON, N. 1977. 'Functional measurement and psychological judgment', *Psychological Review*, **77**, p 153–170.

AXELROD, M.D. 1975. 'Marketers get an eyeful when focus groups expose products, ideas, images, ad copy, etc. to consumers', *Marketing News*, **8**, p 10–11.

BECKLEY, J. and MOSKOWITZ, H.R. 'Databasing the consumer mind: The Crave It!, Drink It!, Buy It! and Healthy You! Databases'. *Institute of Food Technologists Annual Meeting*, Anaheim, CA, July, 2002.

BOX, G.E.P., HUNTER, J. and HUNTER, S. 1978. *Statistics for Experimenters*. New York: John Wiley and Sons.

ENGEL, R. 1928. 'Experimentelle untersuchungen uber die abhangigheit der lust und unlust von der reizstarke beim geschmacksinn', *Pfluegers Archiv fur die Gesamte Physiologie*, **64**, p 1–36.

GALANTER, E., MOSKOWITZ, H.R., and SILCHER, M. 2011. *People, Products, Prices: Sequence the Economic Genome of the Consumer Mind*. London: Bentham Press. In press.

GIL, J.M. and SANCHEZ, M. 1997. 'Consumer preferences for wine attributes: a conjoint approach', *British Food Journal*, **99** (1), p 3–11.

GLADWELL, M. 2009. 'The ketchup conundrum'. In: *What the Dog Saw and Other Adventures*, New York: Little Brown and Company.

GOFMAN, A., and MOSKOWITZ, H. 2010. 'Isomorphic permuted experimental designs and their application in conjoint analysis', *Journal of Sensory Studies*, **25**(1), p 127–145. DOI: 10.1111/j.1745-459X.2009.00258.x

KENNETH, J.H. 1928. 'A few odor preferences and their constancy', *Journal of Experimental Psychology*, **11**, p 56–61.

KRIEGER, A., GREEN, P. and WIND, Y. 2004. *Adventures in Conjoint Analysis: A Practitioner's Guide to Trade-off Modeling and Applications.* Philadelphia, PA: Wharton School of the University of Pennsylvania. Available at: http://marketing.wharton.upenn.edu/ideas/pdf/Green/Monograph/Adventures-Introductory%20Materials.pdf [accessed 21 September 2010].

JOHNSON, L.W., RINGHAM, L. and JURD, K. 1993. 'Behavioural segmentation in the Australian wine market using conjoint choice analysis', *International Journal of Wine Marketing*, **3**(1), p 26–31.

LOCKSHIN, L., JARVIS, W., D'HAUTEVILLE, F. and PERROUT, J.P. 2006. 'Drivers of acceptance: Using simulations from discrete choice experiments to measure consumer sensitivity to brand, region, price, and awards in wine choice', *Food Quality and Preference*, **17**(3–4), p 166–178.

LUCE, R.D. and TUKEY, J.W. 1964. 'Conjoint analysis: A new form of fundamental measurement', *Journal of Mathematical Psychology*, **1**, p 1–36.

MOSKOWITZ, H.R. (1994). *Food Products and Concepts: Just in Time Development.* Trumbull, CT: Food and Nutrition Press.

MOSKOWITZ, H. and GOFMAN, A. 2007. *Selling Blue Elephants: How to Make Great Products Before People Know They Want Them.* Upper Saddle River, NJ: Wharton School Publishing.

MOSKOWITZ, H.R. and GOFMAN, A. (2004). 'System and method for performing conjoint analysis'. Provisional patent application, 60/538,787, filed January 23, 2004.

MOSKOWITZ, H.R., PORRETTA, S. and SILCHER, M. 2005b. *Concept Research in Food Product Design and Development.* Ames, IA: Blackwell Professional Publishing.

MOSKOWITZ, H.R., GOFMAN, A., KATZ, R., ITTY, B., MANCHAIAH, M. *et al.* 2001. 'Rapid, inexpensive, actionable concept generation & optimization – The use and promise of self-authoring conjoint analysis for the foodservice industry', *Foodservice Technology*, **1**, 149–168.

MOSKOWITZ, H.R., GERMAN, B. and SAGUY, I.S. 2005a. 'Unveiling health attitudes and creating good-for-you foods: The genomics metaphor and consumer innovative web-based technologies', *Critical Reviews in Nutrition and Food Science*, **45**(3), 265–191.

MOSKOWITZ, H., GOFMAN, A., BECKLEY, J., and ASHMAN, H. 2006. 'Founding a new science: Mind genomics'. *Journal of Sensory Studies*, **21**(3), 266–307.

MOSKOWITZ, H.R., REISNER, M., LAWLOR, J.B. and DELIZA, R. 2009. *Packages, Ideas and Experience in Food Product Design and Development.* Ames, IA: Blackwell-Wiley.

PERROUTY, J.P., D'HAUTEVILLE, F. and LOCKSHIN, L. (2006). 'The influence of wine attributes on region of origin equity: an analysis of the moderating effect of consumers' perceived expertise', Agribusiness, **22**(3), 1–19.

PINE, J. and GILMORE, J. 1999. *The Experience Economy: Work is Theater & Every Business A Stage.* Boston, MA: Harvard Business School Press.

ROBERTS, K. 2005. *Lovemarks: The Future Beyond Brands* (Expanded edition). New York: powerHouse Books.

SANCHEZ, M. and GIL, M. 1997. 'Drivers of acceptance; consumer preferences for wine attributes in different retail stores: a conjoint approach'. *Journal of Food Products Marketing*, **4**(2), p 63–78.

SYSTAT. 1997. *User Manual for SYSTAT, the System for Statistics.* Evanston, ILL: SYSTAT Corporation Division of SPSS.

WITTINK, D.R. and CATTIN, P. 1989. 'Commercial use of conjoint analysis: An update', *Journal of Marketing*, **53**, 91–96.

21

Preference mapping: principles and potential applications to alcoholic beverages

H. J. H. MacFie, Hal MacFie Training Services, UK and J. R. Piggott, University of Strathclyde, UK

Abstract: Preference mapping is a key tool in portfolio management as it enables marketing personnel to view and understand the marketplace. When linked to descriptive sensory data it also enables new product development (NPD) staff in research and development to understand how to manipulate the sensory properties of their products. This chapter outlines the main techniques in use, and indicates how to design and collect suitable data for preference mapping. The application of various decision-making tools to enable optimal products to be identified and their sensory properties determined is illustrated. Finally some current research areas are described with an assessment of their importance to the practitioner.

Key words: marketing, preference mapping, food, alcoholic beverages, product development, trials, sensory data, cluster analysis, probabilistic modelling, partial least squares.

Note: This chapter is a revised version of Chapter 23, 'Preference mapping and food product development' by H. MacFie, originally published in *Consumer-led food product development*, ed. H. MacFie, Woodhead Publishing Limited, 2007, ISBN: 978-1-84569-072-4.

21.1 Introduction

Preference mapping is a key tool in portfolio management as it enables marketing personnel to view and understand the marketplace and see where the competitors are moving to and where to position their own products. When linked to descriptive sensory data it also enables new product development (NPD) staff in research and development to understand how to manipulate the sensory properties of products to move to desired positions in the marketplace; and both marketing and research and development can understand the key sensory attributes that respondents are using to make assessments about liking. It is important to realize that it is a source of reformulations or hybrids of current products. It is not a source of totally novel products; for those one may apply the tests outlined by van Kleef and van Trijp (2007).

This chapter outlines the main techniques currently in use, their pros and cons and indicates how to design and collect suitable data for preference mapping. The application of various decision-making tools to enable optimal products to be identified and their sensory properties determined is illustrated. Finally some current research areas are described with an assessment of their importance to the practitioner. Given the lack of published examples of preference mapping applied to alcoholic beverages, many of the examples necessarily come from other sectors of the food industry.

Preference mapping really began with a paper by Carroll (1972). He outlined the ideas of internal and external preference mapping, thus confusing generations of workers. To anyone in research and development used to the idea of an internal panel and an external panel, it is clear that with external preference mapping you would be using the external panel to get the map and internal vice versa. Absolutely incorrect! Carroll was a mathematician, not a sensory scientist, and 'internal' refers to the fact that you obtain the map of the products internally from the consumer liking data and external refers to the fact that the map is based on data collected outside of the consumer data.

The first food-based preference mapping paper to our knowledge was on meat (Horsfield and Taylor, 1976). With the wider availability of the internal preference map program MDPREF and the very timely text on multidimensional scaling by Schiffman *et al.* (1981), papers started to appear. In Section 21.5 Sources of further information and advice we have tried to collect all the preference mapping references from the *Journal of Sensory Studies, Food Quality and Preference* and *Journal of Food Science* to provide a useful resource for readers to follow up on methods or particular food categories. This list is not exhaustive. However, the real engine for the dissemination of preference mapping concepts was the sensory and food-based market researchers in the USA and Europe who soon realized that being able to connect up liking and descriptive data and present the so-called sensory drivers and marketplace maps gave them a perceived competitive edge over the competition. Alternative black box methods started to appear and there still remain a large number of hybrid preference mapping techniques that have never been published.

A second reason for the success of preference mapping is that it also satisfies two widespread, but possibly erroneous, assumptions that many workers and new product developers make when working with consumers. These are that the general consumer finds it difficult to articulate and score many perceptual attributes of foods, and that product experts who traditionally make quality assessments of foods are not representative of the general consumer population. If you make these assumptions it naturally follows that the optimum solution is to just ask consumers a minimum set of affective questions and recruit a non-technical consumer with excellent sensory facilities who is trained to recognize and articulate and score sensory attributes of foods. The two data sets can then be related by some statistical linkage method.

The growth of preference mapping has thus been engineered on the basis of conducting simultaneous central location tests and sensory descriptive analysis on the same products. Along the way there have been a number of developments and understandings that have increased the accuracy, precision and reduced bias of the data we collect and we will discuss these first before elaborating on the statistical linkage methods.

21.2 Conducting central location trials (CLTs)

Central location trials (or CLTs as we shall refer to them) are sequential monadic tests conducted in a location such as a church hall, market research facility, etc. The respondents are either pre-recruited or come in directly off the street and assess a number of the products in question. They usually complete a thorough demographic and consumption questionnaire at the end of the test. The foremost advantage of the CLT over the home use test (HUT) is that there is total control over the preparation and presentation of the samples. It is also quick; with enough recruiters and temporary booths it is possible to complete the study in one day. The key advantage over monadic is that the consumer is able to express a number of opinions that enable us to work out which sensory attributes appear to be positively or negatively correlated with their liking and to establish a product map that defines a common space on which we can project liking directions or ideal points.

CLTs are conducted as marketplace screens, for competitive benchmarking and to compare the performance of a number of prototypes with the current product. For preference mapping of any sort then a practical minimum number of products to be assessed is six. However, most practitioners would recommend eight as a realistic minimum and try for 12 or more, so experimenters should try to present a larger number of products than would normally be used in a sensory test. Naturally a reasonable recovery time between assessments is maintained.

The process of selecting which products to include in the test is relatively easy for a control versus prototype test using a statistically-based fractional design. However, the method of selecting samples from the marketplace is usually more tortuous, as marketing and research and development will have some very firm views on which products must be included or excluded. The presence of a sensory panel that can perform sensory descriptive analysis (SDA) is invaluable here. An iron rule for consumer trials should be that it is generally not worth putting two samples in a CLT that are indistinguishable to the sensory panel. Another rule to ensure success in recovering a reasonable map is not to restrict the samples to those that are generally likely to be market leaders as this will give too small a range of variation. The aim should be to select a set that will span the space of the marketplace. A typical scenario for product selection is to collect all the products one can, including any prototypes, and bench test them to select a subset of say 20–30 to go forward to SDA. On completion of the SDA, perform a principal components analysis (PCA) and select a set of 12 or 15 samples that span the space, circle them and send the map over to marketing and research and development for their views. Ben Slama *et al.* (1998) suggested a method for selecting the optimum subset of samples. When the list of samples that absolutely must be included in the trial comes back, make a second iteration, attempting to avoid products with identical sensory properties going into the trial. If there are problems or political battles, then it is relevant to make the point that it will be possible to project samples for which there is an SDA profile into the map to see where they lie, even if they were not submitted to the CLT.

Having selected the samples, the next question will be whether to conduct a complete trial, in which all respondents assess all samples, or conduct an incomplete trial. Preference map practitioners will always push for a complete trial, as many cluster analysis and internal preference mapping algorithms require a complete set, and there is little doubt that this must be the optimum solution. This may be difficult for products such as, say, coffee or alcoholic beverages where there is a limit to the total quantity of product that the respondents can be asked to consume. Multiple tasting sessions can be used, but it may not be possible to ensure that all respondents return for second or subsequent sessions, and over-recruitment may be necessary to compensate for this. If an incomplete design is necessary, then a rough rule for preference mapping trials is that each respondent should assess at least 75% of the total set (Hedderley and Wakeling, 1995). A balanced incomplete block design is used so that every product is used equally often across the respondents. An incomplete design can be analyzed using 'virtual consumers' constructed from small clusters of consumers who behave in a similar way (Callier and Schlich, 1997).

With the samples selected, we can turn our attention to selecting the respondents. Marketing can usually provide profiles of typical users and quotas are set up for age, gender, location and usage level. In some instances previous trials have characterized psychographic profiles and a key set of recruitment questions can be used to select quotas of the different profile groups.

We then come to the thorny question of how many respondents to recruit? One approach is to use previously obtained estimates of residual standard deviations to calculate the number of respondents that will give you the precision you require. Hough *et al.* (2006) have provided a very useful guideline for this. They examined a number of CLT trials across many products and countries and concluded that a typical residual standard deviation for these trials was 0.23 of the scale length. So if one was working with the 9-point hedonic scale there would be a scale length of 8 (9–1) and the RSD would be 0.23×8 (1.84). So using this value and power tables, if we set a Type I error of 5% (5% chance of finding a significant difference between two samples when none exists) and Type II error of 10% (10% chance of concluding there is no difference between two sample when one exists), we can work out the number of respondents needed in a trial to be able to detect a pre-specified difference.

Table 21.1 indicates that for a pre-specified difference of 10% of the scale length, i.e. 0.8 in terms of the 9-point scale, a total of 112 respondents would be required for the Type I and II settings above. This is remarkably close to the minimum number used by many practitioners of 120. However, this number is not

Table 21.1 Number of consumers needed for an acceptability test

RMSL[a]	α%[b]	d[c]	β%[d]		
			20	10	5
0.14	10	0.2	7	9	11
	5	0.2	8	11	14
	1	0.2	12	15	18
	10	0.1	25	34	43
	5	0.1	32	42	52
	1	0.1	47	59	71
	10	0.05	98	135	170
	5	0.05	124	166	205
	1	0.05	184	234	280
0.23	10	0.2	17	23	29
	5	0.2	22	29	35
	1	0.2	32	40	48
	10	0.1	66	91	115
	5	0.1	84	112	138
	1	0.1	124	158	189
	10	0.05	262	363	459
	5	0.05	333	445	551
	1	0.05	495	631	755
0.3	10	0.2	29	39	49
	5	0.2	36	48	59
	1	0.2	53	68	81
	10	0.1	112	155	196
	5	0.1	142	190	235
	1	0.1	211	269	322
	10	0.05	446	617	780
	5	0.05	566	757	936
	1	0.05	842	1072	1284

[a] RMSL = root mean square error divided by scale length. [b] *α*% = probability of Type I error. [c] *β*% = probability of Type II error. [d] d = difference in means that is sought in the experiment (scale 0–1).
Source: Hough *et al.* (2006)

going to be sufficient for a trial in which you may be wishing to analyze certain subgroups or are expecting to detect and study naturally occurring segments in the population. A rough guideline if you are thinking of analyzing groups separately is to recruit at least 50 people per group and ideally more. It is important to realize that some individuals have great trouble in giving numbers to their sensory experiences and their ratings appear to be completely uncorrelated with anyone else's. There is also the frustrating group of non-discriminators that give the same score to every sample. So it is necessary to over-recruit so that we have at least 40–50 respondents in each important subgroup.

Another problem with the 50 respondents per segment rule is that segments do not typically appear to be the same size. It often happens that there is one large group and a few smaller segments. As will be seen later, it is becoming common practice to identify the clusters and then work out products that can satisfy as many segments as possible, rather than psycho-analyze each segment and develop a specific product. In these cases it is not necessary to characterize each cluster too intently and a figure of 150–180 respondents allows for three segments. However, some practitioners routinely identify 6–9 clusters in the population and therefore sampling numbers of 300 plus are recommended. This seems a little high to the current authors, a figure of between 150–200 should suffice for most markets. If an incomplete design is used, the number of respondents must be increased accordingly.

The sampling plan for each respondent is the next task. Following MacFie *et al.* (1989), the Williams Latin squares (Williams, 1949) that balance the occurrence of each sample in each position, and the number of times each sample follows every other sample are now used routinely. If the incomplete design is being used the designs proposed by Wakeling and MacFie (1995) are utilized. Ball (1997) has discussed how to keep the design as balanced as possible throughout the sampling process so that, in the case of an unexpected halt in the flow of consumers into the test facility, a relatively high degree of balance is maintained. For multi-session trials in which the respondent may be returning to the facility two, three or more times, the Design Express program (Design Express version 1.6. 2006, Qi Statistics Ltd, UK) gives easy access to designs that are as balanced as possible, even for incomplete designs. Note that, although these designs are universally used, there are those who maintain that a randomization methodology would be preferable (Kunert J., Dortmund University, personal communication).

These Williams designs do not remove the carry-over effect of one sample on to another from the data. They just distribute any carry-over effects evenly among the samples. One effect that can be removed is the tendency of respondents in sequential monadic trials to score the first product more highly than the rest. The use of a warm-up or dummy sample is recommended here. This sample is presented to the respondents with the full questionnaire and they complete it normally, believing it to be part of the test. However, the results are discarded as being part of the learning process. It is often thought that a warm-up sample is only needed on the first session of a multi-session test. Fig. 21.1 shows the mean

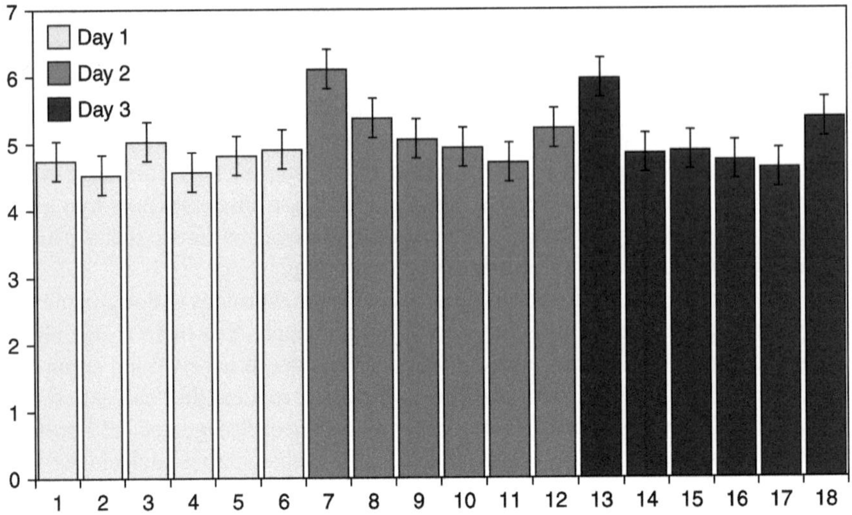

Fig. 21.1 Average liking scores of a CLT test of 18 beers for positions 1–18 across three days of testing. On the first day a warm-up sample was presented prior to the start of the test. Warm-up samples were not presented prior to testing on days two and three. Evidence of a higher score given at the start and end of sessions two and three is clear. The vertical scale is the mean liking on a 7-point hedonic scale.

liking score in each position across a balanced design for some beers across three days of testing. On the first day a warm-up sample was presented and the results discarded. No warm-up was presented on subsequent days. The effects on the data are clear, as is the usual occurrence of a higher score at the end of a session as respondents realize they have finished for the day or are about to collect their recompense. This could be minimized by informing respondents that they are going to score one more sample than is actually the case. The case for a warm-up sample is not so strong when the interest is only on aggregated results across all the respondents as positional effects can be removed as part of the analysis of variance estimation procedure. In cases where each subject is going to be input to a cluster analysis or an internal preference map, or input as a separate y-variable in an external preference mapping procedure, then these effects are difficult to remove by adjustment, and may significantly influence the scores of a respondent. In our view a warm-up sample is an essential component of successful preference mapping.

The next question is: how do we select a sample to be a warm-up? Our advice on this stems from some unpublished data in which we presented 12 fragrances to 60 subjects using the same order of samples with no warm-up. We then repeated this using three other orders with different sets of 60 respondents. Analyzing the data we found that the order that had happened to contain the least liked sample first had given a different order of sample means than the other three orders. We concluded that the halo effect of the first sample had lifted the scores of this sample. Subjects were simply not willing to score every other sample one or two points higher so a re-evaluation of the scale took place with resulting distortion of the final sample means. Our advice is therefore to try to use a sample that is in the middle of the space and will hopefully score in the middle of the liking range. The presence of sensory descriptive analysis data here is again invaluable as one can select a product in the middle of the space defined by the first two principal components.

One further aspect of sample design that needs to be considered is whether to replicate one or two samples in the sequential monadic test. This gives an estimate of respondent reliability and has been used by some practitioners as a way to screen out poorly performing respondents. It also gives extra degrees of freedom to enable a test of respondent by sample interaction in the aggregated analysis of variance. In our experience, clients are usually too eager to put an extra prototype in and therefore do not often include a replicate, but it can on occasion be very useful.

The next step before commissioning the CLT is the design of the questionnaire. It is usual for experimenters to come down with a self-designed, extensive list of questions they would like the respondents to answer. Here we take a fairly tough view and advise clients and researchers to ask a minimum set of questions, ideally just one: How much do you like this product? The rationale for this is that most respondents eat foods, consume beverages, shave, apply make–up, wash, shampoo, drive cars all the time and do not think deeply about the sensory variations in these experiences. By making them complete extensive questionnaires on every product, the likelihood is that they will start to re-evaluate their opinions, often for reasons of so-called impression management (making you think they are someone they are not) and so the resulting scores do not relate to their daily appreciation and the resulting product will not be liked in the 'real world'.

We are particularly opposed to the use of just-about-right (JAR) questions in these preference mapping studies. Earthy *et al.* (1997) reported how the addition of JARS to a questionnaire in a CLT test of chocolate mousses varying in sweetness and ratio of dark to milk chocolate altered the overall liking of products. Popper *et al.* (2004) repeated the trial with extra conditions including intensity of attributes and other liking questions as well as JARS and showed that the presence of attribute liking question and the presence of JARS altered the ranking of the overall liking of products. In our view many respondents do not carry ideal points around in their head but this can be easily imposed on them by the presence of JAR questions. This is not to say that we are implacably opposed to JAR questions. The use of Penalty Analysis to help reformulation is increasing and we refer readers to Rothman (2007) for further details. However, we will maintain that, for an accurate view of consumer liking that is essential for unbiased preference mapping, these JAR questions, and ideally all other attribute questions, should be asked as a separate exercise. Note that Earthy *et al.* (1997) indicated there to be no differences between asking the hedonic question on its own and asking the hedonic question first, so the latter is not a solution.

A very important issue to be considered before completing the preparation of the CLT trial is how the product is going to be presented to the respondent. Here we seek to emulate normal use as far as possible, without losing control of the standardization of presentation that is a key characteristic of CLT trials. After all, if 99.9% of respondents only ever drink gin in the presence of tonic water, what is the point of getting them to assess neat gin? Similarly, if you take sugar in your tea, never eat olive oil spread on its own, or only ever eat sushi with a drop of soya sauce, then the presentation should be modified accordingly. Posri (2001) showed that modifying the presentation of tea to respondents' usual strength, milk and sugar dosage gave improved correlation with subsequent home-use test results. However, the modification had to be standardized for each respondent. When respondents were permitted to prepare the tea themselves, the correlation was reduced, presumably because variation in preparation reduced the perceived differences.

A final issue to be dealt with when commissioning CLTs is who is going to visit on the first day of each series to ensure that the trial is being conducted correctly. This must be someone who understands the trial, is an experienced tester and has the authority to bring the trial to a halt if standards are not being observed. Skipping this operation is not an option unless you have complete trust in the team conducting the trial.

A brief summary of the recommendations is given in Table 21.2. If you observe all of these, you are well on the way to producing the kind of data that will produce

Table 21.2 Recommendations for conducting a successful CLT for a preference mapping exercise

Step	
1	Minimum number of samples is 6, ideal is 8–12.
2	Conduct a descriptive analysis on a larger set first.
3	Do not select samples to go into the CLT that are sensorially indistinct from a sample that has already been selected.
4	Span the space of the marketplace in selecting the samples.
5	Conduct a complete trial so that each respondent assesses all samples.
6	If not complete, each respondent should test at least 75% of the samples.
7	Use target users.
8	Select 120 respondents minimum or if three or more segments are expected add 50 for each additional segment expected.
9	Use a Williams design to order the samples for balance across position of presentation and to balance any possible carry-over effects.
10	Use a warm-up sample at the start of each session. This should be something in the middle of the range.
11	Inform the respondents that they will be tasting one more product than they actually will be.
12	Emulate normal mode of use for sample presentation, e.g. tea with milk, sushi with soy sauce but control so that the treatment of each sample is identical.
13	Arrange to visit the CLT testing facility at the start-up on the first day.

reliable maps that can be used to predict liking of other products within the sensory space you have selected.

21.3 Analyses

21.3.1 Preliminary analyses

Upon receiving the data from the market research company, or from the data preparation team if you have conducted it yourself, it is worth conducting some top line screening analyses to check what has happened. We find the Pivot table option in Excel absolutely invaluable here. The first question is whether the recruiting quotas have been achieved and frequency tables such as that shown in Table 21.3 quickly tell the tale. If a category scale such as the 9-point hedonic scale has been used then it is usual to calculate the frequency table and the top 3 and bottom 3 box percentages as shown in Table 21.4. Note that in this table the two winning products are easily identified. However, there are many products with nearly half the respondents expressing high liking and the other half expressing a high level of disliking. It is clear that a simple overall liking ranking

Table 21.3 Number of respondents: gender by consumption frequency

Consumption frequency	Male	Female	Grand total
1	136	112	248
2	168	240	408
3	616	672	1 288
4	392	248	640
Grand total	1 312	1 272	2 584

Table 21.4 Overall liking. Number of respondents using each category of the 9-point scale

Product name	1	2	3	4	5	6	7	8	9	Grand total	Top 3 box %	Bottom 3 box %
A	7	13	36	52	81	69	43	15	7	323	20.1	17.3
B	19	17	34	66	56	64	46	15	6	323	20.7	21.7
C	12	19	36	56	77	58	47	10	8	323	20.1	20.7
D	27	40	49	58	59	42	30	14	4	323	14.9	35.9
E	5	2	7	27	68	84	80	36	14	323	40.2	4.3
F	4	6	13	46	61	78	75	33	7	323	35.6	7.1
G	10	7	20	39	73	74	64	32	4	323	31.0	11.5
H	4	16	39	50	68	68	43	28	7	323	24.1	18.3
Grand total	88	120	234	394	543	537	428	183	57	2584		

is not going to encapsulate the opinions of these two groups and a more multidimensional approach is called for. These preliminary tables can be sent to the client almost immediately. The winners and losers are identified and the necessity for further analyses is immediately clear.

21.3.2 Cluster analysis

Cluster analysis may be thought of as a preliminary to preference mapping, although we prefer to view it as the lowest level of preference mapping in terms of assumptions about how the population is behaving. With cluster analysis the only assumption is that not all people are displaying the same liking patterns, and with deterministic cluster analysis we hypothesize that each respondent lies in one of several distinct clusters or is considered an outlier. For deterministic analyses there are two algorithms that are provided by all the main statistical packages, hierarchical cluster analyses and non-hierarchical cluster analyses, and we will briefly outline their mode of operation.

The first stage of any cluster analysis must be to define an index of similarity or dissimilarity between respondents. Recall that for this operation we are going to restrict ourselves to the data matrix containing the liking ratings of N respondents on P products. We are going to transform this matrix to a N by N symmetric matrix of (dis)similarity and then use the clustering algorithm to segment the respondents into clusters as shown in Fig. 21.2. Although there are literally dozens

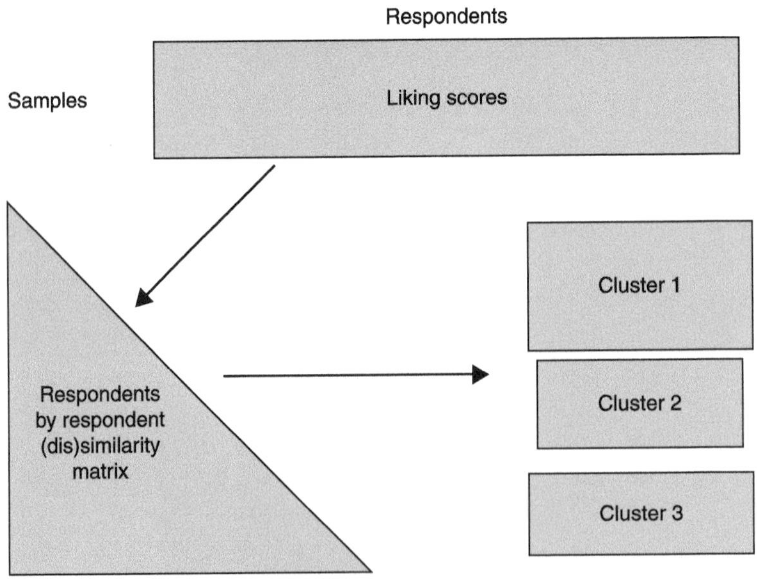

Fig. 21.2 Cluster analysis schemata. Liking data is transformed to a similarity or dissimilarity matrix and a clustering algorithm is used to identify the number and memberships of clusters among the respondents.

of possible indices of (dis)similarity we will discuss only three, and their analogues for the case where ranking data has been collected.

The key question is whether you want to work with the raw data or adjust respondents for the differences in scale usage. The most commonly used raw index is Euclidean distance, which is a multivariate extension of the Pythagorean distance between two points. The ideas are shown in Fig. 21.3 and the distance matrix between four respondents is also shown. A very useful alternative to the Euclidean distance is the city block distance, also termed the Manhattan distance, so named because it is not possible to walk through a block of buildings and so we have to walk around the perimeter. As it is simply the sum of the absolute distances between corresponding scores on a set of samples between two respondents, it is very easy to calculate. The calculations are shown in Fig. 21.4. The key advantage of this measure over the Euclidean distance is that in the latter a single large difference, when squared, can dominate the result. This is not going to be anywhere near as influential in city block.

Since respondents have not been calibrated in any way before performing the rating task, there is a compelling argument that the variation in range usage is nothing to do with perceived product differences and more to do with personality differences. Standardization of scores by subtraction of a respondent mean and division by respondent standard deviation is an obvious method of adjustment. However, the calculation of a Pearson correlation coefficient between two respondents' scores is perhaps even more attractive as the measure has a known

Liking scores

	X_1	X_2	X_3	X_4
Product	Mary	Diana	Jane	Sally
A	5	5	2	8
B	6	7	3	6
C	7	8	5	4
D	8	8	7	2

p products

Euclidean distance

$$D_{ij} = \sqrt{\sum_{k=1}^{p} (X_{ik} - X_{jk})^2}$$

Multivariate extension of Pythagoras' theorem

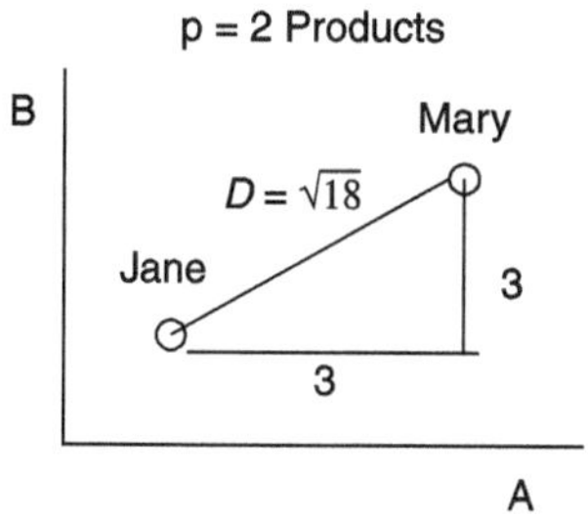

Fig. 21.3 Calculation of Euclidean distance between two respondents' liking scores across all samples. The exact calculation between two respondents for just two samples is shown. For more samples the distance is simply the sum of squared distance between liking scores of corresponding samples in the set.

Liking scores

	X_1	X_2	X_3	X_4
Product	Mary	Diana	Jane	Sally
A	5	5	2	8
B	6	7	3	6
C	7	8	5	4
D	8	8	7	2

p products

City block distance

$$D_{ij} = \sum_{k=1}^{p} \left| X_{ik} - X_{jk} \right|$$

Gives less weight to large distances than Euclidean distance

p = 2 Products

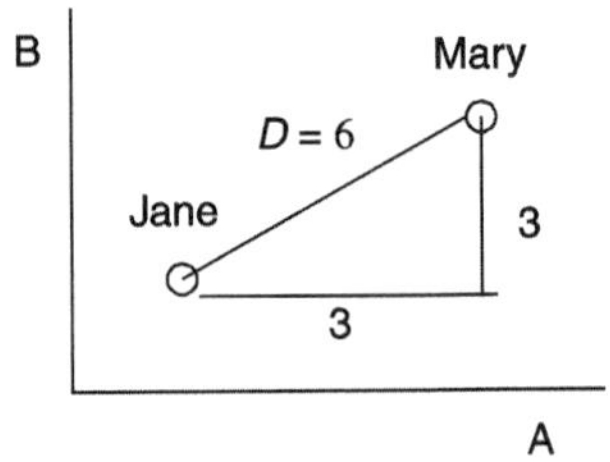

Fig. 21.4 Calculation of city block (Manhattan) distance between two respondents' liking scores across all samples. The exact calculation between two respondents for just two samples is shown. For more samples the distance is simply the sum of the absolute distance between liking scores of corresponding samples in the set.

range and is intuitively understood by experimenters. It also does not require a transformation of the data and so results for cluster means are easily expressed in the raw data. The ideas are shown in Fig. 21.5 and in the resulting correlation matrix in Fig. 21.6 we see that the outlier Sally is in fact simply scoring all products completely inversely to the other respondents.

Having selected a suitable measure of (dis)similarity, we form the symmetric matrix that defines the inter-relationships between the respondents and move on to select a clustering algorithm. We will discuss hierarchical cluster analysis (sometimes termed agglomerative hierarchical classification (AHC)) first. All versions of this algorithm work by first considering each respondent as a separate cluster. Using a single linkage strategy, the two most similar respondents join to form a new cluster at the (dis)similarity level between them, the next two respondents join at the higher similarity level, respondents join clusters of other respondents at the dissimilarity level of their nearest neighbour in the cluster. The process by which respondents join each other, and then clusters join each other is accurately recorded in a dendrogram. The basic ideas are demonstrated in Fig. 21.7.

Early users of single linkage analysis found a tendency for the clusters to join up in chains, and so a system in which the clusters joined at the (dis)similarity level of their two furthest neighbours was devised. Although this technique, appropriately termed furthest neighbour analysis, certainly gave more compact clusters, it still depended very much on the relative positions of just two

Liking scores

	X_1	X_2	X_3	X_4
Product	Mary	Diana	Jane	Sally
A	5	5	2	8
B	6	7	3	6
C	7	8	5	4
D	8	8	7	2

Correlation

$$D_{ij} = Corr(X_i - X_j)$$

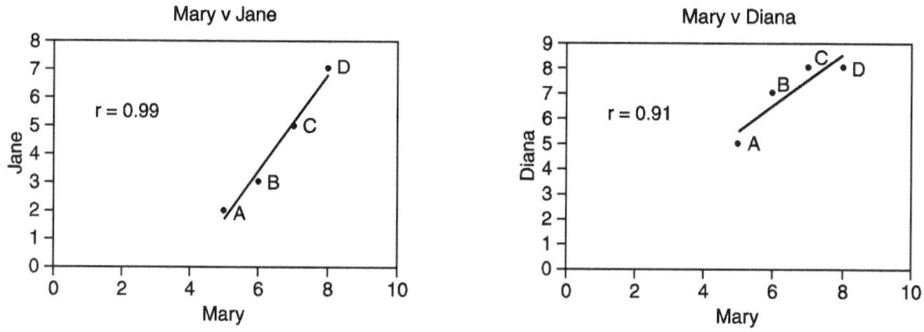

Fig. 21.5 Calculation of Pearson correlation between two respondents' liking scores across all samples. The correlation coefficients and pairwise plots for two pairs of respondents are shown.

Liking scores

	X_1	X_2	X_3	X_4
Product	Mary	Diana	Jane	Sally
A	5	5	2	8
B	6	7	3	6
C	7	8	5	4
D	8	8	7	2

p products

	Correlation between vectors of values			
	Mary	Diana	Jane	Sally
Mary	0	0.91	0.99	–1
Diana	0.91	0	0.85	–0.91
Jane	0.99	0.85	0	–0.99
Sally	–1	–0.91	–0.99	0

Fig. 21.6 The four-respondent correlation matrix that would form the input to a cluster analysis. Respondent Sally clearly likes completely opposite products to the other three.

respondents and was found to give unstable solutions. A preferable, and widely adopted, algorithm joins two clusters at the average (dis)similarity level between all possible cross-cluster member pairings. This technique is termed average linkage, or sometimes unweighted mean pair group. Another widely used algorithm is Ward's method, which attempts to maximise the ratio of between-cluster to within-cluster variance.

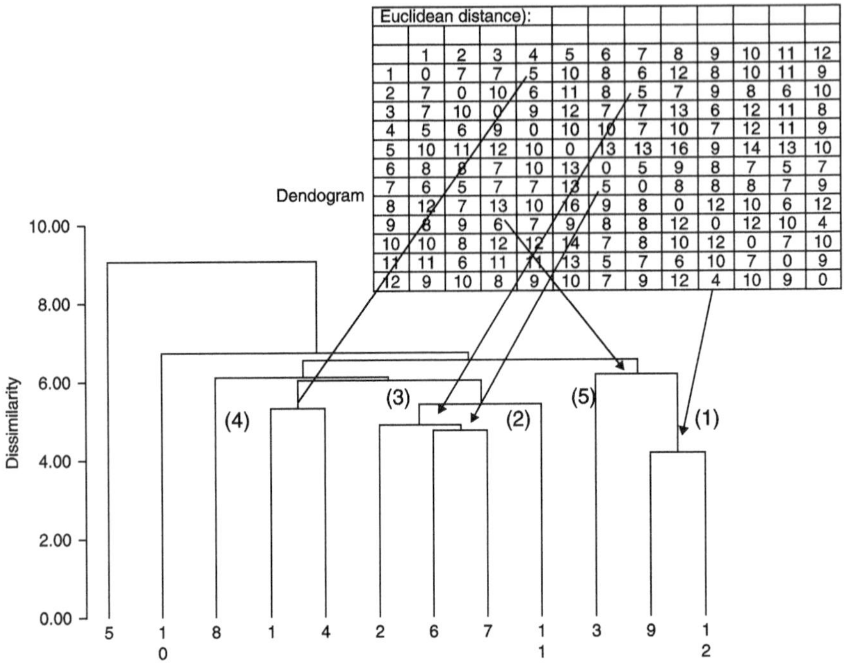

Euclidean distance):												
	1	2	3	4	5	6	7	8	9	10	11	12
1	0	7	7	5	10	8	6	12	8	10	11	9
2	7	0	10	6	11	8	5	7	9	8	6	10
3	7	10	0	9	12	7	7	13	6	12	11	8
4	5	6	9	0	10	10	7	10	7	12	11	9
5	10	11	12	10	0	13	13	16	9	14	13	10
6	8	8	7	10	13	0	5	9	8	7	5	7
7	6	5	7	7	13	5	0	8	8	8	7	9
8	12	7	13	10	16	9	8	0	12	10	6	12
9	8	9	6	7	9	8	8	12	0	12	10	4
10	10	8	12	12	14	7	8	10	12	0	7	10
11	11	6	11	11	13	5	7	6	10	7	0	9
12	9	10	8	9	10	7	9	12	4	10	9	0

Fig. 21.7 Schematic indicating how a single linkage clustering algorithm applied to Euclidean distance matrix of 12 respondents would create the dendrogram illustrated. In step (1) respondents 9 and 12 are identified as having the minimum Euclidean distance between each other. These two respondents are placed on the right of the dendrogram and vertical lines drawn up to be joined at the corresponding distance reading on the vertical scale. In step (2) respondents 6 and 7 are identified as being the next closest to each other and their vertical lines are joined at the respective distance. In step (3) respondent 2 is close to 7 and joins the already-formed cluster of respondents 6 and 7. The dendrogram thus faithfully traces the exact formation of clusters, cluster of clusters until all respondents have joined. Clearly respondent 5 is the outlier in this group.

The number of clusters and their membership is decided by placing a horizontal line across the dendrogram as shown in Fig. 21.8. Most cluster analysis programs have a procedure to detect and identify the number of clusters. However, many users prefer to make that decision themselves. This can be an iterative process, inspecting the clusters at various similarity level cut-offs. In Fig. 21.8 we could take two or three clusters. One point to remember is that if demographic frequency analysis is going to be used by marketing personnel, these clusters should not be too small (i.e. greater than 50). However, sometimes we use cluster analysis as a data reduction process to reduce the number of respondents to a reasonable reduced set for preference mapping and in this case then much smaller clusters can be accommodated.

Once the number of clusters has been determined we can calculate the means and inspecting these enables us to understand what characterizes the difference

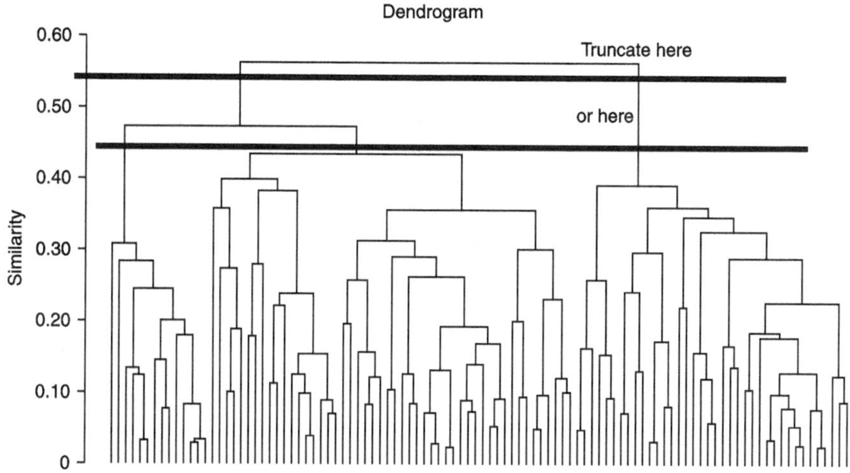

Fig. 21.8 Dendrogram of an average linkage hierarchal cluster analysis based on Euclidean distance of 102 respondents who have rated 10 varieties of apple. The bold horizontal lines represent two possible truncation strategies leading to two or three clusters.

between the clusters, and perhaps give them descriptive labels. Very often they polarize on one or two of the samples. For example, in Fig. 21.9 we show the liking means for the two cluster solution and we can see the green apple liker cluster and red apple cluster that are usually identified in dessert

Composition of the clusters:

Cluster	1	2
Within	4310.306	2309.580
Size	64	38

Cluster centroids:

Cluster	Gibson's Green	Johnson's Red	Golden Delicious	Granny Smith	Pink Lady	Fuji	Top Red	Braeburn	Royal Gala	Sun Gold
Cluster 1	4.875	4.697	6.739	6.794	6.219	4.836	3.459	5.928	3.808	6.989
Cluster 2	4.771	9.137	7.592	3.466	4.605	7.176	6.363	4.374	6.574	3.782

Central observations:

Cluster	Gibson's Green	Johnson's Red	Golden Delicious	Granny Smith	Pink Lady	Fuji	Top Red	Braeburn	Royal Gala	Sun Gold
Cluster 1 (L27)	5.800	5.000	5.700	5.100	3.300	5.100	6.200	5.700	1.700	6.600
Cluster 2 (L80)	3.300	9.900	7.900	2.300	2.100	6.900	7.600	1.100	7.000	2.700

These are the observations that appear to lie in the middle of the cluster space

Fig. 21.9 Typical cluster analysis output after truncation. The size of the cluster in terms of number of respondents and the within groups sums of squares are given. The table of cluster centroids (means) shows the average values for each product using all the respondents in a cluster. It appears that cluster 1 likes the Golden Delicious, Granny Smith, Pink Lady and Sun Gold varieties. Cluster 2 likes Johnson's Red very much and Golden Delicious. Golden Delicious is the best single variety to sell as it is scored well by both clusters. The table of most typical member of each cluster is often given but not used in preference mapping applications.

apple-eating populations in the UK (Daillant-Spinnler *et al,*. 1996). The demographic frequency tables enable us to characterize any unusual patterns. For example in Table 21.5 we see that there are more older people in the green apple liker group.

The case study in Fig. 21.10 demonstrates how useful the demographic frequencies can be. The dendrogram revealed four clusters of likers for prototypes of a new personal product. Inspecting the means in Fig. 21.11 would indicate that the best prototype to invest in would be Sample E. However, the frequency table in Fig. 21.11 indicates that Cluster 1 contains primarily purchasers of the competitor and Cluster 2 contains primarily buyers of our current product. Since the new product is going to be an addition to the portfolio, we will invest in Sample F, liked by all the other clusters except our current purchase group.

In considering which algorithm to use, we believe Nearest neighbour and Furthest neighbour can be discarded as being too dependent on the relative positions of individual respondents. To decide between Average Linkage and Ward's is more difficult as both have their supporters. However, it is helpful to consider the role of two types of outliers in this data. There are respondents who give scores that are totally different to the rest of the population and who should therefore join the other

Table 21.5 Demographic variable by cluster frequency tables for the two cluster solution of the data from Fig. 21.8 and 21.9

Count of row	Cluster		
Age	1	2	Grand total
1	26	20	46
2	38	18	56
Grand total	64	38	102
Count of row	Cluster		
Class	1	2	Grand total
1	40	10	50
2	24	28	52
Grand total	64	38	102
Count of row	Cluster		
Children	1	2	Grand total
1	36	25	61
2	28	13	41
Grand total	64	38	102

Note: Cluster 1 has more old members (Age 2). Cluster 1 has many more A-B working status members (Class 1). There is not much effect due to whether they have children or not.

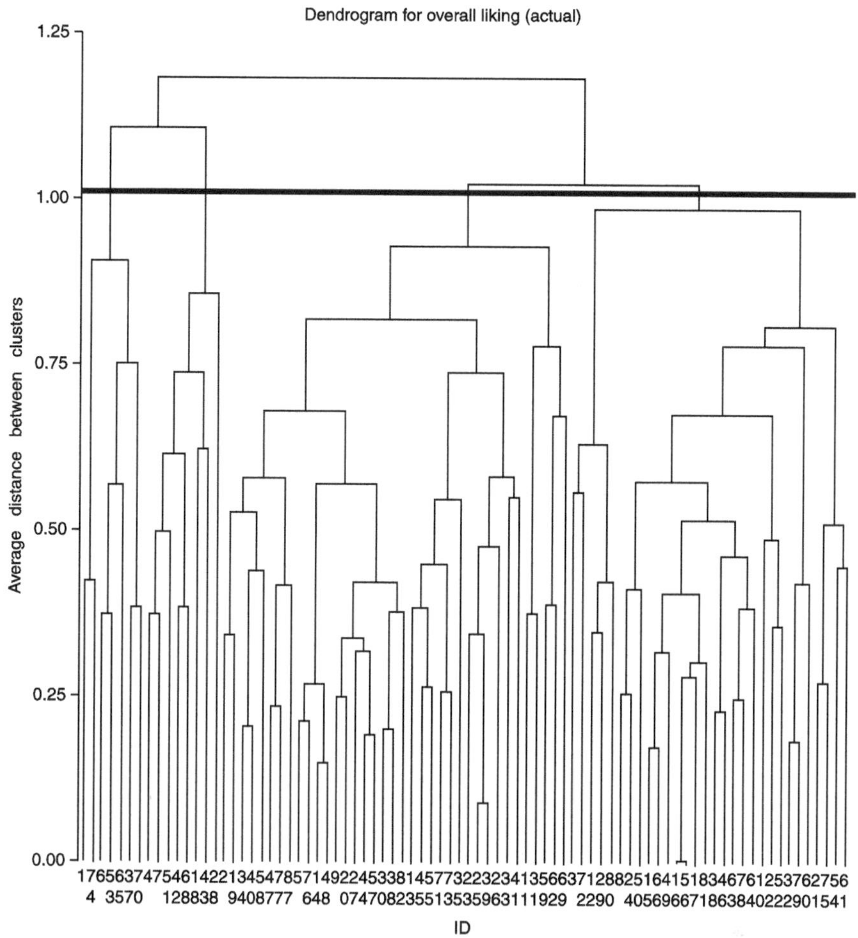

Fig. 21.10 Dendrogram of an average linkage cluster analysis applied to Euclidean distances of respondent liking of a personal product. Although two of the clusters are small, the decision was taken to truncate to establish four clusters.

clusters last in Average Linkage. We can term these 'obvious outliers'. Then there are respondents that do not belong to any particular cluster but lie somewhere between two clusters. We can term these respondents 'in-lying outliers'.

In practice with CLT liking data we have found that Average Linkage coupled with the Pearson correlation coefficient gives reasonable clusters and is resistant to obvious outliers as these show up on the left hand side of the dendrogram. However, we know that in-lying outliers may influence the formation of clusters and it is for this reason that many users are moving towards probabilistic methods such as latent class analysis. We will discuss these further in a later section on new developments.

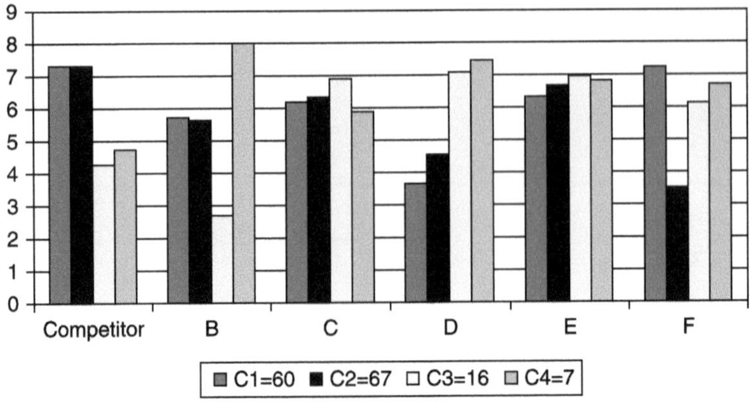

Table of brand by cluster					
Brand (Q7 usual brand)	Cluster				Total
Frequency	1	2	3	4	
Competitor	48	11	7	4	70
Own brand	12	56	9	3	80
Total	60	67	16	7	150

Fig. 21.11 The mean liking scores for the four clusters of Fig. 21.10 are given for the competitor and prototypes B to E. On the surface prototype E looks the best one to invest in. However, the new product is a roll-out and the frequency table indicates that members of Cluster 1 are primarily competitor buyers and Cluster 2 primarily includes our current product buyers. Investment in product F offered the best possibility to take share from the competitor with minimum cannibalization of current product.

With regard to Ward's method, it must be said that this algorithm certainly gives very compact clusters. However, we believe, and others have commented on this (Wakeling I, Qi Statistics, UK, personal communication), that obvious outliers, as well as in-lying outliers, tend to get included in the clusters and can be influential. This makes the solutions unstable. Ward's method also tends to seek clusters of roughly equal size and this is not necessarily true for food- or beverage-based clusters. We have often found one large cluster, with two or three lesser clusters. So our current recommendation for deterministic clustering is to use a Pearson Correlation coefficient (or standardized data) with Average Linkage and use inspection to determine the level at which to truncate. However, we do find that sometimes Ward's method gives the most satisfactory clusters.

21.3.3 Internal preference mapping

In terms of assumptions, the basic tenet of internal preference mapping is that there is a shared perceptual space of a set of foods, beverages or personal products that are perceived to a larger or smaller extent by each respondent. However, the liking of individuals may vary according to which sensory properties they like or do not like. With regard to deterministic internal preference mapping, it is also assumed that individuals focus on a single or highly correlated group of sensory attributes and rate according to how much or how little of those attributes are present. Fig. 21.12 shows an idealized version. This is the so-called vector model and vectors are shown for two individuals in Fig. 21.12. Individual 123 is rating in one direction and individual 37 is rating in a different direction. However, both are seen to be basing their ratings on a common underlying configuration of the samples. The asterisks indicate the ends of the vectors for other individuals. As this is a Principal Component solution the length of the vector is proportional to the percentage of an individual's scores that are accounted for in the picture. In this idealized example we can see that most are well away from the origin and are therefore quite well explained. In most plots we will also plot the unit circle which represents the theoretical maximum length of a vector (all variation projects into the space).

The utility of the model is well demonstrated in this figure. It appears that there are at least two separate groups of respondents, and we could consider launching two products into this marketplace. One is placed at around ten o'clock (NP1) and another at around five o'clock (NP2) as shown in Fig. 21.13.

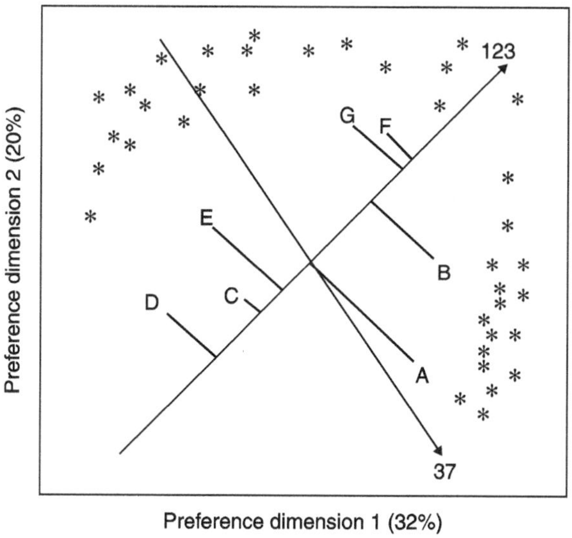

Liking scores

Cons	Sample A	B	C	D	E	F	G
123	5	4	3	2	4	6	7
37	8	7	7	6	6	6	5

Fig. 21.12 Internal preference mapping stylized solution. The positions of the products in the internal preference mapping space indicate samples that are presumably close perceptually as the respondents score the likings of these products to be the same. However, there is a large degree of difference of opinion between respondents. The vectors of increasing liking for respondent 123 and 37 are shown. The asterisks represent the arrow heads of the vectors of remaining respondents.

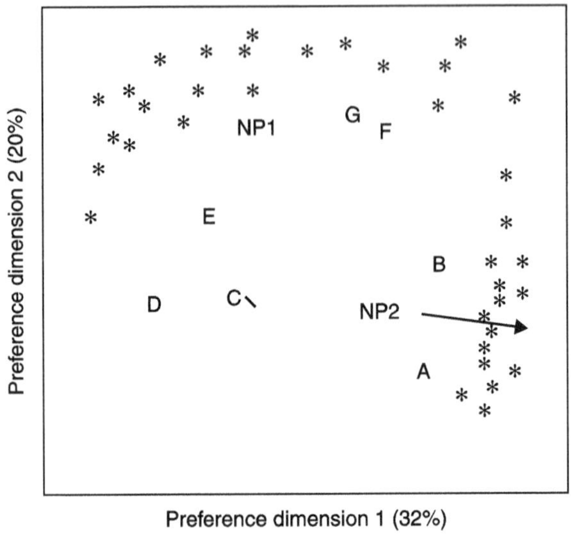

Fig. 21.13 The internal preference map from Fig. 21.12 showing the positions of two possible new products NP1 and NP2 that would compete successfully against the current products.

We obtain the plots of the samples and the vectors of the respondents by applying PCA to the liking data as shown in Fig. 21.14. It is usual to apply PCA to the liking data using the correlation matrix, which is equivalent to removing the mean and dividing by the standard deviation of each individual respondent's scores. In Fig. 21.15 a more typical example of internal preference mapping of eight varieties of Gala apples from a CLT test with 104 respondents is shown. Note that only 40% of the total variance is shown in the first two dimensions. Although a good recovery of the major sample differences, the percentage of variance is never going to be near 100% as there will always be a good deal of individual variance, as is shown by the large scatter of individual respondents. Product developer clients will not be happy with plots representing this proportion of variance. However, suppose we cluster first, find three clusters (Fig. 21.16) and then do internal preference mapping on the clusters as shown in Fig. 21.17. Now 94% of the variance is explained and it is possible to show both samples and clusters on the same bi-plot and clients will feel more comfortable with this solution. Effectively the individual variance has been compressed as part of the clustering step. It is easy to understand why this is done so often in practical applications of preference mapping.

To project the sensory variables on to these plots we simply calculate the correlation coefficient of the panel sensory means with the PC scores and then use these correlations as coordinates to plot the variables on a graph with range -1 to +1 on each axis as shown in Fig. 21.18. These points have the same properties as the points that represent respondents in the loading plots: variables that plot on to the unit circle are completely projected; variables that plot on the same line passing through

	Consumer	Fred	John	Kate	Pat	Sue	Mean
	A	9	6	2	6	2	5
Sample	B	6	9	9	2	6	6.4
	C	2	6	9	6	9	6.4
	D	6	2	2	9	6	5
	E	6	6	6	6	6	6

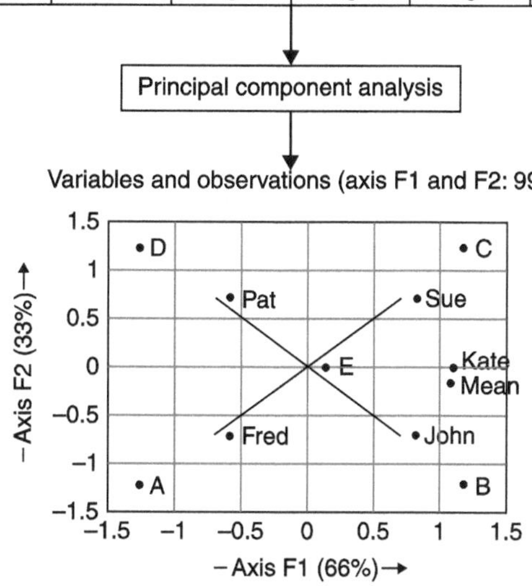

Fig. 21.14 Internal preference mapping is simply principal component analysis of the liking data. The input matrix and the output bi-plot from a principal component analysis of an idealized data set are shown.

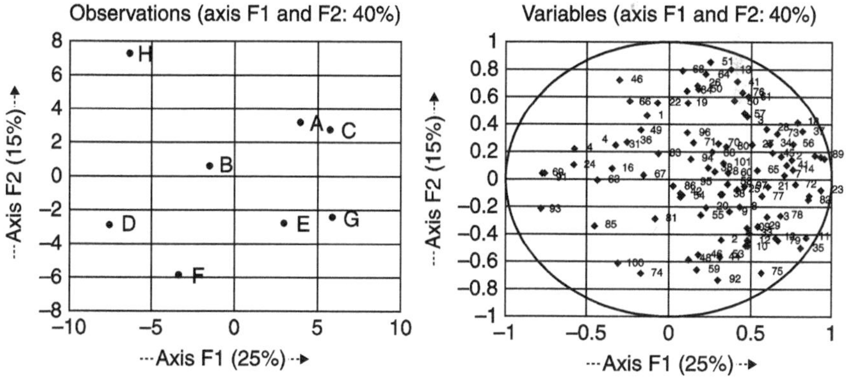

Fig. 21.15 Internal preference mapping of 8 Gala apples rated for liking by 104 respondents. The figure on the left shows the samples A to F plotted against the first two principal components. The figure on the right shows the 104 respondents. Each point is on the best fit vector of increasing liking for a respondent. The distance of each point from the origin is proportional to how much variation in a respondent's scores is projected into this two-dimensional space.

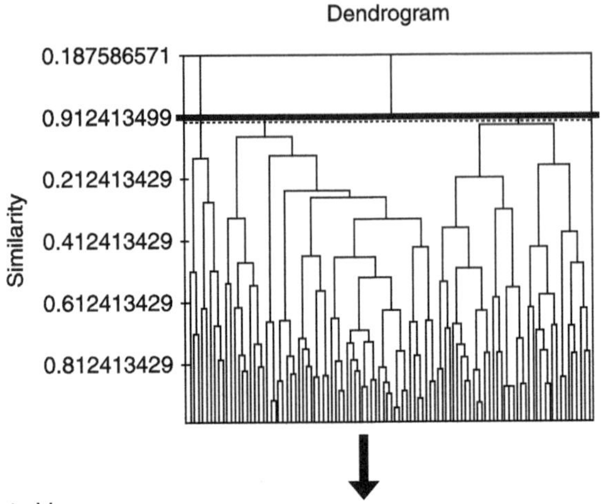

Class centroids:

Class	A	B	C	D	E	F	G	H
1	6.128	5.451	5.992	4.774	6.100	3.674	5.956	6.472
2	6.085	5.553	6.185	4.028	6.085	5.545	6.866	3.709
3	4.478	6.289	5.344	6.967	4.678	6.456	5.633	7.411

Fig. 21.16 Cluster analysis of the Gala apple data. Two main clusters are indicated and the centroids are shown.

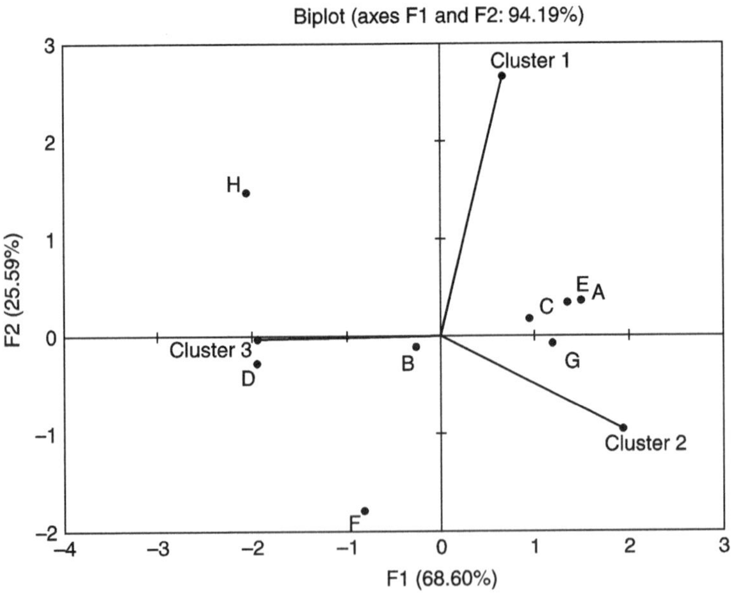

Fig. 21.17 Internal preference mapping of the cluster means (covariance matrix). Notice that the percentage of variation explained in this plot is over 94%, which will be much better received by the clients than the 40% value of Fig. 21.15. A similar configuration of samples is shown, although all the liked samples are compressed.

	Pref 1	Pref 2
• Toughness	–0.9	0.9
• Juiciness	0.8	0.8
• Breaks easily	0.4	–0.7
• Large bits	–0.7	–0.4

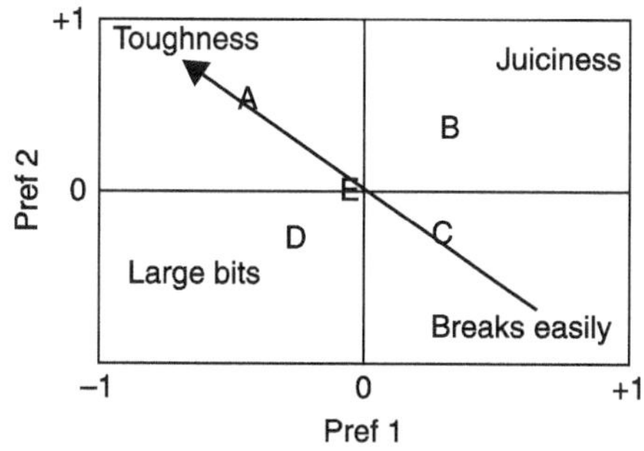

Fig. 21.18 Superimposing the sensory variables on to an internal preference map using correlations with the first two preference dimensions (PCs). The table on the left shows the correlations of four sensory attribute means with the first and second preference scores. The figure on the right indicates the position on rescaled plot with axis endpoints plus and minus one. The arrow indicates a direction of increasing Toughness.

the origin are completely correlated in the space that is shown. Similarly, in a bi-plot such as that shown in Fig. 21.18, we can infer what needs to be done to a product to move it around the space. For example suppose we are currently producing product C in Fig. 21.18 and marketing asks us to move it to match competitor A. We can see that we need to increase *toughness*, decrease the capacity to *break easily* and leave *juiciness* and *large bit* sensory scores the same (as the directions of these attributes from the origin are at right angles to the desired direction of change).

We can use the same rationale to project in the demographic variables such as gender, age, region of origin, etc. We simply calculate the mean scores given to each sample by one level of the variable, such as females, and then correlate this into the plot using the same rationale as in Fig. 21.18.

21.3.4 Identifying the sensory properties of a desired product: reverse regression

The internal preference map with sensory and demographic variables superimposed gives a valuable view of the products from the marketplace that have been selected for trial and sent out to respondents for CLT testing, and the sensory correlates and useful demographic variables. Using cluster means instead of individual respondents simplifies the bi-plot and increases the percentage variance explained. A likely outcome of marketing personnel viewing and understanding this plot is that they identify a position in the space to which they would like to move a current product. The question then arises as to what the sensory properties of this product would be. This problem is solved using reverse regression.

We will illustrate and explain the reverse regression procedure using a CLT test conducted on mozzarella cheese. This data has been reported in Pagliarini *et al.* (1997). The results of the internal preference mapping are shown in Fig. 21.19. The loading plot on the right is a fine example of segmentation. Respondents who plot on the positive side of the first PC like the cow-based cheeses and dislike the

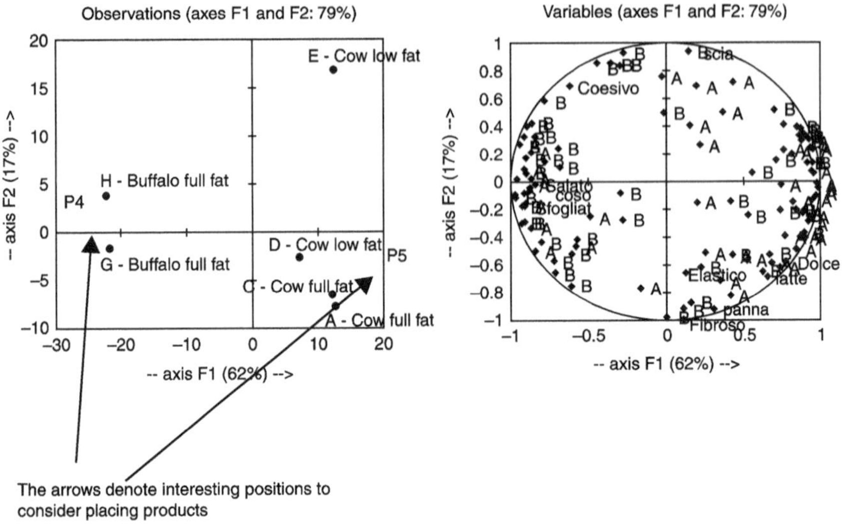

Fig. 21.19 Identifying possible positions that winning products might occupy in the internal preference space. This preference map of Mozzarella cheeses indicates a two-segment population. Product P4 should give a higher liking score for that segment that likes buffalo-based cheese. Product P5 should give a higher liking score to the other segment that prefers cow-based cheese. Placing products as indicated would give a score of −25 for P4 on the first dimension and +20 for P5. On the second dimension P4= 0 and P5 = −5. Note in the figure on the right that the vectors of 12 sensory attributes have been projected.

(authentic) cheeses made from water buffalo. Respondents who plot on the negative side of the first PC have exactly the opposite opinion. Suppose that we decide to enter this marketplace with two products that will score best in each segment. In Fig. 21.19 we have denoted two possible product positions P4 (−25, 0) and P5 (20, −5). In practice one might decide to make a few more test products in each region but for the purpose of illustration we will restrict to these two.

The regression set-up is shown in Table 21.6. The set of sensory panel mean scores is shown on the right and forms the dependent matrix Y. Labelling the first two PCs of the preference mapping PCA as F1 and F2, we see that F1 and F2 are the independent variables in the regression. For each sensory variable we wish to set up a regression model to predict from F1 and F2, and then we can predict the sensory scores that would be obtained if we set (F1, F2) as P4 (−25, 0) or P5 (20, −5). The results are shown together with the significance of the regression in Fig. 21.20. It is important not to make predictions where the regression is not significant, so in this case we do not give much credence to the predicted values for *Elastico* as it is not significant.

21.3.5 External preference mapping

External preference mapping is level 3 of the hierarchy of preference mapping methods. Level 1 – cluster analysis – assumes only that the respondents fall

Table 21.6 The regression set-up to predict the sensory properties of P4 and P5

F1	F2		Acido	Coesivo	Dolce	Elastico	Fibroso	Yogurt	Liscia
12.573	−7.815	A–Cow full fat	2.87	3	5.27	4.97	7.47	3.4	5.63
12.368	−6.576	C–Cow full fat	3	3.2	4.87	5.7	7.13	2.83	6.27
7.055	−2.734	D–Cow low fat	2.9	3.43	5.13	4.23	6.73	3.27	5.73
12.375	16.806	E–Cow low fat	3.6	4	3.57	4.13	3.77	2.87	8.03
−21.650	−1.709	G–Buffalo full fat	4.83	4.13	3.27	4.63	6.03	5.07	6.03
−22.721	2.028	I–Buffalo full fat	5.1	3.77	2.97	4.73	5.87	5.1	6.47
−25.000	0.000	P4	5.09	3.99	3.04	4.63	5.96	5.25	6.18
20.000	−5.000	P5	2.46	3.09	5.43	5.02	7.06	2.62	6.04
		Average	3.72	3.59	4.18	4.73	6.17	3.76	6.36
		RegCoeffs F1	−0.054872	−0.016063	0.045757	0.004154	0.008217	−0.05958	0.007055
		RegCoeffs F2	0.031163	0.034753	−0.066797	−0.041866	−0.146301	−0.01049	0.091888

F1	F2		Salato	Latte	Panna	Sfogliat	Succoso
12.753	−7.815	A–Cow full fat	4.2	5.43	4.27	5.43	5.3
12.368	−6.576	C–Cow full fat	3.57	5	3.73	4.5	4.66
7.055	−2.734	D–Cow low fat	3.3	5.3	3.8	5.4	5.47
12.375	16.806	E–Cow low fat	3.6	3.73	2.3	4.57	4.7
−21.650	−1.709	G–Buffalo full fat	4.7	3.63	3.17	6.23	6.23
−22.721	2.028	I–Buffalo full fat	6	3.83	3	6.87	7.33
−20.000	0.000	P4	5.21	3.85	3.13	6.47	6.69
20.000	−5.000	P5	3.26	5.45	3.98	4.62	4.59
		Average	4.23	4.49	3.38	5.50	5.62
		RegCoeffs F1	−0.048894	0.032057	0.012207	−0.048297	−0.053638
		RegCoeffs F2	−0.001703	−0.063687	−0.072189	−0.016701	−0.009499

Note: F1 and F2, the internal preference mapping score variables are the independent variables and the sensory attribute scores are the dependent (Y) variables. We then regress each of the sensory variables in turn on the two independent variables and use the model to obtain the predicted values for P4 and P5 shown.

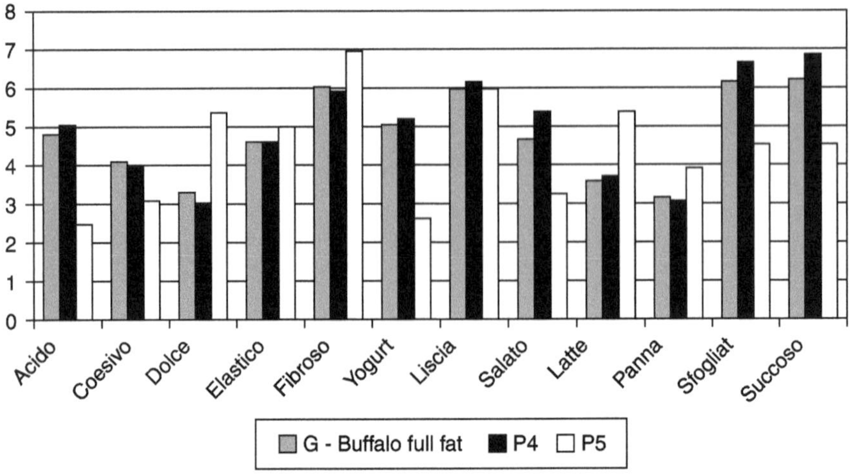

Fig. 21.20 Predicted sensory properties of P4 and P5 plotted against buffalo sample G. This chart shows that P4 (the new buffalo cheese) is predicted to score higher than buffalo on Acido, Salato, Sfogliat and Succoso. P5 (the new cow fat cheese) is very different from the buffalo on these parameters.

into different segments. Level 2 – internal preference mapping – does not necessarily assume different groups but does assume a shared perceptual space. Level 3 – external preference mapping – assumes a shared perceptual space but additionally that we have an external definition of that perceptual space. This is usually the first few principal components of the sensory panel scores, but could also be the first few dimensions of a repertory grid study (Baxter *et al.*, 1998; Carbonell *et al.*, 2007), or the dimensions of multidimensional scaling trial of pairwise difference cores, or simply the perceptual scores from the consumer responses in the CLT trial.

In methodological terms, external preference mapping is most accurately described as principal components response surface methodology (RSM) since effectively we conduct a RSM analysis of overall liking, cluster overall liking, or separately on each respondent's liking scores. Figure 21.21 and Fig. 21.22 illustrate the concepts very well; the sensory data, assumed here to be two-dimensional for illustrative purposes only, form the base and the liking of an individual respondent is plotted in the vertical dimension. It is then possible to fit a wide range of possible surfaces to each respondent as shown on the left of Fig. 21.22. The top surface is an ideal point model. The middle surface is an anti-ideal point model. The bottom surface is a so-called saddle surface with an ideal point model in one sensory direction and a vector in the other.

There is thus no need for specialized software to do external preference mapping. Any multiple regression package will do. The most general form of the model for a two-dimensional sensory space with components F1 and F2 is:

$$\text{Liking} = a + b1 \times F1 + b2 \times F2 + C1 \times F1^2 + C2 \times F2^2 + d \times F1 \times F2 \qquad [21.1]$$

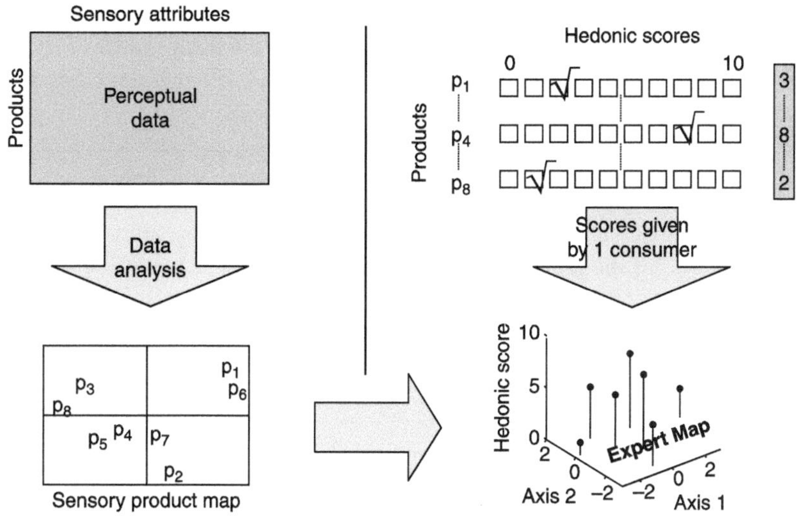

Fig. 21.21 External preference mapping schemata I. The perceptual data on the left is projected down into fewer principal components. These are placed at the base of Fig. 21.21. on the right. Each consumer's scores can then be plotted and a surface visualized.

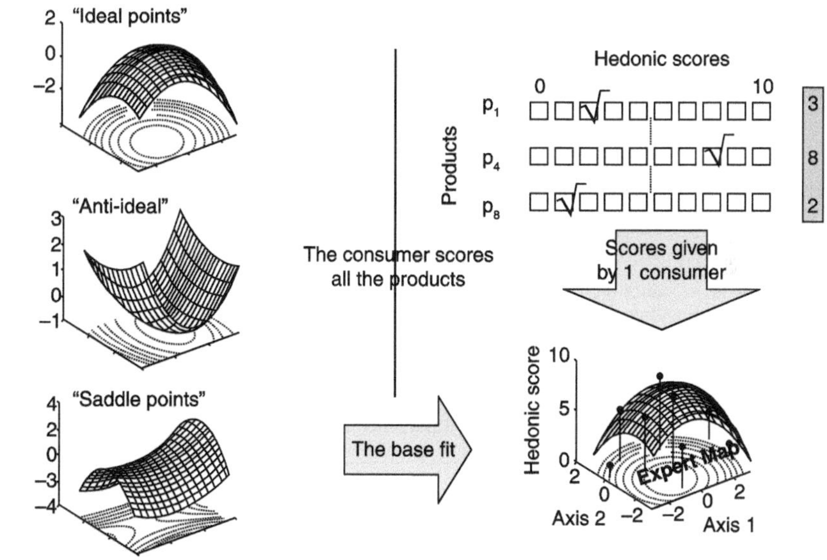

Fig. 21.22 External preference mapping schemata II. The three possible ideal point models that may be used to explain consumer hedonic scores in terms of sensory data are shown. Ideal point models represent consumers where their ideal point is within the current space of the products. Anti-ideal point models represent consumers with a clear idea of what they do not like. Saddle-point models represent consumers who have an ideal point in one of the perceptual directions but have a vector model (ideal point outside the current space) in the other.

The full form of this model is termed the quadratic surface model. However, since F1 and F2 are almost always principal components they are independent of each other and the sum of their cross products is zero. We thus arrive at the elliptical ideal point model shown in Fig. 21.23 and written as:

$$\text{Liking} = a + b1 \times F1 + b2 \times F2 + C1 \times F1^2 + C2 \times F2^2 \quad [21.2]$$

If C1 = C2 we obtain the circular ideal point model shown in Fig. 21.23. Note that if C1 and C2 are positive then it is an ideal point model but if they are negative it is an anti-ideal point model. If C1 and C2 are not significant we arrive at the vector model similar to that used in internal preference mapping. There is thus a simple hierarchy of models moving from vector to circular and elliptical ideal point models that can be tested in an analysis of variance framework.

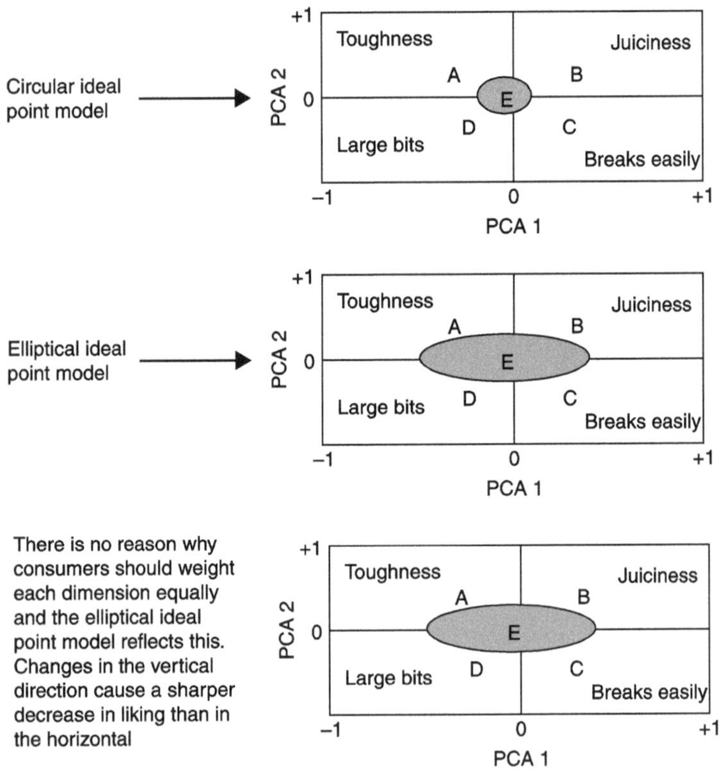

The mathematical expression for a circular idea point model

$\text{Liking} = b0 + b1(X1) + b2(X2) + b3(X1)^2 + b4(X2)^2$

X1 = PCA1 X2 = PCA2

b0 = intercept b1, b2, b3, b4 regression coefficients

The maxima or minima occurs when the first and second derivatives are zero

Quadratic surface model

The general quadratic model allows for covariance between the X variables

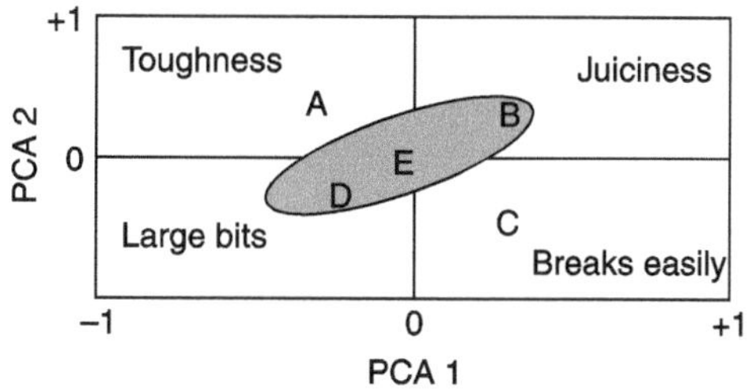

The mathematical expression for a circular idea point model

$$\text{Liking} = b0 + b1(X1) + b2(X2) - b3(X1)^2 + b4(X2)^2 - b5(X1)(X2)$$

$X1 = PCA1$ $X2 = PCA2$

b0 = intercept b1, b2, b3, b4, b5 regression coefficients

The maxima or minima occurs when the first and second derivatives are zero

Fig. 21.23 A circular ideal model looking down on to the sensory space from above. Shaded region is the ideal area for this consumer. As the product moves in any direction away from this area it is liked less. There is no reason why consumers should weight each dimension equally and the elliptical ideal point model reflects this. Changes in the vertical direction cause a sharper decrease in liking than in the horizontal.

The usual method to represent the models on the sensory space is shown in Fig. 21.24. Vector models are represented by an arrow in the direction of increasing liking, where the length of the arrow is proportional to the goodness of fit. Ideal point models are used to calculate the coordinates of the optima in the sensory space and the ideal point is plotted together with an indication of the model type. Thus no brackets indicates an ideal point model, (-) indicates an anti-ideal point model and (±) indicates a saddle point.

21.3.6 Fitting clusters, individuals and contour plotting

After cluster analysis the fitting task is quite reasonable and it is usual to fit the overall liking mean variable as well. If segmentation is severe it is likely that this will not fit significantly. It is also possible to project in the demographic variables on to the plot for interpretive purposes as was done for the internal preference maps.

One problem that will often be encountered is that a cluster will not be fitted significantly by any of the models. One possible reason for this is that the sensory space on which you are fitting may not be representative of the perceptual space that is being used by this cluster. For example, suppose you have done the sensory principal components analysis on data that includes appearance, texture, flavour

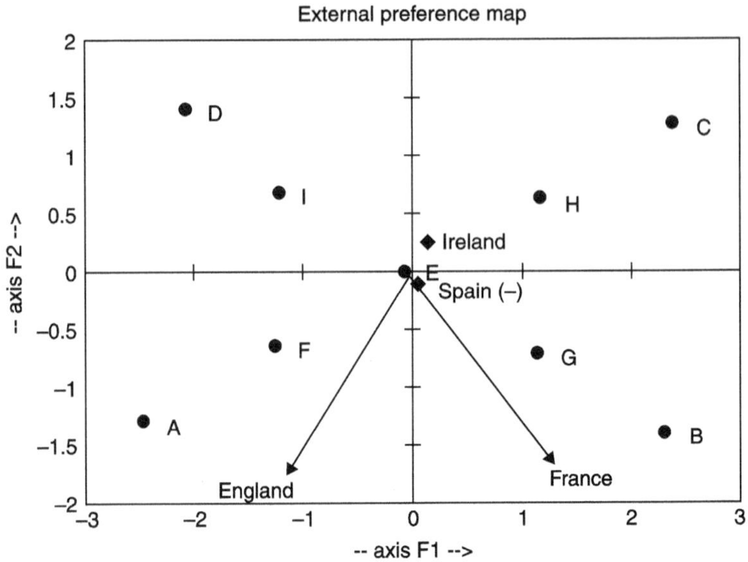

Fig. 21.24 Conventions for displaying the various preference mapping models on to the map. In this map of various countries CLT results on tasting prototypes of a confectionery product, France and England are vector models. Ireland is an ideal point model with the diamond marking the position of the ideal point. Spain is an anti-ideal point.

and aroma data. Typically, assessors are more comfortable with appearance and texture attributes, and these can be highly correlated and tend to dominate the principal components analysis. However, if the non-fitted cluster is only sensitive to differences in aroma and flavour, then repeating the PCA on this subset of attributes will produce a sensory space that is likely to lead to some well-fitting models. It is our current practice to produce spaces separately for appearance, texture and mouth-feel, flavour, aroma and aftertaste and fit each of the clusters to each of these spaces. This gives the maximum possibility to find good fits and makes a lot more interpretive sense. Another possible reason for a non-fitting space is that the clusters are concentrating on a few attributes and a re-weighted PCA is required. This is discussed further by Jaeger *et al.* (1998) who develop a technique of re-weighting based on the correlations of the sensory attributes with the internal preference map space.

Danzart (1998) introduced the idea of defining contours on the sensory space showing the number of consumers who are predicted on the basis of their initially fitted model to score greater than a fixed score, called a cut-off score. This can be very useful to identify regions of suitable prototypes and gives product developers more room in which to innovate, pare costs, etc. On the 9-point hedonic scale a suitable cut-off might be 6.5 to 7.5. Of course not all consumers will have a significant model, so this can mean that many respondents' scores are not used, even when using a liberal $p < 0.2$ significance criterion. Using the cluster models and multiplying by the number of people in the cluster is sometimes more convincing.

21.3.7 Internal versus external preference mapping

The basis of the two approaches is given in Fig. 21.25. The major advantage of internal is that the space is defined from the consumer responses. Many CEOs and marketing personnel would be deeply suspicious of a space defined by ten women who come in three days a week part-time, which is exactly how the external space is often defined. We hope that the two spaces are identical because there are many technical advantages to using external. We have a more sophisticated hierarchy of models than the vector-based model used by PCA on liking. We can build up contours of liking as discussed in the previous paragraph. If a new product comes on to the market, we can have it profiled by the sensory panel and see where it fits into the sensory space and predict its liking overall or within cluster without doing a CLT. If the spaces are not identical, we should attempt to understand why and search for a re-weighting (PCA on the covariance matrix?) as discussed by Jaeger *et al.* (1998).

21.3.8 Preference mapping of lager beers

Guinard *et al.* (2001) reported a preference mapping study of 24 US and imported lager beers, using data from 170 consumers. The beers were first tasted blind in three sessions on different days, and in the same way with the brand and price

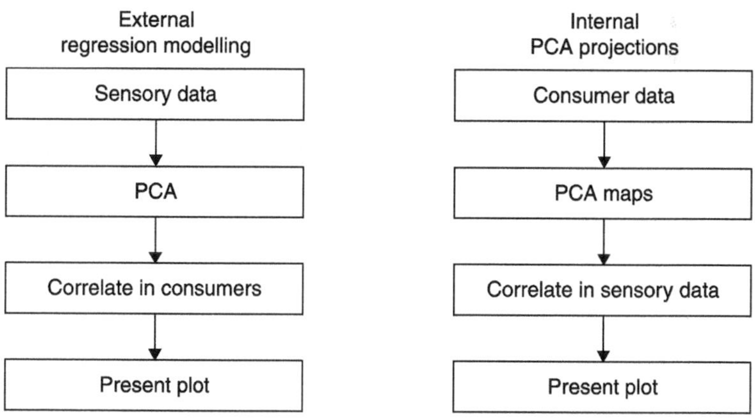

Fig. 21.25 Flow charts denoting the sequence of operations for internal and external preference mapping. The key difference is in the source of the map on which the clusters or respondents will be projected.

displayed. As would be expected, internal preference mapping showed differences in preferences, with some consumers preferring the domestic beers and others the imported ones. Again as might be expected there were significant changes in hedonic scores between the blind and informed conditions, particularly for younger consumers. In an external preference map, 75% of consumers were well-fitted by one of the models described above. The authors attributed this relatively high success rate to the large number of samples, and to the wide variation in sensory properties among the samples.

21.4 Recent developments in preference mapping

21.4.1 Probabilistic modelling

Probabilistic modelling examines the distribution of points in space and identifies areas where there are high concentrations of individuals in the space. These areas, sometimes called latent classes, are defined on the basis of distributional assumptions, and the probability that each individual belongs to each of the classes is calculated. This effectively reveals the in-lying outliers that lie within the space but are not close to any latent class. Although theoretically superior to hierarchical and non-hierarchical cluster analysis, the experience of the authors with the Latent Gold software (Statistical Innovations, 2011) has not been convincing enough for us to switch. Courcoux and Chavanne (2001) have discussed the approach in detail, including a version of external preference mapping. MacKay and Zinnes (1995) developed a version of probabilistic multidimensional scaling that is based on ideal point assumptions and recovers a map of products. The software is freely available at the time of writing (www.proscal.com). In a wine tasting example, Mackay (2005) shows that the ideal point recovery is superior to deterministic multidimensional scaling (this is recovery of a space based on distances between points and may not be the same as PCA-based internal preference mapping). The sensory data that is available is fitted separately using a multidimensional scaling solution based on distances between the products and then a common space of both the sensory and the consumer space is fitted. In an interesting development, Mackay (2006) demonstrates how to use the model with simulation methods to estimate how often a new product would be liked better against a set of products. It must be emphasized that this is a simulation of how well the product would do in a CLT when tested blind and is not an estimate of how well the branded product would do in a marketplace. Nevertheless this is a very useful development and should be adopted in all preference mapping application software. Ennis and Johnson (1994) developed a more refined model that allows for individuals to have wandering ideal points, based on earlier work by De Soete *et al.* (1986). A commercial package directly targeted at the preference mapping application is available with a very nice graphical interface (www.ifpress.com).

21.4.2 Partial least squares (PLS) approaches

Partial least squares (PLS) is an algorithm that is applied to estimate a model based on a set of variables in a matrix X that will predict another set of variables based in set Y measured on the same samples. It is thus a kind of multivariate multiple regression package. An obvious way to do this would be to do a PCA on the X set first and then regress each Y separately on the most important PCs. This method, called Principal Components regression, is widely used but in theory suffers from the fact that the PCs are derived based on the variance in X and do not take any of the Y variation into account. In addition the covariance or correlation between the Y variables is not taken into account. Thus with 100 Y variables that are highly correlated one will have 100 models that are probably all very similar, but this is not explicitly revealed in the analysis. PLS extracts pairs of directions, one from the X set and one from the Y set, that display maximum correlation or covariance and explain as much variation as possible in both sets. The correlations between the variables in both sets are thus acknowledged. Successive pairs are extracted that are independent of previous pairs (i.e. PLS1 direction in X is uncorrelated with PLS2 direction in X). The number of pairs extracted is determined by seeing how well the next pair contributes to the prediction of samples when they are left out of the analysis. With no improvement in prediction the process terminates. The set-up for a preference mapping analysis is given in Fig. 21.26 and the solution for the mozzarella cheese application discussed earlier is given in Fig. 21.27. This is almost identical to the internal preference mapping solution. Helgesen *et al.* (1997) were among the first to apply PLS to a preference mapping application. The method is a linear model so it is

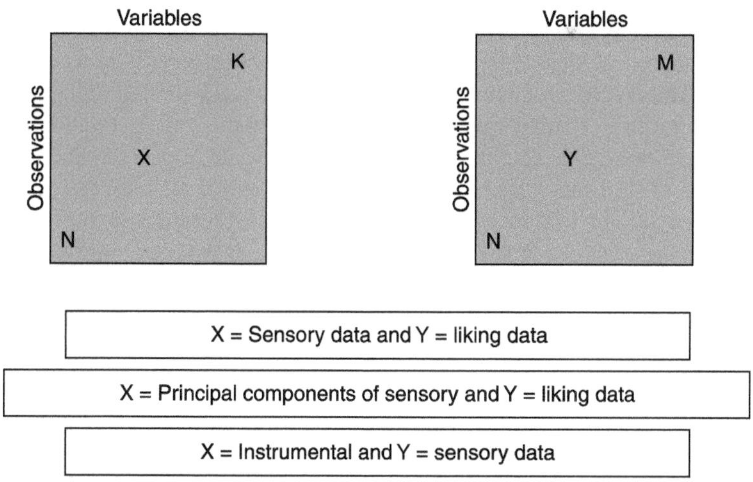

Fig. 21.26 Possible PLS data setups. X is the data that is going to be used to predict Y. With external preference mapping the X data are the sensory panel mean data and Y is the matrix of individual respondent liking scores.

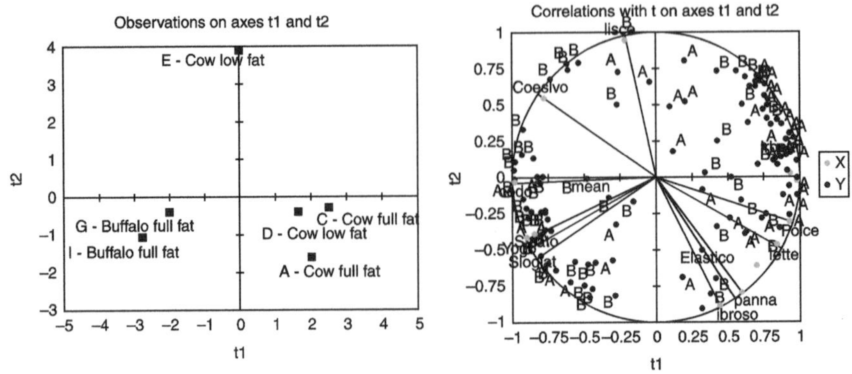

Fig. 21.27 The sample and loading plots of the first two PLS dimensions obtained by fitting the liking data Y to the sensory panel mean data X of the mozzarella data.

effectively fitting a vector model of each person to the sensory data without the need to do a preliminary PCA of the sensory data. If squared terms for the sensory data are added into the X set, then the analysis effectively becomes a quadratic model and ideal points can be incorporated. In our experience this is not often done. PLS is very useful for reverse regression as it is possible to estimate a model for all the sensory attributes in one operation and prediction is quick and easy to set up. It is not so easy to spot which attributes are not being fitted significantly and attention must be paid to this.

There are a number of developments to the basic PLS model that are potentially very useful. L-shaped PLS (Martens *et al.* 2005) uses the set-up shown in Fig. 21.28 and is essentially using the consumer demographic variables as part of the fitting process as well as the sensory variables. This is an extremely attractive concept to marketing personnel, who are often frustrated that the segments extracted by cluster analysis or latent class techniques bear no relation to any demographic cells. See Plaehn and Lundahl (2005) for a preference mapping application. At the time of writing the output given by the technique is complex, and simpler visuals will be needed to convince marketing and research and development personnel that this is a useful development. We need more experience with this technique to find a really convincing example where the fitting process produces a different solution to that obtained by simply fitting the sensory and super-imposing the demographic cell mean attributes (eg female vector and male vector) on to the map.

A second extension of PLS is to use it to model the Y set when there are different blocks of data in the X space. For example, in Fig. 21.29 from Tenenhaus *et al.* (2005) there is a sensory set and a physico-chemical set in the X and we can consider models that might represent how these two components contribute. Although in theory we can apply the structural equation modelling framework developed by Jöreskog and Sörbom (1989), in practice we need over 200 samples

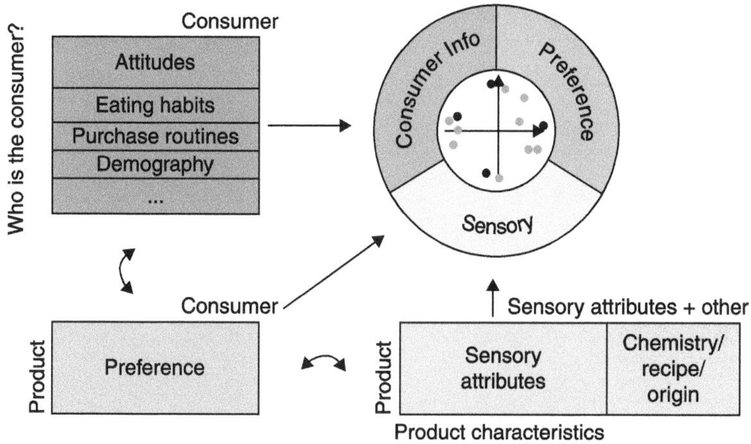

Fig. 21.28 The set-up for L-shaped PLS. In addition to modelling the relationships between liking and sensory or physical chemical, the possible relations between demographic and other personality variables with liking are explored. (Source: presented by Frank Westad of Matforsk, Westad, 2004.)

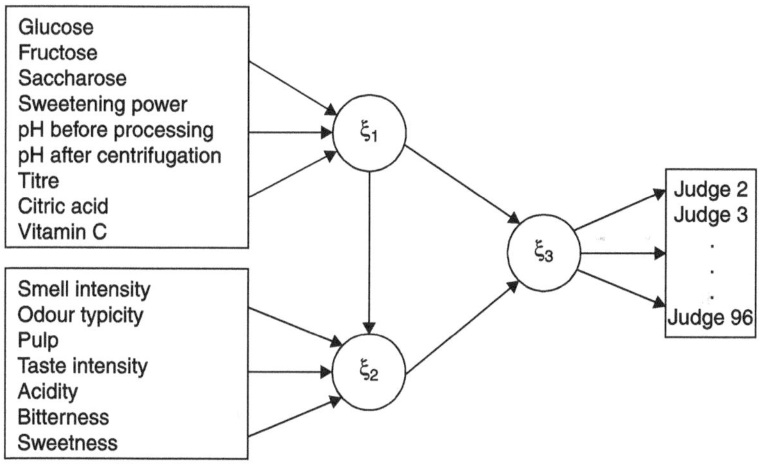

Fig. 21.29 A multi-block set-up in which the X set is composed of two blocks of data. The latent variables ξ that are considered to be underlying the physical relations are connected as shown. (Source: Tenenhaus *et al.* 2005.)

for this and we rarely have so much data so multi-block PLS can be very useful here. Wold *et al.* (1996) proposed the extension to hierarchical multi-block models (Fig. 21.30) and the present authors believe this to be a most important step forward as now we can start to consider how the senses interact and in what order they come together to influence liking.

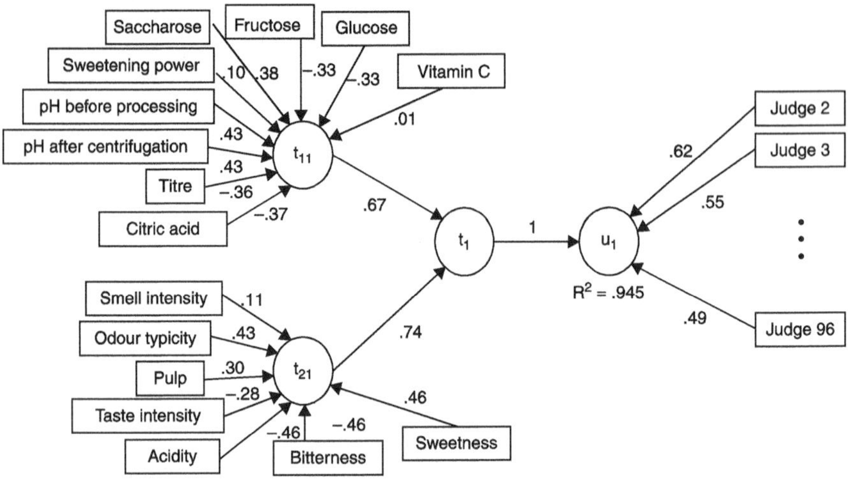

Fig. 21.30 An estimated hierarchical multi-block PLS model. Non-significant loadings are in italics. (Source: Tenenhaus *et al.* 2005.)

21.5 Sources of further information and advice

As mentioned in the text, we have attempted to collect as many publications on preference mapping as we can from *Food Quality and Preference, Journal of Sensory Studies* and other food journals. Hopefully the reader will find an application close to their interest. There are sections in many sensory books including Lawless and Heymann (1998).

ARDITTI S (1997), 'Preference mapping: A case study', *Food Quality and Preference*, **8**, 323–327.

BÁRCENAS P, PÉREZ DE SAN ROMÁN R, PÉREZ ELORTONDO F J and ALBISU M (2001), 'Consumer preference structures for traditional Spanish cheeses and their relationship with sensory properties', *Food Quality and Preference*, **12**, 269–279.

BOWER JA and SAADAT M A (1998), 'Consumer preference for retail fat spreads: an olive oil based product compared with market dominant brands', *Food Quality and Preference*, **9**, 367–376.

BRUECKNER B, SCHONHOF I, SCHROEDTER R and KORNELSON C (2007), 'Improved flavour acceptability of cherry tomatoes. Target group: Children', *Food Quality and Preference*, **18**, 152–160.

CASPIA E L, COGGINS M W, YOON Y and WHITE C H (2006), 'The relationship between acceptability and descriptive sensory attributes in cheddar cheese', *Journal of Sensory Studies,* **21**, 112–127.

CHO H-Y, CHUNG S-J, KIM H-S and KIM K-O (2005), 'Effect of sensory characteristics and non-sensory factors on consumer liking of various canned tea products', *Journal of Food Science*, **70**, 532–538.

ELMORE J R, HEYMANN H, JOHNSON J and HEWETT J E (1999), 'Preference mapping: relating acceptance of "creaminess" to a descriptive sensory map of a semi-solid', *Food Quality and Preference*, **10**, 465–475.

FABER N, MOJET J and POELMAN A A M (2003), 'Simple improvement of consumer fit in external preference mapping', *Food Quality and Preference*, **14**, 455–461.

FAYE P, BRÉMAUD D, TEILLET E, COURCOUX P, GIBOREAU A and NICOD H (2006), 'An alternative to external preference mapping based on consumer perceptive mapping', *Food Quality and Preference*, **17**, 604–614.

GEEL L, KINNEAR M and DE KOCK H L (2005), 'Relating consumer preferences to sensory attributes of instant coffee', *Food Quality and Preference*, **16**, 237–244.

GREENHOFF K and MACFIE H (1994), 'Preference mapping in practice', In: H.J.H. MacFie and D.M.H. Thomson, (Editors), *Measurement of food preferences*, Blackie Academic & Professional, London, pp 137–166.

HERSLETH M, BERGGREN R, WESTAD F and MARTENS M (2005), 'Perception of bread: a comparison of consumers and trained assessors', *Journal of Food Science*, **70**, S95–S101.

HEYD B and DANZART M (1998), 'Modelling consumers' preferences of coffees: evaluation of different method', *Lebensmittel-Wissenschaft und-Technologie*, **31**, 607–611.

HOUGH G and SÁNCHEZ R (1998), 'Descriptive analysis and external preference mapping of powdered chocolate milk', *Food Quality and Preference*, **9**, 197–204.

HUON DE KERMADEC F H, DURAND J F and SABATIER R (1997), 'Comparison between linear and nonlinear PLS methods to explain overall liking from sensory characteristics', *Food Quality and Preference,* **8**, 395–402

JAEGER S R, WAKELING I N and MACFIE H J H (2000), 'Behavioural extensions to preference mapping: the role of synthesis', *Food Quality and Preference*, **11**, 349–359.

JAEGER S R, LUND C M, LAU K and HARKER F R (2003), 'In search of the "Ideal" Pear (*Pyrus* spp.): results of a multidisciplinary exploration', *Journal of Food Science*, **68**, 1108–1117.

JAEGER S R, ROSSITER K L, WISMER W V and HARKER F R (2003), 'Consumer-driven product development in the kiwifruit industry', *Food Quality and Preference*, **14**, 187–198.

KÄLVIÄINEN N, SALOVAARA H and TUORILA H (2002), 'Sensory attributes and preference mapping of muesli oat flakes', *Journal of Food Science*, **67**, 455–460.

LÊ S and LEDAUPHIN S (2006), 'You like tomato, I like tomato: Segmentation of consumers with missing values', *Food Quality and Preference*, **17**, 228–233.

MARTÍNEZ C, SANTA CRUZ J M, HOUGH G and VEGA M J (2002), 'Preference mapping of cracker type biscuits', *Food Quality and Preference*, **13**, 535–544.

MATTILA V (2003), 'Semantic analysis of speech quality in mobile communications: descriptive language development and mapping to acceptability', *Food Quality and Preference*, **14**, 441–453.

MCEWAN J A (1996), 'Preference mapping for product optimization', In: Næs, T. and Risvik, E., (Editors), *Multivariate analysis of data in sensory science*, Elsevier Science BV, Amsterdam.

MCEWAN J A and THOMSON D M H (1989), 'The repertory grid method and preference mapping in market research: A case study on chocolate confectionery', *Food Quality and Preference*, **1**, 59–68.

MEULLENET J.-F, XIONG R, MONSOOR M A, BELLMAN-HOMER T, DIAS P *et al.* (2002), 'Preference mapping of commercial toasted white corn tortilla chips', *Journal of Food Science*, **67**, 1950–1957.

MONTELEONE E, FREWER L, WAKELING I and MELA D J (1998), 'Individual differences in starchy food consumption: The application of preference mapping', *Food Quality and Preference*, **9**, 211–219.

OKAYASUAND H and NAITO S (2001), 'Sensory characteristics of apple juice evaluated by consumer and trained panels', *Journal of Food Science*, **66**, 1025–1029.

PRESCOTT J, YOUNG O and O'NEILL L (2001), 'The impact of variations in flavour compounds on meat acceptability: a comparison of Japanese and New Zealand consumers', *Food Quality and Preference*, **12**, 257–264.

RICHARDSON-HARMAN N, STEVENS R, WALKER S, GAMBLE J, MILLER M *et al.* (2000), 'Mapping consumer perceptions of creaminess and liking for liquid dairy products', *Food Quality and Preference*, **11**, 239–246.

RIVIÈRE P, MONROZIER R, ROGEAUX M, PAGÈS J and SAPORTA G (2006), 'Adaptive preference target: Contribution of Kano's model of satisfaction for an optimized preference analysis using a sequential consumer test', *Food Quality and Preference*, **17**, 572–581.

SERRANO-MEGÍAS M, PÉREZ-LÓPEZ A J, NÓÑEZ-DELICADO E, BELTRÁN F and LÓPEZ-NICOLÁS J M (2005), 'Optimization of tropical juice composition for the Spanish market', *Journal of Food Science*, **70**, 28–33.

SUWANSRI S, MEULLENET J-F, HANKINS J A and GRIFFIN K (2002), 'Preference mapping of domestic/imported jasmine rice for U.S.-Asian consumers', *Journal of Food Science*, **67**, 2420–2431.

THOMPSON J L, DRAKE M A, LOPETCHARAT K and YATES M D (2004), 'Preference mapping of commercial chocolate milks', *Journal of Food Science*, **69**, 1365–2621.

THYBO A K, KÜHN B F and MARTENS H (2004), 'Explaining Danish children's preferences for apples using instrumental, sensory and demographic/behavioural data', *Food Quality and Preference*, **15**, 53–63.

VIGNEAU E, and QANNARI E M (2001), 'Segmentation of consumers taking account of external data. A clustering of variables approach', *Food Quaity and Preference*, **13**, 515–521.

VIGNEAU E, QANNARI E M, PUNTER P H and KNOOPS S (2001), 'Segmentation of a panel of consumers using clustering of variables around latent directions of preference', *Food Quality and Preference*, **12**, 359–363.

WESTAD F, HERSLETH M and LEA P (2004), 'Strategies for consumer segmentation with applications on preference data', *Food Quality and Preference*, **15**, 681–687.

XIONG R and MEULLENET J F (2004), 'Application of multivariate adaptive regression splines (mars) to the preference mapping of cheese sticks', *Journal of Food Science*, **69**, 1365–2621.

YACKINOUS C, WEE C and GUINARD J-X (1999), 'Internal preference mapping of hedonic ratings for Ranch salad dressings varying in fat and garlic flavour', *Food Quality and Preference*, **10**, 401–409.

YOUNG N D, DRAKE M, LOPETCHARAT K and MCDANIEL M (2004), 'Preference mapping of cheddar cheese with varying maturity levels', *Journal of Dairy Science*, **87**, 11–19.

21.6 Acknowledgement

Thanks are due to Anne Hasted of QI Statistics for Figs 21.3 and 21.4.

21.7 References

BALL R D (1997), 'Incomplete designs for the minimization of order and carry-over effects in sensory analysis', *Food Quality and Preference*, **8**, 111–118.

BAXTER I A, JACK F R and SCHRODER, M J A (1998), 'The use of repertory grid method to elicit perceptual data from primary school children', *Food Quality and Preference*, **9**, 73–80.

BEN SLAMA M, HEYD B, DANZART M and DUCAUZE C (1998), 'Plans D-optimaux: une stratégie de réduction du nombre de produits en cartographie des préférences', *Sciences des Aliments*, **18**, 471–483.

CALLIER P and SCHLICH P (1997), 'La cartographie des préférences incomplétes – Validation par simulation', *Sciences des Aliments*, **17**, 155–172.

CARBONELLA L, IZQUIERDO L and CARBONELL I (2007), 'Sensory analysis of Spanish mandarin juices. Selection of attributes and panel performance', *Food Quality and Preference*, **18**, 329–341.

CARROLL J D (1972), 'Individual differences and multidimensional scaling', In R.N. Shepard, A.K. Romney, and S. Nerlove (Eds.), *Multidimensional scaling: Theory and applications in the behavioral sciences* (Vol. 1, pp. 105–155), Seminar Press, New York.

COURCOUX P and CHAVANNE P C (2001), 'Preference mapping using a latent class vector model', *Food Quality and Preference*, **12**, 369–372.

DAILLANT-SPINNLER B, MACFIE H, BEYTS P and HEDDERLEY D (1996), 'Relationships between perceived sensory properties and major preference directions of 12 varieties of apples from the Southern Hemisphere', *Food Quality and Preference*, **7**, 113–126.

DANZART M (1998), 'Cartographie des préférences', In: SSHA, (Editor), *Evaluation Sensorielle: Manuel Méthodologique*, Lavoisier, Paris.

DE SOETE G, CARROLL J D and DESARBO W S (1986), 'The wandering ideal point model: a probabilistic model multidimensional unfolding model for paired comparisons data', *Journal of Mathematical Psychology*, **32**, 449–465.

EARTHY P J, MACFIE H J H and HEDDERLEY D (1997), 'Effect of question order on sensory perception and preference in central location trials', *Journal of Sensory Studies*, **12**, 215–237.

ENNIS D M and JOHNSON N L (1994), 'A general model for preferential and triadic choice in terms of central F distributions', *Psychometrika*, **59**, 91–96.

GUINARD J-X, UOTANI B and SCHLICH P (2001), 'Internal and external mapping of preferences for commercial lager beers: comparison of hedonic ratings by consumers blind versus with knowledge of brand and price', *Food Quality and Preference*, **12**, 243–255.

HEDDERLEY D and WAKELING I (1995), 'A comparison of imputation techniques for internal preference mapping, using Monte Carlo simulation', *Food Quality and Preference*, **6**, 281–297.

HELGESEN H, SOLHEIM R and NAES T (1997), 'Consumer preference mapping of dry fermented lamb sausages', *Food Quality and Preference*, **8**, 97–109.

HORSFIELD S and TAYLOR L J (1976), 'Exploring the relationship between sensory data and acceptability of meat', *Journal of the Science of Food and Agriculture*, **27**, 1044–1056.

HOUGH G, WAKELING I, MUCCI A, CHAMBERS E, GALLARDO I M *et al.* (2006), 'Number of consumers necessary for sensory acceptability tests', *Food Quality and Preference*, **17**, 522–526.

JAEGER S R, ANDANI Z, WAKELING I N and MACFIE H J H (1998), 'Consumer preferences for fresh and aged apples: a cross-cultural comparison', *Food Quality and Preference*, **9**, 355–366.

JÖRESKOG K and SÖRBOM D (1989). *LISREL 7 – A Guide to the Program and its Applications*, 2nd ed., SPSS Publications, Chicago.

STATISTICAL INNOVATIONS, 2011, Latent Gold, http://www.statisticalinnovations.com/products/latentgold_v4.html, [accessed 27 July 2011].

LAWLESS H and HEYMANN H (1998), *Sensory Evaluation of Food: Principles and Practice*, Chapman and Hall, New York.

MACFIE H J H, BRATCHELL N, GREENHOFF K and VALLIS L V (1989), 'Designs to balance the effect of order of presentation and first-order carry-over effects in hall tests', *Journal of Sensory Studies*, **4**, 129–148.

MACKAY D (2005), 'Probabilistic scaling analyses of sensory profile, instrumental and hedonic data', *Journal of Chemometrics*, **19**, 180–190.

MACKAY D (2006), 'Chemometrics, econometrics, psychometrics – How best to handle hedonics?', *Food Quality and Preference*, **17**, 529–535.

MACKAY D and ZINNES J (1995), 'Probabilistic multidimensional unfolding: an anisotropic model for preference ratio judgments', *Journal of Mathematical Psychology*, **39**, 99–111.

MARTENS H, ANDERSSEN E, FLATBERG A, GIDSKEHAUG L, HØY M *et al.* (2005), 'Regression of a data matrix on descriptors of both its rows and of its columns via latent variables: L-PLSR', *Computational Statistics and Data Analysis*, **48**, 103–123.

PAGLIARINI E, MONTELEONE E and WAKELING I (1997), 'Sensory profile description of mozzarella cheese and its relationship with consumer preference', *Journal of Sensory Studies*, **12**, 285–301.

PLAEHN D and LUNDAHL D S (2005), 'An L-PLS preference cluster analysis on French consumer hedonics to fresh tomatoes', *Food Quality and Preference*, **17**, 243–256.

POPPER R, ROSENSTOCK W, SCHRAIDT M and KROLL B J (2004), 'The effect of attribute questions on overall liking ratings', *Food Quality and Preference*, **15**, 853–858.

POSRI W (2001), 'Improving the predictability of consumer preferences from Central Location Test (CLT) in tea', Ph.D. thesis, The University of Reading, UK.

ROTHMAN L (2007) 'The use of just about right (JAR) scales in food product development and reformulation', In: MacFie H (ed), *Consumer-led food product development*, Woodhead Publishing, Cambridge, pp xxx–xxx.

SCHIFFMAN S S, REYNOLDS M L, and YOUNG F W (1981), *Introduction to Multidimensional Scaling. Theory, Methods, and Applications*, Academic Press, New York.

TENENHAUS M, PAGÈS J, AMBROISINE L and GUINOT C (2005), 'PLS methodology to study relationships between hedonic judgements and product characteristics', *Food Quality and Preference*, **16**, 315–325.

VAN KLEEF E and VAN TRIJP HCM (2007), 'Opportunity identification in new product development and innovation in food product development', In MacFie, HJH (ed), *Consumer-led food product development.* Woodhead Publishing, Cambridge, pp xxx–xxx.

WAKELING, I N and MACFIE H J H (1995), 'Designing consumer trials balanced for first and higher orders of carry-over effect when only a subset of *k* samples from *t* may be tested', *Food Quality and Preference*, **6**, 299–308.

WESTAD F (2004), *2004 Sensometrics Conference.*

WILLIAMS E J (1949), 'Experimental designs balanced for the estimation of residual effects of treatments', *Australian Journal of Science Research A*, 2, 149–168.

WOLD S, KETTANEH N and TJESSEM K (1996), 'Hierarchical multi-block PLS and PC models for easier interpretation and as an alternative to variable selection', *Journal of Chemometrics*, **10**, 463–482.**Fig. 21.23** A circular ideal model looking down on to the sensory space from above. Shaded region is the ideal area for this consumer. As the product moves in any direction away from this area it is liked less. There is no reason why consumers should weight each dimension equally and the elliptical ideal point model reflects this. Changes in the vertical direction cause a sharper decrease in liking than in the horizontal.

Index

Kolibri 360 GmbH
Ferdinand-Jühlke-Straße 7
99095 Erfurt, Deutschland
produktsicherheit@kolibri360.de